ETHNOBOTANY OF INDIA

Volume 1
Eastern Ghats and Deccan

ETHNOBOTANY OF INDIA

Volume 1
Eastern Ghats and Deccan

Edited by

T. Pullaiah, PhD
K. V. Krishnamurthy, PhD
Bir Bahadur, PhD

Apple Academic Press Inc. | Apple Academic Press Inc.
3333 Mistwell Crescent | 9 Spinnaker Way
Oakville, ON L6L 0A2 | Waretown, NJ 08758
Canada | USA

First issued in paperback 2021

Exclusive worldwide distribution by CRC Press, a member of Taylor & Francis Group

No claim to original U.S. Government works

ISBN-13: 978-1-77463-119-5 (pbk)
ISBN-13: 978-1-77188-338-2 (hbk)

Library and Archives Canada Cataloguing in Publication

Ethnobotany of India / edited by T. Pullaiah, PhD, K. V. Krishnamurthy, PhD, Bir Bahadur, PhD.

Includes bibliographical references and indexes.
Contents: Volume 1. Eastern Ghats and Deccan.
Issued in print and electronic formats.
ISBN 978-1-77188-338-2 (v. 1 : hardcover).--ISBN 978-1-77188-339-9 (v. 1 : pdf)

1. Ethnobotany--India. I. Pullaiah, T., author, editor II. Krishnamurthy, K. V., author, editor III. Bahadur, Bir, author, editor

GN635.I4E85 2016   581.6'30954   C2016-902513-6   C2016-902514-4

Library of Congress Cataloging-in-Publication Data

Names: Pullaiah, T., editor.
Title: Ethnobotany of India. Volume 1, Eastern Ghats and Deccan / editors: T. Pullaiah, K. V. Krishnamurthy, Bir Bahadur.
Other titles: Eastern Ghats and Deccan
Description: Oakville, ON ; Waretown, NJ : Apple Academic Press, [2016] |
Includes bibliographical references and index.
Identifiers: LCCN 2016017369 (print) | LCCN 2016021535 (ebook) | ISBN 9781771883382 (hardcover : alk. paper) | ISBN 9781771883399 ()
Subjects: LCSH: Ethnobotany--India--Eastern Ghats. | Ethnobotany--India--Deccan.
Classification: LCC GN476.73 .E857 2016 (print) | LCC GN476.73 (ebook) | DDC 581.6/309548--dc23
LC record available at https://lccn.loc.gov/2016017369

Ethnobotany of India 5-volume series

Editors: T. Pullaiah, PhD, K. V. Krishnamurthy, PhD, and Bir Bahadur, PhD

Volume 1: Eastern Ghats and Deccan

Volume 2: Western Ghats and West Coast of Peninsular India

Volume 3: North-East India and Andaman and Nicobar Islands

Volume 4: Western and Central Himalaya

Volume 5: Indo-Gangetic Region and Central India

ABOUT THE EDITORS

T. Pullaiah, PhD
Former Professor, Department of Botany, Sri Krishnadevaraya University, Anantapur, Andhra Pradesh, India

T. Pullaiah, PhD, is a former Professor at the Department of Botany at Sri Krishnadevaraya University in Andhra Pradesh, India, where he has taught for more than 35 years. He has held several positions at the university, including Dean, Faculty of Biosciences, Head of the Department of Botany, Head of the Department of Biotechnology, and Member, Academic Senate. He was President of the Indian Botanical Society (2014), President of the Indian Association for Angiosperm Taxonomy (2013), and Fellow of the Andhra Pradesh Akademi of Sciences. He was awarded the Panchanan Maheswari Gold Medal, the Dr. G. Panigrahi Memorial Lecture Award of the Indian Botanical Society, the Prof. Y. D. Tyagi Gold Medal of the Indian Association for Angiosperm Taxonomy, and a Best Teacher Award from Government of Andhra Pradesh. Under his guidance 53 students obtained their doctoral degrees. He has authored 45 books, edited 15 books, and published over 300 research papers, including reviews and book chapters. His books include *Flora of Eastern Ghats* (4 volumes), *Flora of Andhra Pradesh* (5 volumes), *Flora of Telangana* (3 volumes), *Encyclopedia of World Medicinal Plants* (5 volumes), and *Encyclopedia of Herbal Antioxidants* (3 volumes). He was also a member of Species Survival Commission of the International Union for Conservation of Nature (IUCN). Professor Pullaiah received his PhD from Andhra University, India, attended Moscow State University, Russia, and worked as postdoctoral Fellow during 1976–78.

K. V. Krishnamurthy, PhD
Former Professor, Department of Plant Sciences, Bharathidasan University, Tiruchirappalli, Tamill Nadu, India

K. V. Krishnamurthy, PhD, is a former Professor and Head of Department, Plant Sciences at Bharathidasan University in Tiruchirappalli, India, and is at present an adjunct faculty at the Institute of Ayurveda and Integrative Medicine, Bangalore. He obtained his PhD degree from Madras University, India, and has taught many undergraduate, postgraduate, MPhil, and PhD students. He has over 48 years of teaching and research experience, and his major research areas include plant morphology and morphogenesis, biodiversity, floristic and reproductive ecology, and cytochemistry. He has published more than 170 research papers and 21 books, operated 16 major research projects funded by various agencies, and guided 32 PhD and more than 50 MPhil scholars. His important books include *Methods in Cell Wall Cytochemistry* (CRC Press, USA), *Textbook of Biodiversity* (Science Publishers, USA), and *From Flower to Fruit* (Tata McGraw-Hill, New Delhi). One of his important research projects pertains to a detailed study of Shervaroys, which form a major hill region in the southern Eastern Ghats, and seven of his PhD scholars have done research work on various aspects of Eastern Ghats. He has won several awards and honors that include the Hira Lal Chakravarthy Award (1984) from the Indian Science Congress; Fulbright Visiting Professorship at the University of Colorado, USA (1993); Best Environmental Scientist Award of Tamil Nadu state (1998); the V. V. Sivarajan Award of the Indian Association for Angiosperm Taxonomy (1998); and the Prof. V. Puri Award from the Indian Botanical Society (2006). He is a fellow of the Linnaean Society, London; National Academy of Sciences, India; and Indian Association of Angiosperm Taxonomy.

Bir Bahadur, PhD

Former Professor, Department of Botany, Kakatiya University, Warangal, Telangana, India

Bir Bahadur, PhD, was Chairman and Head of the Department, and Dean of the Faculty of Science at Kakatiya University in Warangal, India, and has also taught at Osmania University in Hyderabad, India. During his long academic career, he was honored with the Best Teacher Award by Andhra Pradesh State Government for mentoring thousands of graduates and post-graduate students, including 30 PhDs, most of whom went onto occupy high positions at various universities and research organizations in India and abroad. Dr. Bahadur has been the recipient of many awards and honors, including the Vishwambhar Puri Medal from the Indian Botanical Society for his research contributions in various aspects of plant Sciences. He has published over 260 research papers and reviews and has authored or edited dozen books, including *Plant Biology and Biotechnology* and *Jatropha, Challenges for New Energy Crop,* both published in two volumes each by Springer Publishers. Dr. Bahadur is listed as an Eminent Botanist of India, the Bharath Jyoti Award, New Delhi, for his sustained academic and research career at New Delhi and elsewhere. Long active in his field, he was a member of over dozen professional bodies in India and abroad, including Fellow of the Linnean Society (London); Chartered Biologist Fellow of the Institute of Biology (London); Member of the New York Academy of Sciences; and a Royal Society Bursar. He was also honored with an Honorary Fellowship of Birmingham University (UK). Presently, he is an Independent Director of Sri Biotech Laboratories India Ltd., Hyderabad, India.

CONTENTS

LIST OF CONTRIBUTORS

S. John Adams
Department of Pharmacognosy, R&D, The Himalaya Drug Company, Makali, Bangalore, India

M. Hari Babu
Department of Botany, Andhra University, Visakhapatnam–530003, India

Bir Bahadur
Department of Botany, Kakatiya University, Warangal–506009, India

V. George
Amity Institute for Herbal and Biotech Products Development, 3 Ravi Nagar, Peroorkada P.O., Thiruvananthapuram–695005, Kerala, India. E-mail: georgedrv@yahoo.co.in

T. P. Ijinu
Amity Institute for Herbal and Biotech Products Development, 3 Ravi Nagar, Peroorkada P.O., Thiruvananthapuram–695005, Kerala, India. E-mail: ijinutp@gmail.com

V. Kamala
National Bureau of Plant Genetic Resources, Regional Station, Rajendranagar, Hyderabad–500030, Telangana State, India. E-mail: kgksvp@gmail.com

S. Karuppusamy
Department of Botany, The Madura College (Autonomous), Madurai–625011, Tamilnadu, India, E-mail: ksamytaxonomy@gmail.com

K. V. Krishnamurthy
Department of Plant Science, Bharathidasan University, Tiruchirappalli–620024, India

K. Sri Rama Murthy
Department of Botany and Biotechnology, Montessori Mahila Kalasala, Vijayawada–520010, Andhra Pradesh, India, E-mail: drmurthy@gmail.com

Ashish Kumar Pal
Formulation Analytical Research Department, Aurobindo Pharma Limited and Research Centre, Survey No. 313, Bachupally Village, Quthubullapur Mandal, R.R. District 500090, Telangana, India, E-mail: ashishkumarhyd@gmail.com

S. R. Pandravada
National Bureau of Plant Genetic Resources, Regional Station, Rajendranagar, Hyderabad–500030, Telangana State, India. E-mail: pandravadasr@yahoo.com

T. Pullaiah
Department of Botany, Sri Krishnadevaraya University, Anantapur–515003, Andhra Pradesh, India, E-mail: pullaiah.thammineni@gmail.com

P. Puspangadan
Amity Institute for Herbal and Biotech Products Development, 3 Ravi Nagar, Peroorkada P.O., Thiruvananthapuram–695005, Kerala, India. E-mail: palpuprakualm@yahoo.co.in

S. Sandhya Rani
Department of Botany, Sri Krishnadevaraya University, Anantapur–515003, Andhra Pradesh, India, E-mail: sandhyasakamuri@gmail.com

J. Koteswara Rao
Department of Botany, Andhra University, Visakhapatnam–530003, India

T. V. V. Seetha Rami Reddi
Department of Botany, Andhra University, Visakhapatnam–530003, India, E-mail: reddytvvs@rediff-mail.com

M. Chandrasekhara Reddy
Department of Botany and Biotechnology, Montessori Mahila Kalasala, Vijayawada–520010, Andhra Pradesh, India, E-mail: chandra4bio@gmail.com

B. Sadasivaiah
Department of Botany, Government Degree and PG College, Wanaparthy–509103, Mahabubnagar District, Telangana, India, E-mail: chumsada@gmail.com

N. Sivaraj
National Bureau of Plant Genetic Resources, Regional Station, Rajendranagar, Hyderabad–500030, Telangana State, India. E-mail: sivarajn@gmail.com

Razia Sultana
EPTRI, Gachibowli, Hyderabad–500032, India. E-mail: emailrazia@yahoo.com

LIST OF ABBREVIATIONS

AICRPE	All India Co-ordinated Research Project on Ethnobiology
AQUAD	analysis of qualitative data
AR	aerial roots
AR	augmented reality
ASHRAM	*Abhayaranya Samrakshan* through Holistic Resource (Array) Management
ASI	Ancient South Indian
AVP	Arya Vaidya Pharmacy
B	bulbs
BDM	Biodiversity Data Management
BMDP	BioMeDical Package
BP	before the present
Br.	bark
BTIS	Biotechnology Information System
C	corms
CADAS	computer-assisted data analysis
CAQDA	computer-assisted qualitative data analysis
CBRs	Community Biodiversity Registers
CCRAS	Central Council of Research in Ayurveda and Siddha
CCRUS	Central Council of Research in Unani Medicines
CFM	Community Forest Management
CPD	Centre of Plant Diversity
CSL	Citation Style Language
DDS	Deccan Development Society
EGMB	Eastern Ghats Mobile Belt
ENVIS	Environmental Information System
EPTRI	Environmental Protection, Training and Research Institute
ERIS	Environmental Resources Information System
FAO	Food and Agricultural Organization
Fl.	flowers
Fr.	fruits
FRIS	Farmers' Rights Information Service
FRLHT	Foundation for the Revitalization of Local Health Traditions
G	gum

IIA	International Institute of Ayurveda
IBIN	Indian Bioresource Information Network
IGCMC	Indira Gandhi Conservation Monitoring Centre
IKS	indigenous knowledge system
In.	inflorescence
INMEDPLAN	Indian Medicinal Plants National Network of Distributed Database
IPBN	Indigenous People's Biodiversity Network
IPC	International Patent Classification
IPR	intellectual property rights
ISMH	Indian Systems of Medicine and Homeopathy
ITDA	Integrated Tribal Development Agency
ITK	indigenous traditional knowledge
IUCN	International Union for Conservation of Nature
JFM	Joint Forest Management
JNTBGRI	Jawaharlal Nehru Tropical Botanical Garden and Research Institute
KKSKT	Kerala Kani Samudaya Kshema Trust
KWIC	key word in context
L	leaves
MAB	Man and Biosphere Programme
MPCAs	Medicinal Plant Conservation Areas
MSSRF	M.S. Swaminathan Research Foundation
MVP	minimum viable population
NBPGR	National Bureau of Plant Genetic Resources
NBRI	National Botanical Research Institute
NBSAP	National Biodiversity Strategy and Action Plan
NIC	National Informatics Centre
NIF	National Innovations Foundation
NTFPs	non-timber forest produces
OCR	optical character recognition
P	pith
PCNM	principal coordinates of neighbor matrices
PGR	Plant Genetic Resources and Commission on Plant Genetic Resources
PIC	prior informed consent
PTG	primitive tribal groups
PVPFR	Plant Variety Protection and Farmers' Rights
QDA	qualitative data analysis
QDAP	qualitative data analysis program

R	root
Rh	rhizome
RRL	Regional Research Laboratory
S	seeds
SAS	statistical analysis system
SPSS	statistical package for the social sciences
SRISTI	Society for Research and Initiatives for Sustainable Technologies and Institutions
St	stem
T	tuber
TDTK	Thiruvananthapuram Declaration on Traditional Knowledge
TK	traditional knowledge
TKDL	Traditional Knowledge Digital Library
TKRC	Traditional Knowledge Resource Classification
TKS	traditional knowledge system
Ts	tender shoots
UNMC	University of Nebraska Medical Center
WHO	World Health Organization
WIPO	World Intellectual Property Organization
WP	whole plant

PREFACE

Humans are dependent on plants for their food, most medicines, most clothes, fuel and several other needs. Although the bond between plants and humans is very intense in several 'primitive' cultures throughout the world, one should not come to the sudden and wrong conclusion that post-industrial modern societies have broken this intimate bond and interrelationship between plants and people. Rather than, plants being dominant as in the 'primitive' societies, man has become more and more dominant over plants, leading to over-exploitation of the latter, and resulting in a maladapted ecological relationship between the two. Hence a study of the relationships between plants and people-ethnobotany and, thus, between plant sciences and social sciences, is central to correctly place humanity in the earth's environment. Because ethnobotany rightly bridges both of these perspectives, it is always held as a synthetic discipline.

Most people tend to think that ethnobotany, a word introduced by Harshberger in 1896, is a study of plants used by 'primitive' cultures in 'exotic' locations of the world, far removed from the mainstream people. People also think wrongly that ethnobotany deals only with non-industrialized, non-urbanized and 'non-cultured' societies of the world. Ethnobotany, in fact, studies plant-human interrelationships among all peoples and among all. However, since indigenous non-western societies form the vast majority of people now as well as in the past, a study of their interrelationships with people becomes important. More than 10,000 human cultures have existed in the past and a number of them persist even today. They contain the knowledge system and wisdom about the adaptations with diverse nature, particularly with plants, for their successful sustenance. Thus, ethnobotanical information is vital for the successful continuance of human life on this planet.

Ethnobotany is of instant use in two very important respects: (i) indigenous ecological knowledge, and (ii) source for economically useful plants. The first will help us to find solutions to the increasing environmental degradation and the consequent threat to our biodiversity. In indigenous societies biodiversity is related to cultural diversity and hence any threat to biodiversity would lead to erosion in cultural diversity. Indigenous cultures are not only repositories of past experiences and knowledge but also form the frameworks for future adaptations. Ethnic sources of economically useful plants have resulted in serious studies on bioprospection for newer sources of food, nutraceuticals, medicines and other novel materials of human use. Bioprospecting has resulted in intense research on reverse pharmacology and pharmacognosy. This has resulted in attendant problems relating to intellectual property rights, patenting and the sharing of the benefits with the traditional societies who owned the knowledge. This has also resulted in serious documentation of traditional knowledge of the different cultures of the world and to formalize the methods and terms of sharing this traditional knowledge. It has also made us to know not only *what* plants people in different cultures use and *how* they use them, but also *why* they use them. In addition it helps us to know the biological, sociological, cultural roles of plants important in human adaptations to particular environmental conditions in the past, present and future.

This series of the five edited volumes on ethnobotany of different regions of India tries to bring together all the available ethnobotanical knowledge in one place. India is one of the most important regions of the old world and has some of the very ancient and culturally rich diverse knowledge systems in the world. Competent authors have been selected to summarize information on the various aspects of ethnobotany of India, such as ethnoecology, traditional agriculture, cognitive ethnobotany, material sources, traditional pharmacognosy, ethnoconservation strategies, bioprospection of ethno-directed knowledge, and documentation and protection of ethnobotanical knowledge.

The present series of five volumes is a humble attempt to summarize the ethnobotanical knowledge of the aborigines of India. The first volume is on Eastern Ghats and adjacent Deccan region of Peninsular India. Published information is summarized on different aspects. Our intention is that this may lead to discovery of many new drugs, nutraceuticals, and other useful products for the benefit of mankind.

Since it is a voluminous subject we might have not covered the entire gamut; however, we have tried to put together as much information as possible. Readers are requested to give their suggestions for improvement of the coming volumes.

ACKNOWLEDGMENTS

We wish to express our grateful thanks to all the authors who contributed their research/review articles. We thank them for their cooperation and erudition. We also thank several colleagues for their help in many ways and for their suggestions from time to time during the evolution of this attractive and readable volume.

We wish to express our appreciation and help rendered by Ms. Sandra Sickels, Rakesh Kumar, and the staff of Apple Academic Press. Above all, their professionalism has made this book a reality and is greatly appreciated.

We thank Mr. John Adams, Senior Research Fellow of Prof. K.V. Krishnamurthy, for his help in many ways.

We wish to express our grateful thanks to our respective family members for their cooperation.

We hope that this book will help our fellow teachers and researchers who enter the world of the fascinating subject of ethnobotany in India with confidence, as we perceived and planned.

ACKNOWLEDGMENTS

CHAPTER 1

INTRODUCTION

K. V. KRISHNAMURTHY,[1] T. PULLAIAH,[2] and BIR BAHADUR[3]

[1]*Department of Plant Science, Bharathidasan University, Tiruchirappalli–620024, India*

[2]*Department of Botany, Sri Krishnadevaraya University, Anantapur–515003, India*

[3]*Department of Botany, Kakatiya University, Warangal–506009, India*

CONTENTS

ABSTRACT

This chapter introduces the scope and contents of this volume, which deals with the ethnobotany of the Eastern Ghats and adjacent Deccan region of India. The physical features and geomorphology of the study region are briefly introduced. The ethnic diversity, worldviews and belief systems, ethnoecology, ethnotaxonomy, ethnonomenclature, traditional crop biodiversity, utilization aspects of plants of different ethnic communities (food, medicine, veterinary medicinal plants, etc.), documentation, conservation and management of ethnoplant resources, etc. of the study region are briefly introduced. The importance of mainstreaming traditional botanical knowledge of the study region is also emphasized.

1.1 PHYSICAL FEATURES AND GEOMORPHOLOGY OF THE STUDY REGION

It is generally agreed upon that the Indian subcontinent was part of the Gondwanaland, got separated from it, drifted northwards and finally collided with the Asian tectonic plate to position itself as we see it today. The drifting took place by the early Cretaceous period and collision around 50–65 million years ago. The present day Indian subcontinent consists of four geomorphic provinces, each of which is structurally and lithologically distinct and physiographically contrasted; the four provinces also have an altogether different evolutionary history (Valdiya, 2010); the four provinces are: (i) The mountainous Himalayan province that girdles the northern border of the subcontinent; (ii) The flat and expansive Indo-Gangetic plains in the middle; (iii) The plateaus and uplands of peninsular India; and (iv) the coastal plains along the seaboard (Arabian Sea on the West, Bay of Bengal on the east and the Indian Ocean on the south). South of the Himalayan mountains is peninsular India, a shield of Archaean antiquity. Four well-defined crustal blocks, called Cratons (containing the oldest granite rocks) make up this mosaic of peninsular shear zones: (i) The Dharwar Craton (3.20 to 3.40 Ga) in south India, covering parts of Maharashtra, Andhra Pradesh, Karnataka, Goa, Tamil Nadu and Kerala; (ii) The Bastor Craton (3.01 Ga) in central India, covering parts of north west Andhra Pradesh, south west Odisha, Chhattisgarh and north east Maharashtra; (iii) the Singhbhum Craton (3.56 Ga) in eastern India, covering northern Odisha, and south West Bengal; and (iv) The Bundelkhand Craton (3.31 Ga) in north western India covering eastern Rajasthan, southwest Uttar Pradesh and north west Madhya Pradesh.

Each of these cratons have undergone "events of crustal rifting and sagging or shrinking of dismembered blocks, with the attendant volcanism, as well as deformations, metamorphosis, and granatization or charnockitization" (Valdiya, 2010). These processes have resulted in the welding of the crustal blocks into composite rigid cratons. During subsequent geological history these craton regions underwent several other geological and geomorphological changes.

Peninsular India is triangular in shape and the apex of the triangle is at the southernmost end of India (at Kanyakumari). It is about 2,200 km long in the N-S direction and around 1,400 km wide in E-W direction (in the region of greatest width). It consists of three physiographic regions: the mountain ranges on its three sides, the uplands and plateaus, the latter two constituting the longer part within the confines of these mountain regions and the coastal plains along the eastern and western seaboards (Figure 1.1). The peninsular Indian region covered in this book is bordered on the north

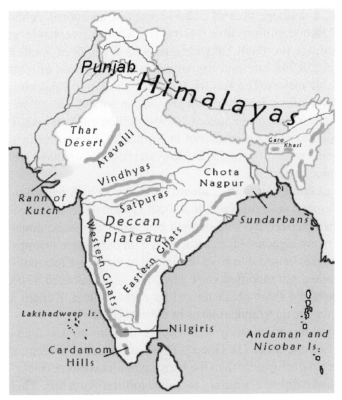

FIGURE 1.1   Major mountain systems of India.

by the Vindya-Satpura hill range that trends in the ENE-WSW direction, the Sahyadri hill range that extends 1, 600 km southwards from the Tapti valley to Kanyakumari on the western side and the East Coast hill range, which forms a series of physiographically discontinuous hill ranges on the East. The plateaus enclosed within these three hill ranges are the Deccan plateau. It encompasses practically the plains of Maharashtra (especially the Vidarba region), Odisha, the adjoining parts of the undivided Andhra Pradesh, Karnataka and Tamil Nadu. The average elevation of Deccan is 600 m above msl. Covering the whole of Karnataka, adjoining Tamil Nadu and the undivided Andhra Pradesh is the Mysore Plateau, which is made up of Archaean gneisses, granites and high-grade metamorphic rocks.

The major rivers of the Deccan region are the following: The Mahanadi, which has its source in Dandakaranya near Sihawan Rajpur district, has a length of 857 km, covers an area of 141,600 km² and the with a volume of average annual flow of 67,000/66,640 million cubic meters. The Godhavari river has its source at Trimbek Plateau near Nasik and has a length of 1,465 km, a drainage area of 312,812 km²and an annual volume flow of 105,000/118,000 million cubic meters. The Krishna river has its source near Mahabaleshwar (northern Sahyadri), runs for a length of 1,400 km, covers an area of 258,948 km² and has an annual volume flow of 62,800/67,670 million cubic meters. The Kaveri river has its origin in Talakaveri at central Sahyadri, runs for a length of 800 km, has an annual water flow of 87,900 km² and covers a drainage area of 20, 950 million cubic meters. The Pennar river originates near Kolar district in Karnataka, runs a length of 910 km, has an annual water flow of 55, 213 km² and covers a drainage area of 3, 238 million cubic meters.

The east coast (=eastern seaboard) stretches from Athagarh in Odisha to beyond Ramnad in Tamil Nadu. It is about 126 km wide. It is a coast of emergence, characterized by well-defined beaches, many sand dunes and sand spits and many lagoonal lakes associated with backwater swamps. The eastern seaboard is believed to have originated in the post-Cretaceous times and has grown and got modified since then. The shore between Visakhapatnam and Ganjam is a shore characterized by cliffs. Pulicat, Kolleru and Chilka lakes are the most prominent lakes in the east coast.

The most important study region covered in this book is the Eastern Hill range or Eastern Ghats (E. Ghats). These are 'tors' of geological antiquity and are geologically older than the Himalayas and Western Ghats. The Ghats orogeny had happened around 1600±100 million years ago. This hill range that lies on eastern side of the Deccan plateau of peninsular India forms a chain of physiographically discontinuous, elevated hill range that does not have any structural unity (Krishnamurthy et al., 2014). Many geographers consider the Khondmal hills in Odisha as the northern extremity of E. Ghats, while others consider the Simlipal massif of northern Odisha as the northern extremity. The E. Ghats traverse through the states of Odisha, undivided Andhra Pradesh, Tamil Nadu and parts of Karnataka and finally meet the Nilgiris of W. Ghats in the Moyar valley (Figure 1.2). BR hills form the southwestern extremity of E. Ghats while the southern extremity is near Ramnad in Tamil Nadu. The discontinuity in this hill range is mainly due to the great rivers, mentioned earlier, and the small rivers like Dahuda, Vamsadhara, Nagavalli, Sarada, Varaha, etc. that cut through the range.

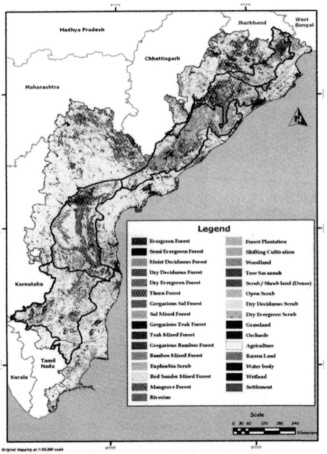

FIGURE 1.2 Four major sections of Eastern Ghats with section 2 divided into two subsections (Source: Murthy et al 2007b [Proc. Nat. Sem. Conserv. Eastern Ghats, 2007, EPTRI]. Used with permission.)

The E. Ghats over an area of about 75,000 km², with an average width of about 200 km in the north and about 100 km in the south. A maximum area of about 48% of E. Ghats falls in the undivided Andhra Pradesh, while its area in Tamil Nadu and Odisha is about 25% each and the remaining 2% pass through Karnataka (see Krishnamurthy et al., 2014). The hill range has a length of 1,750 km. the average elevation is about 700 m, though individual peaks may rise up to a height of 1,675 m msl. Most geologists believe that the E. Ghats is geologically heterogeneous in origin: two distinct kinds of hill ranges make up the so-called E. Ghats. The first one (often called the northern E. Ghats runs parallel to the east coast in a N-SW direction up to Krishna valley (Ongole). This is the true E. Ghats and is also called the E. Ghats Mobile Belt (EGMB). The other part runs south of Krishna valley and has hills that are of heterogeneous character. However, in this volume, E. Ghats is been considered to include both these regions, as many others have done earlier. The composite E. Ghats generally is considered to have three major sections: North (North Odisha to Guntur in undivided Andhra Pradesh), Middle (Krishna river to near about Chennai in Tamil Nadu), and South (the rest of the E. Ghats in Tamil Nadu and Karnataka). In Karnataka E. Ghats runs from Bellary, through Chitradurga, Kolar, Tumkar and ends in BR hills. There are 138 major hills in E. Ghats (Krishnamurthy et al., 2014).

1.2 ETHNIC DIVERSITY OF THE STUDY REGION

India is remarkable for its diversity, both biological and human. The Indian subcontinent has about 427 tribal communities (Singh, 1993) with about 62–65 million people (Vinodkumar, 2007), although others speak of 4635 well-defined groups under 532 tribes (of which 72 are primitive including 36 hunter-gatherer tribes). The great ethnic human diversity of India is due to its position at the tri-junction of the African, the northern Eurasian and Oriental realms, as well as to its great variety of environmental regimes. Its plant (and animal) wealth has been continuously attracting humans in many streams starting from about 70,000 to 50,000 years ago, at different historical times and from different directions. This has brought together a great diversity of human genes and human cultures into India and their subsequent mix-up to various degrees. While in other parts of the world the dominant human culture of those parts has been known to absorb or eliminate other cultures that might enter there, the tendency in India, from historic times, has been to isolate and subjucate the subordinated cultures, thereby segmenting the cultural (and thereby the human) diversity (Gadgil et al., 1996). In the light of this background, the second chapter of this volume contributed by Bir Bahadur et al. provides a detailed account on the ethnic diversity of E. Ghats and Deccan region of India, the ethnobotany of which is the focus of this volume. They have not only explained the origin of the great ethnic diversity of this region, essentially based on the works of Gadgil et al. (1996) and Thangaraj (2011), but also have given details on the various ethnic tribes and their distribution in different parts of the region of study mentioned above.

1.3 WORLDVIEWS AND BELIEF SYSTEMS

Societies, cultures and knowledge systems evolved almost simultaneously with the first establishment of modern human species in different parts of the world all along its migratory routes from Africa. The evolution and establishment of each one of the three was critically dependent on the evolution and establishment of the other two. All the three were not only dependent on the environmental conditions that prevailed in the different places where the human species got settled but also on the threads of culture and knowledge systems that the settling human population already possessed and carried along with it during their migration. Thus, there are two components in their social behavior, culture and knowledge system: one that they had before actually settling down and the other after they got settled to a place. The first component largely explains most, if not all, common aspects of social life, cultures and knowledge systems of the different ethnic societies of the world, however far away they are from each other at present. However, we should not also rule out parallel or independent evolution of certain common components in societal organization, culture and knowledge systems of different ethnic societies of the world.

Hunting-gathering and nomadic society, and cultures and knowledge systems associated with it, were the earliest to evolve as humans were on a constant move during their early migration from Africa, as well as around places of settlement within a reasonably smaller territory until about 10,000 to 12,000 years ago when agriculture and the societies, cultures and knowledge systems associated with evolved and almost replaced the hunter-gatherer culture. It is this hunter-gatherer, short-distance-nomadic populations "settled" on the various migratory routes who have greatly contributed to the evolution of societies, cultures and knowledge systems. These pockets of human population are to be rightly referred to as Indigenous communities. The occupation of new environmental niches by these human societies was initially enabled by the effective tool-making and using abilities, control over fire and the group-gathering and hunting for food. These adaptations fine-tuned the utilization of the natural resources available around them (Gadgil, 1987). The interrelationship between different members of an ethnic/tribal group, which make social life possible, is called social organization. An ethnic society normally has a territory and is endogamous and is often divided into clans, subclasses/subgroups, etc. each pursuing traditionally a well-defined, similar mode of subsistence and similar levels of access to environmental resources like plants and are egalitarian in nature. The social relationship amongst the members is governed by kinship and mutual help and all of them strive for protecting their environment and its resources and sustainably use them (Gadgil and Thapar, 1990).

The world knows (had known in the past also) a great variety of cultures and civilizations) each with their own knowledge, value and belief systems. Biologists define culture as the acquisition of behavioral traits from conspecifics through the process of social learning (Gadgil, 1987) and, thus culture is one that is learnt. Sociologists define it as that complex whole that includes knowledge, belief, art, morals, law, customs and any other attributes acquired by any one as a member of a society (Tylor, 1874) and thus a culture is a man-made component of the environment (Parthasarathy, 2002). Hence, culture of different ethnic communities is largely dependent on the environmental conditions prevailing in any region, particularly the environmental resources available. It is because of this that cultural diversity and biodiversity of any region go hand in hand, particularly in primitive human locations (McNeely and Pitt, 1985; see also Krishnamurthy, 2003).

The knowledge systems known to traditional/indigenous ethnic societies is called *traditional knowledge system (TKS)* or *indigenous knowledge system (IKS)*. TKS is also referred to by terms, such as *worldviews, cosmovisions* or *beliefsystems*. These three terms actually refer to the different ways of perceiving, interpreting and learning about the world around the people. The different worldviews of different ethnic communities have come to gain knowledge about the world and its environmental components (living as well as non-living) around them, thus resulting in different knowledge contents. Invariably, however, worldview is expressed by conceiving life and the knowledge obtained during a life period in terms of three interrelated and inseparable domains (or worlds or spheres): Natural, Human and spiritual. To a great extent, the ability of an ethnic community to use the local environmental resources like plants is determined by the above worldview. The worldview includes knowledge that is not limited to the world that can be perceived with human senses and can be explained in a rational way, but also to a world beyond human perception. Thus, knowledge, according many ethnic societies, is a combination of that which is true and that which is believed, and that truth and belief go together. This knowledge is also qualitative, practical, partial, intuitive and holistic. Information relating to worldviews on plants and their validity and importance has largely been obtained through three main approaches (Cotton, 1996): (i) ecocultural approach (ii) cognitive and socio-cultural approach, and (iii) utilitarian or economic approach. The first invokes certain traditional cultural practices, such as sowing, transplanting and harvesting taboos and/or rituals; their importance can only be realized when the relevant cultural influences are known. Despite their culture-specific nature and importance, many traditional cultural practices involving plants may at first glance appear irrational, but in reality they have important functional consequences. The cognitive approach explains how different ethnic communities perceive plants and vegetative types and how such perceptions are influenced by sociocultural factors and spiritual beliefs. This approach also involves rituals/symbolic behaviors that fall in the realm of society, religion, magic, spiritual and supernatural domains. The utilitarian approach records how different species of plants are used as food, medicine or for other materialistic needs of humans and seeks to explain these uses on the basis of modern scientific methods. In Chapter 3, Krishnamurthy explains in detail the worldviews and belief systems of people of the Indian region covered in this book regarding plants and their importance from an ecocultural and sociocultural perspective; this kind of approach has not yet received adequate attention of those who are interested in Western science and its so-called rational (and reductionistic) approach.

1.4   TRADITIONAL KNOWLEDGE ON ETHNOECOLOGY, ETHNONOMENCLATURE AND ETHNOTAXONOMY

Plants were there on this earth long before the evolution of the modern human species. Once on earth, man had to confront all major groups of plants, which had already adapted themselves to diverse habitats/ecosystems of the world, through his own adaptation to the very different environments around him. The knowledge was gained by him gradually during his prolonged period of interaction with the different environments and the diverse plants associated with them. This has resulted in an enormous body of information on traditional knowledge on plants, both wild and domesticated. As already stated, information relating to traditional knowledge on plants and their validity and importance have largely been obtained through ecocultural, cognitive and utilitarian approaches. Traditional knowledge on Indian plants is both non-codified and codified and cover folk/tribal knowledge. The knowledge on plants as gained in India in historic times is covered under *Vrikshayurveda* (= science of plant life). A thorough understanding of the languages and dialects often are used locally to the correct interpretation of most data/information on plants (Carroll, 1992). A variety of linguistic techniques have been employed so far (Martin, 1995).

The foremost and very important traditional knowledge on plants concerns knowledge on ethnoecology, ethnonomenclature and ethnotaxonomy, the former referring to naming of plants (and animals) around them by ethnic societies, while the third refers to a study of traditional systems of classification of plants (and animals). A critical study reveals that plants were named and classified by almost all traditional societies of the world. According to Malinowski (1974), and many others, satisfying the needs (for food, medicine, etc.) of traditional people is more important in recognizing naming and classifying plants (a materialistic view point). On the contrary, for Levi-Strauss (1966) and some others, the outlook of traditional people towards plants (and animals) is primarily intellectual and cognitive (and a natural urge) and divorced from pragmatic concerns, such as the one mentioned above. Both these perspectives are important in understanding ethnonomenclature and ethnotaxonomy. Ethnoecology concerns the ideas and concepts of ethnic communities on ecology and ecosystems that prevailed around their communities.

Ethnonomenclature and ethnotaxonomy require a deep and critical knowledge on the life of plants (animals) around the different ethnic communities. It also requires several technical/descriptive terms to denote the different structural and functional characteristic of the different plants. Also needed are practical and intuitive knowledge on characters of primary importance, especially identifying sets of contrasting characters, for the diagnosis and classification of plants. In naming plants, linguistic definition has often attained importance. Patterns revealed through linguistic analysis of plant names and categories have often provided clues to the used and other characteristics of plants (Martin, 1995). It is now widely recognized that not all conceptual categories of taxa received linguistic recognition and hence folk taxonomists are now beginning to look only at the *structural* details of linguistically defined nomenclatural and taxonomic systems, but also at their substantive nature (Ellen, 1994). Perhaps the first and detailed ethnonomenclatural and ethnotaxonomic analysis was done by Conklin (1954, 1974). This was followed by a synthesis of general principles by Berlin and his co-workers (Berlin et al., 1973; Berlin, 1992), and a statement on other cross-language patterns in ethnonomenclature and ethnotaxonomy (Brown,

1984, 2000). These aspects are described in detail by Krishnamurthy and John Adams in Chapter 4. These authors have tried to fit in details of nomenclature and classification of ethnic Tamil people into the general principles proposed by Berlin and his group and in the cross-language patterns detailed by Brown.

1.5 TRADITIONAL AGRICULTURAL CROP DIVERSITY

Today, most people depend on agriculture for their daily sustenance. Yet, it is a very recent development in the history of humanity. It is commonly believed that only around 10,000 radio-carbon years ago agriculture started first in South West Asia. The transition from foraging to farming drastically changed the relationship of humans with their environment. Because agriculture allowed more people to be sustained per unit area of cultivable land, it paved the way for a settled human life as well as for the development of towns, crafts, trade, scripts and technologies; there was a drastic cultural, social, and political change as well (Harris, 2005). By about 1500 CE, when Europeans were beginning to colonize other continents, most people of the world were already dependent for their sustenance on agriculture practiced in a variety of environmental systems, except in Australia.

Tracing the exact origin of agriculture (the transition from foraging to farming), at least in some parts of the world, such as Africa, is very difficult, particularly because lack of clinching archaeobotanical records/data. Neither is clear the causes of origin of agriculture. Although several hypotheses, such as the Oasis hypothesis, the Natural habitat hypothesis, the Population pressure hypothesis, the Edge-zone hypothesis, etc. have been proposed in the past to explain the origin of agricultural activity [see Harris (2005) for detailed literature], it is now generally agreed that climate changes, population pressure and technological advancement involving people-plant interaction all have resulted in the origin of agriculture. Once initiated agriculture evolved through the involvement of cultivation, domestication and the establishment of agricultural economies.

The facts relating to the origin, antiquity and sources of agriculture in India are very vexed, hazy and controversial problems (Srivastava, 2008). The earliest evidence of agriculture in the Indian subcontinent is from 8000 years ago from the present (Srivastava, 2008; Jarrige and Meadow, 1980). It started in the Indo-Gangetic region and then got spread to south a little later (see Fuller et al., 2001). Traditional biodiversity knowledge is being studies at three levels: genetic, specific and ecosystem. Traditional communities throughout the world have been promoting efforts to maintain all these three levels of biodiversity. Maintenance of genetic diversity within a species implies maintenance of that plant species. India is one of the eight centers of origin of crop plant species accounting for about 115 food species and for about 110 non-food species out of about 325 domesticated in India (Krishnamurthy, 2009) out of which about 45 are believed to have been domesticated in Eastern Ghats and adjacent regions (Krishnamurthy et al., 2014). E. Ghats is believed to be a primary/secondary /diffuse center of origin, diversity and spread for rice, pigeon pea, some cucurbits, black gram, banana, mango, jamun, some millets, cow pea, sesame, Okra, green gram leafy amaranths, etc. It also has around 460 out of about 2000 domesticated medicinal plants. Besides, this region is very rich in wild relatives of useful plants. In Chapter 5, Pandravada et al. discusses all aspects related to the diversity of agricultural crops in the region under consideration in this volume.

1.6   UTILITARIAN ETHNOBOTANICAL KNOWLEDGE

It was discussed earlier in this chapter that there are three approaches to the study of ethnobotanical knowledge, one of which is the utilitarian or economic approach (see Cotton, 1996). Harshberger (1896) was perhaps the first to emphasize the strictly utilitarian aspects of ethnobotanical knowledge. This approach involves the collection of knowledge about the uses and management of different plant species as well as about the identification of useful species possessed by the tribal communities although it generally fails to take into account the cultural perception of plants used by different tribal communities, Most protagonists of western science are interested only in the utilization aspects of ethnobotany. The physical structures, the chemical contents and their nutritional, therapeutic and other effects on humans have all enabled the use of plants for various human requirements, such as food, nutraceuticals, medicines, cosmetics, etc. In an age when population is exponentially increasing, man has to find additional/alternate sources of plants for meeting various human needs. Hence it is not surprising that research is mainly aimed at identifying novel plants and plant products that have untapped economic potential and to conserve traditional plant sources. This type of research is known as bioprospecting or geneprospecting (Reid et al., 1993; Sittenberg and Gamez, 1993; Krishnamurthy, 2003). Essentially three methods have been used until now for bioprospecting. The first is the random method, which randomly selects a plant for its probable economic potential. The second is the phylogenetic method, which selected a plant species for analysis if its related species is already known to have economic value. The third is the ethno-directed method. In this method, attention is specifically focused on plants, which are known to be used by tribal people/ ethnic communities, but not yet received wider attention. It is a ready-made knowledge that is sure to yield the desired results in addition to involving less research and development costs; it is also less time-consuming. Hence it is ideal to know and document available information on the various plants used by ethnic communities.

1.7 FOOD PLANTS

The most important utilitarian botanical knowledge relates to food plants. Food is considered the very stuff of life. Along with water and heat, food is the substance and agent of operation and driving force of life. Most, if not all, of the food plants that humans use today have been identified as food plants by the ancient human communities in different parts of the world. It has been estimated that about 70,000 species of plants of the world have been found to be edible by indigenous human societies (Krishnamurthy, unpublished information). However, over the history of humankind on this earth, food and food systems (and their plant and animal sources) have got changed depending on the life-style of people. Hunter-gatherers depended on a tremendous variety of plant food, from tubers and grass seeds to the pith of palms and fleshy fruits. Agricultural-pastoral community tended a much smaller number of plant (and animal) species (around 3,000 species of plants). However, the level of production per unit area from these species is far greater. Such food plant harvests imply an intensification of the outflow of materials from small areas of cultivated and intensely grazed lands. The number of plants depended on as food got drastically reduced as transition from hunting-gathering to specificalized agriculture took place (Gadgil and Thapar, 1990). Thus, the humans gradually became more dependent and specialized for their food supply on a small selection of crops, grains in some parts of the world and on tuber crops and roots in other parts. The traditional foods and food systems, although still survive in many primitive ethnic societies of the world, have been mostly corrupted or altogether replaced in several world communities, gradually in most cases, but quite abruptly in certain others. Globalization and homogenization have replaced local food plants and diet-related chronic disorders and other forms of malnutrition (Kuhnlein et al., 2009). It should, however, be emphazied that the traditional food systems of indigenous communities touch the full spectrum of life in ways that the present day food systems do not. If something bad happens to the present day food systems and their source plants, we have no other option except to go back to our traditional food sources.

Food is close to the hearts of indigenous peoples and is often suited to their local environmental and cultural conditions. Social networks and cohesion were often involved in traditional food systems. Indigenous peoples' food systems contain treasures of knowledge from the long-evolved cultures and patterns of living in their respective ecosystems. The dimensions of nature and culture that define a food system of an indigenous society contribute to the whole health picture of the individual and the community-not only physical health but also the emotional, mental and spiritual aspects of health, healing and protection from diseases. An impressive array of food species and varieties has been documented and some of these still require botanical identification and nutrient-composition analysis. Equally impressive are the diverse methods that traditional societies have employed in food collection, preservation, processing and cooking. Not all the foods that the ethnic societies used are equal. Some are relished, others only tolerated and still others are loathed, being eaten only when absolutely necessary (Minnis, 2000). Although the greatest scientific attention has been paid so far on the most common and relished food and its source plants, attention need also to be focused on the less desirable foods, frequently called 'famine foods,' 'starvation foods,' 'emergency foods,' or 'queer foods.' In Chapter 6, Sadasivaiah and Pullaiah discuss on the various aspects of ethnic food systems and their source plants. It is evident from their discussion that many tribal food plants have been domesticated; yet there are several others that need to be further researched on their potential to be added to the principal food sources of mainstream people. Also needs to be researched is whether 'famine' or 'starvation' food plants mentioned in this article are once commonly used food plants, as per the suggestions of Minnis (2000).

1.8 MEDICINAL PLANTS

Plants are known not only to have nutritive value and to providing a long life, but also are known to be important allies in the curing of ailments and as antidotes for poisons. Plants also save humans from ailments. The traditional people's wisdom is that there is no herb (=plant), which does not possess medicinal properties (Zimmer, 1935). Most, if not all, human diseases now known were also known to ancient people, although in different names. Ancient medical knowledge is available in the form of both non-codified folk medicine and of codified medicine. Traditional (also often called alternate or complementary medicine) medicine was particularly well developed in India, China, Tibet, Greece, Egypt and other Arab countries and in many traditional societies of other parts of the world. Although the power of plants is certainly due to the actual physiological processes of healing, for traditional societies it equally seems that their ritual principles of similarity, for example, to "sympathetic" transformations, wherein the appearance, location, or other properties of plants (for example the *panch-abhuta* or five-element nature) are directly related to the medical problems at hand.

The tribal communities have a tremendous depth of knowledge regarding the use of natural medicinal resources, including plants. The tribals, in India alone, use over 7,500 to 9,000 species of plants as medicines and nutraceuticals; similarly the codified traditional medical systems of India, such as Ayurveda, Siddha, Unani, etc. use around 8,000 species of plants. An integral component of tribal medicine is its association with shamanism. A *shaman* or *poojari* is a person regarded as having access to, and influence in, the world of benevolent and malevolent spirits; he typically enters into a trance state during a ritual, and practices divination and healing. Shamans are said to treat ailments/illness by mending the soul, while at the same time prescribing ritually sanctified herbals or their extracts/juices. Shamanism cannot be strictly defined as medicine, although healing is its main objective. Shamans have a vast knowledge on the medicinal properties of plants. Thus, tribal medical system effectively combines belief systems and medicinal properties of plants.

In Chapter 7, Karuppusamy and Pullaiah have given an elaborate account on the ethanomedicinal plants of Eastern Ghats and the adjacent Deccan region of south India. They have not only given a detailed list on medicinal and aromatic plants but also the ailments for which these medicinal plants are used. Also given in this chapter are the accounts on the indigenous medicinal systems followed in this region as well as on the major types of medical formulations used.

1.9 PLANTS OF ETHNOVETERINARY IMPORTANCE

India is blessed with a very rich animal biodiversity. It is also one of the important centers of animal domestication. Animal domestication, particularly of cattle, was perfected predominantly by the pastoral people, generally called *Yadavas* (in ancient Tamil country these people were known as *Idaiyars* or *Konars*), who were generally nomadic and pastoral initially but were settled subsequently. India is known to have 30 indigenous cattle breeds, 12 buffalo breeds, 20 breeds of goat, 40 breeds of sheep, 6 breeds of horse, 8 breeds of camel, 3 breeds of pig, 18 breeds of poultry (Aruna Kumara and Anand, 2006), and many breeds of dogs and cats. Farmers use livestock and many other animals as a source of milk, manure and fuel, as draft animals for plowing and carting, and as a source of animal protein; some are treated as pets. The most important breeds of cattle in the region covered in this book are Amruthmahal, Hallikar, Krishna valley, Ongole, Punganur, Baragur, Kangayam, Manapparai, Malaimadu, Pulikkulam, Toda buffalo, etc.

Traditional communities paid a good deal of attention to animal husbandry which included breeding, feeding and maintenance, and preventing and curing diseases that may afflict domesticated animals or livestock. Fodder from wild and cultivated resources was given importance in feeding domesticated and in attracting game animals. Indeed, many traditional communities demonstrated a considerable knowledge of both the nutritional quality of different grazing or forage plant species and the ecological interactions between particular wild species of plants and animals (Krishnamurthy et al., 2014). Traditional communities also paid very great attention to prevention, control and eradication of diseases of domesticated and game animals. Ethnoveterinary medicine is as old as the domestication of animals, and in India and many other tribal localities of the world this medicine is rich and efficient and plays an important social, religious and economic role in the life of traditional societies. It comprises of belief, knowledge, practices and skills pertaining to healthcare and management of livestock (Nair, 2006). More than 250 diseases and their preventive and curative herbs have been known. However, there is a great need for documenting local ethnoveterinary practices as well as to assess these practices and knowledge for their efficacy and safety. There is also a need to revitalize these practices. In Chapter 8, Hari babu et al. discuss and summarize our knowledge of the traditional societies of Eastern Ghats and adjacent Deccan region on local ethnoveterinary practices involving plants.

1.10 PLANTS THAT ARE USED FOR PURPOSES OTHER THAN FOOD AND MEDICINE

Species of plants provide an array of products, other than food and medicine, used by people. Certain of these plants are exploited from the wild, while others sustain humanity through cultivation. In spite of vast over-all development, plant biodiversity as a resource largely remains poorly understood, underexploited and inadequately documented. Knowledge on plant use from indigenous people has not yet been translated into wider use largely because of its non-availability to most people. Other than for food and medicine ethnic communities throughout the world have been exploiting, sustainably, plants around them as sources of horticultural and ornamental plants, timber, fiber, dyes, fuel and other renewable energy and a host of other products used in industry and commerce (Krishnamurthy, 2003). In Chapter 9, Chandrasekhara Reddy et al. have provided a de-tailed account on the traditionally used plants of Eastern Ghats and the adjacent Deccan region for purposes other than for food and medicine. It is evident from their account that many traditionally-used plants of this region should be brought into mainstream use not only through popula-tionization and cultivation but also through biotechnological tools and techniques.

1.11   CONSERVATION, DOCUMENTATION, AND MANAGEMENT OF TRADITIONAL KNOWLEDGE ON PLANTS

The idea that ethnic knowledge on plant biodiversity is worth conserving rests on several fundamental arguments including nostalgia and human ben-efits and needs. The innate desire to experience the great pleasure that ethnic plant biodiversity knowledge has given us is part of the nostalgic argument for its conservation, although this nostalgic argument should not push us into construing conservation as an act aimed at considering tribal biodiver-sity knowledge an untouchable entity. Conservation, on the contrary, should be considered as a philosophy of managing plants and other environmental resources in a sustainable way so that it does not despoil, exhaust, or extin-guish the resources and the values and uses they have.

The first and foremost effort towards conservation of traditional knowl-edge on plants is the serious and effective documentation of such knowl-edge since most, if not all, such knowledge have all along been passed on from one generation to another, mostly orally. It was possible that valuable information might have been already lost in this process. Documentation should begin at the most local level, for example, at each of the traditional communities of the world. An important example is the initiation of com-munity Biodiversity Registers by Gadgil and his coworkers (Gadgil et al., 1995; see details in Krishnamurthy, 2003). The creation of databases and

Networks on indigenous knowledge is another. As examples we may cite the Indigenous people's Biodiversity Network (IPBN) and SRISTI. Yet another effort is the running of newsletters like the ones on Eastern Ghats, medicinal plants, Seshaiyana, etc. with the help of Ministry of Environment and Forests, Government of India. Mention must also be made on the All India Coordinated Project on Ethnobotany and the National Agricultural Technology Project on germplasms of useful Indian plants as well as books written by various authors on ethnobotanical information pertaining to Indian indigenous communities. Bir Bahadur et al. have given an excellent account on the documentation efforts so far undertaken in Chapter 10.

Traditional communities themselves have been excellent conservators of their own knowledge and resources. Their approach is based more on cultural and social perspectives as well as on belief systems than on political and reductionist western scientific systems. A truly sustainable developmental approach is followed by them with the involvement of the entire community in benefit sharing of the common resources. Participatory resource management and use of the resources owned commonly by the entire community is one such approach. A detailed account on conservation strategies, documentation and management of ethno-knowledge followed for Eastern Ghats and the adjacent Deccan region is also provided in Chapter. These activities are made easy through developments in computer science and information technology in the last two to three decades. Chapter 11 written by Pal and Bir Bahadur deal in detail with computer applications in ethnobotany.

1.12   MAINSTREAMING TRADITIONAL BOTANICAL KNOWLEDGE

Attempts to make a wider use and application of indigenous knowledge system, because of the superiority of the ethno-directed approach, have begun to revolutionize the food, agriculture, health and other consumer sectors. Hence there is an increasing effort to mainstream traditional knowledge, popularize it and to exploit it through Bioprospecting. Such an effort has often resulted in biopiracy and deprivation of the legitimate rights to benefits of traditional societies which own this knowledge (see discussion in Krishnamurthy, 2003). Bioprospecting looks for every valuable traditional genetic and/or biochemical resource that finds use in pharmaceutical, food, cosmetic, agricultural and biotechnological industries either through bioprocesses unique to the resource or through novel end or byproducts that can be obtained from it. The medical formulations found in codified and non-codified systems of medicine are now getting gradually subjected to phytochemical and therapeutic analysis to understand the chemical bases of their activity through reverse pharmacognosy and reverse pharmacology. The number of plants investigated so far in this way is very few and more traditionally used plants should be prioritized for analysis on a war-footing (Krishnamurthy, 2009). In Chapter 12, Pushpangadan et al. discuss some aspects related to bioprospecting, ethnopharmacology and patenting. The demand for ethnobotanicals is often not with adequate supply of genuine raw materials and is topped up, often, with substitutes or adulterants. As a result, we need standardization of ethnobotanicals through proper authentication using techniques like microscopy, phytochemistry and molecular biology.

1.13   CONCLUSIONS

There is no doubt of the considerable potential benefits that arise from ethnobotnaical research. There is also no doubt that these benefits have considerable importance to the sustainable economic development, particularly of rural areas, in spite of problems that may be associated with ethno-directed developmental projects. India, as discussed in the chapters of this volume, is an ethnobotanically rich country and, therefore, can develop into an economically sound country, if its ethnoresources are adequately conserved and sustainably exploited. It is also very clear from the chapters of this volume that attention needs to be focused in future not only on a utilitarian approach but also on a cultural and cognitive social-cultural approach, so that the development that would be achieved would be definitely holistic and not reductionistic. It is certain that ethnobotany-based development in India would indirectly contribute to the overall development of the whole world. One of the fundamental problems which, however, remains is the protection of the ethnic tribes themselves and also their bundle of rights, including the benefit-sharing rights. The other problem pertains to the standardization of ethnobotanicals using adequate techniques including techniques of ethnogenomics.

KEYWORDS

- **Belief Systems**
- **Deccan**
- **Eastern Ghats**
- **Ethnocrop Diversity**
- **Ethnofood Plants**
- **Ethnogenomics**
- **Ethnomedicinal Plants**
- **Ethnotaxonomy**
- **Worldview**

REFERENCES

Aruna Kumara, V.K. & Anand, A.S. (2006). An Initiative Towards the Conservation and Development of Indian Cattle Breeds. In: A.V. Balasubramanian & T.D. Nirmala Devi, (Eds.) Traditional Knowledge Systems of India and Sri Lanka. Centre for Indian Knowledge Systems, Chennai, India. pp. 104–113.

Berlin, B. (1992). Ethnobiological Classification: Principles of Categorization of Plants and Animals in Traditional Societies. Princeton, USA: Princeton University Press.

Berlin, R. Breedlove, D.E. & Raven, P.H. (1974). Principles of Tzeltol Plant Classification. New York: Academic Press.

Brown, C.H. (1984). Language and Living Things. Uniformities in Folk Classification and Naming. USA: Rutgers University Press.

Brown, C.H. (2000). Folk Classification. New Jersey, USA. In: P.E. Minnis (Ed.). Ethnobotany a Reader. Norman, USA: University Oklahoma Press, pp. 65–68.

Carroll, M.P. (1992). Allomotifs and the psychoanalytic study of folk narratives. *Folklore 103,* 225–234.

Conklin, H.C. (1954). The relation of Hanunóo Culture to the Plant world. PhD Thesis. USA: Yale University.

Conklin, H.C. (1974). The Relation of Hanunóo Culture to the Plant World (Yale University, PhD 1954). High Wycombe, USA: University Microfilms Ltd.

Cotton, C.M. (1996). Ethnobotany: Principles and Applications. John Wiley & Sons, Chichester.

Ellen, R.F. (1994). Putting plants in their place: anthropological approaches to understanding the ethnobotanical knowledge of rain forest populations. Presentation in UBD-RGS Conference.

Fuller, D.Q., Korisettar, R. & Venkatasubbiah, P.C. (2001). Southern Neolithic Cultivation Systems: A Reconstruction on Archaeobotanical Evidence. *South Asian Studies 17,* 171–187.

Gadgil, M. (1987). Diversity: Cultural and Biological. *TREE 2*, 369–373.

Gadgil, M., Devasia, P. & Seshagiri Rao, P.R. (1995). A comprehensive framework for nurturing practical ecological knowledge. Centre for Ecological Science, Indian Institute of Science, Bangalore, India.

Gadgil, M., Joshi, N.V., Manoharan, S. Patil, S. & Shambu Prasad, U.V. (1996). Peopling of India. In: D. Balasubramanian & N. Appaji Rao (Eds.). The Indian Human Heritage. Universities Press, Hyderabad, India. pp. 100–129.

Gadgil, M. & Thapar, R. (1990). Human Ecology in India. Some Historical Perspectives. *Interdisciplinary Sci. Rev. 15*, 209–223.

Harris, D.R. (2005). Origins and Spread of agriculture. In: G. Prance & M. Nesbitt (Eds.) The Cultural History of Plants. Routledge, New York. pp. 13–26.

Harshberger, J.W. (1896). The purposes of ethnobotany, *Bot. Gaz. 21*, 146–154.

Jarrige, J.F. & Meadow, R.H. (1980). The antecedents of civilization in the Indus Valley. *Scientific American 243*, 120–123.

Krishnamurthy, K.V. (2003). Text Book of Biodiversity. Science Publishers, Enfield (NH), USA.

Krishnamurthy, K.V. (2009). Ancient roots and modern shoots. Indigenous Biodiversity knowledge and its Relevance in Modern Science. Professor A. Gnanam Endowment Lecture, Madurai Kamaraj University Madurai, India.

Krishnamurthy, K.V., Murugan, R. & Ravikumar, K. (2014). Bioresources of the Eastern Ghats Their Conservation and Management. Bishen Singh Mahendra Pal Singh, Dehra Dun.

Kuhnlein, H.V., Erasmus, B. & Spigelski, D. (Eds.) (2009). Indigenous People's Food Systems: The Many Dimensions of Culture, Diversity and Environment for nutrition and Health. Centre for Indigenous People's Nutrition and Environment. Rome: FAO.

Levis-Strauss, C. (1966). The Savage Mind. London: Weidonfeld and Nicolson.

Malinowski, B. (1974). Magic, Science and Religion. London: Souvenir Press (reprinted 1925 edition).

Martin, G.J. (1995). Ethnobotany—A Conservation Manual. London: Chapman & Hall.

McNeely, J.A. & Pitt, D. (Eds.) (1985). Culture and Conservation. Dublin: Croom Helm.

Minnus, P.E. (2000). Famine Foods of the North American Desert Borderlands in Historical Context. In: P.E. Minnis (Ed.) Ethnobotany—A Reader. Norman, USA: University Oklahoma Press. pp. 214–239.

Murthy, M.S.R., Sudhakar, S., Jha, C.S., Sudhakar Reddy, C., Pujar, G.S., Roy, A., Gharai, B., Rajasekhar, G., Trivedi, S., Pattanaik, C., Babar, S., Sudha, K., Ambastha, K., Joseph, S., Karnatak, H., Roy, P.S., Brahmam, M., Dhal, N.K., Biswal, A.K., Mohapatra, A., Mohapatra, U.B., Misra, M.K., Mohapatra, P.K., Mishra, R., Raju, V.S., Murthy, E.N., Venkaiah, M., Venkata Raju, R.R., Bhakshu, L.M., Britto, S.J., Kannan, L., Rout, D.K., Behera, G. & Tripathi, S. (2007b). Vegetation land cover and Phytodiversity Charaterization at Landscape Level using Satellite Remote Sensing and Geographic information system in Eastern Ghats, India. *EPTRI-ENVIS Newsletter 13(1)*, 2–12.

Nair, M.N.B. (2006). Documentation and Assessment of Ehtnoveterinary practices from an Ayurvedic viewpoint. In: A.V. Balasubramanian & T.D. Nirmala Devi (Eds.) Traditional Knowledge Systems of India and Sri Lanka. Centre for Indian Knowledge Systems, Chennai, India. pp. 78–90.

Parthasarathy, J. (2002). Tribal People and Eastern Ghats: An Anthropological Perspective on Mountains and Indigenous Cultures in Tamil Nadu. In: Proc. Nat. Sem. Conserv. Eastern Ghats. ENVIS Centre, EPTRI, Hyderabad. pp. 442–450.

Reid, W.V., Laird, S.A., Gamez, R., Sittenfeld, A., Janzen, D.H., Gollin, M.A. & Juma, C. 1993. A New Lease on Life. In: W.V. Reid, S.A. Laird, C.A. Meyer, R. Gamez, A. Sittenfeld, D.H. Janzen, M.A. Gollin & C. Juma (Eds.). Biodiversity Prospecting, World Resources Institute, Washington, DC. pp. 1–52.

Singh, K.S. (1993). Peoples of India (1985–92). *Curr. Sci. 64,* 1–10.

Sittenberg, A. & Gamez, R. (1993). Biodiversity prospecting by INBIO. In: Reid, W., Laird, S.A. et al. (Eds.). Biodiversity Prospecting using genetic Resources for Sustainable Development. World Resources Institute, Washington, DC.

Srivastava, V.C. (2008). Introduction. pp. xxix–xxxiv. In: L. Gopal & V.C. Srivastava (Eds.) History of Science Philosophy and Culture in India (up to 1200 AD). Centre for Studies in Civilization, New Delhi.

Thangaraj, K. (2011). Evolution and migration of modern human: Inference from peopling of India. In: Symposium volume on 'New Facets of Evolutionary Biology.' Madras Christian College, Tambaram, Chennai, India. pp. 19–21.

Tylor, S.B. (1874). Primitive Culture. New York.

Valdiya, K.S. (2010). The Making of India: Geodynamic Evolution. New Delhi: MacMillan Publishers Ltd.

Vinodkumar (2007). Sustainable development perspectives of Eastern Ghats, Orissa. In: Proc. Natl. Sem. Conserv. Eastern Ghats. ENVIS center, EPTRI, Hyderabad, India. pp. 558–575.

Zimmer, H. (1935). The Art of Indian Asia. New York: Pantheon Books.

CHAPTER 2

ETHNIC TRIBAL DIVERSITY OF EASTERN GHATS AND ADJACENT DECCAN REGION

BIR BAHADUR,[1] RAZIA SULTANA,[2] K. V. KRISHNAMURTHY,[3] and S. JOHN ADAMS[4]

[1]*Department of Botany, Kakatiya University, Warangal–506009, India*

[2]*EPTRI, Gachibowli, Hyderabad–500032, India*

[3]*Department of Plant Science, Bharathidasan University, Tiruchirappalli–620024, India*

[4]*Department of Pharmacognosy, R&D, The Himalaya Drug Company, Makali, Bangalore, India*

CONTENTS

ABSTRACT

This chapter deals with the ethnic diversity of Eastern Ghats and the adjacent Deccan region. Emphasis is laid on the major ethnic tribes of Odisha, Undivided Andhra Pradesh, Tamil Nadu, and Karnataka. Ethnic tribal communities form a fairly a dominant percentage of the population of this area. These communities are the sources and holders of great knowledge on plants of cultural, social and utilitarian value. Not only the evolutionary origin of these ethnic communities, but also their social and cultural life are dealt with.

2.1 ORIGIN OF ETHNIC TRIBAL DIVERSITY

The term "tribe" means a group of people that have lived at a particular place from time immemorial. Anthropologically the tribe is a system of social organization which includes several local groups on lineage and normally includes a common territory, a common language and a common culture, a common name, political system, simple economy, religion and belief, primitive law and own knowledge system. India is culturally, linguistically religiously and ethnically a very diverse country. Hence, "Tribals" are found in almost all the States of India. Tribals constitute 8.14% of the total population of India, numbering 84.51 million (as per 2001 census) and cover about 15% of the country's area. Currently about 540 scheduled tribal communities exist. In terms of geographical distribution about 55% of tribals live in Central India, 28% in west, 12% in north-east India, 4% in South India and 1% elsewhere. These communities are actively working to preserve their rich cultures through broad institutional efforts. The strength of these communities varies from 31 people of Jarwa tribe to over 7 million Gonds. Thus, the Gonds form a very big tribal community, whereas the small communities comprising less than 1000 people include the Andamanese, Onge, Oraon, Munda, Mina, Khond and Saora. India is one among the top few countries with respect to its ethnic diversity (Singh, 1993; Vinod Kumar, 2002). Besides the ancient tribal communities, there is great ethnic diversity even among the mainstream people of India. It is derived from both the ANI and ASI populations (see later in this section for details). Although Hindus constitute the majority, there are also Muslims, Christians, Jains, Buddhists and Parsis. While ancient tribals invariably occupy forested and hilly tracts, the mainstream people occupy plains of India.

It is now more or less clear that the modern human species (*Homo sapiens*) originated in East African near Ethiopia around 200,000 years ago. It is also now known that the modern humans must have lived in Africa twice as long as anywhere else in the world. These details were evident from a study of mitochondrial genome (mt DNA) of females and Y chromosome of males of diverse primitive ethnic tribes of the world. As in the case of the origin of modern human species, its spread to different parts of the world is also deciphered by a study of mt DNA and Y chromosome. The earliest known mutation, which is found in all non-Africans, that helps to detect the human spread outside Africa is M168 in Y chromosome. This mutation had happened around 70,000 to 50,000 years ago (see Carney and Rosomoff, 2009). This mutation was followed by M9, which is common in all Eurasians and which first appeared in Middle East/Central Asia around 40,000 years ago. This was followed by M3 mutation, which arose in all Asian human populations that reached the Americas around 15,000 to 10,000 years ago. What made them to migrate out of Africa when they did so is still an unresolved mystery, although a few hypotheses have put forward (Scholz et al., 2007)

Which route did the modern humans take when they migrated out of Africa? Two paths lay open to Asia: (i) the path that led up the Nile valley, across the Sinai Peninsula and north into the Levant in Middle East. However, genetic data do not support this migration route; and (ii) from the horn of Africa via the mouth of Red Sea into Arabia and from there to central Asia, particularly Kazakhstan. From Kazakhstan humans got spread to other parts of Asia, Europe and Australia. Once in Asia, genetic evidence suggests that the population got split, one moving to Middle East, second to Europe, third to South East Asia and China (eventually reaching Siberia and Japan) and the last to Australia via India. Genetic data also indicate that humans in north Asia migrated eventually to Americas.

Migration of modern human species into India is the most complicated and discussed aspect of human spread. It is a well-known fact that India is remarkable for its rich ethnic diversity, as also for its plant (and animal) diversity. The ethnic diversity is due to India's geographical position at the tri-junction of African, the north Eurasian and the Oriental realms, as well as due to its great variety of environmental regimes. India's biological wealth has been attracting humans in many streams, at different times and from diverse directions of the old world. This had resulted in bringing together a great diversity of human genes as well as human cultures. This had also resulted in the mixing up between different ethnic groups. Hence, it is vital to focus on early human migrations into India in order to correctly understand the present day ethnic diversity that is seen in any region of the Indian subcontinent.

Gadgil et al. (1996) have made a fairly detailed discussion on the major migrations of humans into India. They speak of four major migrations: (i) the Austric language speakers soon after 65,000 years before the present (BP), probably from the north east; (iv) the Dravidian speakers in several waves after 46,000 years BP; (iii) the Indo-European speakers in several waves after 6,000 years BP; and (iv) the Sino-Tibetan speakers in several waves after 6,000 years BP.

Fairly recently, Thangaraj and his co-workers (see detailed literature in Thangaraj, 2011) have analyzed nearly 16,000 individuals from different ethnic populations of India (including several tribes) with genetic evolutionary markers (medinas and Y chromosome) to understand the genetic origins and structure of the ethnic Indian populations. In another genetic study the same group screened 560,123 SNPs across the genomes of 132 individuals belonging to 25 diverse groups from 14 Indian States (including Andaman) and six language groups. From these studies it was concluded by them that relatively small groups of ancestors founded most Indian groups, which then remained largely isolated from one another with limited cross-gene flows for long periods of time, perhaps 45,000 years BP. They have identified two main ancestral groups in India: (i) an "Ancestral North Indian (ANI)," and (iii) an "Ancient South Indian (ASI)." The first one is directly related to the Middle East, central Asia and Europe, while the second one is fairly indigenous (either not related to groups outside India or had some connection which is not yet established). Based on their studies, these authors have suggested three early major migrations from Africa to India: (i) *via* sea to Andaman; (ii) *via* land to South India through west coast (ASI population); and (iii) *via* land to North India (ANI population). From both ANI and ASI populations, the remaining parts of India were then populated.

2.2 ETHNODIVERISTY

Nearly one tenth of the total population of India lives in Eastern Ghats. According to Chauhan (1998) and Ratna Kumari et al. (2007) 54 tribal communities (nearly 34% of the total population of Eastern Ghats region) occur in Eastern Ghats but according to another estimate there are

62 tribal communities (Swain and Razia Sultana, 2009). In the northern Eastern Ghats alone (Odisha and undivided northern Andhra Pradesh north of Godavari river) 54 tribal communities with about 60,00,000 people have been reported (see Krishnamurthy et al., 2014). According to other estimates there are 63 ethnic communities in Odisha State alone, of which several live in the Eastern Ghats region (Merlin Franco et al., 2004; Sandhibigraha et al., 2007). There are 33 tribes (27 according to Pullaiah, 2001) in the undivided state of Andhra Pradesh with a population of 42 lakhs (Sastry, 2002). Of these 33 tribals, 27 inhabit Eastern Ghats (Ratna Kumari et al., 2007). Some of these tribes are common to Odisha and northern Tamil Nadu. There are 36 tribal communities in Tamil Nadu of which about 10 communities are associated with E. Ghats and adjacent regions. There are five tribals communities that are associated with the E. Ghats of Karnataka. Thus, there is no uniformity in past reports with reference to the number of ethnic communities in Eastern Ghats and the adjacent region and the problem is at least partly due to the fact that some tribes are known by more than one name in different parts of this study region. Most, if not all, traditional ethnic communities of the Deccan region are the earliest inhabitants and autochthonous people of the forest tracts. It is needless to emphasize here that all the ethnobotanical knowledge of this study region are the result of the contributions of these various ethnic tribals groups due to their long interaction with nature. The tribals use a variety of plant species in their daily life and are well-versed with knowledge of edible greens, vegetables, fruits, seeds, medicines and other materials.

This section deals with the most important tribals communities of E. Ghats and the adjacent Deccan region. A detailed list of ethnic tribals associated with E. Ghats is given in Krishnamurthy et al. (2014).

2.2.1 TRIBALS OF ODISHA

Odisha accounts for 3.47% population of India with a population density of 269 as against the national 342 per km^2. Tribals form a major share of Odisha population, have many sociocultural similarities and together they characterize the notion of tribalism. Although as many as 75 tribals have been reported 11 are the most important. Some of these are described here.

2.2.1.1   GONDS

Gonds or Gondi people are a Dravidian people of Central India, spread over the states of Madhya Pradesh, Eastern Maharashtra (Vidarbha), Chhattisgarh, Uttar Pradesh, Telangana and Odisha. With over seven million people, they are the largest tribe in Central India. They are also the most important tribe of Odisha. They are a designated Scheduled Tribe. The Gonds are also known as Raj Gonds. The term "Raj Gond" was widely used in 1950s, but has now become almost obsolete, probably because of the political eclipse of the Gond Rajas. The Gondi language is related to Telugu and other Dravidian languages. About half of Gonds speak Gondi language or 'kui' language while the rest speak Indo-Aryan languages including Hindi. According to the 1971 census, their population was 51.54 lakhs (5,154,000). By the 1991 census this had increased to 93.19 lakhs (9,319,000) and by 2001 census this was nearly 110 lakhs.

The Pardhan Gonds are a clan of the large Gond tribe inhabiting Central India. They traditionally served the larger tribal community as musicians, bardic priests and keepers of genealogies and sacred myths. With declining support for their traditional role, the Pardhan Gonds have adapted to making auspicious designs on the walls and floors of mud huts, acrylic paintings on canvas, pen and ink drawings, silkscreen prints and large-scale murals. Traditionally the Gondi people had a social institution (school) known as Ghotul, a kind of mixed dormitory system for the unmarried youth, which was the main means of education and introduction to adult life.

Gonds go out for collective hunts and eat fruits and roots they collect. They usually cook food with oil extracted from sal and mahua seeds. They also use medicinal plants. These practices make them mainly dependent on forest resources for their survival. Their religion is animistic, and their pantheon of gods includes 83 gods. Kandhamal district in Odisha has a 55% Gond population, and was named after a subtribe.

Dongria Gonds inhabit the steep slopes of the Niyamgiri Range of Koraput district and over the border into Kalahandi and work entirely on the steep slopes for their livelihood. The Niyamgiri Range provides a wealth of perennial springs and streams, which greatly enrich Dongria cultivation. Gonds also occupy northern parts of Telangana and Andhra Pradesh (Figure 2.1).

FIGURE 2.1   Two Gond Women.
(Source: http://indianholocaustmyfatherslifeandtime.blogspot.in/2010/02/unseen-lessons-of-history-taught-by.html)

2.2.1.2   SAVARAS

The Savaras are found inhabiting the Eastern Ghats of Odisha and undivided Andhra Pradesh. Their population is 1,05,465 (1991 census). The total literacy rate among Savara is 13.68. They build their settlements on hill slopes and near hill streams to facilitate easy access to terrace fields, and for fetching water. The most significant feature of the social organization of the Savaras is the absence of clan organization. For all practical social purposes, such as marriage, the group having a common surname is exogamous.

2.2.1.3   BHILS

Bhils or Bheels are primarily an aborigine Adivasi people of Central India, particularly of Odisha. They speak the Bhil languages, a subgroup of the Western Zone of the Indo-Aryan languages (Figure 2.2).

FIGURE 2.2   Bhils tribe.
(Source: http://tribes-of-india.blogspot.in/2008/09/bhils-tribes-of-india.html)

Bhils are divided into a number of endogamous territorial divisions, which in turn have a number of clans and lineages. The Ghoomar dance is one well-known aspect of Bhil culture.

2.2.1.4   BAGATA

The Bagata tribe is regarded to be one of the aboriginal tribes of India. Tribal communities reside in different parts of Odisha and in Northern Andhra Pradesh. Festivals, dance as well as musical bonanza make the culture of these Bagata tribes quite exquisite. Special mention must be made about the Dhimsa dance that has been practiced in the Bagata tribal society. It is a dance form where Bagata tribes of all ages participate quite energetically.

2.2.1.5   MUNDA

Munda tribe mainly inhabits Odisha. Hunting is the main occupation of the Munda tribe. Originally they were living in core forest areas of Odisha but now have been pushed to buffer zones (Figure 2.3).

FIGURE 2.3 Munda tribal woman.
(Source: http://tribes-of-india.blogspot.in/2008/10/munda-tribes-of-india.html)

2.2.1.6 SANTHAL

The Santhal tribe is the third largest tribes in India. Belonging to pre-Aryan period, and have been the great fighters, this tribe is found, Odisha, Santhal Tribe take pride in their past. Santhali is the prime language spoken by the Santhal tribe. This tribe also has a script of its own called Olchiki. Apart from Santhali they also speak Bengali, Oriya and Hindi.

2.2.1.7 GADABA

The Gabada tribe is one of the oldest and jovial tribes in India and are located in the southern fringes of the Koraput district. Gadabas are very friendly and hospitable. Their villages are with square or circular houses and conical roofs. The women are well-dressed and are fond of wearing ornaments generally made out of brass or aluminum.

2.2.1.8 JATAYA

Jataya tribe of Odisha is named after the mythological figure Jatayu of Ramayana epic.

2.2.2   TRIBALS OF UNDIVIDED ANDHRA PRADESH

Nearly 70% of the total population of undivided Andhra Pradesh lives in rural and forested areas. More than 35 ethnic tribes have been reported and the most important are discussed in the following subsections.

2.2.2.1   SAMANTHAS

The Samanthas of Visakhapatnam agency are one of the few traditional agricultural communities living in the Eastern Ghats of Andhra Pradesh and Odisha (Sathya Mohan, 2006). They speak "Kuvi," a language which is a brand of Telugu language. Samanthas clear the jungle on hill slopes, burn the trees and grow the crops in the ashes. Podu—the Slash and burn cultivation—is the major livelihood for these tribals. They used to cultivate a plot for six or seven years and leave the land fallow for about 10 years, by shifting their cultivation to another hill slope thus enabling the soil fertility of the old plot. Now-a-days, the fallow period is reduced to two or three years. The podu cultivation is simple and uses only hoe and human labor. Though the crop output is poor and not profitable, slash and burn cultivation is meant for their own survival. The remarkable feature of podu cultivation is that many varieties of cereals and vegetables can be grown in one plot (Figure 2.4).

FIGURE 2.4   (a) Samantha women; (b) Samantha community enjoying festivities through drinking, and smoking (Sathya Mohan, 2006; used with permission from ENVIS Division of EPTRI).

The Samanthas have a strong sense of community living. Each and every activity of the village including festivals is carried out by all the families working in close co-operation with each other and every household contributes for it. Slash and burn cultivation is initiated with a religious ritual. In February, during the seed festival known as "Biccha Parbu," the Samanthas worship the village Goddess "Jakiri Penu" by offering animal sacrifices. The Samanthas believe that sowing seeds mixed with the sacrificial blood will impress the fertile powers of Nature. They mainly grow dry paddy, ragi (*Eleusine coracana*), *sama* and *oliselu* (*Guizotia abyssinica*) in these fields.

Every family also cultivates kitchen garden crops like chillies, tobacco and vegetables in a small piece of land near their hamlets. Women and children collect minor forest produce of various types, such as edible and herbal roots, tubers and creepers, leaves and fruits. The Samanthas sell most of these products at the weekly shandies and buy commodities like kerosene, oil, salt, ornaments and clothing. Traditionally, the shandies have provided the people with an opportunity to barter their surplus produce. The distribution system earlier was limited to the tribal communities in the shandies of this area. But today, these market places have become the centers for commercial exploitation of the tribals by the traders from the plains.

Their economic activity is interdependent with their religious life, which consists of various Gods and Goddesses, who are symbols of various forces of Nature. They believe in absolute surrender of human spirits to the Natural forces. The availability of food in the jungle, the fertility of the Mother Earth, the rainfall and also the outbreak of epidemics are supposed to be dependent on the mercy /wrath of the respective Gods and Goddesses. In the event of an epidemic the Samanthas propitiate the Goddess of the disease known as "Ruga Penu". After worshipping they ceremonially send the Goddess out of the village. The religious sense of archaic oneness with Nature has characterized the many generations of traditional life among the Samanthas.

2.2.2.2   KOYAS

The Koyas are one of the few multi-lingual and multi-racial tribal communities (Sathya Mohan, 2006). They are also one of the major peasant tribes of Andhra Pradesh and Telangana numbering about 3.60 lakhs in 1981. Physically they are classified as Australoid. The Koyas call themselves "Koithur." The lands of Koithur includes those near the Indravati, Godavari, Sabari, Sileru rivers and the thickly wooded Eastern Ghats, covering parts of Bastar, Koraput, Warangal, Khammam, Karimnagar and the East and West Godavari districts. Most Koyas speak the Koyi language which a blend of Telugu (Figure 2.5).

FIGURE 2.5   Koya men with traditional head gear.
(Source: http://www.storypick.com/27-beautiful-photos-from-different-ethnic-tribes-of-india/)

The story of the Koyas dates back to pre-historic times. They seem to have had a highly evolved civilization in the past in which they were a ruling Tribe. According to the Koya mythology, life originated from water. The friction between the fourteen seas resulted in the emergence of moss, toads, fish and saints. The last saint was God and He first created *Tuniki* (*Diospyros melanoxylon*) and *Regu* (*Ziziphus mauritiana*) fruits. During the 18th century, the Marathas invaded and subverted the Koyas along with the Gonds. The continuous pillage and harassment by the non-tribals resulted in the loss of the vestige of Koya civilization. The Koyas were driven to take refuge in the inaccessible highlands. In this period they were depicted by travelers as treacherous savages.

There are many endogamous sub-divisions among the Koyas of Bhadrachalam agency, such as Racha Koya, Lingadari Koya, Kammara Koya and Arithi Koya. Each group is vocationally specialized having a separate judiciary system, which ensures group endogamy. There are also differences in food habits. Lingadari Koyas do not eat beef and do not interdine with others. They perform purificatory rites to depollute the effects of inter group marriages. The Racha Koyas are village administrators. They also perform rituals during festivals. Kammara Koyas make agricultural implements. They are blacksmiths and are generally paid in kind. Arithi Koyas are bards. They narrate the lineages. They are the oral custodians of Koya mythology. Each of these sub-divisions among the Koyas has exogamous phratries having separate totems, which are again split into a number of totemistic sects, which form the lineage ("velpu") pattern. For example, in Chinthur mandal of Bhadrachalam agency, the Paderu Gatta (phratry) of Racha Koyas worship "Dhoolraj" and their totem is wood. These phratries have a number of totemistic sects each denoted by a name, totem and worshipped by a group of families having separate names. For instance, Gatta worshippers of Bheemraj are further classified into three groups on the basis of their "Ilavelpulu" (family deities).

The Koyas have a patrilineal and patrilocal family called "Kutum." The nuclear family is the predominant type. Usually, sons in a family live separately after marriage, but continue to do joint cultivation (*Pottu Vyavasayam*) along with parents and unmarried brothers. Monogamy is prevalent among the Koyas. The preferential marriage rules favor mother's brother's daughter or the father's sister's daughter. Generally, the mate is selected through negotiations. But other practices of capture and elopement also exist, involving a simple ritual of pouring water on the girl – the water being the symbol of fertility. There is bride price involved in arranged marriages. Marriage is celebrated for three days. It is not simply an affair between two families. It is an occasion for two villages and all the relatives. Every person carries grain and liquor to a marriage to help the bridgroom's family. Marriages take place in summer when palm juice is abundantly available. The Bison-horn dance is a special feature on the occasion of a marriage ceremony among the Koyas. Birth, marriage and death are the three important aspects of life and each event is celebrated on a grand scale in Koya society. The funeral ceremony among the Koyas is strikingly peculiar. The corpse is carried on a cot accompanied by the kinsmen and villagers including women. They symbolically offer material objects like grains, liquor, new clothes, money and a cow's tail by placing them on a cot besides the corpse and the whole cot is placed on the pyre with the feet towards the west. They generally burn the corpse. The corpses of pregnant women and children below five months old are buried. They have a ceremony on the eleventh day after the death, which is called "Dinalu." At this time they believe that the spirit of the dead comes back and resides in the earthen pot called "Aanakunda." The occasion of death is a common concern in which all the relatives share the burden and expenditure of the family of the deceased. After the ceremony is over, they sing, dance and have a feast.

The major forest species exploited by Koyas are teak, bamboo, maddi (*Terminalia alata*) and cashew. The minor forest produce includes beedi leaves (*Diospyros melanoxylon*), gum, honey and tamarind. Sorghum is the staple crop and rice and tobacco are grown along the river banks.

There are 89 Koya villages and a small town in Chintur mandal with a population density of 123 persons per sq.km. Agriculture is totally dependent on rains. Owing to small land holdings (the average land-holding per family is 2.0 acres wet and 4.1 acres dry land) and no irrigation facilities, about 55% of the families continue practicing slash and burn (podu) cultivation, while 10% of the population is landless. Due to the limited availability of land for cultivation, total dependence on rain for irrigation and the growing population pressure over the Koyaland, the agriculture activity of the Koyas has become predominantly a subsistence way of farming. The ecological surroundings – especially forests – provide the Koyas with food, beverages, fodder, shelter and medicinal herbs.

Though the Koyas are farmers by occupation, most of their food supplies are drawn from the forest. Roots and fruits form their subsidiary food. They eat *Keski dumpa* and *Karsi dumpa*, which are the common roots available in this region. They cut these roots into pieces, keep them in running water for three days and boil them to make them edible. During drought years the Koyas go in groups into the forest to collect these roots in large quantities. Their staple diet is sorghum. They grow several varieties of sorghum (*Konda jonna, Pacha jonna*, etc.) and a few pulses. Rice is also grown in a few wetlands.

The Koyas also collect various forest products to supplement their meager agricultural returns. They sell these products in the weekly shandy and buy other required commodities. There is no other monetary transaction among the Koyas except in the shandy. On the whole only 0.4% of the agricultural produce is sold.

Joint cultivation, known as "Pottu Vyavasayam" is a common practice among the Koyas. Landless families go with their agricultural implements and join those who own land. The yield is shared between the landowner and others who have contributed labor. This practice ensures unity within the group and avoids further division of land holdings.

The Koyas are expert hunters and the good hunters are looked upon as heroes. For the Koyas, hunting is an essential skill for food as well as for defense from wild animals in the forest. On the occasion of the "Vijja Pandum" (the festival of seeds), Koyas go hunting in groups. Fish is another important food for the Koyas. In villages near rivers, quite often fish is a meal for every family. They ensure fair share of fish to all. The Koyas use various types of nets tied to bamboo poles, which are used in still waters.

During the toddy palm season, every Koya family lives mainly on palm juice for almost four months. For them palm juice is not just a beverage, but also a complete food. On an average, every Koya family owns at least four to eight palm trees. Palm juice is consumed three to four times a day in large community gatherings known as "gujjadis." The Koyas consider the palm tree as a gift of nature and to secure this gift they worship the village Goddess "Muthyalamma." On all social and religious occasion, liquor plays an important role among the Koyas. The "Ippa Sara" or the mohuva drink

(*Madhuca longifolia*) is purely an intoxicating beverage. The Koyas consume mohuva liquor to get relief from the physical hardship of the day and to withstand extreme variations in the climate.

The houses are built within one's own agricultural land. These are rectangular and are built of the material that is available from the forest. These houses are constructed on an elevation of two to three feet with walls made of bamboo, plastered with mud and roofed with palm leaves. They are leakproof, quite warm during winter and cool during summer. Most of their festivals are related to agricultural operations. Kolupu is one such occasion, which comes during November. The Koyas worship the Earth-Goddess "Bhudevi" and they enlist the co-operation of the Goddess by offering animal sacrifices during the festival. They believe that sowing seeds that are soaked in sacrificial blood brings them good crops. The Koyas deify their ancestors and worship them on all social occasions. The entire clan members join together to worship their ancestors. The Koyas believe in four guardian deities who are supposed to control the four directions. The Koya pantheon consists of various gods and goddesses who are the symbols of various forces. Among them Bhima, Muthyalamma, Sammakka and Sarakka are worshipped by non-tribals of the surrounding regions as well. The sense of supernaturalism is strongly rooted in the Koya's concept of nature. They worship personal spirits, which are thought to animate nature. They also believe in evil spirits that are dangerous to the harmony of group life. The traditional medicine man "Buggivadde" and the sorcerer "Vejji" are supposed to ward off all kinds of evil spirits. The Koyas celebrate festivals indicating the onset of particular seasons for tapping palm juice, collecting mohuva flowers, beginning agricultural operations, hunting and fishing. Every koya village is a socio-political unit and also a part of a larger social and territorial unit called "Mutha," a cluster of villages linked by economic, political and kinship ties. The customary law of the Koyas ensures communal ownership of natural resources administered by the village headman known as "Pedda." The pedda is the senior-most person who first settled in the village and established the village Goddess. This position is held by descendents of the same family. Pedda controls the social, political and religious activities in the village. The village panchayat consists of the other members (Pina pedda, Vepari, Pujari, etc.) who deal with minor problems. Sometimes the pedda holds two or three positions in a panchayat. The village panchayat is the final authority over all issues in a village. The overall judicial system of a cluster of villages is maintained by the "Samithi Poyee," a judicial head who is assisted by the people known as "Veparis." Issues are dealt with in co-operation with the village panchayat and this makes every village a part of a wider cluster known as mutha and is held by tribal norms.

2.2.2.3 BANJARA

The Banjara are a class of nomadic people from the Indian State of Rajasthan, now spread all over Indian sub-continent. They claim themselves to have descended from Rajputs, and are also known as Lakha Banjara means Lakhapati, Banjari, Pindari, Bangala, Banjori, Banjuri, Brinjari, Lamani, Lamadi, Lambani, Labhani, Lambara, Lavani, Lemadi, Lumadale, Labhani Muka, Goola, Gurmarti, Dhadi, Gormati, Kora, Sugali, Sukali, Tanda, Vanjari, Vanzara, and Wanji. According to J.J Roy Burman's book titled, "Ethnography of a Denotified Tribe the *Laman* Banjara," the name Laman is popular long before the name Banjara and the Laman Banjaras originally came from Afghanistan before settling in Rajasthan and other parts of India. According to Motiraj Rethod also, the Lamans originally hailed from Afghanistan. Together with the *Domba*, they are sometimes called the "gypsies of India". They are known for colored dress, folk ornaments and bangles. The Lambadas are one of the largest tribes in Andhra Pradesh and Telangana and live in exclusive settlements of their own called Tandas, usually away from the main village. They tenaciously maintain their cultural and ethnic identity. The Lambadas believe that the world is protected by a multitude of spirits benign and malign. Hence the malignant spirits are periodically appeased through sacrifice and supplication called Tanda.

Banjara people celebrate *Teej* festival in a grand scale. The festival, which is celebrated during Shravan in the month of August, is considered as a festival of unmarried girls who pray for better grooms. Girls sow seeds in bamboo bowls and water them three times a day, for nine days and if the sprouts grow thick and high, it is considered as good omen for a better future groom. The bowls with seedlings are kept in a prominent place and girls sing and dance circling the bowl.

Folk art of Banjara people includes Dance, Rangoli, Embroidery, Tattooing, Music and Painting, of which embroidery and tattooing have special significance in the community. Lambani women specialize in preparing lepo embroidery on clothes by interweaving glass pieces in colorful clothes. The craft known as Sandur Lambani Craft made by Lambani people has got Registered Geographic Indication tag in India, enabling the community people to exclusively market them in that name.

Banjara people are generally classified as Hindus. They worship Hindu gods like Krishna, Balaji, Jagadamba Devi, Hanuman, etc. They also worship Sati Aayi, Seva Bhayya (or Sevalal), Mithu Bhukhiya, Banjara Devi, etc., which are specific to the community. Banjara Devi is located usually in forests in the form of a heap of stones. Mithu Bhukhiya was known as an expert dacoit of the tribe and the community pays high respect to him who is worshipped in a hut built in front of Tanda or village with a white flag on top. Nobody sleeps in the special hut built for Mithu Bhukhiya. Seva Bhaya or Seva Lal is another historic person who draws high respect from Banjara people. He became a saint and protector of women of the community. They speak Banjari language, also called Goar-boli, which belongs to Indo-Aryan group of languages and the language has no script. In India, Banjara people were transporters of goods from one place to other and the goods they transported included salt, grains, firewood and cattle.

2.2.2.4   KONDA REDDY

Konda Reddy is a community that prefers to remain unmarried than to stir out of its habitation mostly in Khammam, Telangana. They are concentrated in Pusukunta. They preferred to live in hills for decades by stonewalling the influence emanating from the civilized plains. Living far from the maddening crowd, the clan is still primitive. Apprehensive of losing their rights over minor forest produce, which is their source of livelihood, they turned down the offer of land and pucca houses as part of the rehabilitation programmers. The 'Konda Reddy' tribe is against exchanging Pusukunta for any kind of marriage proposal.

2.2.2.5   VALMIKI

It is a Dalit sect of Hinduism centered on the sage Valmiki. The community is found in Punjab, Rajasthan, Andhra Pradesh, Karnataka, and Gujarat and Punjab, Rajasthan (Figure 2.6).

FIGURE 2.6   Valmiki composes the Ramayana.
(Source: https://en.wikipedia.org/wiki/Valmiki)

They worship Valmiki as their ancestor and God. They consider his works, the Ramayana and the Yoga Vasistha, as their holy scripture. In the state of Andhra Pradesh, Valmikis are referred to as Boyas. The titles of the Boyas are said to be Naidu (or Nayudu), Naik, Dora, Dorabidda (children of chieftains), and Valmiki.

2.2.2.6   KONDA DORA

Visakhapatnam district of Andhra Pradesh is known for Konda Dora tribe. Konda Dora tribe is divided into a number of clans, such as Korra, Killo, Swabi, Ontalu, Kimud, Pangi, Paralek, Mandelek, Bidaka, Somelunger, Surrek, Goolorigune, Olijukula, etc., Konda Dora are very dominant in the district. These tribal communities are considered to be forest dwellers living in harmony with their environment. They depend heavily on plants and plant products for making food, forage, fire, beverages and drinks, dye stuff and coloring matters, edible and non-edible oils, construction of dwellings, making household implements, in religious ceremonies, magico-religious rituals, etc. A close association with nature has enabled these tribal people to observe and scrutinize the rich flora and fauna around them for developing their own traditional knowledge and over the years, they have developed a great deal of knowledge on the use of plants and plant products as herbal remedies for various ailments.

2.2.2.7   CHENCHUS

The Chenchus are a designated and conservative Scheduled Tribe community found in the Indian States of Andhra Pradesh, Telangana, Northern Karnataka and Odisha but predominantly undivided Andhra Pradesh. They are an aboriginal tribe whose traditional way of life is based on hunting and gathering, particularly in the Nallamalai forests or Andhra Pradesh. The Chenchus speak the Chenchu language, a member of the Telugu branch of the Dravidian language family. Some Chenchus have specialized in collecting forest products for sale to non-tribal people. Many Chenchus live in the dense Nallamala forest of Andhra Pradesh for hundreds of years.

The Chenchus are unfazed by their natural surroundings. The bow and arrow and a small knife is all the Chenchus possess to hunt and live. The slender build of their bodies is deceptive and express little of their strong and resilient nature.

The dark-complexioned Chenchus are one of the primitive tribal groups that are dependent on forests and do not cultivate land but hunt with bow and arrow for a living. The Chenchus have responded rather unenthusiastically to government efforts to induce them to take up farming themselves. They prefer to live in the enclosed space and geography leading a life of an unbroken continuity (Sathya Mohan, 2006). Their meal is simple and usually consists of gruel made from jowar or maize, and boiled or cooked jungle tubers. The Chenchus collect jungle products like roots, fruits, tubers, beedi leaf, mohua flower, honey, gum, tamarind and green leaf and sell them. They have hardly developed any technique for preserving their food. The Chenchu village is known as 'penta' with a few huts scattered here and there. The village elder is the authority. The Chenchus are a broad exogamous group and are basically Hindus. The marriage ceremony is performed with traditional rituals in front of the community and the priest or *Kularaju* officiated over the marriage rites. The Chenchus have a strong belief system and worship their deities like Lingamayya, Maissamma/Peddamma, Particularly during July/August. Their celebrations are austere, serene, simple and sometimes can be wild, intoxicating and mystical.

2.2.2.8 MALIS

Tribes are predominantly found in tribal areas of Visakhapatnam, Vizianagaram and Srikakulam districts. They are also called Mahali and Malli. Their population according to 1991 census is 2925. The total literacy rate among Mali is 17.47. The traditional dormitories, known as 'Kuppus' were once popular in this community. Marriage by negotiations, marriage by mutual love and elopement, marriage by service are different ways of acquiring mates.

2.2.2.9 KOTIA

Kotia is chiefly found in the tribal areas of Visakhapatnam district of Andhra Pradesh and adjoining Odisha; their population as per 1991 census is 41,591 and their total literacy rate was 17.83. Four types of acquiring mates are in vogue in this community. They marry by negotiations, by mutual love and elopement, by capture and marriage by service. Divorce is permitted. Widow or widower re-marriages are permissible (Figure 2.7).

FIGURE 2.7   A Kotia woman.
(Source: https://www.pinterest.com/pin/24629129183141977/)

2.2.2.10. KOLAMS

Kolam tribe inhabits areas of Telangana (and Maharashtra). They mainly speak Kolami language, but are also fluent in Marathi, Telugu and Gondi languages. They follow Hindu based rituals and ceremonies. Agriculture and working in the forests are the occupations. Kolam tribe return barefoot after performing a pilgrimage. Drum or tappate and bamboo flute or vas are the musical instruments used in festive occasions.

2.2.2.11 ANDH

The Andh are one of the tribes of India living in the hilly tracts of Adilabad in Telangana. They are further subdivided into the Vertali and the Khaltali. Marriage by negotiations is common among Andhs but marriage by intrusion is also prevalent. Widow remarriages are permitted among Andhs. Divorce is permitted. They mainly subsist on agriculture followed by agricultural labor. They partly subsist on collection of forest produce, hunting and fishing.

2.2.2.12 YANADIS

Yanadis are one of the most primitive tribes and occupy the forest areas of East Godavari in Andhra Pradesh and a part of Khammam District in Telangana state. They occupy of the Godavari river in the forests. The Yanadis of Nallamalais are essentially hunters and foragers and their settlement patterns reflect this type of activity. Their huts are made of bamboos. They frequently migrate from one place to another when resources are exhausted, but this migration is only within a short radius. They use simple traps and snares and catch small rodents for their food (Parthasarathy, 2002).

2.2.2.13 YERUKALA

The Yerukala tribe is considered as one among the 33 scheduled tribes of Andhra Pradesh and Telangana (India census 2006). It is a semi-nomadic tribe inhabiting the plains. The people of this tribe are traditionally basket makers and swine herders. Though live mostly in multi-caste villages, maintaining symbiotic relations with non-tribals, they cultivate their unique beliefs and practices. They are considered to be the native of Southern Andhra Pradesh but now largely occupy Telangana, particularly in Warangal district. The traditional healers of Yerukala ethnic community have been using around 30 plant species for various formulations to cure chronic disorders.

2.2.3   TRIBALS OF KARNATAKA

A blend of culture, religion and ethnicity is represented by the tribes of Karnataka. More than 55 ethnic tribes have been reported. The most important tribes of Eastern Ghats of Karnataka State includes Bedar, Hakkipikki, Kadu kuruba, Kattunayakan, Konda kapus, Sholaga, etc. These tribes of Karnataka have built their settlements in several hilly and mountainous areas like Sandur, Chitradurga, Nandi and BR hills. As far as the languages are concerned, the tribes of Karnataka state converse with each other in different languages. Kananda language is the main language. Following the tradition of most of the tribes of the whole country, these tribes of Karnataka too have taken to diverse religions, although Hinduism is the most prevalent religion. Several other tribal communities of Karnataka possess their distinct tradition and ethnicity. They communicate in their local dialect and they also maintain their own tradition. Some of them are also reckoned as being originated from the warrior race.

The tribes of Karnataka are also known for their costumes, cultural habits, folk dances and songs, foods and their way of celebrating different festivals and occasions. A renowned dance format of the tribal communities of Karnataka is the open-air folk theater, better known as Bayalata. The theme of this dance drama centers around several mythological stories. This dance is executed at religious festivals and various social and family occasions. Generally these festivals start at night and carry on till quite a long period of time until day break.

The Bedar tribes belong to the Dravidian language family group. The Bedar tribal community can be found in several places of Karnataka. They are also known as Beda, Berad, Boya, Bendar, Berar, Burar, Ramoshi, Talwar, Byadar, and Valmiki. The word 'bedar' has an etymological significance and is derived from the word bed or bedaru, which signifies a fearless hunter. The ancestors of Bedar tribes may be the Pindaris or Tirole Kunbis. Within the Bedar tribal community, there are few Hindus and are called Bedar while the Muslims are referred to as Berad. The societal structure of the Bedar tribal community is quite significant. The Bedar tribe has six social groups. They have their indigenous customs and traditions. They eat meat and also drink liquor. Just like many of the tribal communities, the institution of marriage is given prime importance in Bedar tribal communities. The proposal of marriage usually comes from the parents of the bridegroom. Although child marriage is prevalent in the Bedar society, the bride does not reside with her husband till her puberty. Marriage within the subgroup of the Bedar community is not allowed. Widow re-marriage and divorce are permitted. In matters of administration, especially in case of disputable matters, the Bedar tribes take the help of the village headmen, popularly called Kattimani. Bedar tribal community has developed immense faith on various practices related to religion and spiritualism like fortune telling, magic and astrology.

Kadu Kuruba tribals of Karnataka live in forests as their name indicates. Cultural excellence is widely being depicted in all its aspects like dance, language, religion, festival, etc. by this community. Just like many of the tribal communities of Indian subcontinent, Kuruba tribes also are the ardent followers of Hinduism. To top of it, these kadu Kuruba tribes practice Halumatha, also known as palamatha by many people of the Indian Territory. The peculiar ritual of this Kadu Kuruba tribal community is that they revere 'Almighty Source' in a stone, which has been identified as Linga. According to the beliefs of these tribes, stone is the source for the soil, which in turn nourishes all the plants. Some anthropologist go to the extent of saying that the worship of stone as well as Shakti, in the form of deities like Yellamma, Renuka, Chowdamma, Kariyamma, Chamundi, Bhanashankari, Gullamma, etc. have originated from the tradition of Kadu Kuruba tribal community. However ancestral worship too has been incorporated in the religion of the Kadu Kuruba tribal community. Kadu Kuruba tribe is one of the significant tribes who have got the rich tradition of worshiping stone and also their predecessors with lots of festivity and enthusiasm. Apart from these tribal groups, the Kattunayakan tribe is said to be the descendants of the Pallavas. Collection of food is one of the chief professional activities of the Kattunayakan tribe.

Another important tribal group is the Sholaga tribe. Members of the scheduled Sholaga tribe converse with each other in the beautiful language of Sholaga. They are known by different names like Kadu Sholigau, Sholigar, Solaga, Soligar, Solanayakkans, Sholanayika. They follow of Hindu religion. They occupy the BR hills besides being present in Tamil Nadu's Timban-Sathyamangalam hills. As per 1981 census their population was 4828. They worship Madeshwara, the god of hunting. They are nonvegetarians. They were originally described to sleep naked around a fire lying on a few plantain leaves and covering themselves with others. They live chiefly on the summits of mountains where tigers do not frequent. Their huts are made on bamboos. The sholaga society is divided into 12 exogamous, partrilineal clans called *kulams* which regulate their marriage system. They bury the dead. They are hunters and food gatherers and were earlier known to practice shifting cultivation (Parthasarathy, 2002). Sholaga tribe are scattered in different parts of Karnataka, such as Mysore and BR hills. Sholaga tribal community of Karnataka is basically settled in several parts of this state and has enriched the culture and heritage of this state with their own distinctness.

Quite a number of this Sholaga tribal community collects various products from the forest areas. That this Sholaga tribal community is very much pious and the people of this group are religious minded. Hinduism is the main religion. However, many of the members of this Sholaga tribal community have still retained the local practices and customs of this community.

Lingayat Mathpatis often act as their priests. Janai, Jokhai, Khandoba, Hanmappa, Ambabai, Jotiba, Khandoba are some of the supreme deities of the Bedar tribal community. Images from deities like Durgava, Maruti, Venkatesh, Yellamma and Mallikarjun are made from silver, copper or brass. Cultural exuberance of the whole of the Bedar tribal community has nicely being depicted in all its aspects like festivals, language, jewelries, etc. The people communicate in Bedar language. Both Bedar males and females are very fond of wearing ornaments that are made up mainly from silver and gold. In addition, Bedar females place their hair in loose knots, wear several other ornaments like nose-rings and a gold necklace. Moreover there are quite a handful of Bedar tribes who also shave their heads, according to the custom. Tattooing also is a special custom of these Bedars. Males and females of the Bedar tribe do tattooing on the several parts like forehead, corners of the eyes and forearms. Rites, rituals and customs are part of the Bedar community.

2.2.4 TRIBALS OF TAMIL NADU

According to the 2001 census the tribal population of Tamil Nadu is 6,51,321. There are around 38 tribes and sub-tribes in Tamil Nadu. The tribal people are predominantly farmers and cultivators and most of them are much dependent on forest lands.

2.2.4.1 KURUMBA

Kurumbas are jungle-dwellers. They are generally believed to be the descendants of the Pallavas. They have settled in scattered settlements. The tribe is divided into five groups. Shola Nayakkars, Mullu, Urali, Beta and Alu. They have flat noses, wedge-shaped faces, hollow cheeks and prominent cheekbones, slightly pointed chins, and dark complexion. The women wear a waist cloth and sometimes a square cloth that comes up to the knees. Ornaments play a major part in the costumes. They have their distinct culture, tradition, religious customs and social practices. Kurumbas are one of the backward tribes of Tamil Nadu with no accessibility to modern amenities. Many of them are illiterate. Many Kurumbas were known for their black magic and witchcraft in the past. They speak the distinctive Kurumba language. The Kurumba art uses four colors traditionally: red (from red soil), white (from both soil), black (from the bark of Karimaram), and green (from the leaves of Kattu Aavaarai).

The Uraly subtribe of Kurumba tribe Tamil Nadu is found only in the Eastern Ghats in the districts of Erode and Salem. Their population as per 1981 census was 9224. The term Uraly means a village person. According to Parthasarathy (2002) they occupy the low land region between Eastern Ghat peaks of Salem and Erode. They speak Kannada and Tamil. They are non-vegetarians although ragi is their staple food. The Uraly society is divided into seven exogamous class called Kulams (Kanar, Kodiyari, Ennayari, Onar, Thuriyar, Vethi and Vayanar). Married women wear a Thali around their neck. The society is patrilineal. They bury the dead and perform ancestral worship. They practiced shifting cultivation in the past but no longer do it now. For their living they collect fruits, trap wild animals and birds. Women are skilled at making fine mats and baskets out of reed and bamboo.

2.2.4.2 IRULAS

Irulas are an important tribal community of Tamil Nadu and occupy Coimbatore, Erode and part of Nilgiri districts. Their total population is around 25,000. Nearby 100 Vettakad Irula settlements are found in forest areas and in the deep mountainous jungles. Traditionally Irulas main occupation has been snake and rat catching, but also work as laborers in agricultural fields during sowing and harvesting. They also are involved in fishing. This tribe produces honey, fruits, herbs, roots, gum, dyes, etc. and trades them with the people in the plains. They speak the Irula language, a member of the Dravidian family. The Irulas are related to the mainstream Tamils, Yerukalas and Sholagas.

2.2.4.3 MALAYALIS

One of the most important and largest scheduled ethnic tribes of Tamil Nadu is the Malayali tribe. They occupy the Eastern Ghats hills of Tamil Nadu, such as Kolli, Shervaroys, Pacha malais, Kalrayans, Javad, Pudurnadu and Yelagiri. According to 1981 census the population of Malayalis was 2,09,040. They are Hindus. The tribes consists of cultivators, woodmen and shepherds and are not uncivilized. They migrated to hills in fairly recent times. They are believed to be Vellalars who have migrated to hills due either to Muslim rule or due to fearing for conversion to Vaishnavites. They speak the Tamil language. They are divided into two endogamous groups on the basis of their tattooing marks on their bodies (Parthasarthy, 2002): Karalar gounders without tattooing marks and Vellalar gounders with tattooing. The society is patrilineal. The entire composite extended families are involved in cultivation and share the produce. Malayali marriages are by elopement, force or by negotiation. The Malayali traditional political organization has close affiliation with their territory within the hills. The territories are called *nadus* and are headed by Nattans.

KEYWORDS

- **Cultural Diversity**
- **Eastern Ghats**
- **Ethnic Diversity**
- **Ethnic Societies**
- **Origin of Ethnic Tribes**

REFERENCES

Carney, J.A. & Rosomoff, R.N. (2009). In the Shadow of Slavery. Berkeley, USA: University California Press.

Chauhan, K.P.S. (1998). Framework for conservation and sustainable use of biological diversity: Action plan for the Eastern Ghats region. pp. 345–358. In: The Eastern Ghats—Proc. Natl. Sem. Conserv. Eastern Ghats. Hyderabad, India: ENVIS Center, EPTRI.

Gadgil, M., Joshi, N.V., Manoharan, S., Patil, S. & Shambu Prasad, G.V. (1996). Peopling of India. pp. 100–129. In: D. Balasubramanian & N. Appaji Rao (Eds.). The Indian Human Heritage, Hyderabad: Universities Press.

Krishnamurthy, K.V., Murugan, R. & Ravi Kumar, K. (2014). Bioresources of the Eastern Ghats. Their Conservation and Management. Bishen Singh Mahendra Pal Singh, Dehra Dun.

Merlin Franco, F. Narasimhan, D. & William Stanley, D. (2004). Patterns of Utilization of natural resources among tribal communities in the Koraput region. In: Abstracts- Natl.

Sem. New Frontiers in Plant Biodiversity Conservation. TBGRI, Tiruvanandapuram, India. pp. 149.

Parthasarathy, J. (2002). Tribal people and Eastern Ghats: An anthropological perspective on mountain and indigenous cultures in Tamil Nadu. pp. 442–450. In: Proc. Natl. Sem. Conserv. Eastern Ghats. ENVIS Centre, EPTRI, Hyderabad, India.

Pullaiah, T. (2001). Draft Action Plan-Report on National Biodiversity Strategy and Action Plan: Eastern Ghats Ecoregion, Anantapur, India.

Ratna Kumari, M., Subba Rao, M.V. & Kumar, M.E. (2007). Tribal groups and their Podu type of Cultivation in Andhra Pradesh. In: Proc. Natl. Sem. Conserv. Eastern Ghats. ENVIS center, EPTRI, Hyderabad, India. pp. 384–385.

Sandhibigraha, G., Dhal, N.K. & Mohapata, B. (2007). Preliminary survey on folklore claims of anti-arthritic medicinal plants of Gandhmardan-Harishankar hill ranges of Orissa, India. In: Abstracts-Int. Sem. Changing Scenario in Angiosperm Systematic. Shivaji University, Kolhapur, India. pp. 162–163.

Sastry, V.N.V.K. (2002). Changing tribal economy in Eastern Ghats: Problems and Prospects. In: Proc. Natl. Sem. Conserv. Eastern Ghats. ENVIS center, EPTRI, Hyderabad, India. pp. 49–495.

Sathya Mohan, P.V. (2006). People. ENVIS-SNDP Newsletter. Special issue. pp. 10–17.

Scholz, C.A., Johnson, T.C., Cohen, A.S. et al. (2007). East African megadroughts between 135 and 75 thousand years ago and its bearing on early-modern human origins. *Proc. Natl. Acad. Sci. USA. 104,* 16416–16421.

Singh, K.S. (1993). Peoples of India (1985–92). *Curr. Sci. 64,* 1–10.

Swain, P.K. & Razia Sultana. (2009). Tribal Communities of Eastern Ghats. *EPTRI-ENVIS Newsletter 15*(2), 3–6.

Thangaraj, K. (2011). Evolution and migration of modern human: influence from peopling of India. In: Symp. Vol. on New Facets of Evolutionary Biology. Madras Christian College, Tambaram, Chennai, India. pp. 19–21.

Vinod Kumar (2007). Sustainable development perspectives of Eastern Ghats-Orissa. In: Proc. Natl. Sem. Conserve. Eastern Ghats. ENVIS center, EPTRI, Hyderabad, India. pp. 558–575.

CHAPTER 3

ETHNOBOTANY OF WORLDVIEWS AND BELIEF SYSTEMS OF EASTERN GHATS AND ADJACENT DECCAN REGION

K. V. KRISHNAMURTHY

Department of Plant Science, Bharathidasan University, Tiruchirappalli–620024, India

CONTENTS

ABSTRACT

This chapter deals with plants that are associated with the worldviews and belief systems of the ethnic communities of Eastern Ghats and the adjacent Deccan region of India. Plants have played a vital role in the evolution of societies and cultures of these ethnic communities. They are connected to the natural, spiritual and human domains and their interactions. The information on plants and their validity and importance in traditional worldviews and belief systems have been obtained through three approaches: utilitarian, ecocultural and cognitive. Plants often symbolize rituals and the correct interpretation of rituals provides insight into various aspects of human behavior while using a plant or its product. This chapter covers aspects relating to plants involved in life cycle, social, religious, agricultural and food rituals. It many cases, the origins of such plant symbolisms associated with various rituals have become largely obscured and unfortunately, only their so-called 'superstitious' and 'occasional irrational' tags remain, leading to the questioning of the rituals themselves. A critical reexamination of the importance of this plant symbolism in rituals is urgently needed.

3.1   INTRODUCTION

It is generally believed that societies, cultures, worldview and belief systems were simultaneously evolved in different regions of the world which formed parts of various migratory routes of modern human species (*Homo sapiens*) when it left the African continent around 70,000 years ago. It is also generally believed that the evolutions of all the above is inextricably related to one another and that this evolution was also dependent on the environmental conditions that existed in the different regions of original human occupation. It was also dependent on the threads of social ideas, cultures, worldviews and belief systems that the settling human populations already possessed on its long journey before settling down in a region. Hence, it is natural to find two distinct components in all the four: (i) the components that are common to all ancient traditional ethnic communities; and (ii) the components that are unique to each one of these communities. It is also possible that some of the first category components might have had a parallel or independent evolution in these communities.

The pockets of human populations that got settled on specific locations of the world and have been associated with these locations for a very long time are called indigenous or traditional communities. The interrelations

between different members of an indigenous community that enable better social life constitute social organization or society. All ethnic societies
normally have a distinct territory, are endogamous and are often subdivided
into clans, subclasses, classes, etc. The social relationship amongst members
of such endogamous ethnic groups is governed by kinship and mutual help.
Their beliefs extend these interrelationships from the social to the natural environment. Each of these societies pursue tradionally well-defined modes of
subsistence, have similar levels of access to various environmental resources
and are egalitarian (Gadgil and Thapar, 1990). A very important component
of the ethnic society is the *shaman*, the spiritual leader, and *shamanism* is
central to the well-being and the continued conservation of the ethnic society
and its culture, worldviews and belief systems.

It is often difficult to define culture and many definitions are available. It
is defined by Gadgil (1987), from a biological perspective, as the acquisition
of behavioral traits from conspecifics through the process of social learning.
These behavioral traits, which include traits related to knowledge, belief systems, arts, morals, laws, customs and any other capability and habit acquired
by people as members of a society, are socially transmitted from one generation to another (Tylor, 1874). Culture, thus, is a man-made component of the
environment (Parthasarathy, 2002).

The knowledge systems known to indigenous/traditional societies are
known as indigenous/traditional knowledge systems (IKS/TKS). TKS is
also known by terms, such as Worldviews and Cosmovisions. All these
terms refer to the different ways of perceiving, ethnic societies/communi-
ties. The different worldviews of various ethnic societies have come to gain
knowledge about the environment and its living and non-living components
around them. However, more often, worldviews are expressed by conceiving
life's knowledge obtained during a life time in terms of three inter-related
and often inseparable *domains* (or spheres or worlds): Natural (or Material),
Human (or Social) and Spiritual (Figure 3.1). The natural world provides the
message from the spiritual world to the human world.

It is to be emphasized here that, to a very large extent, the ability of an
ethnic community to sustainably use the environmental resources available
around it is determined by the above worldview. This worldview always
includes knowledge that is not only limited to the world which can be perceived with human senses and which can be explained in a rational/scientific
way, but also to a world beyond human perception and existing rational/
scientific explanation. Thus, knowledge, according to traditional worldview,
is a combination of that which is true and rationally explained by science,
and that which is believed by humans (=belief systems); thus, truth and

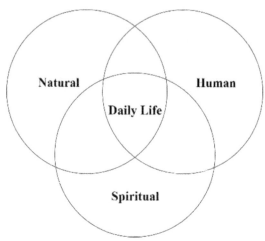

FIGURE 3.1   Interactions of three domains that results in worldviews.

belief go together. This worldview is also qualitative, practical, intuitive and holistic. TKS in the natural domain includes thematic areas related to specific resource generation/collection, agricultural, health and other practices. It includes knowledge about the physical world and the biological world and their constituents and material resources. The human domain implies the social life of people and includes knowledge about local organization, community life, family ties, local spiritual leadership, management of local resources, mutual help, conflict resolution, gender relations, and language and communication. The spiritual domain includes knowledge and beliefs about the invisible world, divine beings, spiritual forces and ancestors and transmitted through values and practices, such as rituals, festivals, etc. None of these three domains exists in isolation; all the domains together reflect the expressions of a unity. Some ethnic societies consider the confluence of these three domains as giving rise to a fourth domain/sphere called daily life domain. It is in this fourth domain that all the shared practices, such as the necessary techniques and technologies for the continuity of life and the social, material and spiritual reproduction take place for all humans. When a man respects his natural sphere and adapts himself to it spiritually and socially, nature will maintain its equilibrium and supply him what he needs. Thus, "all is related to all."

The ability of the ethnic communities to use the local resources to a large extent is determined by the above worldview. There are six categories of resources: (i) Natural resource (landscapes, ecosystem, climate, plants and animals); (ii) Human resources (Knowledge, skills, local concepts, learning ways and experiments); (iii) Human-made resources (buildings, infrastructure, tools, equipments, etc.); (iv) Economic and financial resources (markets, incomes, ownerships, price-relations, credits, etc.); (v) social resources (family, ethnic organizations, social institutions and leaderships); and (vi) Cultural resources (beliefs, norms, values, rituals, festivals, art, languages, lifestyle, etc.). This division is a further elaboration of the concept of three domains mentioned above.

Information relating to worldviews on plants and their validity and importance have largely been obtained through three main approaches (Cotton, 1996) in the ethnic societies of the study region, as in most other ethnic societies of the world: (i) Ecocultural approach; (ii) Cognitive and Socio-cultural approach; (iii) Utilitarian or Economic approach. The first approach invokes certain traditional cultural practices involved not only in foraging but also in settled agriculture, such as those involved in sowing, transplanting, and harvesting taboos and/or rituals. Many traditional cultural practices involving plants may at first appear irrational, but in reality these practices have very important and significant functional consequences. The cognitive approach tells us how the various ethnic societies perceive plants and also how such perceptions are influenced by sociocultural factors and spiritual beliefs. This approach invariably involves rituals/symbolic behaviors that fall in the realms of society, religion, magic, spirituality and supernatural domain. Utilitarian approach records how different ethnic cultures use the plants for their various materialistic needs. All the old ethnic communities in the study region are Dravidians. In the account that follows only ecocutural, cognitive and sociocultural approaches and plants associated with these approaches are dealt with; utilitarian approach to plant utilization is given importance in a few other chapters of this volume.

3.2 RITUALS AND PLANTS ASSOCIATED WITH THEM

3.2.1 DEFINITION AND TYPES OF RITUALS

Rituals may be defined as any serious and voluntarily conducted act/event at the individual, family or community level with the appropriate and correct behavioral and procedural formalities to gain impetus to lead a purposeful life in a serious and symbolic manner. Thus, rituals make people to lead life with serious and virtuous goals, as they add auspiciousness and sanctity to life (Eliade, 1959). Rituals help people to accept, without questioning, the importance and sanctity of energy and the myths and belief systems associated with this energy. Rituals are seen in all ethnic communities of the world, including those studied in this chapter, and are characteristic of all religions, whether basic or formal, primitive or advanced. According to basic religious and Hindu religious philosophies of India, which are mainly followed by the ethnic communities in the Eastern Ghats and adjacent Deccan region, human life is intimately connected to the cycle of life, right from birth until death and that between these two phases there is a sequential change from childhood to senescence. Since, human life in these ethnic societies is also culture-related, the biological phase changes are also subjected to the influences of culture. In other words, life cycle changes are also socio-cultural changes. In fact more than life cycle changes, it is at the socio-cultural level that life-phase changes get more polished. Each phase change gets culturally entangled to very specific ritualistic phenomena or gets itself ritualized. The ritualized phase changes become life cycle rituals. Birth, naming of the child, ear-boring, birthday celebrations, attaining puberty, marriage, nuptials, pregnancy, child-bearing, childbirth, and ultimately the death of the individual, etc. are the most important life phases which are ritualized in the ethnic societies of Eastern Ghats and adjacent Deccan region.

In addition to these life cycle rituals, there are other rituals that are associated with social, economic, religious, agricultural, food and arts/crafts activities of these ethnic communities. Thus, rituals are multifaceted. The nature of rituals changes with many factors, such as time, place and environment at which they are conducted, as well as on the aim and need for the conduct of a ritual. Although the rituals of each ethnic community have unique characteristics, we can also find common characteristics between them. The common ones are the following: Thought-related action-oriented, conduct-related, voluntary, visible, esthetic, periodic, repetitive, auspicious, collective, social, non-entertainmental and having rational and instrumental components. Each ritual also has a unique role, unique character, unique action and unique way of conduct. Cults like *totemism* (see a subsequent page for more details) followed in many ethnic communities in the study region depend on belief systems and practices associated with them. These belief systems and rituals are inseparable, not only because rituals are often the role manifestation of otherwise imperceptible ideas but also because they react upon and, thus, alter the nature of the ideas themselves. The rituals serve and can serve to sustain the vitality of beliefs and to keep the beliefs from being effaced from memory. Rituals help to protect the integrity of the society and its culture, keep them intact, remove all contradictions and oxymorons, bring about cooperation among people of the community, protect all the values that the society should preserve, and conserve societal relations, carry out the aspirations of the individuals of the society in order to make him a valuable component of the society, etc. Rituals of ethnic communities in the study region can be classified in various ways: technical *vs* ideological, positive *vs* negative, conformatory *vs* transformatory, separatory *vs* incorporatory, auspicious *vs* inauspicious, prescriptive *vs* performatory, etc. Ritual is known in ancient Tamil language as *Karanam*.

With all the rituals, plants or plant products are invariably associated. They often symbolize rituals and an analysis of these symbolic items basically involves the recurring themes of all rituals, the correct interpretation of which can provide insight into certain aspects of human behavior while using a plant or a plant product. In many cases the origins of such plant symbolism have become largely obscured and only their so-called 'superstitious' and occasional 'irrational' tags remain, leading to the questioning of the rituals themselves.

3.2.2   LIFE CYCLE RITUALS

3.2.2.1   BIRTH RITUALS

Birth rituals are very special for certain ethnic societies. Birth indicates the addition of one body and a soul to the earth. During this ritual the umbilical cord and the uterine liquid that comes out of the vagina of the mother are considered as very sacred and auspicious. These two are buried behind the house and the burial mound is considered sacred until 10th or 16th day after child birth depending on the community. The mound is enclosed by a fence made of Palmyra or coconut leaves and rachis. The mother takes bath within this fenced enclosure until the 10th or 16th day as the case may be. On that day three full plantain leaves are put in front of the mound on which are served cooked rice. *Sambar* prepared with salted dried fish or black pepper (depending on non-vegetarian or vegetarian family) is poured on the rice; also put are either boiled eggs or balls made of palmyra jaggery mixed with *Alpinia* rhizome paste. Then a wooden twig/stick is inserted vertically on the rice. Also kept near the banana leaves is a vessel filled with water and sprinkled with floral petals. This is followed by worship to God to protect both the mother and the newborn child. Then the food on the three leaves is eaten respectively by mother, the woman who helped her during childbirth, and the nurse who helped her medically to yield the child. The child is kept over the burial mound. The whole ritual is conducted by women only and men are not allowed. The burial pit plus the mound over it are considered as equivalent to the *Yoni* on vagina, the stick/ twig the male organ and the liquid *sambar* poured over cooked rice is the fluid that comes out of uterus along with the child (Paramasivan, 2001).

Couple who do not have a child and who want to adopt a child follow an interesting ritual. The couple gives a cup/vessel full of powered paddy husk to the person who gives her child for adoption by the couple. This means that the child for adoption is not got free but by paying husk in return.

3.2.2.2   MARRIAGE RITUALS

The most important ritual connected to marriage in many ethnic communities of the study region is the tying of the *mangala sutra* (*Thaali* in Tamil and Telugu) or the sacred cotton thread soaked in turmeric dye on the neck of the bride by the bridgegroom. Wearing this sacred thread until the death of husband is a must for the married women. Once this thread is tied, all relatives of both the bride and bridegroom (both patrilineal and matrilineal) bless the newly-married couple by showering on them with rice grains stained with turmeric dye. In the most ancient Dravidian, particularly Tamil, culture, there existed only secret love affair (called *Kalavu* in Tamil) that was either permitted or not by the parents and there was no marriage *per se*. However, sooner or later this *kalavu* system gave rise to the marriage system that was more and more tagged with chastity through wearing of *Thaali*. No remnant of Thaali was discovered in Adichanallur excavations conducted in the southern part of Tamil Nadu. Initially the thaali thread was tied along with flowers of *Jasminum auriculatum* (which represent chastity) before it is put on the bride. The most ancient foraging communities wear a black thread to which a collar bone (Clavicle) was tied. The dying of *thaali* with turmeric was done much later in Dravidian history but not earlier than 10th century CE according to many Tamil scholars (Paramasivan, 2001; Krishnamurthy, 2007). In the marriage hall/stage are kept many colored (often yellow/red) pots (called *Arasaani* pots) filled with turmeric water. These indicate that the marriage should lead to wealth, prosperity and fertility, as turmeric symbolizes auspiciousness. Some tribes also keep a pair of winnowing pan (*muram* in Tamil), made of bamboo, again tinged with turmeric dye. Betel leaves, and coconuts, plantain fruits, areca nuts, etc. are all kept on a pan in the marriage stage always in even numbers. Banana trees one on either side are kept on the entrance to the hall or on the marriage stage, often along with the inflorescence, again in pairs, of coconut, palmyra, phoenix species or very rarely in northern Eastern Ghats region *Corypha* species tied to the banana stems. The keeping of all the above in even numbers signifies *bilateral symmetry* that is reflected in the life of the people of these ethnic communities. This is in contrast to the death rituals (see later) where things including plants/ plant parts used are always in *odd numbers*.

The marriage hall/stage specially houses laticiferous plants or their twigs in the marriages of many ethnic communities of the study region. This symbolizes fertility of the woman to get married, especially the lactating ability of her breasts after child birth and the child-bearing fertility of her womb. This ritualistic symbol is characteristic of the matriarchal society, according to Turner (1963), although it should be mentioned that this ritual is followed in ethnic societies which are also patriarchal. Probably, all the ethnic societies were originally matriarchal and then some of them became patriarchal. Ritualistic plants, such as banana, coconut, palmyra, phoenix, *Corypha*, *Ficus*, Mango, turmeric and a few others used in the marriage venue are believed to convert the mortal bodies of the marriage couple into divine couple and godly status in the luminal space of the marriage hall. The wearing of floral garland and new cotton clothes and new cotton thread by the bride also helps to bring about this change.

3.2.2.3 DEATH RITUALS

In the study region, the ethnic communities either bury or burn the dead people depending on the community (Paramasivam, 2001; Rajan, 2004). The dead body is buried in specially devised structures in the ground or is put in a huge mud pot (*Thazhi* in Tamil) which is then buried. Along with the dead body grains of rice, sorghum or minor millets are put inside the pot or in the pit. Such pots have been discovered in several parts of ancient Tamil country through archaeological excavations (Rajan, 2004). Burning the dead is considered by some as more ancient than burying since the latter is often believed to have arisen along with the evolution of agriculture. Many death rituals are quite opposite to the ones that are conducted during the auspicious phases of the life cycle like birth or marriage. This is indicative of bilateral symmetry, as already mentioned. The death rituals are intimately connected to the concept of rebirth after death. The cycle of rebirth includes humans and animals, but does not include plants (i.e., plants after death do not have rebirth) in all the ethnic societies. However, as early Hindu texts analyze about what may happen to a dead person, plants are brought into the process. Burying the entire dead body or burying the bones of the dead after burning are considered as equivalent to sowing seeds of a crop plant and the emergence of seedlings from them as equivalent to rebirth/resurrection after getting a new life. Plants are also believed to feature as a destination or a way-station for parts of the dead person on journey after his death in the life cycle process. For example, his hairs go to the herbs and his head to the trees. In other words, the dead are born on earth again as "rice and barley, herbs and trees, as seasame plants and beans" (Radhakrishnan, 1993). Just before burying or burning the body of the deceased the patrilineal relatives stuff the mouth of the dead with rice grains. Thus, plants and their seeds are centered to the ritual mechanics that cycle the deceased into his next existence.

In the *Saanthi* or peace-making death ritual performed by many ethnic communities, after collecting the bones of the burnt dead person, flowers, wood, grass and butter/ghee are used by the survivors of the dead person to renew his life after the sweeping and eleaning the site of the mound using special ritual twigs of *Palasa* (*Butea frondosa*) tree, plowing the site and then sowing seeds of various herbs before pouring out the bones. Over the top of the mound, the worshipper sows paddy grains and covers the mound with *Avaka* plants for moisture and *Dharba* grass (*Imperata cylindrica*) for softness (Kane, 1990–1991).

Many offerings of cooked rice and vegetables are made throughout the death rituals. One particular instance that exemplifies the role of plants in death rituals is the *Sraardha* ritual. In this ritual offerings are made for the *pitroos*, the immediate paternal ancestors of the worhipper so as to enable the deceased to be received by them to be among them. This ritual uses *pindum* or ball of cooked rice, which represents the dead body. This *pindum* is worshipped using incense, flowers, ghee, white threads/clothes and a cup of water with sesame seeds. This worship is done normally for 10 days, although in some communities this period may be shortened or prolonged by a few days. Each day the number of cups is increased by one. It is believed that the use of *pindum* ritually creates the new body of the deceased for life in the community of ancestors and eventually for rebirth for himself among the world of humans. Thus, this ritual places plants at the center of the ritual's transformatory process. The material that makes up the *pindum* is invariably rice, although some minor millets are used in some ethnic communities of the study region, and barley in some communities in northern-most Eastern Ghats.

The Koyas of Eastern Ghats symbolically offer material objects like grains, palm liquor (from *Caryota urens*), new clothes, money and a cow's tail by placing them on a cot beside the corpse and the whole cot is placed on the funnel pyre with the feet of the corpse towards the west. They have a ceremony on the 11th day after death called 'Dinalu.' At this time they believe that the spirit of the dead comes back and resides in the earthen pot called '*anakunda*'

3.2.3 RELIGIOUS RITUALS

All the traditional ethnic societies of the study area follow Hinduism, which believes in the existence of many gods. It is well-known that Hinduism is a very diverse, varied and broad-based religion and has no identified founder. It originated in India perhaps as an amalgamation of vedic religion and the various basic religions of the different ancient ethnic groups. However, under the broader umbrella of Hindu religion, it is the worldviews of basic religions that are still dominant in most, if not all, ancient ethnic groups of the study region. As per basic religions, people considered themselves dependent on various natural forces, which are supposed to be controlled by Gods/spirits to whom they can pray, worship or make sacrifices for support (Krishnamurthy, 2005). The basic religions include animism, bongaism, totemism, naturalism, ancestral worship and polytheism (Hopfe and Woodward, 1998; Pushpangadan and Pradeep, 2008). Although basic religious customs, rituals and taboos still prevail in many cultures, there is a gradual transition from these basic aspects to those of mainstream Hinduism. Village/family gods/ goddesses are still the most popular among the tribal groups than formal and mainstream gods. The village gods/goddesses are believed to protect crops, irrigation water, harvested produce, etc.

The ultimate goal of all religions is reaching god and thereby truth. Religious show the way as to how humans must relate themselves to god and truth. Praying, worshipping and offerings are manifestations of this relationship. Because of their centrality in human lives, plants are often the objects of sacred attentions and religious activities. Besides being the objects of worship themselves, plants come to play important roles in religious practices, such as fuel for the sacred fire, as wood for ritual implements, as the materials (such as leaves, flowers, fruits, seeds and even whole plants) used for worship of god, as food offered for gods and priests, and as *prasaadams* given to devoters (Findly, 2008). For example, Lord Shiva himself is conceived as a *yupa* fashioned in *Khadira* or *sami wood*, and both woods are believed to have fire in them.

All the basic religious, and particularly totemism, involve some form of identification between tribes or between clans of a tribe and an animal or plant species (Zimmer, 1935; Whitehead, 1921; Subramanian, 1990) or a natural phenomenon/object. It is a belief in the mysterious relationship between plants (and animals) on the one hand and humans on the other and their veneration and propitiation as totem objects. The totem is often considered by many tribes as the ancestor of the clan and its relationship implies certain common characteristics as well as taboos in eating or foraging/hunting. Such a relationship is articulated as the concerned plant (or animal) is believed to have helped or protected the respective ancestors of a clan, or it had been proved to be of some use. The totem plant is held in great respect and reverence and it should not be hurt or destroyed. In some cases, its utilization is not normally permitted. The dead totemic plants (and animals) are given full honor and are attended to in their last rites. The best way thought by these ancient ethnic groups to protect plants (and animals) was to create a sense of fear, respect or faith in human minds towards the plants as abodes of gods/spirits. Deities living in various trees are not all the same, however, and, they reside there for different purposes. Tree deities are spirits of fertility, prosperity and protection. The early understanding of the nature of a deity inhabiting a plant is how the deity relates to the health of humans. That is, those plants with healing properties house deities of health and prosperity, while those that cause symptoms of disease house deities of wickedness and debilitation. Concerning the latter, many examples of plants are being used to protect humans from curses, witchcraft, sorcery and black magic-inflicting enemies. Trees can also house the spirits (*Preta* or ghosts) of the recently departed (dead) who, while still in states of transition, are unsatisfied and need human attention. Thus, plants are considered as sentiment beings equal to humans in that they, too, can become *pretas* –spirits in the process of being remembered on their way to rebirth. Services by the deity living in a tree are much like the services offered by the tree itself.

The concept of *Sthalavriksha* (=temple trees) is an extension of tree worship practiced in basic religions of ancient S.E. India, when these religions got transformed into formal religions, particularly when institutions of worship like temples developed. Even today many ethnic societies in this region worship trees. The names of presiding deities in many temples are based on or related to tree/plant names (Varadarajan, 1965; Kalyanam, 1970). In 270 temples (but 275 according to Sundara Sobitharaj, 1994) of Tamil Nadu, nearly half have one of the 80 species of trees/ plants as the *sthalavriksha*. These are mostly trees, although there are shrubs, subshrubs, herbs, climbers and even grasses (Swamy, 1978; Sundara Sobitharaj, 1994; Krishnamurthy, 2007). The concept of *sthalavriksha* must have been born just before 7th century CE in Tamil country as a result of the evolution of formal religion and the development of temples as major religious institutions. The idea that each temple should have a temple-plant seemed to have been formalized gradually. Krishnamurthy (2007) has given a comprehensive list of all the temple plants of Tamil Nadu and adjoining regions of Andhra Pradesh, Karnataka and Kerala, which formed part of the ancient Tamil country. These temples include not only *Saivite* (74 species of plants) but also *Vaishnavite* (18 plants) temples and 12 plants were common to both. The same plant species may serve as the temple plant of more than one temple. There are also a few temples with more than one temple tree. For instance, Bilva (*Aegle marmelos*) is the temple tree of 26 temples, *Vanni* (*Prosopis spicigera*) of 26, *Konrai* (*Cassia fistula*) of 22, *Punnai* (*Calophyllum inophyllum*) of 18, Jack tree (*Artocarpus integrifolia*) of 17, Mango (*Mangifera indica*) of 12, and palmyrah (*Borassus flabellifer*) of 9 temples. Another way in which trees/ plants were attached religious sanctity by the ethnic societies of this region has been to associate them with the 27 stars recognized and named by ancient Hindus. It is to be stated here that even before a temple was built, its temple tree was decided by the temple committee. The builder first mentions the place where the temple has to be built and the presiding deity to be enshrined in the temple. Subsequently, the architect and the priest identify the star related to this deity and accordingly, the shape and size of the idol are decided. The details on the plants assigned to each of the 27 stars are found in Krishnamurthy (2007).

Sacred groves, *Nandavanas* (temple gardens) and temple tanks that are associated with temples form another important aspect of sanctity and importance given to plants by the ethnic communities of the study region. Sacred groves are parts of natural vegetations that are always associated with the so-called minor temples that are given importance to by basic religions and are seen in village/rural/forested regions, in contrast to *Nandavanas* which are newly established in association with larger temples (of Hinduism) in towns/cities. In the former case, sacred groves come first and 'temples' are established subsequently, while the reverse is true in the latter. The concept of sacred groves can be said to have evolved even at the hunter-gather stage (Kosambi, 1962), while that of *Nandavanas* is of post-agricultural evolution. There are more than 50,000 sacred groves in India of which in the study region covered by this book there are about 3,000 groves; 500 of these are fairly well-known. The sacred groves are places of great power and people have traditionally set them aside as places of physical and spiritual benefits to humans and non-humans alike. They are one of the finest instances of traditional ways of reverence to plants. Plants in sacred groves cannot be cut and several other taboos are associated with sacred groves (Krishanmurthy, 2003). The deity in the temple associated with a sacred grove may be male or female. No one owns the sacred grove and it belongs to everyone in the community.

Temples and the sacred groves/*nandavanas*/tanks associated with them have strengthened the tie between religions and plants. In all cultural, educational and religious activities and rituals associated with temples, plants and their parts (leaves, flowers, fruits, seed, etc.) obtained from *nandavanas*/sacred groves and tanks are used; particularly for the daily worship of the deity and the offering of *Prasadams* to the devotes (see details in Krishnamurthy, 2007).

Specific species of plants are associated with specific gods in the formal Hindu religious practices of the study region, as is evident from literary and epigraphic evidences as well as from temple murals/paintings. Generally *Saivite* gods were worshipped with leaves of *bilva* (*Aegle marmelos*) and flowers of *Leucas aspera* and *Cassia fistula*, while *vaishnavite* gods with *tulsi* (*Ocimum* spp.) leaves, flowers of *Mimusops elengi*; Lord Ganesa is worshipped with the grass *Cynodon dactylon*. Certain plants or their parts were offered to all gods: coconut, betel leaf, betel nut, lotus flowers, *Nerium* flowers, etc. Each god was also assigned specifically a plant and its flower or leaf: *Cassia fistula* for Lord Shiva, *Neolamarckia cadamba* for Lord Muruga, *Nymphaea* spp. For goddess Lakshmi, *Nelumbo nucifera* (white variety) for Goddess Saraswati, *Mimusops elengi* for Lord Vishnu, *Saraca asoca* for Mahavir, *Nerium oleander* for goddess Kali, *Bauhinia* spp. for Lord Indra and so on. Later on, this specificity was corrupted and all the above were used for all gods/goddesses. Sandal paste is continued to be used for anointing the idols as also coconut water for holy bath of idols. Camphor was beginning to be used in the *Sanctum Sanctorium* of temples so as to enable the devotees to see the idol, while oil cloth wrapped on woody twigs was lighted in village temples (=*Teeppandam* or torch of fire) for the same purpose. The oil used for burning temple lamps was originally obtained from the seeds of *Madhuca longifolia*, although this was subsequently replaced by other oils, such as sesame oil.

Prepared plant food items were being offered to god, often as substitutes for live animals like chicken, goats, etc., or along with them as sacrifice. One of the oldest and most popular food items offered to village deities is *pongal*, a preparation made of rice, pulses and ghee. It is often prepared at the community level (Paramasivam, 2001; Krishnamurthy, 2007). The offered *pongal* was known as *Madi* or *Kulirthi* in Tamil. During the *Thaipoosam* religious festival *Pongal* is offered to God, by the whole community, along with *Colocasia*, elephant foot yam, palmyra seedling tuber, etc. *Pongal* is also a ritual food throughout the Tamil *Margazhi* month (December–January) for most communities. *Pongal* represents an offering to God from the raw plant to the cooked state of the same, a trend that was common to all ethnic societies. Some consider the pot in which *pongal* is prepared as the goddess, and the boiling rice as the destructions of male chauvinism. When *pongal* offering is done at the community level, each family prepares it by using only palmyra leaves as the fuel. Another food item offered by some traditional communities in the arid regions is *Paanakam* (a mixture of palmyra jaggery, tamarind, ginger and water). Yet another item is *ulundu sundal* (a preparation of boiled and salted black gram). In some societies, cooked rice is not offered, but only rice soaked in water and germinated pulses. These communities, on critical analysis, were found to be originally foraging nomadic communities. In S.E. Indian Hindu temples food items offered as sacrifice (*Padayal* in Tamil) to God and given as *prasaadams* to devotees were always vegetarian. In all likelihood this practice began in the 7th century CE, after the *Bakthi* movement. A partly dilapidated epigraph discovered in 1927 (epigraph number 127) mentions ten different items of food as dear to Lord Shiva, but details could not be obtained. A 9th century CE epigraph of king Varagunapandian found in Ambasamudram in Tirunelveli distirct, Tamil Nadu mentions a list of food items offered as ritual sacrifice including *Pulikkari, Pulugookkari, Porikkari*, Kummayam, etc. (Epigraphica Indica IX. No.10). A 11th century CE epigraph of king Veerarajendra Cholan (Epigraphica indica XXI. No.30) mentions, in addition to the above, *Milagukkari*. A 16th century Vijayanagara kingdoms's epigraph reveals that different kinds of food items were offered to Gods and these included vegetables, *Kariamudu, Vaaikkamudu, Neyyamudu, Kootu, Pachchadi, Kadukkorai, Aappam, Vedhuporiaappam, Vadai, Idli, Sugian, Daddiyannam* (Curd rice), etc. Some of these have either been lost, or have had name changes, so as not enabling us to identify them correctly.

Sacrifice to Gods/Spirits arose in Totemism which recognizes totems or emblems which are usually plants or animals. In all ethnic societies, animal sacrifice to Gods/Spirits was offered to begin with during cultural revolution and it is continued to be done even today in many societies at least on special religious rituals. Sooner or later animal sacrifice was replaced by plant sacrifice. Initially both animal and plant sacrifices were done and finally plants replaced animals totally in formal religious institutions like big temples. The blood of chicken or goat is mixed with rice and offered in cane baskets to Lord Muruga, according to ancient Tamil literary sources. Along with *Tinai* rice (a millet) and flowers, goat and chicken were also offered. However, blood was replaced by red dyes obtained from plants and these were mixed with rice or stuffed into lemon or ash gourd fruits and offered to Gods. Such a transition is also seen in many ethnic communities in other parts of the world like the Jews and Nuers. Evolution from animal sacrifice to plant sacrifice did not change the ideology and theoretic basis of a ritual but only brought about an operational and technical chance (Evans Pritchard, 1956).

3.2.4 AGRICULTURAL RITUALS

As in many other ethnic cultures of the world, most ethnic societies of Eastern Ghats and adjacent Deccan region follow agriculture for their sustenance and food security. The agricultural cycle in this region is hedged around with ritual and ceremonious activities directed towards furthering the powers of fertility manifest in soils and in various crops to ensure a good harvest. Hence, agricultural activities, such as planting, harvesting, weeding, etc. in these societies are associated with belief systems, taboos and rituals. Agriculture is treated as a pious and virtuous act by all these societies, as in ethnic societies elsewhere in the world (Salas, 1994). These rituals are related to the type of agricultural activity and the time of the year, and take place at various levels (individual/family/community), and include social conduct based on the concept of reciprocacity. At family level simple rituals are performed more or less on a daily basis at home or in the agricultural field to appease the natural spirits and to attract good luck and abundance of crop yield. Many rituals are combined with a sacrifice or offering of an animal or a plant food preparation, such as *pongal*. For example, in February during the seed festival known as '*Biccha Parbu*' the Samantha tribe of Eastern Ghats worship the village Goddess '*Jakiri Penu*' by offering animal sacrifices and then they sown seeds mixed with sacrificial blood of animals. Offering of *pongal* (also called *pallayam* offering or *Padayal*) signifies the fact that the space inside the pot for *pongal* preparation is symbolically converted into a space in the agricultural field and the prepared *pongal* is symbolically equivalent to the crop yield. Most often this *pongal padayal* is carried out at the community level.

3.2.5 FOOD RITUALS

Human subsistence in all traditional societies signified more than the food that fed commerce (Carney and Rosomof, 2009). It asks us to engage the broader relationship of food to culture, and culture to identity. According to the belief systems of many traditional societies, including those of the region studied here, food is provided by Gods/Spirits, often in response to prayers and gratitude, expressed through a variety of ceremonial rituals. The Hindu texts, which have been the source of inspiration to many ethnic societies of this region, say clearly that plants are most often necessary to make food for Gods and priests to eat. Some of these texts specify ten cultivated grains that are to be used in foods associated with rituals: rice, barley, sesame, beans, millet (probably black gram) and Vetch. When used for sacrificial purposes, these grains are ground and soaked in curd, honey or ghee before being offered as part of on oblation. Cakes made of different grains are offered in different directions, often wrapped up in leaves of *Udumbara* (*Ficus glomerata*) tree. Rice cake, for example, is placed on the eastern direction. Even trees were worshipped with ritual plant foods, often at night. The foods offered include milk, porridge, sweets, rice, curd, sesame and "eatables of various kinds." Food offerings are done to trees also before they are cut down for some purpose.

The food taken by many ethnic communities in the ancient Tamil country is ritualistic. In many communities food taken on Tuesdays, Fridays and/ or Saturdays are purely plant-based; so also throughout the Tamil month of *Purattasi* (September–October). In many villages, people consume only vegetrarian diet during the village religious festivals, particularly after the temple flag has been hoisted (=*Kaappukattal* in Tamil) (Paramasivam, 2001). During these days, in many families, the following vegetable items are not cooked: bitter gourd, some pulses (especially those that are not native) and *Sesbania* leaves. The food that is consumed includes thick porridge of black gram, puffed rice, cooked rice flour sweetened and added with coconut shreds, *Kummyam*, sesamum seed preparations, etc.

3.2.6   SOCIAL RITUALS

A number of social rituals are conducted. These rituals are meant to keep the social harmony intact and to bond the members of a society together. Social rituals are related to festivals, pilgrimage, etc. Rituals practiced to remove the effects of 'evil eye' (removing *Drishty*) are very important in the social context. By 'evil eye' we mean the jealousy/hatred/cold war that is shown towards a person or family by some other member/family of an ethnic community for various reasons; the affected person/family believes that the evil eye is the reason for the suffering (in terms of health, finance or otherwise). Sometimes sorcery is also involved in creating the evil effects. In order to remove these evil efforts, the affected person or family follows certain rituals which it believes will not only remove the ill-effects but also will not reveal the identity of the person responsible for fixing the 'evil eye in public. The latter preserves social harmony. The rituals conducted are either preventive or curative and take many forms. Hanging of an ash gourd (often with a ghost face drawn on it), an *Agave* plant upside down on a black-colored thread, or a bunch of chilly fruits sewn together along with a lemon fruit on a black thread in front of a house is done to remove the evil eye. Crushing of a lemon or ash gourd fruit stuffed with red dye in front of the house is done in some instances. Sometimes chilly fruits with or without betel leaves kept on water dyed red in a pan is spilled in front of the house or in the street corner. The reason why these particular plants are used for these rituals is not clear.

Another type of social ritual is related to the act of thanking God for fulfillment of vows, promises, oaths, etc. During these rituals the concerned person/family consumes plant food that is different from their daily food. Social rituals also include rituals connected to education. The *Gurukula* system of education was the prevalent system that existed among the ethnic communities of the study region. On the first day of initiation the child is taught to write the first alphabet on paddy/rice spread on a plate. The teachers or *Gurus* are paid *Gurudakshina* for educating the child. This concept arose in Buddhism and then got extended to Hinduism (Paramasivam, 2001). The students are given leave on the newmoon and fullmoon days and these days were called '*Uva*' days. On these days the students give their teachers *Dakshina* (=fees) in the form of rice, vegetables and fruits. This rice is called '*Uva rice*' which later on got corrupted as '*Vavarisi*.' This practice is still continued in some ethnic societies,

Many social rituals involve laticiferous plants. It is under laticiferous trees like banyan and peepul (species of *Ficus*) that the meeting of ethnic communities including *Panchayats* are held. Women go around a *Ficus religiosa* tree praying for a child. Marriages and puberty functions are conducted in halls/venues kept with twigs of laticiferous/resinous/or other juicy trees like mango, banana, palms, *Ficus*, etc. These latex-secreting plants probably symbolize a matriarchal nature of most, if not all ancient ethnic societies of this region. These symbols are archetypal symbols. *Panchayats* below *Ficus* trees (symbolizes mother's rule), worshiping of *F. religiosa* trees by women wanting to have a baby (matrilineal tradition), tying the umbilical cord onto a laticiferous tree (matrilineal reproductions), etc. are all represented by the same symbol to indicate a multifaceted and collective meaning and content. These symbols connect womenhood and fertility.

Rituals connected to festivals remove social problems and bring about social harmony. In village festivals, the idol of the village temple is brought from the temple to the place where it was originally residing and from there it is brought back to the temple. The whole place is decorated with festoons of plant leaves (particularly of mango, coconut, etc.) and flowers. Invariably *pongal* is offered along with sweetened rice flour cakes over which a ghee lamp (*Maavilakku* in Tamil) is lit. The idol is worshipped also with incense stick, betel leaf and nut, banana fruits, etc. The Banjara tribe celebrates the *Teej* festival in the month of August in a ritualistic manner. This festival is considered as important for unmarried girls who pray for better bride-grooms. The girls sow seeds in bamboo bowls and water them three times a day for nine days. The bowls are kept in a prominent place and girls sing and dance around the bowls. If the sprout from the seeds grow thick and tall, it is considered as a good omen for getting a better groom.

3.3   HEALTH CARE RITUALS

Rituals connected to disease prevention and cure in the primitive ethnic communities of the study region are intimately related to a concept similar to *Shamanism* followed in some Latin American and African ethnic societies (Thaninayagam, 2011). The *Shaman* or spiritual leader (male or female) in this region is often known by the Tamil terms *Poojary* or *Pandaram* The shaman of *Koya* tribe is called '*Buggivadde*'; this tribe also has a sorcerer called '*Vejji.*' The shanman is not only the village doctor but also the priest of the village temple and caretaker of the sacred grove associated with it. It is from the sacred grove that he often obtains the medicinal plants needed for the village community. He often practices medicine by combining with it spirituality, magic, godly faith and fear and sometimes even sorcery. Very often curing of diseases is ritualistic and combines one or more of the above. The *poojary* is regarded as having access to, and influence in, the world of benevolent and malevolent spirits; he typically enters into a trance state during a ritual through ritual dances and practices divination and healing. The *poojaries* are often considered as intermediaries/messengers between the human world and the spiritual world. They operate primarily within the spiritual world, which in turn affects the human world. They are said to treat ailments/diseases by mending the soul. They also enter supernatural realms or dimensions to obtain solutions to problems afflicting the ethnic community. The restoration of balance results in the elimination of the disease (Eliade, 1972). Shamanism cannot be strictly defined as medicine, although healing is its main objective. The *poojary* has a vast knowledge on the medicinal properties of plants around him. Strange ceremonies, rites, rituals, chants, dances, colorful outfits, perfuming with incense and invocations are part of the shamanic world. Consumption of trance-giving plants is regarded as sacred is not hallucinogenic. Naokov (2000), for example, has elaborated his observations on shamanism after extensive fieldworks in Villupruam district of Tamil Nadu. The Nattar people, according to him, get cured through religion and magic-related actions, or mental and physical illness. He mentions of rituals related to *Kuri*-telling (=divine indication or forecasting/fortune-telling), black magic, counter black magic, sorcery, driving away of ghost/ evil spirits, offering live sacrifice, etc. which are important in disease prevention/curing in this community. On getting subjected to these rituals, the person who gets a disease shows slow changes towards betterment while being watched by his relatives. Both prescriptive and performative rituals are conducted in this connection.

One ritual that is followed is tattooing (*Akki*-writing in Tamil) or suitable make ups on the body, aimed at disease curing. Anointing parts of body with clay or acquisition/control of body energies in an intelligent manner are also done through rituals and associated activities mentioned in the previous paragraph. Most of these rituals are often classified under technical rituals. Specific plants are used in this connection and the nature of relationship of these plants to actions like black magic and sorcery is not very clear. However, many of these plants contain tranquilizing or hallucinating principles in them.

KEYWORDS

- **Cosmovision**
- **Domain Concept**
- **Eastern Ghats**
- **Rituals**
- **Traditional Knowledge System**

REFERENCES

Carney, J.A. & Rosmof, R.N. (2009). In the Shadow of Slavery. Berkeley: University California Press.

Cotton, C.M. (1996). Ethnobotany. Principles and Applications. Chichester: John Wiley & Sons.

Eliade, M. (1959). The Sacred and the Profane: The Nature of Religion. New York: Harcourt Brace Jovanovich.

Eliade, M. (1972). Shamanism. Archaic Techniques of Ectasy. pp. 3–7. In: Bolligen Series LXXVI. Princeton: Princeton University Press.

Evans Pritchard, E.E. (1956). Nuer Religion. Oxford: Clarendon Press.

Findly, E.B. (2008). Plant Lives: Borderline Beings in Indian Traditions. New Delhi: Motilal Banarsidas Publishers Pvt. Ltd.

Gadgil, M. (1987). Diversity: Cultural and Biological. *TREE 2*, 369–373.

Gadgil, M. & Thapar, R. (1990). Human Ecology in India. Some Historical Perspectives. *Interdisciplinary Sci. Rev. 15*, 209–223.

Hopfe, M. & Woodward, R. (1998). Religious of the World. New Jersey: Prentice Hall.

Kalyanam, G. (1970). *Sivalayam Ratnagiri Thalavaralaru* (in Tamil). Ratnagiri, Tamil Nadu, India.

Kane, P.V. (1990–1991). History of Dharmasastra. Vols. 1–5. (Revised and enlarged edition). Bhandarkar Oriental Research Institute, Poona.

Kosambi, D.D. (1962). Myth and Reality: Studies in the Formation of Indian Culture. Bombay: Popular Press.

Krishnamurthy, K.V. (2003). Text Book of Biodiversity. Enfield (NH), USA: Science Publishers.

Krishnamurthy, K.V. (2005). History of Science. Tiruchirappalli, India: Bharathidasan University Publications.

Krishnamurthy, K.V. (2007). *Tamilarum Thavaramum* (in Tamil). 2nd Edition. Tiruchirappalli, India: Bharathidasan University Publications.

Nabokov, I. (2000). Religion Against the Self: An Ethnography of Tamil Rituals. New Delhi: Oxford University Press.

Paramasivam, T. (2001). Panpattu Asaivugal (in Tamil). Nagercoil, India: Kalachuvadu Publications.

Parthasarathy, J. (2002). Tribal People and Eastern Ghats: An Anthropological Perspective on Mountains and Indigenous Cultures in Tamil Nadu. pp. 442–450. In: Proc. Nat. Sem. Conserv. Eastern Ghats. ENVIS Centre, EPTRI, Hyderabad.

Pushpangadan, P. & Pradeep, P.R.J. (2008). A Glimpse at Tribal India—An Ethnobiological Enquiry. Thiruvanandapuram: Amity.

Radhakrishnan, S. (1993). The Principal Upanisads. London: George Allen & Unwin Ltd.

Rajan, K. (2004). *Tholliyal Nokkil Sangakalam* (in Tamil). Chennai: International Institute of Tamil Studies.

Salas, M.A. (1994). "The Technicians only believe in science and cannot read the sky." The cultural dimension of the knowledge conflict in the Andes. pp. 57–69. In: I. Scoones & Thompson, J. (Eds.). Beyond Farmer First: Rural People's Knowledge, Research and Extension Practice. London: Intermediate Technology Publications.

Subramanian, N. (1996). The Tamils. Institute of Asian Studies, Chennai.

Sundara Sobitharaj, K.K. (1994). *Thala Marangal* (in Tamil). Sobitham, Chennai.

Swamy, B.G.L. (1978). Sources for a History of Plant Sciences in India. IV. "Temple Plants" (*Sthala-Vṛksas*)—A Reconnaissance. Trans. Arch. Soc. S. India.

Thaninayagam, X.S. (2011). The Educators of Early Tamil Society pp. 181–196. In: Sivaganesh, D. (Ed.). The Collected Papers on Classical Tamil Literature in the Journal of Tamil Culture. Chennai: New Century Book House (P) Ltd.

Tylor, S.B. (1874). Primitive Culture. New York.

Varadarajan, M. (1965). The Treatment of Nature in Sangam Literature. Second Edition. Tirunelveli Saiva Siddhanta Nool Pathippagam, Chennai.

Witehead, H. (1921). The Village Gods of India. Madras: Oxford University Press.

Zimmer, H. (1935). The Art of Indian Asia. New York: Pantheon Books.

CHAPTER 4

ETHNOECOLOGY, ETHNOTAXONOMY, AND ETHNONOMENCLATURE OF PLANTS OF ANCIENT TAMILS

K. V. KRISHNAMURTHY[1] and S. JOHN ADAMS[2]

[1]*Department of Plant Science, Bharathidasan University, Tiruchirappalli–620024, India*

[2]*Department of Pharmacognosy, R&D, The Himalaya Drug Company, Makali, Bangalore, India*

CONTENTS

4.1 INTRODUCTION

It is generally agreed that the first land plants evolved on this earth during the late Silurian and early Devonian periods. Hence, when the modern human species, *Homo sapiens*, arose around 200,000 to 250,000 years ago, he had to confront all major groups of plants, including angiosperms, which were already there on this earth. By that time, plants had adapted themselves to the diverse habitats/ecosystems of the world through structural and functional modifications. The early human populations also had to adapt themselves to these diverse habitats/ecosystems, made possible largely through the generation and application of knowledge, both ecological and technological, which they gradually gained during their long years of hunter-gatherer experience (Cotton, 1996). This ethnic knowledge system is often called traditional knowledge system (TKS). When one critically analyzes the ethnic cultures and their non-codified and codified TKS of different parts of the world, he would be greatly impressed on knowing that these cultures placed great emphasis on the value and importance of their environment and its resources. Hence, it is not surprising that Merculieff (1994) had shown how traditional concepts on environment and ecology have preempted modern ecological ideas of western science. It is now possible to get a fair idea on the traditional systems of classifications of ecosystems, vegetation types and plants (and animals). This chapter deals with ethnoecology, ethnotaxonomy and nomenclature of the ancient Tamil people belonging to the Dravidian race that occupied the major part of the study region covered in this volume. The Tamils are one of the most ancient ethnic peoples of this world with a history of around 50,000 years. The Tamil TKS is also a well-codified system of knowledge in the form of ancient literary works that belong to the *Sangam* (200 BCE to 250 CE) and post-*Sangam* (250 BCE to 600 CE) periods. This article summarizes the knowledge that belonged to these periods, although it should be stated that most of this knowledge continued to be there till the British occupation of India in the 16th century.

4.2 ECOSYSTEM CLASSIFICATION

The ecosystem classification proposed by ancient Tamils is one of the most significant and oldest of all traditional ecosystem classifications. There are very strong evidences in *Tolkappiam*, *Sangam* and post-*Sangam* Tamil literature to show that this classificatory system was in existence even by about 250 BCE in the Tamil country (Krishnamurthy, 2007). The Tamils classified their landscape into five ecosystems or *Thinais*: *Kurinji, Mullai, Marutham, Neithal* and *Paalai*. These names were believed by many to be based on the most characteristic plants (type plants) of the respective ecosystems: *Strobilanthes kunthianus* (*Kurinji*), *Jasminum auriculatum* (*Mullai*), *Lagerstroemia speciosa* (*Marutham*) (wrongly denoted earlier as *Terminalia arjuna*), *Nymphaea nouchali* (*Neithal*) and *Wrightia tinctoria* (*Paalai*). It was however, believed by Varadarajan (1965) that these names of ecosystems were first applied to the flowers of the type plants belonging to the different ecosystems, subsequently to the landscapes and finally to indicate the differences in the love-related virtuousness and behavior of the peoples of these ecosystems. On the other hand, Nedunchezian (2003) did not agree with this opinion. He argued that these names were based, beyond flowers, on the habits of plants. However Krishnamurthy (2007) believed that the names of ecosystems were based on flowers, landscapes, love-related virtuousness and behavior and habits of plants, in a holistic manner since this viewpoint is in agreement with the traditional concept of biodiversity (see details on a subsequent paragraph of this article).

The five ecosystems, respectively, represent the mountainous ecosystem (*Maivarai Ulagam*), Scrub savanna ecosystem (*Kaaduraiulagam*), the predomianatly agricultural ecosystem with abundant lotic and lentic water bodies (*Theempunal Ulagam*) the marine and coastal ecosystem (*Perumanam Ulagam*), and the degraded mountain and scrub savanna ecosystem (*Vanpulam*). Each ecosystem was assigned a system of primary components or *MudalPorul*, of core components or *Karupporul* and of the personal love component or *Uripporul* (See Table 4.1). Each ecosystem has two primary components: Landscape type or *Nilam*, as mentioned above and time or *Pozhudu*. The latter is spoken in terms of different periods of a day (*Siru-pozhudu*) (six periods were recognized in a day) and of different periods or seasons of a year (*Perumpozhudu*) (six seasons were recognized in a year). Each ecosystem was related to a particular *Pozhudu* (both *Siru-* and *Perum-Pozhudu*), which was considered significant to that ecosystem. Fourteen items were assigned to constitute the core components of each ecosystem. Among these the most important are the principal God, plants, animals, birds, harvested food crops, foods, the main substances used, generic names of people, their profession, the hamlets (Place of living), etc. The third component, *Uripporul*, is the most important and it added a great deal of intrinsic, moral and personal importance and significance to this ecosystem classification.

TABLE 4.1 *Mudal, Karu* and *UriPoruls* of the Different Ecosystems Recognized by Ancient Tamils

Sl. No	Ecosystem	Mudal Porul	Karupporul (most important only given)	Uripporul
1	*Kurinji*	1) Landscape Mountainous area 2) Time a) Season Windy season & Early winter b) Day Night time	1) God *Seyon* (=Murugan) 2) Plants-*Kurinji* (*Strobilanthes kunthianus*), *Venkai* (*Pterocarpus santalinus*), *Moongil* (*Bambusa* sp.) 3) Animals-Bear, Tiger 4) Birds-Peacock, Parrot 5) Harvested crops-*Thinai* (millet), mountain paddy. 6) Foods-Rice, fruits, honey. 7) People-*Koravas* 8) Profession-Collecting honey, plucking tubers. 9) Hamlet-*Chirukudi* 10) Leader-*Kundranadan*	Union of lover (or husband) and the loved (or wife).
2	*Mullai*	1) Landscape Scrub savanna (=*Puravu* or *Vianpulam*) 2) Time a) Season Rainy season b) Day Evening time	1) God-*Maayon* (=*Maal, Tirumaal*) 2) Plants-*Mullai* (*Jasminum auriculatum*) *Konrai* (*Cassia fistula*) *Kaanthal* (*Gloriosa superba*) 3) Animals-Dear, Rabbit 4) Birds-Peacock, Hen/cock 5) Harvested Crops-*Varagu, Muthirai, Saamai* (all minor millets) 6) Foods-Millets 7) People-*Aayar, Kovalar, Idaiyar* 8) Profession-Cattle-rearing and feeding 9) Hamlets-*Paadi*, Cheri, *Palli* 10) Leader-'*Kon*'	Separation of lovers (or husband and wife); women maintaining chastity.

TABLE 4.1 *(Continued)*

Sl. No	Ecosystem	Mudal Porul	Karupporul (most important only given)	Uripporul
3	*Marutham*	1) Landscape-cultivated agricultural lands with a lot of lentic and lotic water bodies (called in Tamil *Pazhanam*, *Kazhani* and *Cheru*)	1) God-Vendan	Tiffs between lovers or between husband and wife and their subsequent removal.
			2) Plants-*Marutham* (*Lagerstroemia speciosa*)	
		2) Time	*Ambal* (*Nymphaea pubescens*) *Pakandrai* (*Operculina turpethum*)	
		a) Season All six seasons of the year	3) Animals-Buffalo, Bull	
			4) Birds-Stork, Duck	
		b) Day Very early morning and sunrise time	5) Harvested crops-Paddy, Plantain, sugarcane	
			6) Foods-Rice, Banana, Jack fruit, Sugarcane	
			7) People-Uraar	
			8) Profession-Agriculture Hamlets-*Perur*, *Mudur*	
			9) Leader-Uran	
4	*Neithal*	1) Landscape	1) God-*Varunam*	Wife/lover waiting for the separated husband/lover to return home
		a) Coastal land and sea (Kaanal and Kadal)	2) Plants-*Neithal* (*Nymphaea nouchali*),	
		2) Time	*Thazhai* (*Pandanus* sp.),	
		a) Season All six seasons of the year	*Punnai* (*Calophyllum inophyllum*)	
			3) Animals-Crocodile, Crabs, Fishes	
		b) Day Morning time	4) Birds-Swan, Egret	
			5) Harvested Food source-Fishes, Prawns	
			6) Foods-Fish, Prawns, Crabs	
			7) People-*Bharathavar* (Fisherfolks)	
			8) Profession-Fishing	
			9) Leader-*Thuraivan*, *Pulamban*	

TABLE 4.1   *(Continued)*

Sl. No	Ecosystem	Mudal Porul	Karupporul	Uripporul
			(most important only given)	
5	*Paalai*	1) Landscape-Degraded *Kurinji* and *Mullai* lands (=*Vanpulam*) 2) Time a) Season Late winter and summer b) Day Midday time	1) God-*Durgai* 2) Plants- *Panai* (*Borassus flabellifer*) *Omai* (*Anogeissus latifolia*) Kura (*Tarenna asiatica*) 3) Animals-Dogs 4) Birds-Vultures 5) People-*Ainer* 6) Leader-*Midalai, Kaalai* 7) Profession-Robbery 8) Hamlet-*Kurumbu*	

Although the Tamil concept of ecosystem has certain features of certain other ethnic ecosystem classifications, it has certain unique features not known to other systems. The common features concern the distinction of local vegetation types on the basis of factors, such as location, dominant life forms and predominance of particular plant/animal species of cultural or utilitarian value. This, for example, is the virtue of Hanuno's culture in the Mindora island of Philippines also (Conklin, 1974). The unique feature of the Tamil system is that a social and cultural dimension has been intricately added to the ecosystem concept, for example, the assignment of separate *Mudal, Karu* and *UriPoruls* to each ecosystem. Hence, the Tamils may be considered as one of the earliest ethnic societies that included social and cultural diversity aspects in their concept on biodiversity. This inclusion is only recently being suggested and emphasized by UNESCO and UNEP (see Krishnamurthy, 2003). Cultural diversity recognizes the pivotal role of sociological, ethical, religious and ethnic values in human efforts concerning biodiversity classification (UNEP, 1995).

4.3   ETHNONOMENCLATURE AND CLASSIFICATION OF PLANTS

4.3.1   TRADITIONAL ETHNIC APPROACHES

A critical examination of traditional cultures of various parts of the world reveals that plants were recognized, named and classified. Ethnonomenclature refers to the recognition and naming of plants (and animals) around them by the ethnic societies, while ethnotaxonomy is a study of the traditional system of classification of plants (and animals). Ethnotaxonomy requires the identification and naming of plants that need to be classified. For ethnic societies satisfying their basic needs, such as food and medicine is more important in properly recognizing and naming plants, a materialistic view point (Malinowski, 1974). Hargreaves (1976) has, in fact, shown that plants were named according to their uses and that many plants in Chitipa have no local name because they have no use. However, others like Levi-Strauss (1966) argued that the outlook of the ethnic societies towards the natural world in general, and its resources like plants in particular, is primarily intellectual an cognitive and divorced from pragmatic concerns. Some plants in Malawi, as elsewhere, have names but no apparent utility and hence there appears to be no correlation between ethnic nomenclature and plant use. In his early theories of structural anthropology, Levi-Strauss showed the universal human tendency or urge to organize and classify perceived phenomena, experience or things (Seymour-Smith, 1986). These two different viewpoints respectively of Malinowski (1974) and Levi-Strauss (1966) naturally advocate different kinds of intellectual and classificatory modes. For Levi-Strauss (1966), ethnic societies are concerned with a mode of thinking that unifies through symbolic logic the various aspects of their cultures, while for Malinowski (1974), Berlin and his associates (1974) and others ethnic peoples are protobotanists who are concerned with ordering the natural world through criteria based on structural morphology of plants. Although both these perspectives are necessary and are not mutually exclusive, they have limited our understanding of ethnic taxonomies. To consider ethnotaxonomics simply as taxonomics, abstracted from utilitarian, environmental and cultural concerns, greatly limits our understanding of how human societies are intimately related to the natural environment. The approach of Levi-Strauss focuses on symbolic logic and over systematizes the social communities. On the contrary, the approach of ethnotaxonomists tends to underplay the relevance of practical interests in structuring ethnotaxonomies. A critical analysis of ethnotaxonomy of the ancient Tamils shows that it is a mainly based on Levi-Strauss' intellectual and cognitive approach. It does not appear to have a materialistic basis.

4.3.2   PERFECTION OF DESCRIPTIVE TECHNICAL TERMS

The fact that both ethnonomenclature and ethnotaxonomy have developed very well in many, if not all, ethnic societies of the world must have required a deep and critical knowledge on the life of plants on the part of these societies. This, in fact, is true for the ancient Tamil culture. These people have also coined specific technical terms numbering to around 150 to denote the various characters and character states of plants. A list of the more important technical terms used in ancient Tamil literature to denote the various characteristics is given in Table 4.2. These technical terms have been used to not only distinguish one taxon from another, but also to name and classify them. The ancient Tamils have also used several metaphoric similes to compare/describe a number of plant characteristics of ethnotaxonomic importance. Some examples are given Table 4.3.

TABLE 4.2   Representative Botanical Terms Used by Ancient Tamil Community

Sl. No	Botanical terms in Tamil	Equivalent English terms
1.	Maram, Maran	Plant
2.	Marundu	Plant
3.	Maram	Tree
4.	Kodi	Climber/twiner
5.	Pul	Grass
6.	Poodu	Bulbous Plant
7.	Pudal	Clump of herbs
8.	Pavar	Straggler
9.	Payir	Runner, also as crop.
10.	Paruookkodi	Woody climber (Liane)
11.	Punkodi, Menkodi, Nunkodi	Herbaceous climber
12.	Aambi, Kaalaambi	Fungus/Mushroom
13.	Paasi	Alga
14.	Kandu	Trunk
15.	Sinai	Branch, Twig
16.	Kaambu	Stalk
17.	Thandu	Stem
18.	Thaal	Slender stalk
19.	Kizhangu	Tuber

TABLE 4.2   *(Continued)*

Sl. No	Botanical terms in Tamil	Equivalent English terms
20.	*Mul*	Thorn/spine
21.	*Kodu/kottu*	Branch
22.	*Ilai*	Leaf
23.	*Siriilai*	Leaflet/small leaf
24.	*Perilai*	Large leaf
25.	*Nettilai*	Lanceolate leaf
26.	*Olai*	Palm like leaf
27.	*Idazh*	Lamina/blade
28.	*Eerkku*	Midrib
29.	*Paalai*	Spathe
30.	*Madal*	Folded blade
31.	*Thalir*	Young leaf
32.	*Sethil*	Scale
33.	*Thol*	Skin
34.	*Koththu*	Flower bunch
35.	*Thoththu*	Hanging inflorescence/flower bunch
36.	*Kulai*	Inflorescence of palms
37.	*Manjari*	Inflorescence with distinctly seen flowers
38.	*Thunar*	Spike
39.	*Inar*	Catkin
40.	*Poo*	Flower
41.	*Malar*	Open Flower
42.	*Nani*	Floral primordium
43.	*Arumbu*	Young flower bud
44.	*Mugai*	Older flower bud
45.	*Podu/Podhi/Pugil/Pogil*	Flower at anthesis
46.	*Alar*	Pollinated open flower
47.	*Vee*	Pollinated flower with abscising floral parts
48.	*Pulli*	Calyx
49.	*Alli*	Corolla
50.	*Adazh*	Perianth
51.	*Makaram*	Stamen
52.	*Poguttu/Kannigai*	Ovary
53.	*Taadhu*	Pollen
54.	*Then*	Nectar
55.	*Nara/Narai*	Honey
56.	*Kal/Mattu/Theral*	Fermented honey
57.	*Nirai poo*	Complete flower
58.	*Kurai Poo*	Incomplete flower
59.	*Sool*	Ovule

TABLE 4.2   *(Continued)*

Sl. No	Botanical terms in Tamil	Equivalent English terms
60.	*Kai*	Unripe fruit
61.	*Kani/Pazham*	Ripe fruit
62.	*Vidai/Kazh*	Seed
63.	*Kural*	Cob
64.	*Koli*	Non-Flowering tree
65.	*Umi*	Lemma/palea
66.	*Thoombu*	Air canal
67.	*Vazhumbu*	Mucous
68.	*Naar*	Fiber
69.	*Akakazh*	Wood
70.	*Vayiram*	Heartwood
71.	*Pararai*	Trees with heartwood
72.	Veezh	Aerial root

TABLE 4.3   Some Examples of similes of Ethnotaxonomic Importance Used by Ancient Tamils (Krishnamurthy, 2006)

Sl. No	Similies used
1.	*Mullai* flower buds like teeth of women
2.	*Neithal, Kaavi, Kuvalai* and *Senkazhuneer* flowers like the eyes of women
3.	*Sesamum* flowers like the nose of women
4.	*Vallai* leaves similar to the ear lobe of women
5.	Flower buds of *Kongu* akin to the breast of young women
6.	Perianth lobes of *Gloriosa superba* like the fingers of women tinged with *Lawsonia* dye
7.	*Nochi* flower buds similar to the eyes of a crab
8.	*Vagai* flower similar to peacock's headtuft
9.	The young flower bud of *Kuravu* similar to the teeth of snake
10.	The young flower bud of *Punnai* similar to the egg of house lizard
11.	*Thalava* flower looking like the beak of Kingfisher bird
12.	*Thalava* flower bud similar to the nail of *kauthari* bird
13.	*Iluppai* flower similar to the foot of a cat
14.	*Adappam* flower similar to the young one of *kurumpoozh* bird
15.	*Mulli* flower similar to the tooth of Squirrel
16.	*Avarai* flower similar to the beak of a parrot
17.	*Agaththi* flower similar to the tooth of a pig
18.	The leaf of *nochi* looking like the foot base of peacock
19.	*Ambal* leaf like the ear lobe of a rabbit
20.	*Oogu* is similar to the tail of *anil*
21.	*Kundri* seed looking like the eye of a white rat
22.	*Adumbu* leaf looks like a deer's foot.

TABLE 4.3 *(Continued)*

Sl. No	Similies used
23.	The truck of *Omai* tree is similar to the skin of crocodile
24.	*Perupoolai* leaf hairs similar to the hairs of a kitten
25.	Leaflet margin of Neem looking like a saw
26.	*Iluppai* flower similar to arrowhead
27.	*Makilam* flower similar to the wheel of a cart
28.	*Padiri* flower looking like a painter's brush
29.	*Pungam* flower bud similar to puffed rice
30.	*Karumbu* flower similar to a upright spear
31.	The inflorescence of *Venkadambu* and the fruit of *vila* similar to a ball
32.	*Paalai* unripe fruit similar to a tong
33.	The corm of *Kanthal* looking like a plow
34.	Bunch of *Konnai* fruit similar to the beard of a sage

4.3.3 ETHNONOMENCLATURE

It is interesting to know how different ethnic societies of the world named and classified the different plants (and animals) and, thus, conceptualize the natural world. Like any other sub-discipline of anthropology, ethnobotany has heavily borrowed from linguistic analysis, both in the emphasis placed on recording and studying those categories of taxa which are linguistically defined and in the focus on identifying sets of contrasts. Patterns revealed through taxonomic labels often provide clues, not particularly cultural, of local plants (Martin 1995; Cotton 1996) see also Ellen (1994). The transcribed traditional names can then be translated either by the use of a gloss or free translation, in which case the closest equivalent word available in English is used, or, the terms may be translated literally, in which case the world is translated word-for-word.

Between 250 BCE and 600 CE, the Tamil ethnic community had around 350 generic plant names. Subsequent to this period, this number gradually increased (Krishnamurthy, 2007). There are two unique aspects about these generic plant names: (i) These names cannot be translated either by the use of a gloss or free translation, as discussed in the previous paragraph, and hence, the closest equivalent English word cannot be found, or these generic names cannot also be translated literally and, hence, the name cannot be translated word-for-word, as in many other ethnonomenclatural systems of the world. The Tamil generic names for plants can only be transliterated. Since these generic names and the information as to which plants they refer to have been passed on through successive generations of Tamil people, we now know what Tamil generic name indicates which plant. In spite of this, we are not able to find the correct Tamil botanical names for some of the ancient Tamil generic names found in literature since enough descriptions of these plants are not available in the literature, which could facilitate their correct identification. In some cases the descriptions are not enough to help in accurate identification and only tentative identifications have been given. As examples of the former we may cite the following: *Anicham, Asakam, Arai, Aravu, Ingulam, Iram, Kaduvu, Kalmitham, Kaavithi, Kusappul, Koovai, Sengurali, Puzhagu, Vayalai, Visai*, etc. (ii) The generic names used by ancient Tamils to denote the different plants are very unique in their own rights (Shanmugasundaram, 1970; Krishnamurthy, 2007). Almost all generic names are very short and consist of one (for example, *Che-Alangium salvifolium*), two (Ari-*Bambusa* species), three (*Aacha-Hardwickia binata*), four (*Adumbu-Ipomoea pes-caprae*), or five Tamil letters (*Iranthai, Zizyphus mauritiana*). Six and seven letter generic names are extremely rare (e.g., *Kannikaram, vellothiram*). The names are easy to pronounce and are very attractive. The names often end with soft and rhythmic sound with the following Tamil letters only: Aa, E, U, I, Il, Im or Ir (representative examples are *Ukaa, Inge, Kamugu, Panai, Vel, Aaram* and *Aar*). However, from the *Bakthi* Literature period (late half of 7th century CE) onwards such a type of naming of generic categories of plants declined very rapidly. Many Sanskrit generic names started to be in prevalent use, as is evident from Tamil *Nigantus* (Dictionaries) (Examples: *Athimaduram, Kanaveeram, Aravindam*, etc.). There was also an influence of Ayurveda and Siddha medicines on the naming of generic taxa of plants, particularly at the village/rural level. In villages new names for plants were coined (sometimes even for the existing Tamil generic names) so as to enable lay people to easily recognize and identify the plants of medicinal value. As examples, we may mention *Kalappai Kizhangu* (tuber shaped like a plow) for *kaanthal* (*Gloriosa superba*), *Santhanam* (for Aaram, *Santalum album*), *Kudaivelam* (for *Udai, Acacia planifrons*) and *Udumbaram* (for *Atthi, Ficus glomerata*). The method of naming generic taxa by ancient Tamils also allows us to distinguish original plants that were available at the time of such naming from the plants (and their names) that were subsequently introduced into the Tamil country from other parts of India or from other parts of the world. As examples we may cite native *Malli* (*Jasminum* sp.) from the introduced *PavazhaMalli* (*Nyctanthes arbortristis*) (Probably introduced from Odisha) and *Panai* (*Borassus flabellifer*) from *Koonthal Panai* (*Caryota urens*).

4.3.4 ETHNOTAXONOMY

Traditional societies have been classifying objects around them, including plants and animals. It is interesting to record that "a culture itself amounts to the sum of a given society's folk classifications" (Sturtevant, 1964). The earliest critical study of ethnic taxonomy/classificatory system was that of Conklin (1954, 1957, 1974), who investigated the classification system followed by a tribe in the Philippine island of Mindoro. Subsequently, the plant classification systems of several ethnic cultures of the world have been reviewed and a synthesis of their features, in the form of some general principles has been proposed (Berlin et al., 1973). These principles are discussed below in reference to the classificatory system that was followed by the ancient Tamil culture.

The first general principle is that in all languages it is possible to identify groups of taxa which are recognized linguistically and that these taxa are based on varying degrees of inclusiveness (like in English language *Casuarina*, tree and plant respectively represent taxa of increasing inclusiveness). In ancient Tamil language also this was seen (for example, *Mullai*, *Kodi* and *Maran* or *Maram*, respectively, represent the increasing hierarchy of generic name, lifeform and plant).

The second general principle is that biological taxa are grouped into a number of ethnobiological categories, similar to the modern plant/animal taxonomic ranks. In 1972, these categories were respectively designated as unique beginner, life form, intermediate, generic specific and varietal, in a decreasing order of inclusiveness. This categorization was also noticed in the Tamil classification system. Folk taxonomic hierarchies are considered to be relatively shallow and that the term hierarchy is almost is misnomer. It is especially true, for instance when about 20% of Tzeltal plant categories are unaffiliated to any life-form taxa, and that 85% of the generics are monotypic. Similarly, plants in Bunaq taxonomy appear to be classified more according to a complex web of resemblances rather than forming, neat hierarchy.

The third principle states that mutually exclusive ethnobiological categories mentioned above are arranged hierarchically, each encompassed by the single unique beginner taxon, which is almost equivalent to the plant kingdom suggested in modern biological classificatory system. This principle is also seen in the Tamil ethnic classification system.

The fourth principle states that the taxa of the same ethnobiological rank commonly occur at the same taxonomic level. This is true of Tamil ethno classificatory system also.

The fifth principle states that the unique beginner taxon (i.e., plant) is not normally named with a single, habitual level, but at the next level, there normally exist between five to ten life forms, which are normally labeled the Tamil classificatory system the unique beginner taxon is not labeled by the term *Thaavaram* which is used in modern Tamil as equivalent to plant until around the 7th century CE (Krishnamurthy, 2007). On the contrary, the unique beginner taxon was labeled as *Maram* or *maran*, which in fact was a term used at the next habitual level also to denote trees. In other words, all plants were called by ancient Tamil culture as *Maram* or *Maram* and at the same time to denote a habit category also, for example, tree. The term *Marundu* was also used by the ancient Tamils to denote plants at the unique beginner taxon level since they believed that plants formed the source of medicines (see discussion in Nedunchezian, 2003; Krishnamurthy, 2007).

The sixth principle states that most ethnotaxonomies appear to contain around 500 taxa at the *generic* level and that these taxa invariably represent the basic building blocks of these taxonomies. These taxa are also the most salient psychologically. This principle further states that most of these *generic* taxa are included within a given life-form category. There are also a few morphologically unique or economically very important and often aberrant taxa and also taxa that are not conceptually regarded as affiliated to any life-form categories in those ethnotaxonomies. The Tamil ethnotaxonomy contains around 350 generic taxa; like in other ethnotaxonomies these taxa represent the basic building blocks of the Tamil taxonomic system. These taxa are also the most salient. A critical analysis shows that all these taxa are included within the life-form categories recognized by this system. Eight life-form categories have been identified by the ethnotaxonomy of Tamils: of these seven refer to land life-forms and one to aquatic life form. The ligneous life-form constitutes the trees which are characterized by wood, often with a region of heartwood, and perennial life span. This life-form was indicated by the terms *Maram*, or *Maran* in Tamil, as already mentioned. Grasses form the next life form category. They are indicated by the word '*pul*' in Tamil (although in a broader sense, this Tamil term indicated the 'monocots' also, see discussion later). *Pudal* forms the third life-form category. In modern Tamil this word refers to a bush. According to some Tamil scholars *Pudal* includes all life forms other than trees, but this is not accepted by most other Tamil scholars. A critical study of literature shows that Pudal refers to a close clump of herbs. It is interesting to note here that the modern Tamil term for herb is *chedi* but this term was not used for herb until around the 7th century CE. The fourth category of life-form is *poodu*, which in modern Tamil got corrupted into *Poondu*. This word indicates plants with underground bulbs as in onion. The fifth category is *Kodi*, which denotes a climber/twiner. There are herbaceous climbers recognized by the use of terms, such as *Menkodi* (weak/slender climber), *Nunkodi* (small climbers) and *Punkodi* (weak-stemmed climbers), and there are woody climbers or lianes, which from the sixth category. These lianes are indicated by the term *Paruookkodi*. The seventh category is referred to by the Tamil word *pavar*. This word describes plants that grow adpressed to another plant, and form literary description and examples cited in literature (like *Calamus*) it is likely to indicate a straggler. The eighth category refers to aquatic plants. There are also references to fungi and algae in the early Tamil literature. These respectively are indicated by the Tamil words *Aambi/Kalambi* and *Paasi*. Whether these should be treated as intermediate categories or as life-form categories is not clear. Some, with whom the authors of this paper had discussion with, even suggest these terms as *generic*.

According to the seventh principle, specific and varietal taxa are less numerous than the generic taxa. The members of a given *contrast set* differ from the other members in a few, often 'verbalisable,' characters. In the Tamil ethnotaxonomy there are around 350 *generic* level taxa. Out of these few alone have specific and varietal level taxa; these constitute about 10% of the total. This number is much higher than the average of less than 2% of such taxa in world ethnotaxonomies (Martin, 1995; Berlin, 1992). The species and varietal level taxa always have prefixes to the original *generic* labels in Tamil ethnotaxonomy, such as *Karu* (indicating black), *Sen(g)* (indicating red), *Ven* (indicating white), *Siru* (indicating 'smaller'), *Peru* (indicating 'bigger'), *kaattu* (indicating 'forest'), etc. These prefixes are often adjectives or another noun.

Special discussion must be made here about the '*intermediate*' taxa recognized in only some ethnotaxonomies and hence are rare or difficult to detect. These are also seldom named in such ethnotaxonomies and hence are called covert or unnamed taxa. These generally encompass a number of related generic level taxa. As far as Tamil ethnotaxonomy is concerned two intermediate level taxa were recognized: *Akakkazh* plants and *Purakkazh* plants. The former refer to plants that have a true wood while the latter to plants that lack true wood. These two groups are respectively equivalent to modern day taxa dicotyledons and monocotyledons. It is very interesting to note that the supposedly earliest Tamil grammar text *Tolkappiam* classified plants into trees and grasses, which respectively indicated dicotyledons and monocotyledons. The latter included palmyrah and other palms and bamboo under the life-form category *Pul* (=grass), although they are arborescent and 'woody' (Rajeswari, 2005). Tolkappiar, the author of *Tolkappiam*, has listed the following structures as belonging to the grasses (i.e., monocots): *Thodu* (Leaf or inflorescence sheath), *Madal* (folded leaf blade), *Olai* (Palm leaf), *Eadu* (lamina without petiole), *Idazh* (perianth lobe), *Paalai* (spathe), *Eerku/Eadu* (linear) and *Kulai* (inflorescence like in palms). Similarly, he had listed the following appendicular structures as characteristic of *Maran* (=the dicotyledons). *Ilai* (leaf), *Muri* (twig), *Thalir* (sprout), *Sinai* (branch), *Kuzhai* (twig with leaf bunch), *Poo* (flower), *Arumbu* (young flower bud), *Nanai* (floral primordium, etc.). He has also listed the following as belong-ing to both these groups: *Kaai* (unripe fruit), *Pazham* (fruit), *Thol* (Skin), *Setil* (scale), Veezhr (aerial root).

The seven principles discussed above have been accepted by many eth-nobiologists, although other cross-language, based on a survey of folk classi-fication in 188 languages for plants, patterns have become apparent in ethno-botany (Brown, 1984, 2000). Some pertinent to Tamil Ethnotaxonomy have been detailed above. Brown has assembled evidence from a large number of globally distributed languages and suggested that plant life-form categories are typically added to languages (i.e., lexically encoded in more or less fixed sequences as shown in Figure 4.1.

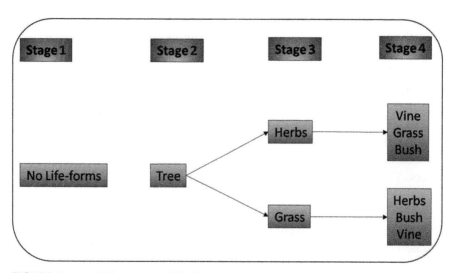

FIGURE 4.1   Addition of plant life-form categories to ethnic languages fixed sequences (based on Brown, 2000).

In the figure, stage 1 languages lack names for botanical *life-form* cat-egories and in the successive stages these languages have added further life-form categories. Ancient Tamil language is one language that has all life-form categories except herb (in Tamil, *Chedi*). This life-form category was added in the language only by about the 5th century CE.

In 1997 Berlin has revised and expanded the seven principles discussed above and proposed a total of seven principles of ethnobiological categorization and, five principles of ethnobiological nomenclature. These principles were summarized by Martin (1995) as a general purpose ethnotaxonomy. This is summarized, with modifications, in order to explain the ancient Tamil ethnotaxonomy of plants (Table 4.4).

TABLE 4.4   Major Features of General Purpose Classification of Plants of Ancient Tamils*

Sl. No	Plant Category	Nature of label	Label proposed in Tamil culture	Numbers
1.	Kingdom (Plant)	Not covert	Labeled as *Maram/Maran* or as *Marundu*	1
2.	Life-forms	Primary		8
	Ligneous and with wood		*Maram*	
	Grass		*Pul*	
	Clump of herbs		*Pudal*	
	Bulbous plants		*Poodu*	
	Herbaceous Climbers		*Kodi*	
	Liana		*Paruookodi*	
	Straggler		*Pavar*	
	Aquatic plants		*Neerpoo*	
3.	Intermediate 'Dictos'	Not covert	*Akakkazh*	2**
	'Monocots'		*Purakkazh* or *Pul*	
4.	Generic	Primary	Basic plant name	350
5.	Specific and Subspecific	Secondary	Basic plant name plus attribute(s)	35(10%)

*This table has been made on the basis suggested by Berlin (1992).

**Whether to include fungi and algae here is a matter of debate and still undecided.

4.4 CONCLUSIONS

The above account clearly demonstrates the deep knowledge that ancient Tamils had on plants, their naming and classification. The Tamil classificatory system is of great convenience to the originator and user of this system. This system is not utilitarian but basically based on symbolic logic and culture, but at the same time allows for a communication between members of the Tamil society. This chapter also falsifies statements, such as the following: "Great in accuracy and a general absence of scientific system obtains in the Tamil botanical nomenclature."

KEYWORDS

- Ancient Tamil People
- Mudal Porul
- Sangam Literature
- Similes
- Thinais

REFERENCES

Berlin, B. (1992). Ethnobiological classification: Principles of Categorization of Plants and Animals in Traditional Societies. Princeton: Princeton University Press.

Berlin, R., Breedlove, D.E. & Raven, P.H. (1974). Principles of Tzeltol Plant Classification. New York: Academic Press.

Brown, C. (1984). Language and Living Things. Uniformities in Folk Classification and Naming. New Jersey, USA: Rutgers University Press.

Brown, C.H. (2000). Folk Classification. pp. 243–246. In: P.E. Minnis (Ed.). Ethnobotany, A. Reader. Univ. Norman, USA: Oklahoma Press.

Conklin, H.C. (1954). The Relation of Hanunóo Culture to the Plant World. PhD Thesis Yale University, USA.

Conklin, H.C. (1957). Hanunóo Agriculture, a Report on an Integral System of Shifting Cultivation in the Philippine. FAO, Rome.

Conklin, H.C. (1974). The Relation of Hanunóo Culture to the Plant World (Yale University PhD 1954). High Wycombe, USA: Univ. Microfilms Ltd.

Cotton, C.M. (1996). Ethnobotany. Chichester, UK: John Wiley & Sons.

Ellen, R.F. (1994). Putting plants in their place: Anthropological approaches to understanding the ethnobotanical knowledge of rainforest populations. Presentation, UBD-RGS Conference.

Hargreaves, B.J. (1976). Killing and Curing: Succulent use in Chitipa. *Catus and Succulent J. 48*, 190–196.

Krishnamurthy, K.V. (2007). Tamils and Plants (in Tamil). Bharathidasan Univ., Tiruchirappalli, India.

Levi-Strauss, C. (1966). The Savage Mind. Weidenfeld and Nicolson, London.

Malinowski, B. (1974). Magic, Science and religion. Souvenir Press, London. (reprinted 1925 edition).

Martin, G.J. (1995). (2nd Edition) Ethnobotany: A Conservation Manual. Chapman & Hall, London.

Merculieff, I. (1994). Western Society's linear systems and aboriginal cultures: the need of two-way exchanges for the sake of survival pp. 405–415. In: E.S. Burch & L.J. Ellanna (Eds.). Key Issues in Hunter-Gatherer Research. Oxford: Berg Publishers.

Nedunchezian, V. (2003). Botany as seen by the Tamils (in Tamil). World Tamil Research Institute, Chennai.

Rajeswari, R. (2005). Scientific thoughts in Tolkappian. pp. 1–11. In: The Wealth of Tamils-Science, Technology. Vol. 1. World Tamil Research Institute, Chennai.

Seymour-Smith, C. (1986). MacMillan Dictionary of Anthropology. MacMillan Press Ltd. London.

Shanmugasundaram, L. (1970). Tamil and Plants (in Tamil). Tenkasi, Tamil Nadu.

Sturtevant, W.C. (1964). Studies in Ethnoscience. Amer. Anthro. pp. 174–222. In: J.W. Berry & P.R. Dasen (Eds.). Culture and Cognition. London: Metheun.

UNEP (1995). Global Biodiversity Assessment. Cambridge: Cambridge University Press.

Varadarajan, M. (1965). The Treatment of Nature in Sangam Literature. Second Edition. Madras: Tirunelveli Saiva Siddhantha Noolpathippukazhagam.

CHAPTER 5

ETHNIC PLANT GENETIC RESOURCES DIVERSITY OF EASTERN GHATS AND DECCAN

S. R. PANDRAVADA, N. SIVARAJ, and V. KAMALA

National Bureau of Plant Genetic Resources, Regional Station, Rajendranagar, Hyderabad–500030, Telangana, India. E-mail: pandravadasr@yahoo.com; sivarajn@gmail.com; kgksvp@gmail.com

CONTENTS

ABSTRACT

Eastern Ghats is a rich abode and a treasure trove for ethnic diversity in plant genetic resources consisting of different agri-horticultural crops, their wild/ weedy relatives, medicinal, aromatic and dye yielding plants. However, due to socio-economic developmental programs and other biotic pressures the endemic crop genetic diversity accumulated through years of evolution under domestication and natural selection by the tribal groups is being wiped out from the nature. Concerted and systematic efforts have to be made now and in future as there is a tremendous urgency and scope for collection and conservation of ethnic plant genetic diversity for sustainable utilization from the Eastern Ghats. This chapter describes the nature and spectrum of plant genetic resources diversity with special reference to ethnic agri-diversity and their wild relatives in the Eastern Ghats region. Conservation strategies and utilization of these ethnic genetic resources and the factors contributing to genetic erosion of native wealth are also discussed.

5.1   INTRODUCTION

The Eastern Ghats of Indian sub-continent is immensely rich in ethnic plant genetic resources diversity of crop species, wild relatives and medicinal plants. It constitutes our invaluable assets to meet the growing demands to increase crop production and productivity. Ethnic plant genetic diversity is fundamental to crop improvement programs and the key to establishing future food and nutritional security. Ethnic plant diversity in the form of seeds and plants provide the raw materials that scientists use to address crop production challenges, develop new crops and identify new uses for existing crops. Scientists use these resources to develop knowledge or products valuable in coping with inadequate water or nutrient supplies, diseases or insect pests, heat and cold tolerance, understand their nutritional properties and for many other purposes. The importance of ethnic plant genetic resources has increased significantly in the recent years with the changing global scenario in material ownership and the legal regimes with respect to access to plant genetic resources under the International Agreements.

Plant genetic resources are the genetic material of plants, which determines their characteristics including their ability to adapt and survive. The ethnic plant genetic resources diversity profile of a crop, therefore, includes its wild species, weedy companion species, sub-species, botanical varieties, landraces, ancient and heirloom cultivars, that make up the part of total gene pool of the crop.

The Eastern Ghats, one of the major hill ranges of India, located between 77°22' and 85°20' E and 11°30' and 21°0' N form an assembly of discontinuous ranges, hills, plateaus, escarpments, narrow basins and spread in an area of about 75,000 km². The Eastern Ghats stretching from Orissa, Chhattisgarh, through Andhra Pradesh to Tamil Nadu and parts of Karnataka are endowed with a large variety of biological species, geological formations and indigenous tribal groups. For Eastern Ghats, the Mahanadi basin marks the northern boundary while the southern boundary lies in the Nilgiri hills. While the tips of Bastar, Telangana, Karnataka plateaus and Tamil Nadu uplands form the boundary in the West, the coastal belt forms the boundary in the east.

The Eastern Ghats region is inhabited by nearly 54 tribal communities, which constitute nearly 30% of total population (Chauhan, 1998). The major tribes in the Eastern Ghats are *Arondhan, Irular, Kota, Kotanayakam, Kurmar, Puniyan, Pulayan, Sholaga, Tuda* and *Malayali* in the southern region, *Bagata, Chenchu, Gadaba, Jatapu, Kammara, Kondadora, Kondakapu, Kondareddy, Kandha, Kotiobenthu Oriya, Koya/ Goud, Kulia, Mali, Mannedora, Nayaka, Nukadora, Paraja, Reddidora, Savara, Valmiki, Yenadi* and *Yerukala* in central region and *Bathudi, Birjhal, Bhuiyan, Dhuma, Bhumis, Bhuttada, Gond, Khana, Kisan, Kolba, Munda, Oraon, Soarha* and *Sounti* in the northern region. The variations in altitude and climatic conditions, especially in rainfall have immensely contributed in the evolution of rich ethnic floristic diversity in the Eastern Ghats. This region is very rich in terms of natural wealth, which is manifested in its greatest biological diversity. Out of 2,500 species of flowering plants belonging to Angiosperms, Gymnosperms and Pteridophytes known to occur in Eastern Ghats, about 77 species (67 Dicots, 9 Monocots and 1 Gymnosperms) are endemic.

5.2   ETHNIC AGRICULTURAL DIVERSITY

In the Eastern Ghats, the natural flora includes many economic plant species that offer food, fiber and shelter. The sowing of selected seeds of different crop plants in limited pockets or under some trees and harvesting them later was the practice over a number of years initially. The farming community in the Eastern Ghats constitute only the tribal population initially. With the increasing requirements of quantity and the spectrum of food, the pressure to bring in more land for organized cultivation came in to existence. As the gathering of wild seeds, beans and tubers stopped, the hill slopes are brought in to cultivation with the slash and burn (*podu*) cultivation. As the soil is virgin, the ash and the plant debris acted as the needed organic fertilizer. In this process, the forest felling, degradation and denudation paved a way for Agriculture in the Eastern Ghats.

Agriculture is practiced by all the tribal groups among others concentrated in the forest areas like *Kondaporas, Kondareddy, Chenchus, Gonds* (Andhra Pradesh), *Sholingars, Kurumbas, Thodas, Irulas, Nari kuravas* (Tamil Nadu) and *Porjas, Gadabas, Bondas, Savaras, Samanthas* and *santhals* (Odisha) mainly by *podu* cultivation. Due to intensified Integrated Tribal Development Agency (ITDA) extension programs the tribals have been given some hill slopes for cultivation and all the required inputs to divert them to agriculture. There are instances very recently where the PTG (Primitive tribal groups, for example, *Kondareddis* in Khammam district) have been persuaded by the ITDA officials to come down the hills and settle down near the foothills and join the mainstream. Ethnic plant genetic resources from Eastern Ghats with more emphasis on medicinal plants had been dealt earlier by many researchers (Banerjee, 1977; Dikshit and Sivaraj, 2014; Krishnamurthy et al., 2002, Pandravada and Sivaraj, 1999; Pandravada et al., 2000; Pullaiah, 2002; Rama Rao and Henry, 1996, Ravisankar and Henry, 1992; Rao and Harasreeramulu, 1985; Reddy, 1980; Reddy et al., 2002; Sandhya Rani and Pullaiah, 2002; Saxena and Dutta, 1975; Sivaraj et al., 2006, 2015; Sudhakar Reddy et al., 2002; Thammanna and Narayana Rao, 1998; Varaprasad et al., 2006, 2010; Vedavathy et al., 1997).

5.2.1 NATURE AND SPECTRUM OF ETHNIC PGR DIVERSITY

Given the history of development of modern agriculture especially that of improved varieties, the importance of agri-diversity and germplasm encompassing landraces, primitive cultivars, wild and weedy relatives of crop species need not be emphasized, which is the basic material in any crop improvement program. Out of the existing native ethnic floristic wealth many plant species are yet to be utilized/explored by man especially with regards to the future needs/ requirements. The list of some of the landraces existing in several ethnic agri-horticultural crops in Eastern Ghats region from parts of Andhra Pradesh, Chhattisgarh, Odisha, Tamil Nadu and Telangana is provided in Table-1.

The contribution of the tribal groups in the domestication and enrichment of the genetic variability in different agri-horticultural crops is immense and indispensable in the crop improvement programs. The way the tribal groups utilize different crops as food material is very interesting and throws light on the trials and errors, permutations and combinations of culinary processes

TABLE 5.1 Ethnic Crop Diversity in Eastern Ghats, India

S. No.	Botanical Name	Family	Common English name	Landraces/Local Name
Cereals and Pseudocereals				
1	*Oryza sativa* L.	Poaceae	Rice, Paddy	*Akasavai vadlu, Akkullu, Anjan, Attedlu, Bandabudamalu, Battadhanyam, Bodadhanyam, Badshabhog, Bagni dhan, Baiganmangi, Bakti chudi, Bandabudamalu, Bandhichudi, Barad-hai, Barangi, Bayahunda, Bhayagunda, Bhus katia, Budama, Budimvanji, Chittimutyalu, Chittiporu, Chudi dhan, Dasarabhogalu, Dasara vadlu, Davaralvanji, Dawalakavanji, Dengichudi, Desidhan, Dhagruvanji, Dhanwakitaka, Dhavaralvanji, Dhonanellu, Dhubraj, Doddodlu, Ekhlo, Errabuddama, Erra mallelu, Errodlu, Erravadlu, Garadavadlu, Gowrani vadlu, Isukaravvalu, Jajati, Kalajeera, Kakirekkala vadlu, Kondadhanyam, Kondabuda-malu, Kukumbanthulu, Lalat, Meher, Malainellu, Mettadhanyam, Modugarikalu, Mullodlu, Nalla vadlu, Nallamettadhanyam, Nallasathikalu, Nevarivari, Nimmalosari, Pandadivanji, Pandarisada, Pisodi vadlu, Polala vadlu, Puraval, Ragalvanji, Regativadlu, Sannadhanyam, Seeraga samba, Seethammasavaralu, Tellabudamalu, Tella vadlu, Umri chudi, Vattodlu, Voodasannalu, Yerrakondadhanyam, Yerramallelu*
2	*Triticum aestivum* L.	Poaceae	Bread wheat	*Metta goduma, Kodumai*
3	*Triticum durum* Desf	Poaceae	Durum wheat	*Erra goduma, Kodumai*
4	*Zea mays* L.	Poaceae	Maize, corn	*Chinna makka, Erra makka, Gundu makka, makka cholam, Pelala makka*
Millets				
5	*Sorghum vulgare* Pers.	Poaceae	Sorghum	*Aragidi jonna, Badigi jonna, Chikkati jonna, Chinna jonna, Dambral, Deyam jonna, Darawat jonna, Gadda jonna, Gattumalle jonna Gunjidi jonna, Gunjidipedda jonna, Guvvi jonna, Jalleda jonna, Jinghri jowar Kathani jonna, Kempujola, Kondajonna, Kondamud-dajonna, Konakadala jonna, Leh jowar, Markandi jonna, Moddu jonna, Motitura, Pachcha-jonna, Pachchaboda jonna, Padijonna, Palepujonna, Podujonna, Pala jonna, Pelala jonna, Pandari jonna, Pasupu jonna, Pasupupachcha jonna, Potiki jonna, Potithimmalu, Purabo-daka jonna, Sai jonna, Sanna jonna, Sevata jonna, Sivira jonna, Tellaboda jonna, Varagadi jonna, Vubiripatti jonna, Vullipitta jonna, Billi jola, Chiru talavalu, Moti jowar, Natujonna, Seethammavarijonna, Tella jonna, Tellamalle jonna, Yerrajonna*

TABLE 5.1 *(Continued)*

S. No.	Botanical Name	Family	Common English name	Landraces/Local Name
6	*Echinochloa frumentacea* Link	Poaceae	Barnyard millet	*Badasuan, Bonta sama, Burada sama, Chinna samalu, Koyya sama, Maghi sama, Malle sama, Nalla samalu, Pedda sama, Sama, Peru sama, Samai, Kudirai vali*
7	*Eleusine coracana* Gaertn.	Poaceae	Finger millet	*Burada chollu, Chinna chollu, Garuvu chodi, Metta chodi, Mudda chodi, Nadipi chollu, Pedda chodi, Punasa chollu, Pyru chollu, Tella ragulu, Tholakari chollu*
8	*Panicum miliaceum* L.	Poaceae	Common, Proso millet	*Badi, Variga, Varagu*
9	*Panicum scrobiculatum* L.	Poaceae	Kodo millet	*Bonthalu, Burka sama, Chinna oodalu, Konda voodalu, Voodalu, Pedda voodalu, Punasa voodalu*
10	*Pennisetum typhoides* (Burm.) Stapf & C.E. Hubbard	Poaceae	Pearl millet	*Cumbu, Gantlu, Pedda ganti, Pitta ganti, Podu ganti, Punasa ganti*
11	*Setaria italica* (L.) Beauv.	Poaceae	Italian or fox-tail millet	*Chinna korra, Erra korra, Jada korra, Konda korra, Kukka toka korralu, Nakka toka gaddi, Punasa korra, Thinai, Perunthinai*
Pulses				
12	*Cajanus cajan* (L.) Millsp.	Papilionaceae	Pigeon pea, Red gram	*Erra kandulu, Kandulu, Mabbu kandi, Natu kandulu, Srikandi, Tellakandulu, Chirukandi, Konda kandi, Parimi kandulu, Pedda kandi, Podu kandi, Siri kandulu, Tandur*
13	*Cicer arietinum* L.	Papilionaceae	Chick pea, Bengal gram	*Buta, Senegalu, Konda kadalai*
14	*Cyamopsis tetragonoloba* (L.) Taub	Papilionaceae	Cluster bean	*Kottavarai, Gorechikkudu*
15	*Dolichos lablab* L.	Papilionaceae	Hyacinth bean	*Avarai, Chikkudu, Nalla chikkudu, Gane chikkudu*
16	*Glycine max* Merril	Papilionaceae	Soybean	*Soya mochai*

TABLE 5.1 *(Continued)*

S. No.	Botanical Name	Family	Common English name	Landraces/Local Name
17	*Lathyrus sativus* L.	Papilionaceae	Grass pea	*Kesari*
18	*Lens culinaris* Medic	Papilionaceae	Lentil	*Masur pappu/Chirisenegalu*
19	*Macrotyloma uni-florum* (Lamk.) Verdc	Papilionaceae	Horse gram	*Kollu, Ulavalu, Kulthi*
20	*Mucuna utilis* Wall ex.wight	Papilionaceae	Velvet bean	*Dulagondi, Pilliadugu, Poonai pidukkan, Poonai Kali, Kainchow*
21	*Pachyrrhizus ero-sus* (L.) Urban	Papilionaceae	Yam bean	—
22	*Psophocarpus teragonolobus* DC.	Papilionaceae	Goa bean	*Morisu avarai*
23	*Vicia faba* L.	Papilionaceae	Broad bean	
24	*Vigna aconitifolia* (Jacq.) Marechal	Papilionaceae	Dew gram, moth bean	*Tuluka payir; Kuncumapesalu*
25	*Vigna mungo* (L.) Hepper	Papilionaceae	Black gram	*Bhema urad, Biri, Biri kandalu, Bunadiri, Gaju minapa, Giddu minumulu, Koloth, Konda minumulu, Kuppa minumulu, Malbiri, Mettu minumulu, Minimulu, Nalla Minumulu, Muga dali, Pacha minumulu, Sada minumulu, Teega minumulu, Toppa Minumulu, Tudu minumulu, Barre minumulu, Moddu Modda minumulu, Lal urad*
26	*Vigna radiata* (L.) Wilczek	Papilionaceae	Green gram	*Nalla pesarlu, Paccha pesaru, Pesarulu, Teega pesarulu, Chinna pesarulu, Ganga pesarlu, Chamki pesarlu, Kotta pesarlu, Nelala pesarlu, Konda pesalu, Balintha pesarlu*
27	*Vigna umbellata* (Thunb.) Ohwi & Ohasi	Papilionaceae	Rice bean	*Thattam payiru*

TABLE 5.1 *(Continued)*

S. No.	Botanical Name	Family	Common English name	Landraces/Local Name
28	*Vigna unguiculata* (L.) Walp	Papilionaceae	Cowpea	*Alasandalu, Judumulu, Controma, Bijunumulu, Junumulu, Barabatti, Batta judumulu, Baragudi chhoin, Bobbarlu, Boijudumulu, Bobbiri judumulu, Challa bobbarlu, Chinna bobbarlu, Chitti bobbarlu, Dhota judur, Erra bobbarlu, Judum, Judumulu, Karri bobbarlu, Konda bobbarlu, Konda judumulu, Matcha bobbarlu, Nalupu judumulu Nalla bobbarlu, Natu alasandhalu, Pedda bobbarlu, Siri bobbarlu, Tella bobbarlu, Thoppa junumulu*
Oilseeds				
29	*Brassica juncea* (L.) Czern. & Coss	Brassicaceae	Indian mustard	*Malai kadugu, Sorisa*
30	*Brassica nigra* (L.) Koch	Brassicaceae	Black mustard	*Malai kadugu, siru kadugu, Nallaavalu*
31	*Carthamus tinctorius* L.	Asteraceae	Safflower	*Sendurakam, Tella kusuma, Kusumalu*
32	*Guizotia abyssinica* Cass.	Asteraceae	Niger	*Verrinuvvulu, Paiyellu, Uchellu, Alisi*
33	*Linum usitatissimum* L.	Linaceae	Linseed	*Avise, Alivirai, Javas*
34	*Madhuca longifolia* (Koening) Macbr.	Sapotaceae	Butter tree	*Ippa, Moha*
35	*Madhuca indica* J.F. Gmel.	Sapotaceae	Mahua	*Ippa, Iluppai, Iluppa, Mahula, Moha, Madigi*
36	*Pongamia pinnata* Pierre	Papilionaceae	Pongam oil tree	*Kanuga, Pungu, Ponga, Koranjo*
37	*Ricinus communis* L.	Euphorbiaceae	Castor	*Amanakku, Kottai muthu, Amudamu*
38	*Seamum indicum* L.	Pedaliaceae	Sesame	*Nuvvulu, Ellu, Khasa, Rasa*

TABLE 5.1 *(Continued)*

S. No.	Botanical Name	Family	Common English name	Landraces/Local Name
Vegetable and Tuber Crops				
39	*Abelmoschus esculentus* (L.) Moench	Malvaceae	Okra, Lady's finger	*Benda, Naatu benda, Bhendi, Vendai, Eedakula benda, Pasara benda, Patcha benda, Ssudibenda*
40	*Allium cepa* (L.) var. *aggregatum*	Alliaceae	Multiplier onion	*Chinna vengayam*
41	*Allium cepa* L.	Alliaceae	Onion	*Vengayam, Ulligadda, Nirulli, Piyaz*
42	*Allium sativum* L.	Alliaceae	Garlic	*Rasuna, Tella gadda, Velluli, Vellai poondu*
43	*Alocasia indica* (Roxb.) Schott	Araceae	Taro	*Charakanda, Merukan kilangu*
44	*Amaranthus viridis*	Amaranthaceae	Green Amaranth	*Siru keerai, Kuppai keerai, Chailakathottakara*
45	*Amorphophallus campanulatus* Bl.ex Decne	Araceae	Elephant foot yam	*Kanda, Karunai kizhangu*
46	*Asparagus officinalis* L.	Liliaceae	Asparagus	*Shimai shadavari, Challa gadda, Pilli tegalu teegalu, Satavari*
47	*Basella alba* L.	Basellaceae	Indian spinach	*Batsala*
48	*Basella rubra* L.	Basellaceae	Indian spinach	*Erra Batsala*
49	*Benincasa hispida* (Thunb.) Cong.	Cucurbitaceae	Ash gourd	*Buditha gummadi, Gummadikaya, Poosanikkai, Pani kakharu*
50	*Beta vulgaris* L.	Chenopodiaceae	Sugar beet	*Chukandar*
51	*Canavalia gladiata* (Jacq.) DC.	Papilionaceae	Sword bean	*Thammakaya, Adavi thamma, Kozhi avarai, Vellai Tambattai, Vellai tamma, Tella tamma, Sigapu tambattai, Erra tamma*
52	*Chenopodium album* L.	Chenopodiaceae	Pig weed	*Paruppu keerai, Pappu kura*

TABLE 5.1 *(Continued)*

S. No.	Botanical Name	Family	Common English name	Landraces/Local Name
53	*Coccinea cordifo-lia* Cogn.	Cucurbitaceae	Ivy gourd	*Dondakaya, Donda*
54	*Colocasia escu-lenta* (L.) Schott	Araceae	Elephant ear yam	*Sharu, Chama dumpa, Chema gadda, Cehppa kizhnagu*
55	*Cucumis melo* L. var *Conmon*	Cucurbitaceae	Pickling melon	*Dosakaya*
56	*Cucumis sativus* L.	Cucurbitaceae	Cucumber	*Budama, Adavi dosakaya, Keera Dosakaya, Channekakide, Vellarikkai*
57	*Lagenaria siceraria* (Mol.) Standl.	Cucurbitaceae	Bottle gorud	*Anamkap kaya, Ekathari kaya, Lau, Sorakaya, Suraikkai*
58	*Luffa acutangula* (L.) Roxb.	Cucurbitaceae	Ridgegourd	*Beera, Donda beera, Beerakaya, Chedu beera, Janni, Pedda beera, Peerkai*
59	*Luffa hermaphrodita*	Cucurbitaceae		—
60	*Luffa cylindrica* (L.) M.J. Roem.	Cucurbitaceae	Sponge gourd	*Neti beera, Guthi beera, Ghiya tori, Mezhugu peerkai*
61	*Lycopersicon esculentum* Mill	Solanaceae	Tomato	*Naatu Thakkali, Takkali*
62	*Lycopersicon pimpinellifolium*	Solanaceae	Cherry tomato	*Ramankaya*
63	*Manihot escu-lenta* Crantz	Euphorbiaceae	Tapioca	*Mara valli kizhangu, Karra pendalamu, Kappa*
64	*Momordica charantia* L.	Cucurbitaceae	Bitter gourd	*Chedu kakara, Kakara, Kakara kaya, Pagarkkai*
65	*Momordica dioica* Roxb. ex Willd.	Cucurbitaceae	Spinegourd, Kakora	*Agakara, Tholoopavai, Paluppakal, Golkandra*

TABLE 5.1 *(Continued)*

S. No.	Botanical Name	Family	Common English name	Landraces/Local Name
66	*Moringa oleifera* Lam.	Moringaceae	Drumstick	*Munakaya, Munaga, Mulaga, Murungai, Shobanjana*
67	*Phaseolus lunatus* L.	Papilionaceae	Lima bean	*Butter beans, Manjal beans, Khasi kollu*
68	*Phaseolus vulgaris* L.	Papilionaceae	French bean	*Beans, Mittai beans, Kakki bean*
69	*Solanum melongena* L.	Solanaceae	Brinjal, egg plant	*Bangda, Baigana, Katthiri, Tellacharakaya, Nallavanga, Namalakaya, Tellakaya, Guttivanga, Neetivanga, Manda saapa, Mettavanga, Mulla vankaya, Pandiri vnakaya, Chigurukotavanga, Jegurupaduvanga, Tantikondavanga, Medichinta, Vankaya*
70	*Spinacia oleracea* L.	Chenopodiaceae	Spinach	*Palak, Palanga saga, Palakkeerai*
71	*Trichosanthes anguina* L.	Cucurbitaceae	Snake gourd	*Potlakaya, Chetipotla, Kooshi, Chhachindra, Pudalai*
72	*Trigonella foenum-graecum*	Papilionaceae	Fenugreek	*Methi, Vendayam, Methi saga*
73	*Vigna unguiculata* (L.) Walp. var. *sesquipedalis* Aschers. & Schweinf.	Papilionaceae	Yard long bean, Asparagus bean	*Alasanda, Jhudanga, Kampa chikkudu, Podugu chikkudu*
Fruits and Nuts				
74	*Achras sapota* L.	Sapotaceae	Sapota	*Safeta, Sapota*
75	*Aegle marmelos* Correa ex Roxb.	Rutaceae	Bengal Quince	*Maredu, Bilvam*
76	*Anacardium occidentale* L.	Anacardiaceae	Cashewnut	*Jidi, Jidi-mamidi, Muntha-mamidi, Mindri, Hijli-badam, Munthiri*

TABLE 5.1 *(Continued)*

S. No.	Botanical Name	Family	Common English name	Landraces/Local Name
77	*Ananas comosus* (L.) Merril	Bromeliaceae	Pineapple	*Annaci*
78	*Annona reticulata* L.	Annonaceae	Bullocks heart	*Rama sita phal, Nona, Ramphal*
79	*Annona squamosa* L.	Annonaceae	Custard apple	*Aata, Sita phal, Seetha pazham*
80	*Areca catechu* L.	Arecaceae	Arecanut	*Paaku, Vakka, Poogiphalam*
81	*Artocarpus altilis* (Park.) Fosberg	Moraceae	Bread fruit	*Seema panasa, Seema pala, Seema pila*
82	*Artocarpus heterophyllus* Lam.	Moraceae	Jack fruit	*Panasa pandu, Pala pazham, Pilapalam*
83	*Averrhoa carambola* L.	Averrhaceae	Star fruit, Carambola	*Arai nelli*
84	*Borassus flabellifer* L.	Arecaceae	Palmyrapalm	*Tadi chettu, Panai, Tal*
85	*Buchanania lanzan* Spreng.	Anacardiaceae	Almandette	*Sara pappu, Morala, Pival, Chironji*
86	*Canarium ovatum* Engl.	Burseraceae	Pilu	*Jangli badam, Karapu kongiliam, Nalla rojanamu*
87	*Capparis decidua* Edgew.	Capparidaceae	Ket	*Kariramu, Sengam, Karira*
88	*Carica candamarcensis* Hook.f	Caricaceae	Mountain papaya	*Kondapapaya, Malaipappalli*
89	*Cairica papaya* L.	Caricaceae	Papaya	*Amrutabhanda, Boppayi, Pappalli*
90	*Carissa congesta* Wight	Apocynaceae	Karonda	*Vaka, Kalakkai*

TABLE 5.1 *(Continued)*

S. No.	Botanical Name	Family	Common English name	Landraces/Local Name
91	*Carissa spinarum* L.	Apocynaceae	Karamadika	*Chiru, Kila, Kalivi, Karamadika*
92	*Citrus grandis* (L.) Osbeck	Rutaceae	Pummelo	*Pambalimasu, Pampalamasam, Chakotra*
93	*Citrus limon* (L.) Burm.f.	Rutaceae	Lemon	*Bijapuram, Periya elumichai, Nimma*
94	*Citrus madurensis*	Rutaceae	Edapandu	*Edapandu*
95	*Citrus sinensis* (L.) Osbeck	Rutaceae	Sweet orange	*Sathugudi, Chini, Narinja, Sini, Satghudi, Musambi*
96	*Elaeocarpus floribundus* Blume	Elaeocarpaceae	Jalpai	*Rudrakhyo, Rudraksha*
97	*Emblica officinalis* Gaertn.	Euphorbiaceae	Emblic myrobalan	*Usirikai, Nelli, Amlaki*
98	*Feronia limonia* (L.) Swingle	Rutaceae	Wood apple	*Velaga, Vilanga, Kait*
99	*Ficus carica* L.	Moraceae	Fig	*Anjuru, Anjira, Manjimedi, Simayatti, Simaiyatti, Tenatti*
100	*Grewia subinaequalis* DC.	Tiliaceae	Phalsa	*Jana, Nallajana, Phutiki, Pharasakoli, Palisa, Tadachi*
101	*Mangifera indica* L.	Anacardiaceae	Mango	*Amba, Mamidi, Maangai, Mavi, Mau, Am*
102	*Manilkara hexandra* (Roxb.) Dubard	Sapotaceae	Khirni	*Khirni, Khirni pazham, Pala, Manjipala, Palla, Palai, Rayan*
103	*Morus nigra* L.	Moraceae	Black mulberry	*Shah-tut*
104	*Musa paradisiaca* L.	Musaceae	Banana	*Arati, Kura arati, Konda arati, Malai vazhai, vazhai*

TABLE 5.1 *(Continued)*

S. No.	Botanical Name	Family	Common English name	Landraces/Local Name
105	*Psidium guajava* L.	Myrtaceae	Guava	*Jama, Koyya, Pijuli*
106	*Punica granatum* L.	Punicaceae	Pomegranate	*Mathulai, Dalimba, Danimma, Dalim*
107	*Pyrus communis* L.	Rosaceae	Pear	*Berikkai, peri*
108	*Spondias pinnata* (L.F.) Kurz	Anacardiaceae	Indian Hogplum	*Ambula, Amodo, Amaratoko, Kondamammidi, Adavimammidi, Kotamara, Amabalam, Eginam*
109	*Syzygium cuminii* (L.) Skeels	Myrtaceae	Java plum	*Neredam, Naval, Sambal, Neredu, Jamun*
110	*Tamarindus indica* L.	Caesalpiniaceae	Tamarind	*Chintha pandu, Puli, Tentuli*
111	*Terminalia catappa* L.	Combretaceae	Indian almond	*Badamuchettu, Vedam, Natvadom, Deshi-badam*
112	*Ziziphus jujuba* Mill.	Rhamnaceae	Chinese date	*Pitni-ber, Ilandai palam, Kandika*
113	*Ziziphus mauritiana* Lam.	Rhamnaceae	Jujube	*Ber, Barkoli, Bodokoli, Bodori, Elandai, Elladu, Regu, Gangaregu, Karakandhavu*
114	*Ziziphus nummularia* (Burm.f.) Wight & Arn.	Rhamnaceae	Jharber	*Neelaregu, Korgodi, Bhukamtaka*
Spices and Condiments				
115	*Capsicum annuum* L.	Solanaceae	Chilli	*Mirapa, Miri, Milagai, Lanka*
116	*Capsicum frutescens* L.	Solanaceae	Bird chili	*Lanka, Seema mirapa*
117	*Cinnamomum tamala* Nees & Eberm	Lauraceae	Indian Cassia Lignea	*Talisapatri, Tejpatra, Tamalaka, Tejpat*

TABLE 5.1 *(Continued)*

S. No.	Botanical Name	Family	Common English name	Landraces/Local Name
118	*Cinnamomum zeylanicum* Breyn	Lauraceae	Cinnamom	*Dalchini, Lavangapattai, Ilayangam*
119	*Cuminum cyminum* L	Apiaceae	Cumin	*Jilakara, Siragam, Jiraka, Zeera*
120	*Curcuma longa* L.	Zingiberaceae	Turmeric	*Manjal, Pasupu, Haldi, Kasturipasupu, Hinga, Woldi, Kodipasupu*
121	*Foeniculum vulgare* Mill.	Apiaceae	Fennel	*Shombu, Sopu, Pedda jilakarra, Mauri*
122	*Mentha piperita* L.	Lamiaceae	Mint/peppermint	*Pudina, Paparamina, Gamathi phudina*
123	*Murraya koenigii* (L.) Spreng.	Rutaceae	Curry leaf	*Bhrusanga patra, Karepak, Kariveppilai*
124	*Nigella sativa* L.	Ranunculaceae	Black cumin	*Kalnonji, Nela Nella jeelakarra, Karunjiragam*
125	*Piper betle* L.	Piperaceae	Betel leaf	*Tamalaku, Vettrilai kodi*
126	*Piper longum* L.	Piperaceae	Long pepper	*Pippalu, Thippili*
127	*Piper nigrum* L.	Piperaceae	Black pepper	*Milagu, Kari milagu, Gol maricha*
128	*Trachyspermum ammi* (L.) Sprague	Apiaceae	Carum	*Vaamu, Omum, Ajowan, Ajwain*
Dye Yielding plants				
129	*Bixa orellana* L.	Bixaceae	Annatto	*Japhara, Latkan*
130	*Caesalpinia sappan* L.	Caesalpiniaceae	Sappan wood	*Bakam, Bakamu, Patungam*
131	*Indigofera tinctoria* L.	Papilionaceae	Indigo	*Nil, Aviri, Nili*

TABLE 5.1 *(Continued)*

S. No.	Botanical Name	Family	Common English name	Landraces/Local Name
132	*Lawsonia inermis* L.	Lythraceae	Henna	*Benjati, Maruthani, Goranti*
133	*Mallotus philippensis* (Lam.) Muell.Arg	Euphorbiaceae	Kumkum tree, Red Kamala	*Kapli, Kungumam, Sinduri, Kunkuma, Kapilogundi*
134	*Rubia cordifolia* L.	Rubiaceae	Indian madder	*Barheipani, Manjistha, Shevelli, Manjitti, Chiranji, Taamaravalli*

perfected over a period of time. The diversity of plants under cultivation include an array of crops belonging to cereals, millets, legumes, tubers, vegetables by the tribal groups who inhabited the Eastern Ghats in six states of the country. Pandravada et al. (2004) and Sivaraj et al. (2009) have given an account of the spectrum of agri-biodiversity that is available in the Eastern Ghat areas.

5.2.1.1   CEREALS AND MILLETS

Significant ethnic diversity is reported in rice, sorghum, pearl millet, finger millet, italian millet, proso millet, little millet, kodo millet and barnyard millet from Eastern Ghats (Pandravada et al., 2008b).

Rice is known in India from about 2,000 BC or even earlier (Vishnu-Mittre, 1968). It can be deduced from the available evidences that, eastern India is the area of domestication of Asian rices (Harlan, 1975). Tremendous diversity in both the cultivated and wild *Oryza* species occur in Koraput (Odisha), which could be the place of origin and domestication for the Asian rices. Rich diversity occurs in paddy germplasm especially for scented ness, plant height, seed dormancy, grain slenderness and resistance to pests and diseases, resistance to lodging, moisture stress, maturity, panicle structure, grain size, glume color, kernel color and endosperm. The endemic scented rice diversity (*isuka ravvalu, kampusannalu, vasanodlu*) occurs in Visakhapatnam (Araku hill tracts), pockets of Vizianagaram, Srikakulam and Nizamabad districts. Some of the tall cultivars like *basangi, krishnakatu-kalu, bayahunda, gottedlu* and *akkullu* are grown in the coastal districts specifically for the straw to thatch the huts and as a feed for cattle. Landraces from Srikakulam (*pottibasangi, soppiri* and *jajati*) are reported to have seed dormancy and do not germinate on standing. Landraces of upland rainfed ecosystem with very fine-grained medium/long slender seed (*voodasannalu, chinnakondengi, mettabudagalu*) with good potential either for direct selection and/or for crossing are found in the tribal belts of Srikakulam district. In the Telangana and Rayalaseema regions also there are diffused pockets in which indigenous variability occurs in upland rice. These landraces have high degree of resistance to pests and diseases.

Most of the above landraces are being fast replaced with the high yielding varieties ignoring the specific qualities for high yield/income particularly in the tribal areas of north coastal districts of Andhra Pradesh. The practice of cultivating early maturing (3 months) upland rice in mixed cropping mainly with sorghum, pearl millet and small millets and sometimes even with pigeon pea is disappearing and currently improved varieties of rice are mostly grown as a sole crop. Cultivation of upland rice itself is replaced by the high-income crops, such as tomato, cauliflower, cabbage, etc. in high altitude tribal pockets.

The important areas of diversity for *kharif* sorghum are districts of Adilabad and Kurnool in Telangana and Andhra Pradesh states respectively. Many distinct landraces (*chiru talavalu, moti jowar, motitura, jinghri jowar, leha jowar, pottithimmalu, tellamalle jonna*) belonging to *durras* and *roxburghiis* are the main races grown under *kharif* sorghums. Mainly variability occurs in plant height, peduncle shape and size, ear head shape and size, glume and grain color. Landraces of *rabi* sorghums (*markandi jonna, gattumalle jonna, sai jonna, mudda jonna, pelala jonna, erra jonna*, etc.) are grown to a great extent especially in the tribal pockets of Adilabad, Khammam, Kurnool and Cuddapah. Sorghum belonging to races *durra, bicolor, guinea, caudatum* and their intermediate types occur in these areas (Figure 5.1). Potential diversity occurs for panicle compactness/shape and glume covering. Sources of resistance for drought and birds (fully covered glumes) occur in the northern Telangana areas.

FIGURE 5.1   Ethnic sorghum (*Sorghum bicolor*) diversity in Eastern Ghats region (Courtesy: NBPGR, RS, Hyderabad).

Pearl millet germplasm variability is known to occur in Visakhapatnam, Vizianagaram, Srikakulam, Nalgonda, Mahaboobnagar, Kurnool and Prakasam districts. *Pittaganti*, a popular landrace of north coastal districts is an early maturing and highly tillering type. Variability occurs in plant height, stem thickness, tillering, spike length, size and shape and seed characters.

In small millets, the important crops are finger millet, italian millet (Figure 5.2), proso millet, kodo millet, little millet and barnyard millet. Variability in plant height, tillering, finger compactness, number of fingers/ ear and ears/ plant exists in finger millet in the districts of Visakhapatnam, Vizianagaram, Srikakulam, Ranga Reddy and Mahaboobnagar. Endemic local variability occurs in the other small millets.

FIGURE 5.2 Ethnic diversity in Italian millet from Eastern Ghats region (Courtesy: NBPGR, RS, Hyderabad).

However, the area under small millets has been coming down very alarmingly because of the introduction of subsidized rice scheme in Andhra Pradesh. It has insidiously contributed in the replacement of small millets with improved varieties of crops, such as chilli and other vegetables in north coastal districts and sunflower and castor in Telangana and Rayalaseema regions, there by losing the endemic diversity of small millets.

5.2.1.2   PULSES

The important pulse crops in which significant endemic diversity occurs are Pigeon pea (Figure 5.3), Lima bean, French bean, Cowpea, Hyacinth bean and Rice bean. In Pigeon pea variability exists in the north coastal tribal belt and Telangana region in Andhra Pradesh and Koraput and Gajapati districts of Orissa for days to maturity, flower color, plant height, pod size, seed shape/color etc. Local landraces, which are tolerant to pod borers *viz.* 'Konda kandulu' are perennial with very bold white/creamish-white seeds, exists in the above areas. The landraces with red/brownish-red seed are cultivated in the northern Eastern Ghats region where as white/creamish-white seeded landraces are preferred in the central Eastern Ghats region.

In Cowpea (*chittibobbarlu, bobbarlu*) the variability mainly includes cultivars with bushy/viny forms, days to maturity, pod character and seed color.

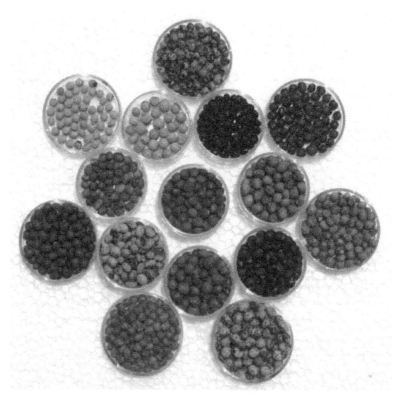

FIGURE 5.3   Ethnic diversity in Pigeon pea from Eastern Ghats region (Courtesy: NBPGR, RS, Hyderabad).

5.2.1.3   OIL SEEDS

The important oil seed crops in which good variability still occurs are sesame and niger. Sesame is an important ancient and traditional oil seed crop for which the sub-continent is the secondary center of diversity. There is significant variability in the cultivars with regards to specific planting season and region specificity due to photosensitivity. Variability is available in plant height, days to maturity, seed size, color, oil content and the extent of tolerance/resistance to different biotic/abiotic stresses.

Among the oilseeds, niger is the crop of the tribals, by the tribals and for the tribals and plays a significant role in the tribal economy. Rich variability exists in Visakhapatnam, Medak and Mahaboobnagar districts of Andhra Pradesh, Koraput of Orissa and Jagdalpur of Chhattisgarh especially for height, stem color, leaf size, branching habit, days to maturity, capitulum size and number of florets, achene size/shape, yield and oil content. The landraces do possess drought tolerance also. Germplasm accessions having high oil content were found and collected from the Koraput region.

5.2.1.4   VEGETABLES

The important vegetables for which good native diversity occurs are brinjal, chillies, okra, cucumber, gourds, onion, beans, tuber crops, and leafy vegetables.

The districts of Srikakulam, Visakhapatnam, East Godavari, West Godavari, Krishna, Guntur, Chittoor and Ranga Reddy, Koraput and Gajapati are important for brinjal diversity. Variability occurs for plant height, fruit shape, color, spininess of the pedicel and clustering and presence of blotches/ stripes on the fruit surface. In Koraput the *pottangi* variety of very primitive landrace which is the progenitors of the present day cultivars of *Solanum melongena* occur. The important landraces that are under the cultivation include *tellacharakaya, nallavanga, namalakaya, tellakaya, guttivanga, neetivanga, mettavanga, chigurukotavanga, jegurupaduvanga, tantikondavanga, medichinta* etc.

In chilli the local diversity for growth habit, fruit color, bearing, shape and size and pungency occur all along the Eastern Ghats. Chillies with small round cherry types, small oblong stout and small conical stout, extra long broad and tapering fruits occur in the north coastal districts of Andhra Pradesh. Cultivation of chilli local landrace which bear yellow fruits (*pachchamirapa*) is confined to Gollaprolu area in the East Godavari district. Indigenous paprika chilli landraces with low pungency are mostly confined to Warangal (*doddukaya, warangalkaya, tomato chilli*) and Rayalaseema region (*byadige* types) which are characterized by folds after drying. *Capsicum frutescens* (Bird's pepper) occurs in all the tribal areas in the Eastern Ghats.

In Okra diversity is mainly concentrated in the districts of Kurnool, Visakhapatnam, East Godavari and Adilabad. The important landraces that occur are *edakula benda, pasara benda, patcha benda* and *sudibenda*. Native diversity occurs for plant height, pigmentation and fruit characters. Very tall, robust, purple pigmented, tolerant to cold and late maturing types (*chalibenda*) occur in the Adilabad district of Telangana region.

The important gourds for which local ethnic diversity especially for fruit size and shape occur are bottle gourd, pumpkin, snake gourd, ridged gourd and bitter gourd (Sivaraj and Pandravada, 2005; Sivaraj et al., 2010; Sunil et al., 2014). The important districts under gourds cultivation where local diversity occurs are East Godavari, Warangal, Khammam, Adilabad, Karimnagar, Visakhapatnam, Koraput and Gajapati districts. Bottle gourd with different fruit shapes occurs in the tribal pockets in the above districts (Figure 5.4). The wild forms are very bitter and trailed on the huts and harvested after drying to use them for carrying water and arrack and also for storage. Extra-large fruited types (*Nelabeera*) almost to one meter are found in the Adilabad district.

FIGURE 5.4   Ethnic bottlegourd (*Lagenaria siceraria*) diversity in Eastern Ghats region (Courtesy: NBPGR, RS, Hyderabad).

In tuber crops the important districts are Srikakulam, Visakhapatnam Vizianagaram, East Godavari, West Godavari, Krishna, Nellore, Koraput and Gajapati. Local cultivars and wild tubers which are well adapted and are being grown/ collected from the forests occur in Elephant foot Yam (*pedda kanda, chinna Kanda, theepi kanda*), *Colacasia, Dioscorea, Xanthosoma* and Tapioca (Figure 5.5).

In the leafy vegetable the important types for which local variability occur are amaranth, spinach, *Rumex, Basella, Trigonella*, etc. in all the districts of Eastern Ghats. In Amaranth types which are green, completely/ partially purple and spiny generally occur.

FIGURE 5.5 Ethnic community with tuber crops diversity in Eastern Ghats region (Courtesy: NBPGR, RS, Hyderabad).

5.2.1.5 FRUIT CROPS

The important fruits which have native diversity are mango, banana, citrus fruits, custard apple, ber, pine apple, wood apple, bael and jamun etc.

In Mango the main variability occurs in the coastal areas, Khammam district in the Telangana and eastern Rayalaseema regions. Diversity occurs for tree size, maturity, bearing, fruit color, size, flesh characters, keeping quality and yield. Also cultivars specific for table and pickling and resistant to hoppers and which can withstand wind and drought are also found. The wild species occur in the tribal dominated hilly areas of Visakhapatnam and Vizianagaram districts.

In banana the districts which are important for germplasm variability are East Godavari, West Godavari, Kurnool, Vizianagaram, Visakhapatnam, Cuddapah, Koraput and Gajapati. The traditional and popular varieties (*amruthapani* and *chakkarakeli*) are alarmingly replaced by the improved varieties. Good local varieties occur both in table and vegetable types. Diversity occurs in plant height, maturity, bunch size, fruit shape, size and aroma. *Musa ornata* and *Ensete glaucum* occur in the hills of Eastern Ghats interspersed in East Godavari and Visakhapatnam districts.

Among the *Citrus* species, *Citrus madurensis* has been naturalized in the hills of East Godavari and Srikakulam. The other fruits in which significant local diversity exist is pine apple in the Visakhapatnam (Simhachalam). Custard apple, which occurs in semi-wild state in the hills has good diversity for habit, plant type, fruit and reticulation size, seed size and number in Mahaboobnagar, East Godavari and Anantapur. In jack, the important local types include *kharja panasa* and *tene panasa*. Good local diversity for fruit size and shape occur in the coastal areas. In the hills, good variability occurs in wood apple, bael and jamun.

5.2.1.6 SPICES

The crops of importance in spices include turmeric, ginger, and Indian long pepper, etc.

In turmeric, landraces are still popular because of their adaptability and significant levels of tolerance/resistance to biotic/abiotic stresses. Native and wild types occur in Araku (*kasturipasupu, hinga, woldi, kodipasupu*) and Mahendragiri hills with good variability for days to maturity, rhizome shape and size, inside color, surface color, aroma and yield.

In ginger the important areas of diversity are Araku hill tracts and other areas of Visakhapatnam, Medak, Koraput and Gajapati districts. The native types (*pandimallelu, rellakommalu*) possess variability for rhizome shape, size, in side and surface color, aroma and resistance/tolerance to rhizome rot and other diseases.

5.2.1.7   FIBER CROPS

The important fiber crops in Eastern Ghats are cotton and mesta. Among the old world diploid Asiatic cottons rich genetic diversity occurs in *Gossypium arboreum* and *Gossypium herbaceum*. The landraces under *G.arboreum* include *srisailam, errapathi, pandapur, mudhole, nandyal, mungari* etc. and *javvari* and *jayadharu* in *G.herbaceum*. Diversity occurs in the cotton germplasm in plant height, boll shape, size and surface, lint color and in seed characters. The perennial types of *G.arboreum* (*jadapathi, pagadapathi, pydipathi* etc.) are grown in temples, backyards of the households and as escapes throughout the ghats. Also it is amazing that both the annual and perennial types of *G.arboreum* cottons are disease/pest free and are sources of resistance for black arm, boll warm and sucking pests. However, with the introduction of superior long stapled new world cottons of *Gossypium hirsutum,* the desi cottons are being replaced at an alarming rate.

In *Mesta* diversity in *Hibiscus sabdariffa* and *H.cannabinus* occurs in the north coastal districts. Variability could be observed in stem pubescence, stem color, branching habit, fruit pubescence, stem diameter, seed color and other seed characters.

5.2.1.8   WILD RELATIVES OF CROP PLANTS

The ethnic plant genetic resources diversity of Eastern Ghats includes wild/ weedy relatives and related taxa of crop plants which are crucial for the improvement of crops. These plant species constitute a part of the crop genepool and possess, a big reservoir of untapped genes that have potential to be utilized in improvement of crops.

At least 91 wild related species of crop plants are reported in the Eastern Ghats region by Arora and Nayar (1984). Pandravada et al. (2008a) enumerated the genetic resources of wild relatives of 73 major cultivated crops belonging to crop groups of cereals, millets, small millets, pulses, oil seeds, vegetables, leafy vegetables, tuber crops, fruit crops and spices in Eastern

Ghats region of Andhra Pradesh which are represented by 71 genera and 203 species of 36 plant families. Important examples of wild relatives from Eastern Ghats ethnic pockets are: *Abelmoschus ficulneus, Abelmoschus moschatus, Amorphophallus paeoniifolius, Cajanus cajanifolius, Capsicum frutescens, Citrullus colocynthis, Cucumis hardwickii, Cucumis pubescens, Curcuma angustifolia, Dioscorea oppositifolia, Luffa acutangula* var. *amara, Luffa cylindrica, Luffa tuberosa, Lycopersicon pimpinellifolium, Momordica dioica* (Figure 5.6), *Moringa concanensis, Moringa pterigosperma, Mucuna monosperma, Murraya paniculata, Musa ornata, Ocimum americanum, Ocimum basilicum, Oryza malampuzhaensis, Oryza nivara, Oryza rufipogon, Pennisetum hohenackeri, Piper longum, Sesamum alatum, Solanum incanum, Solanum indicum, Solanum nigrum, Solanum pubescens, Solanum surattense, Solanum torvum, Solanum trilobatum, Sorghum halepense, Trichosanthes bracteata, Trichosanthes cucumerina, Vanilla wightiana, Vigna trilobata, Zingiber cassumunar, Zingiber roseum, Ziziphus oenoplia, Ziziphus xylopyrus* etc. Identification of wild relatives of many crop plants and establishing their close genetic affinities have made it possible to utilize them as potential source of genetic variation by the breeders (Kalloo and Berg, 1993; Sharma et al., 2003; Pandravada et al., 2008a).

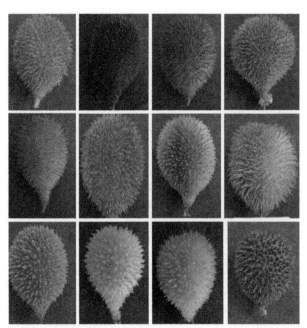

FIGURE 5.6   Ethnic diversity in spine gourd (*Momordica dioica*) in Eastern Ghats region (Courtesy: NBPGR, RS, Hyderabad).

FIGURE 5.7   Wild relatives of crop plants species diversity from Eastern Ghats region (Courtesy: NBPGR, RS, Hyderabad).

A. *Luffa tuberosa* B. *Citrus madurensis* C. *Sesamum alatum* D. *Vanilla wightiana*

5.2.1.9   ETHNIC LIFE SUPPORT SPECIES OF EASTERN GHATS (NON-TIMBER FOREST PRODUCE)

Ethnic life support species have an important role to play for the betterment of mankind, especially under situations of abiotic stresses related to soil, water, nutrients and energy, particularly in view of increased human and livestock population. Some of the important NTFP species recorded in the Eastern Ghats region are *Andrographis lineata, Amorphophallus paeoniifolius, Artocarpus heterophyllus, Bambusa bambos, Borassus flabellifer, Caesalpinia bonduc, Canarium strictum, Ceiba pentandra, Dioscorea bulbifera, Entada pursaetha, Phyllanthus emblica, Ficus hispida, Gloriosa superba, Helicteres isora, Gymnema sylvestre, Hemidesmus indicus, Jatropha curcas, Madhuca indica, Myristica fragrans, Myristica dactyloides, Ocimum gratissimum, Pongamia pinnata, Strychnos nux-vomica, Syzygium cumini, Sterculia urens, Tamarindus indica, Terminalia bellirica, Terminalia chebula, Urginea indica, Ziziphus mauritiana, Ziziphus xylopyrus* etc. The dependence of tribal communities on these plant resources is inevitable. In this context, more important is exploitation of plants occurring in extreme environmental conditions of Eastern Ghats region. These life support species are expected to provide new sources of food, fiber, fuel, fodder, hydrocarbon and industrial products.

FIGURE 5.8   Palmyra palm diversity, a life supporting species of ethnic communities (Courtesy: NBPGR, RS, Hyderabad).

5.3   ETHNIC PLANT GENETIC DIVERSITY CONSERVATION

The conservation of ethnic plant genetic diversity involves two basic strategies (i) *in-situ* and (ii) *ex-situ*. *In-situ* conservation of ethnic plant genetic diversity has to be carried *on-farm,* where landraces and locally improved/ adapted material are cultivated, evaluated, utilized and conserved as part of traditional farming systems. These farming systems are of particular importance in maintaining local genetic diversity and providing food for local consumption and local markets. *Ex-situ* conservation requires collection and systematic storage of seeds/propagules outside the natural habitats of species for short-medium and long-term after proper characterization and evaluation.

The ethnic plant diversity including agro-biodiversity is a vibrant and indispensable component in the overall conservation strategies for Eastern Ghats. The seed material of different agri-horticultural crops of orthodox nature is stored at $-20°C$ with the seed moisture brought down to 5–8% and RH being maintained at 25–32% at National Gene Bank, New Delhi. In some difficult species, which are recalcitrant, pollen and seed material is stored at $-180°C$ in liquid nitrogen in the cryo tanks at the National Gene Bank.

For medium term conservation, the seed material is stored at 5°C with the seed moisture brought down to 5–8% and RH being maintained at 30–35% in the cold storage modules at National Bureau of Plant Genetic Resources (NBPGR) Regional Station, Hyderabad. The crops which are multiplied by vegetative means and medicinal plant species which are non-seed bearing (stem cuttings/root cuttings/whole plant) are being maintained in the Glass house/ Field gene bank at NBPGR Regional Station, Hyderabad in live condition.

Genomic resources of ethnic plant diversity, such as cloning vectors, expression vectors, binary vectors, RFLP probes, Cloned genes, promoters fused to reporter genes, sub-genomic, cDNA, EST, repeat enriched libraries, BAC, YAC, PAC clone set from sequencing projects, genomic, mitochondrial or chloroplast DNA, cloned DNA from wild and weedy species produced exclusively for the repository can be stored by the following methodologies:

- 1–2 years at 4°C; 4–7 years at –20°C and greater than 5 years when stored at –70°C;
- ESTs, full-length cDNAs, BACs, PACs and YACs, are maintained in 96-well or 384-well micro plates at –80°C;
- cDNA clones as plasmid DNA at –20°C;
- Lyophilized DNA for long-term storage;
- Ambient temperature storage.

A total of 1,87,439 germplasm accessions of various crops including ethnic plant diversity of Eastern Ghats were characterized and preliminarily evaluated at NBPGR and its regional stations. These have been documented and published as several crop catalogs for the utilization by breeders in various crop improvement programs in the country.

5.4 UTILIZATION OF ETHNIC PLANT GENETIC RESOURCES DIVERSITY IN CROP IMPROVEMENT

Keeping in view, the importance and potential of the crop genetic diversity, NBPGR Regional Station, Hyderabad has been making earnest efforts to explore, collect and conserve this ethnic plant genetic diversity (Agri-diversity) in the Eastern Ghat areas in Andhra Pradesh, Telangana and adjoining areas of other states.

A paddy accession *Voodasannalu*, a super fine grained upland drought resistant landrace collected jointly by NBPGR, Hyderabad and ANGRAU from the *Jatapu* tribal group from Seethampeta Mandal, Srikakulam district was released as a variety as *Maruteru Sannalu* in Andhra Pradesh. A coriander accession collected from Ongole Mandal, Prakasam district was released as a variety *Sudha* by ANGRAU as per the state varietal release committee recommendations. A Roselle landrace which was found to be quite promising as a leafy vegetable and designated as UJWALA is in the minikit trials of ANGRAU for release as a variety. '*Arka Mangala*' a yardlong bean variety was released at Institute level by IIHR for extra long pods. This variety was developed as pure line selection from accession IC582850 from Eastern Ghat region of Jeypore, Orissa. The variety, evaluated at IIHR over two seasons, recorded a yield of 24.7 t/ha with an increase of 24% and 30% pod yield over the check varieties Lola and Vyjayanthi respectively. The accession is a pole type, photo insensitive variety with green smooth pods. Significant number of Sorghum accessions found to be promising got included in the ICRISAT core collections (*kharif* and *rabi*) for utilization in crop improvement programs. Some collections in Paddy and Chillies were found to be promising against biotic stresses as well. In addition to these, 17 accessions with special traits were registered with the ICAR – Cowpea accession INGR 08084 resistant to *Black eye cowpea mosaic virus;* Chilli accession INGR 08097 for resistance to thrips and powdery mildew; Chilli accession INGR 08095 for resistance to thrips and mites; Jatropha accessions INGR 08087 and INGR 08088 for high oil content; Linseed accessions INGR 10027 high oleic acid content (32%), INGR 10028 High oil content (42.6%); Dolichos bean INGR 110311 field tolerant to Anthracnose and mites; Bottle gourd INGR 10064 unique spindle shape; Chilli accession INGR14040 Purple genotype; Sorghum accessions INGR 09103, INGR 09104, INGR 09105, INGR 09106 Source of resistance to multiple foliar diseases (Rust, zonate leaf spot, sooty stripe, downy mildew and ergot); Greengram accession INGR 11031 Photosensitive line; Blackgram accession INGR 13057 Photosensitive line; Pongamia accession INGR 10134 four seeded pod.

5.5 FACTORS CONTRIBUTING TO GENETIC EROSION OF ETHNIC PLANT GENETIC RESOURCES DIVERSITY

The Eastern Ghats is a vibrant habitat for ethnic diversity in different agri-horticultural crops, their wild/ weedy relatives, medicinal, aromatic and dye yielding plants. However, due to degradation of forests as a result of *podu* cultivation by the tribals, encouragement for raising plantations by the departments of agriculture/horticulture, increase in population and the need to produce more food and non-food agricultural commodities, changing food habits and initiation of other socio-economic developmental programs by the Government and NGOs, the endemic genetic diversity accumulated through years of evolution under domestication and natural selection by the tribal groups is being wiped out from nature. The local landraces/ traditional cultivars and even some crops are gradually being replaced by improved HYV/ Hybrids and other profitable crops.

The nutritional balance of the soil and the ecological foundations of the Ghats has been affected by inappropriate land use (shifting cultivation cycles), changes in agricultural systems, overgrazing of grasslands by herbivores, deforestation due to over exploitation of forest resources, land clearance for developmental activities, such as mining, thermal and hydro-projects and the lack of pollution control measures are adding to the problem. With the increase in the population, the demand of land for agricultural purposes has been increasing, thus resulting in the encroachment of large forest area by the people.

The subsidized rice scheme especially in north coastal Andhra Pradesh has made the hitherto subsistence farming being practiced by the tribal groups in to commercial farming looking for remuneration and profits in cultivation as they were getting rice at a very cheap rate for consumption. This has resulted in unforeseen changes in the cropping patterns and replacement of traditional crops like sorghum, pearl millet and small millets with coffee/ tea/cashew or mango plantations there by losing the diversity in those crops. The practice of raising crops in the kitchen gardens and seed storage for next season sowing being discontinued made them to rely on market forces and middle men. This also contributed towards erosion of ethnic diversity in several agri-horticultural crops. The reduction/abandonment of utilization of wild tubers, fruits, millets etc. as sources of food also has resulted in the loss of diversity and indigenous traditional knowledge (ITK). The dependence on modern medicine is gradually leading to erosion of ethno botanical knowledge.

5.6 FUTURE THRUSTS

Approaches to better management of ethnic plant genetic resources diversity need to focus on the following major elements (i) looking for more species and genes to provide bio-alternatives and (ii) using both traditional breeding approaches and modern technologies. It has to be addressed in order to match the increasing human and animal interventions in Eastern Ghats region of India. Following are some of the future thrusts for effective management of ethnic plant genetic resources diversity on sustainable terms:

- Harnessing the eco-regional (Eastern Ghats) potentials, to meet the climate change.
- Need to have geo-referenced and time series data on ethnic plant genetic resources diversity to take informed decisions.
- Need to develop methodologies and tools to make the dynamic conservation of the genetic diversity of multiple ethnobotanical species compatible with poverty alleviation and increased wellbeing for its keepers.
- Development of Genomic Resources from ethnobotanically valued genetic resources and their conservation.
- *In-situ/on-farm* conservation.
- Developing cost-effective *in-vitro* multiplication and conservation protocols for ethnic plant genetic resources diversity.
- Mapping of ethnic genetic diversity for quality conservation.
- Developing ability, appropriate institutional arrangements and policy framework for handling intellectual property rights related issues.
- Need for safeguarding the regional and national interests in order to meet the challenges of the new legal regimes of the CBD, the WTO & TRIPS, UPOV and ITPGRFA.

5.7   CONCLUSIONS

There is a tremendous urgency and scope for collection and conservation of Ethnic plant genetic diversity in general and medicinal plants, wild relatives and endemic tree species in particular for sustainable utilization from the Eastern Ghats. As the replacement of local cultivars has become very fast and alarming, rapid efforts have to be resorted to collect and conserve the ethnic crop genetic resources diversity from Eastern Ghats. Another concern is the collection of wild/weedy relatives of crop plants and endemic diversity in medicinal and aromatic plants before the natural habitats are destroyed. Concerted and systematic efforts have to be initiated involving all the agencies/institutes concerned to avoid duplication of efforts. It should be ensured that, participation of the communities and stakeholders in comprehensive documentation of tribal life systems including ITK, folklore and domestication of plant species etc. In view of changing scenario in both agricultural and vegetation diversity, a suitable conservation strategy should be evolved for addressing sustainable development to save the Eastern Ghats for the people of present and future generations.

KEYWORDS

- **Conservation**
- **Eastern Ghats**
- **Germplasm**
- **Plant Genetic Resources**
- **Tribal Groups**

REFERENCES

Arora, R.K. & Nayar, E.R. (1984). Wild relatives of crop plants in India. NBPGR Sci. Monogr. No.7. NBPGR, New Delhi. pp. 1–90.

Banerjee, D.K. (1977). Observation on ethnobotany of Araku valley, Visakhapatnam district, Andhra Pradesh. *J. Sci. Club, 33*, 14–21.

Chauhan, K.P.S. (1998). Framework for conservation and sustainable use of biological diversity: Action plan for the Eastern Ghats region. In: Proceedings of Seminar on Conservation of Eastern Ghats. pp. 345–357.

Dikshit, N. & Sivaraj, N. (2014). Folk medicinal plants, uses and claims of tribal peoples of Similipal Biosphere Reserve, Odisha. In: Proceedings of National Seminar on Ethnobotany, Traditional Knowledge and Access and Benefit Sharing (NSEBTK-2014), Tirupati, pp. 71–77.

Harlan, J.R. (1975). Crops and Man. American Society of Agronomy, Crop Science Society of America, Madison, Wisconsin. 295 pp.

Kalloo, G. & Bergh (Eds.) (1993). Genetic Improvement of Vegetable Crops. Oxford, New York, USA: Pergamon Press Ltd.

Krishnamurthy, K.V., Siva, R. & Senthilkumar, T. (2002). Natural dye yielding plants of Shervaroy hills of Eastern Ghats. In: Proceedings of the National Seminar on Conservation of Eastern Ghats, Tirupati (ed.) Anonymous. Hyderabad: Environment Protection Training and Research Institute,, pp. 151–153.

Pandravada, S.R. & Sivaraj, N. (1999). Diversity and collection of germplasm of spices, medicinal, aromatic and dye yielding plants from Andhra Pradesh, South India. In: Biodiversity, Conservation and Utilization of Spices, Medicinal and Aromatic Plants. Ravindran et al. (Eds.). Calicut, Kerala: Indian Institute of Spices Research. pp. 219–228.

Pandravada, S.R., Sarath Babu, B., Sivaraj, N., Maheswara Rao, G. & Satyanarayana, Y.V.V. (2000). Species diversity and germplasm collection of medicinal plants from Eastern Ghats. *Indian Forester, 126,* 1191–1203.

Pandravada, S.R., Sivaraj, N. & Varaprasad, K.S. (2004). The changing pattern of plant bio-diversity in the Eastern Ghats. In: B.S. Dhillon, R.K. Tyagi, Arjun Lal & S. Saxena (eds.). *Plant Genetic Resource Management.* New Delhi: Narosa Publishing House, pp. 136–152.

Pandravada, S.R., Sivaraj, N., Kamala, V., Sunil, N. & Varaprasad, K.S. (2008a). Genetic resources of wild relatives of crop plants in Andhra Pradesh – Diversity, Distribution and Conservation. *Proc. A.P. Academy of Sciences. Vol.* 12 (1 & 2): 101-119.

Pandravada, S.R., Sivaraj, N., Kamala, V., Sunil, N., Sarath Babu, B. & Varaprasad, K.S. (2008b). Agri-biodiversity of Eastern Ghats – Exploration, Collection and Conservation of Crop Genetic Resources. Proceedings of the National Seminar on Conservation of Eastern Ghats., Hyderabad. pp. 19–27.

Pullaiah, T. (2002). Medicinal Plants in Andhra Pradesh. New Delhi: Regency Publications, 226 p.

Rama Rao, N. & Henry, A.N. (1996). The Ethnobotany of Eastern Ghats in Andhra Pradesh. India. Kolkata: Botanical Survey of India.

Rao, K.P. & Harasreeramulu, S. (1985). Ethnobotany of selected medicinal plants of Srikakulam district, Andhra Pradesh, *Ancient Sci. Life, 4,* 238–244.

Ravisankar, T. & Henry, A.N. (1992). Ethnobotany of Adilabad district, Andhra Pradesh, *Ethnobotany, 4,* 45–52.

Reddy, T.A. (1980). Notes on some medicinal plants of Polavaram agency tracts, West Godavari district, Andhra Pradesh. *J. Indian Bot. Soc., 59,* 169.

Reddy, K.N., Sudhakar Reddy, Ch. & Raju, V.S. (2002). Ethnobotany of certain orchids of Eastern Ghats of Andhra Pradesh. In: Proceedings of the National Seminar on Conservation of Eastern Ghats, Tirupati (ed.) Anonymous. Hyderabad: Environment Protection Training and Research Institute, pp. 154–160.

Sandhya Rani, S. & Pullaiah, T. (2002). A taxonomic survey of trees in Eastern Ghats. In: Proceedings of the National Seminar on Conservation of Eastern Ghats, Tirupati (ed.) Anonymous. Hyderabad: Environment Protection Training and Research Institute, pp. 5–15.

Sharma, H.C., Pampathy, G. & Reddy, L.J. (2003). Wild relatives of pigeonpea as a source of resistance to the pod fly (*Melanagromyza obtusa* Malloch) and pod wasp (*Taraostinodes cajanianae* La Salle), *Genetic Resources and Crop Evolution, 50,* 817–824.

Saxena, H.O. & Dutta, P.K. (1975). Studies on the ethnobotany of Orissa. *Bull. Bot. Surv. India, 17,* 124–131.

Sivaraj, N. & Pandravada, S.R. (2005). Morphological diversity for fruit characters in bottle gourd (*Lagenaria siceraria* (Mol.) Standl.) germplasm from tribal pockets of Telangana region, Andhra Pradesh, India. *Asian-Agri History Journal, 9*(4), 305–310.

Sivaraj, N., Pandravada, S.R., Varaprasad, K.S., Sarath Babu, B., Sunil, N., Kamala, V., Babu Abraham & Krishnamurthy, K.V. (2006). Medicinal Plant Wealth of Eastern Ghats with special reference to indigenous knowledge systems. *The Journal of the Swamy Botanical Club, 23,* 165–172.

Sivaraj, N., Varaprasad, K.S., Pandravada, S.R. & Sharma, S.K. (2009). Agrobiodiversity hotspots in Eastern Ghats – Issues and challenges. In: Agrobiodiversity Hotspots: Access and Benefit Sharing. In: S. Kannaiyan (ed.), New Delhi: Narosa Publishing Company, 325 pp.

Sivaraj, N., Pandravada, S.R., Kamala, V., Sunil, N., Sarath Babu, B., Babu Abraham & Varaprasad, K.S. (2010). Bottle gourd diversity in tribal pockets of Andhra Pradesh, India – A potential livelihood component for rural folk. In: U.V. Sulladmath & K.R.M. Swamy (Eds.). Proceedings of International Conference on Horticulture (ICH-2009), PNASF, Bangalore. pp. 1311–1315.

Sivaraj, N., Kamala, V., Pandravada, S.R., Sunil, N., Elangovan, M., Sarath Babu, B., Chakrabarty, S.K., Varaprasad, K.S. & Krishnamurthy K.V. (2015). Floristic ecology and phenological observations on the medicinal flora of Southern Eastern Ghats. *Open Access Journal of Medicinal and Aromatic Plants* 5(2), 5–24.

Sudhakar Reddy, Ch., Murthy, M.S.R. & Dutt, C.B.S. (2002). Vegetational diversity and endemism in Eastern Ghats, India. In: Proceedings of the National Seminar on Conservation of Eastern Ghats, Tirupati (ed.) Anonymous. Hyderabad: Environment Protection Training and Research Institute, pp. 109–134.

Sunil, N., Thirupathi Reddy, M., Hameedunnisa, B., Vinod, Pandravada, S.R., Sivaraj, N., Kamala, V., Prasad, R.B.N., Rao, B.V.S.K. & Chakrabarty, S.K. (2014). Diversity in bottle gourd (*Lagenaria siceraria* – (Molina) Standl.) Germplasm from peninsular India. *Electronic. J. Plant Breeding*, 5(2), 236–243.

Thammanna & Narayana Rao, K. (1998). *Medicinal Plants of Tirumala*. Tirumala and Tirupati Devasthanams, Tirupati.

Varaprasad, K.S., Abraham, Z., Pandravada, S.R., Latha, M., Divya, S., Raman, Lakshminarayanan, S., Pareek, S.K. & Dhillon, B.S. (2006). Medicinal plants germplasm of Peninsular India. New Delhi: National Bureau of Plant Genetic Resources, 203 p.

Varaprasad, K.S., Sharma, S.K., Sivaraj, N. & Sarker, A. (2010). Integrated gene resource management of underutilized legumes in India. *Euphytica*, *180*, 49–56.

Vedavathy, S., Mrudula, V. & Sudhakar, A. (1997). Tribal Medicine of Chittoor District, A.P. (India). Herbal Folklore Research Centre, Tirupati.

Vishnu-Mittre (1968). Protohistoric Records of Agriculture in India. Bose Res. Inst. Calcutta, Trans., 31, 87–106.

CHAPTER 6

ETHNIC FOOD PLANTS AND ETHNIC FOOD PREPARATION IN EASTERN GHATS AND ADJACENT DECCAN REGION

B. SADASIVAIAH[1] and T. PULLAIAH[2]

[1]Department of Botany, Government Degree and PG College, Wanaparthy–509103, Mahabubnagar District, Telangana, India, E-mail: chumsada@gmail.com

[2]Department of Botany, Sri Krishnadevaraya University, Anantapur–515003, Andhra Pradesh, India, E-mail: pullaiah.thammineni@gmail.com

CONTENTS

ABSTRACT

The present study provides information about 785 species of edible plants under 449 genera and 134 families from various sources and published literature (Table 6.1). Among 134 families Helvellaceae (Fungi), Marseliaceae, Polypodiaceae (Pteridophyta) and Cycadaceae (Gymnospermae) are from other groups while 130 families belong to Angiosperms. Fabaceae is the dominant family with 51 edible plant taxa, followed by Amaranthaceae (32 taxa), Poaceae (29 taxa), Euphorbiaceae (28 taxa), Rubiaceae (26 taxa) and Cucurbitaceae (25 taxa). A total of 47 families like Begoniaceae, Bromeliaceae, Caricaceae, Nelumbonaceae are represented with single species; 24 families are represented by 2 taxa, 41 families are represented by 3–10 taxa and 22 families are representing with more than 10 species. Among the all edible plant parts a total of 320 are fruits, 294 leaves, 122 underground parts (including roots, bulbs, corms, tubers), 113 seeds, 67 stems, 36 flowers, 8 whole plants and 19 other parts like gum, bark, etc. are consumed by the tribal people of Eastern Ghats and adjacent Deccan region.

6.1 INTRODUCTION

All the familiar vegetables and fruits of our kitchen gardens, as well as the cereals of our fields, were once wild plants; or to put it more accurately, they are the descendants, improved by cultivation and selection, of ancestors as untamed in their way as the primitive men and women who first learned the secret of their notoriousness (Saunders, 1920).

There is sufficient capacity in the world to produce enough food to feed everyone adequately; nevertheless, in spite of progress made over the past two decades, 850 million people still suffer from chronic hunger. Among children, it is estimated that 161 million under five years of age are chronically malnourished (FAO, 2012). In human history, 40–100,000 plant species have been regularly used for food, fibers, industrial, cultural and medicinal purposes. Now a day's only 30 plant species are used to meet 95% of the world's food energy needs (Anon., 1996). About 50% of the world's food dry weight is derived from four cereals: rice, wheat, maize and barley. These crops are widely and intensively cultivated and have been selected from a large agro biodiversity basket containing more than 7,000 food species (Wilson, 1992). In many developing countries millions of people, particularly tribal and rural communities still collect and consume a wide variety of wild plant resources to meet their food requirements (FAO, 2004; Balemie and Kebebew, 2006; Bharucha and Pretty, 2010). In India, a considerable proportion of rural population, particularly in remote areas do not produce enough food grains to meet yearly food requirement. Therefore, most of the rural population is meeting their nutritional requirement through unconventional means, by consuming various wild plants (Singh and Arora, 1978).

There are 45,000 species of wild plants in India, out of which 3,900 plant species are used by tribals as food (out of which 145 species comprise of root and tubers, 521 species of leafy vegetables) (Kamble and Jadhav, 2013). About 7000 plant species have been cultivated for consumption in human history. Presently, only 30 crops provide 95% of human food energy needs, four of which (rice, wheat, maize and potato) are responsible for more than 60% of our energy intake (FAO, 2012).

Forests among the most important repositories of terrestrial biological diversity, and play a vital role in the daily life of rural communities in many areas, as source of timber and non-timber forest produce, as contributors to soil and water conservation, and as repositories of aesthetic, ethical, cultural and religious values. Losing forest diversity means missing opportunities for medicines, food, raw materials and employment opportunity in one word: welfare. The traditional knowledge and consumption of wild edible plants of Indian ethnic communities are rich and unique in the world (Panda, 2014).

The Eastern Ghats is inhabited by about 54 tribal communities, which constitute about 30% of tribal population of India (Verma, 1998; Chauhan, 1998; Karuppusamy and Pullaiah, 2005; Pandravada et al., 2007; Sanyasirao et al., 2014). They depend mostly on various forest resources available locally for their livelihood. The common tribal people living in the hills of Eastern Ghats are as follows:

Orissa: Bathudi, Bhottoda, Bhumia, Bhumiji, Dharua, Gadaba, Gond, Khanda, Kandha, Gouda, Kolha, Koya, Munda, Paroja, Omanatya, Santal, Saora, Birhor, Bonda, Didayi, Dogaria Kondh, Hill Kharia, Juang, Kutia Kondh, Lanjia Saora, Mankirdia, Paudi Bhuyan and Saora.

Andhra Pradesh and Telangana: Chenchus, Gadabas, Savaras, Konda Reddis, Koyas, Khonds, Kolamis, Nayakpods, Valmikis, Bhagatas, Jatayus, Yanadis and Yerukalas, Nookadora, Kotiya, Kondakammari, Muliya, Kondadora.

Tamil Nadu: Malayalis, Paliyans.

The forest related tribal groups are confronted with deteriorating livelihoods due to a declining resource base, population increase, and the impact of economic policies. There are nevertheless a number of contemporary Forest Related Tribal Groups who, after contact with other societies, continue their ways of life with very little external influence, these include the

Chenchus, the Jarawa, the Sentinelese, the Onge and the Shompen. They face great problems with respect to health, nutrition, control over resources and participation in decision making processes. Due to commercial logging operations, immigration and conversion of forest into agricultural land, the resource base has come under increasing pressure over the past century (Appalanaidu, 2013).

According to Pushpangadan (1994) in India, 68 million people belonging to 227 ethnic group and comprising of 573 tribal communities derived from six racial stocks namely – Negroid, Proto – Australoid, Mongoloid, Mediterranean, West Breachy and Nordic exists in different parts of the country. The population of tribal people in India is 8,43,26240, mainly concentrated in the forest and high altitude zones of Eastern Ghats, Western Ghats, Central, North-Eastern and Himalayan mountains. These tribal people are spread throughout the country. India is the second largest tribal populated country after Africa continent. United Andhra Pradesh itself has 6.6% tribal population of total population (Census of India, 2001). Haimendorf (1943, 1945, 1979) published accounts of life-styles, customs, socio-economic conditions and, to some extent, the crops raised and plants used by Chenchus, the Reddis of Bison hills and the Gonds of Adilabad.

Forests have a large and indispensable role to play in improving food security of tribes. Wild edible plants are important in the livelihood strategies of forest dwellers/tribal populations because they help the people to meet one of their most important basic needs the food. While these foods are not widely accessible, locally they are great relevance for nutrition and food security in many countries. India has a tribal population of 42 million of which some 60% live in forest areas and depend on forest for various edible products. Wild edible plants are much important than is generally assumed in the food supplies of many countries. Some wild foods are used as staples or as basic components of substantial meals. Many plants used in industrialized countries today were originally identified and developed through indigenous knowledge.

Wild fruits and tubers are available throughout the year; these wild edible plants grow in organic rich soil and in pollution free area. They are not exposed to any artificial chemical fertilizers and are enriched with natural nutrients. Consumption of these wild edible plants makes the tribal people less prone to diseases and they have more strength when compared to people in plains. Useful wild plants in ethnic ecosystems show a trend of utilization of locally available resources both in areas with high plant diversity and marginal habitats. The oral transfer of the indigenous knowledge of conventional uses of wild plants between elder and younger generation is not always ensured. Now-a-days, the traditional knowledge is declining due to lack of interest in the present generation and also absence of records about the useful plants. Hence, the truthful indigenous knowledge is immediately to be documented and validated for serving future generations and their nutritional values should be analyzed.

6.2 REVIEW OF LITERATURE

Due to rich Biodiversity; India has large ethnic society and immense wealth. There are 45,000 species of wild plants out of which 9,500 species are ethno botanically important species. Of these 3,900 plant species are used by tribals as food; out of which 145 species comprise of root and tubers, 521 species of leafy vegetables, 101 species of bulbs and flowers, 647 species of fruits (Arora, 1991).

Krishnamachari (1900) has reported the use of the leaves of *Erythroxylum monogynum* (Devadari) and the roots of *Aloe vera* (Kalabanda) as a food during famine. Pal and Banerjee (1971) reported the less-known plant-foods among the tribals of Andhra Pradesh and Orissa. According to Mathew (1983) Malayali tribes in Jawadhu hills of Tamil Nadu are using 24 plant species for food, religious purposes, rituals, decorative purposes, insect repellents, bio fertilizers, construction purposes, making household implements and hedge and fuel. Uma and Singh (1987) have explored 15 edible fruit species of *Diospyros* with distribution, fruit morphology and way of utilization observed in India. The tribal belts of Orissa were studied for wild edible plants by Girach et al. (1992) and Girach and Aminuddin (1992). Ansari et al. (1993) worked on less known edible plants of Shevoroy and Kolli hills of Eastern Ghats. Goud and Pullaiah (1996) ethnobotanically surveyed Kurnool district of Andhra Pradesh and stated that some of the wild edible plants observed in their study. Alagesaboopathi et al. (1996) studied about 30 wild edible fruits used by Malayali tribes in Shevaroy hills in Tamil Nadu. Umashankar et al. (1996) explained about importance of food processing of *Phyllanthus emblica* fruits. Four wild edible plants used by the local people of B.R. Hills were studied by Murali et al. (1998). Girach et al. (1997) have compiled 31 plant species consumed as wild edible plants by tribal and rural people of Bhadrak district of Orissa. Sudhakar and Vedavathy (1999) studied wild edible plants used by the tribal people of Chittoor district of Andhra Pradesh. Prasad et al. (1999) documented the food plants of Konda Reddis of Rampa Agency, East Godavari district. Subramanyam and Rama Mohan

(2001) studied on the ecology and food security of tribes lived in Visakha agency area and noted that tribes following shifting cultivation. The tribal farmers grow mixed crops of millets, pulses and oil seeds.

Murthy et al. (2003) reported a total of 419 wild edible plants from undivided Andhra Pradesh, of them most of the plants used tribal people lived in the hills of Eastern Ghats. Hebbar et al. (2003) have recorded 29 wild edible fruit yielding plants under 21 different families from Dharwad district of North Karnataka. Ganesan and Setty (2004) have enumerated the importance and uses of amla fruits in the preparation of pickles, jam and juice. Rajasab and Isaq (2004) enlisted 51 wild food plants from North Karnataka. Out of 51 plants most of them are trees, followed by herbs, shrubs and climbers. Sinha and Lakra (2005) studied about three tribal dominated districts of Orissa namely Kheonjhar, Mayurbhanj and Dhenkenal for plant consumption pattern in five tribal groups (Gond, Sounti, Bhumiz, Kol and Juang) and a total of 126 wild edible plants have been reported, of them majority are leaves (50) and fruits (46). Reddy et al. (2006) worked on ethno botany of 28 endemic plants of Eastern Ghats used by local ethinic gropus namely Bagatas, Chenchus, Gonds, Kondareddis, Koyas, Lambads, Nukadoras, Valmikis, Yanadis, Yerukalas of Andhra Pradesh and Kondhas, Gadabas, Sauras, Didayas, Kolhas of Orissa and reported 28 endemics; of them the seeds of *Cajanus cajanifolius* used to prepare curry, the seeds of *Cleome chelidonii* var. *pallai* are used as condiment, pith of *Cycas beddomei, Cycas sphaerica* used to prepare "*Sago*" and fruits of *Phyllanthus indofischeri* are edible. Mamatha et al. (2006) have described 100 medicinal plants used by Soliga community of B.R. Hills, among them 10 are edible plants. Devaraj et al. (2006) have enumerated 11 edible plants along with common name, part used, harvesting season and uses from households of tribal communities from Kollegal taluk.

Reddy et al. (2007) reported the traditional knowledge on wild food Plants in Andhra Pradesh and listed 156 wild food plants. Rout (2007) did a wonderful work on Ethnobotany of diversified wild edible fruit plants in Similipal Biosphere Reserve of Orissa. Behera et al. (2008) documented wild edible plants of Mayurbhanj district of Orissa. Gayatri and Srividya (2008) made a note on ethno medicinal knowledge of traditionally used edible leaves, seeds flowers among women. Dhole et al. (2009) studied on ethno medicinal properties of weeds in crop fields of Marathwada region and reported 57 problematic weeds, of them 18 are used as medicinal plants to cure many diseases and some of them are edible like *Alternanthera sessilis* and *Portulaca oleracea*.

Among the various types of plants, edible plants received the earliest attention of mankind and reflect man's search for knowing more and more about their nutrients (Jain, 1981). The seeds of *Paracalyx scariosus* are rich in sodium, phosphorus, calcium, zinc, manganese and iron. Antinutritional factors, such as total free phenols (5.56%) tannins (2.78%), L-DOPA (0.63%), hydrogen cyanide (0.065%) and phytic acid (0.85%) are present in variable quantities. From the results these plants have a good potential as food crops in Andhra Pradesh (Murthy and Sambasiva Rao, 2009). Even though Rice is a major crop in the Agrobiodiversity of Jeypore tract, the other cultivated crops include *Zea mays, Eleusine coracana, Vigna radiata, V. mungo, Brassica juncea, Sesamum indicum, Arachis hypogaea, Panicum miliaceum, Setaria italica, Guizotia abyssinica, Cajanus cajan, Macrotyloma uniflorum, Saccharum officinarum, Solanum tuberosum* and *Zingiber officinale* (Sharma et al., 1997; Misra, 2009). The study was done on the forest patches of Orissa in some districts and it revealed that a total of 15 wild edible plants recorded from Maliparbat area of Koraput district, 40 species from Khandualmali of Kalahandi district, 13 species from Kutrumali and 40 species from Baphlimali of Raygada district. *Morchella esculenta* is an important edible mushroom belonging to the family Helvellaceae also reported from Orissa in Baphlimali village, Rayagada district (Vasundhra, 2009).

Sadasivaiah (2009) in his work on diversity, quantification and conservation of herbaceous plant resources of Nallamalais recorded and quantified 36 wild edible plant resources. Of the 36 wild edible herbs of Nallamalais, *Ceropegia spiralis* is a vulnerable species and 12 are medicinal plants, 3 are wild relative plants. Basha (2009) recorded 102 wild edible trees from the forests of Nallamalais and they are consumed by human as well as animals, most of them are eaten by human beings.

The estimated 2800 species of vascular plants of Orissa state (India), about 150 wild edible fruit species occurring in different parts of eastern India's deciduous forests are consumed in various quantities by rural communities (Mahapatra and Panda, 2009). Naidu and Khasim (2010) reported 118 ethno botanical plants from Eastern Ghats of Andhra Pradesh and among them a good number of wild edible plants also mentioned in their work. According to Sahu et al. (2013) Boudh district of Orissa is famous for cultivation of minor millets. The tribal communities cultivate various millet species like Jav (*Hordeum vulgare* L.), ragi (*Eleusine coracana* (L.) Gaertn.), Maka (*Zea mays* L.), Kangu (*Setaria italica* (L.) P. Beauv.), Mandia (*Eleusine coracana* (L.) Gaertn.), Suan (*Panicum sumatrense* Roth. ex Roem. & Schult.) and pearl millet (*Pennisetum typhoides* (Burm. f.) Stapf & C. E. Hubb.).

Recently Rao et al. (2011) discovered a new species of *Brachystelma* and named after Pullaiah as *Brachystelma pullaiahii* and the tubers of the species are locally eaten. An ethnobotanical survey undertaken by Xavier et al. (2011) in Kolli hills of Eastern Ghats of Tamil Nadu revealed that the Malayali tribes of Kolli hills are using 50 plant taxa to treat various diseases while *Moringa oleifera, Solanum nigrum* and *S. torvum* are edible. The chemical and nutritional composition of *Cajanus albicans*, an underexploited tribal pulse in Eastern Ghats of Andhra Pradesh was determined and proved that it is very rich in crude protein than other commonly consumed legumes like *Cicer arietinum* (Murthy, 2011). Murthy and Emmanuel (2011) studied the nutritional properties of a wild legume *Rhynchosia bracteata* and concluded that it may be further exploited in breeding programs and popularized for mass cultivation and consumption in third world countries, such as India to alleviate hunger and poverty. A variety of trees are grown by the Adivasis in Eastern Ghats on their land, the produce (fruits, nuts, oil from seeds) of which are used for household consumption as well as for augmenting income through sale. The common trees grown on land include jackfruit, orange, guava, mango, gooseberry, java plum, Indian fir/mast tree, tamarind, myrobalan and marking nut/black cashew. The other trees grown include silver oak, custard apple, Indian beech tree, beech, fishtail/toddy palm, lemon, eucalyptus, teak, pomegranate and banana (Seema Mundoli, 2011).

Launea procumbens, a wild edible plant of Asteraceae is found as a weed in crop fields and wastelands and it is rich in Calcium and Iron and commonly found in black cotton soils. Local women used this as leafy vegetable and sold in market in Gulbarga district of Karnataka (Rajasab and Rajshekhar, 2012). Dhore et al. (2012) explored Digras Tahsil, of Yavatmal district of Maharashtra for wild edible plants and reported 25 species. Mahapatra et al. (2012) identified superior/identical nutritional status in terms of carbohydrate, sugar and protein and mineral contents in non cultivated indigenous forest species, for example, *Eugenia rothii, Mimusops elengi, Ziziphus oenoplia, Zizipus rugosa, Bridelia tomentosa* and *Carissa spinarum* comparable to the cultivated fruits like mango, pomegranate, sapota, grapes, guava, cherry, banana and lemon, etc. The analysis indicates the scope of using wild edible fruits for dietary supplement since it has valuable ingredients as Iron, Sodium, Potassium and Calcium. Kumar et al. (2012) reported that 11 edible wild *Dioscorea* species are available in the Simlipal Biosphere Reserve forest, Orissa. Reddy (2012) reported a total of 61 wild edible plants from Chandrapur district of Maharashtra state. Samydurai et al. (2012) surveyed Kolli hills of Eastern Ghats for wild edible tubers and revealed that a total of

38 tuberous plants are being used by tribal people of Kolli hills. Kumar et al. (2012) studied on wild edible plants of Simlipal Biosphere Reserve of Orissa and they specially focused on the tuberous species of *Dioscorea*. Sadasivaiah and Ravi Prasad Rao (2012) worked on tribe Ceropigeae in Eastern Ghats of Andhra Pradesh and listed 19 taxa under 6 genera. Of these, the stems of *Boucerosia, Caralluma,* the tubers of *Brachystelma, Ceropegia* are edible by the tribal people of Eastern Ghats.

The species of *Dioscorea* play a vital role among the tribal communities by serving as a food and as a traditional medicine to cure different types of diseases during critical period (Kumar et al., 2013). A total of 216 numbers of ethnobotanical interested species were collected with the help of medicine man (Baidya or elder village people) from different tribal populated village areas at Nabarangpur district of Odisha. Out of 270 uses recorded 153 are as medicinal, 54 as food, 7 veterinary, 8 fodder, 7 rope making, 2 herbal dye, 9 tooth brush, 3 insect repellent 4 hair oil and 25 other purposes (MoEF, 2013). Mukesh Kumar et al. (2013) reported 21 plant species belonging to 19 families used as wild edible plants by the tribal and rural communities of Balasore, Bhadrak, Jajpur, Keonjhar and Cuttack districts of Odisha. Bagul (2013) enlisted 27 medicinal plants from Jalgaon district of Maharashtra and some of them are edible. Deshpande and Kulkarni (2013) studied on *Theriophonum indicum* a leafy vegetable commonly consumed by Gondia tribe in Vidarba region of Maharashtra and explained a new technique developed by Gondia tribes for the preparation of food material with *Theriophonum indicum* because, the tubers will give irritation. Misra and Misra (2013) emphasized on leafy vegetables available in South Odisha and listed 106 leafy vegetables. Out of 106 leafy vegetables 78 grow in wild and 28 are cultivated. Singh (2013) on his work on probable Agricultural Biodiversity Heritage Sites in the South-Central region of Eastern Ghats mentioned many edible and wild edible plants. The Malayali tribals cultivate edible plants, like tapioca, pineapple, banana and cash crops, such as pepper, coffee, jack fruit, clove and cereals like ragi, thinai, makkasolam, samai and panivaraku (Vaidyanathan et al., 2013). Prabakaran et al. (2013) worked on wild edible forest products of Chitteri Hills of Southern Eastern Ghats collected by Malayali tribes which included wild fruits, leafy vegetables, tubers, commonly used for self subsistence. A total of 38 species of wild edible fruits, 11 different leafy vegetables were listed. The Bagatas, Koyas, Gonds, Manne Doras, Malis, Reddi Doras, Nooka Doras and Valmikis living in the valleys and nearby streams where plain landscape prevails, have totally adopted to settled cultivation but the same tribes inhabiting near the hill tracts and interior forests are resorting to shifting cultivation (Subramanyam and Veerabhadrudu, 2013). However, a few of them currently growing commercial crops like turmeric, maize, tobacco, chillies, cotton, cashew, orange, ginger, pippallu (*Piper longum*), different varieties of beans, etc., mixed cropping like pulses, millets and oil seeds is the dominant feature in the dry and Podu cultivation. Misra (2013) documented 26 wild edible plants and stated that they are the common property of villagers, of them 13 are fruit yielding plants and 13 are leafy vegetable. Vaidyanathan et al. (2013) studied ethnomedicinal plants used by Malayali tribes in Kolli hills of Eastern Ghats and recorded 250 ethno medicinal plants, among them some are edible by the tribal people.

Misra and Misra (2014) reported 38 wild edible plants with underground parts from South Odisha. Nagalakshmi (2014) reported 54 species of wild leafy vegetables from rural households of Anantapur district of Andhra Pradesh. Among 54 wild edible leaves, 11 are coming from the family Amaranthaceae. Tripathy et al. (2014) stated that *Trichosanthes cucumerina* possesses sound ethnobotanical values and its parts are used in various disorders and microbial infections and also young fruits are used as vegetable. Fruits are rich with carbohydrate (26.24%), lipid (2.20%), protein (1.50%), fiber (1.96%) and good amount of moisture. Panda (2014) enlisted 86 wild edible plants of 51 families from Odisha state and stated that and Dioscoreaceae and Amatanthaceae are the dominant families.

Investigation on uncultivated vegetables in the Dumbriguda agency region of Visakhapatnam District was done by Sanyasi Rao et al. (2014) and listed 55 indigenous food plants and stated that most of them are leafy vegetables. Around 64% of the people in the North-western Deccan Plateau Region are employed in agriculture and allied activities. Field crops include wheat, rice, sorghum, pearl millet, minor millets, pulses, and oilseeds, while horticultural crops are dominated by tropical fruits, such as mango, grape, banana, orange, pomegranate, etc., and diverse vegetables. Cash crops include cotton, sugarcane, turmeric, groundnut, and tobacco in the North-western Deccan Plateau Region (Singh, 2014). Mukesh Kumar et al. (2014) studied 43 Ethno medicinal plants in the Khordha forest division of Khordha District, Odisha, which are used by Kondh, Sabra, Naik tribes of the area, among them some of the plants also used as edible, such as the rhizomes of *Amorphophallus paeoniifolius* and the fruits of *Averrhoa carambola*.

Misra and Misra (2014a) did nutrient analysis of 27 wild leafy vegetables available in South Odisha and concluded that all the edible leafy vegetable plants contain appreciable amount of nutrients which are readily available.

Hence they could be consumed to supplement the scarce or non-available sources of nutrients to the tribal and poor rural people. Pandravada et al. (2014) reported the distribution and cultivation practices of *Luffa hermaphrodita* for the first time in Andhra Pradesh from Adilabad district. Deepa et al. (2014) studied wild edible plants used by Malayali tribes of Bodha Hills, Southern Eastern Ghats, Namakkal district of Tamil Nadu and identified 95 wild edible plants belonging to 75 genera and 48 families. Out of 95 wild edible plant species, 43 are leafy vegetables, 38 wild fruits. Rekka and Senthil Kumar (2014) published a note on ethnobotanical wild edible plants used by Malayali tribals of Yercaud Hills, Eastern Ghats, Salem District of Tamil Nadu and identified 42 wild edible plant species under 36 genera and 29 families. Among these, 27 are fruits, 9 are leafy vegetables. Among 42 plant species, trees (13 species) were found to be most used plants followed by herbs (11 species), shrubs (8 species), climbers (6 species) and small trees (4 species). These products are collected from both wild and cultivated plants. Documentation of wild edible plants in old Mysore district by Nandini and Shiddamallayya (2014) encompasses 105 plant species belonging to 77 genera of 47 families. The wild edible plant species are composed of 6 climbers, 28 shrubs, 30 herbs and 41 Trees. Of enlisted 62 plants used as raw, 30 are used as boiled and 13 plants are used as raw and boiled form in rural and tribal population of old Mysore. Satyavathi et al. (2014) gathered information on Bogata tribes of Paderu division of Andhra Pradesh and listed a total of 30 angiosperm plant taxa for ethno botanical uses, among them some are used as edible plants.

Nayak and Basak (2015) evaluated some of the nutritional properties in 8 wild edible fruits reported to be consumed by rural people and tribals of Odisha. Among them, *Dillenia pentagyna* showed highest total carbohydrate (18.5%), total sugar (16.8%) and iron content (16 mg/100 g) making it a good competitor against other popular cultivated fruits like mango (17.00%), pomegranate (17.17%) with reference to carbohydrate content and grapes (16.25%) with regard to total sugar content. Similarly, wild edible fruits like *Streblus asper* (12.7%) and *Carmona retusa* (11.8%) also recorded high sugar content in comparison to cultivated fruits, viz., apple (10.39%), pineapple (10.8%) and pears (7.05%). The protein content in *Melastoma malabathricum* (5.48%) and *Carmona retusa* (4.1%) were found higher in comparison to cultivated fruits like banana (1.09%), mango (0.51%) and guava (2.54%). With regard to the mineral content, like iron, manganese, copper and calcium, *Dillenia pentagyna*, *Streblus asper*, *Melastoma malabathricum*, *Calamus guruba* were also found at par with apple, mango, banana and guava.

6.3   GROWTH FORM ANALYSIS

The growth form of the edible plants used by the tribal people of Eastern Ghats as observed in the review includes herbs are dominating with 362 taxa followed by trees (255 taxa), climbers (113 taxa) and shrubs (55 taxa). Sanyasi Rao et al. (2014) enlisted wild food plants consumed by tribals of Dumbriguda area of Vizag and stated that most of them are trees which occupied highest position with 21, followed by herbs (14 species), climbers (9 species), Shrubs (6 species), each one species of fern and Vine.

6.4   EDIBLE PART ANALYSIS

Rekka and Senthil Kumar (2014) recorded 42 wild edible plants from Yercaud Hills, Eastern Ghats, of which 13 trees, 11 herbs, 8 shrubs, 6 climbers and 4 small trees. The edible parts consumed as food by the tribal people of Eastern Ghats are Aerial roots (AR), Bark (Br.), Bulbs (B), Corms (C), Gum (G), Whole Plant (WP), Root (R), Tuber (T), Rhizome (Rh), Leaves (L), Stem (St), Flowers (Fl.), Fruits (Fr.), Seeds (S), Pith (P), Tender shoots (Ts) and Inflorescence (In.) and these plants still share a good proportion of tribal dishes all over world (Anonymous, 1970–1988; Samant and Dhar, 1997; Sasi et al., 2011). Most of the collected edible parts of plants eaten as raw as well as cooked as vegetables, snacks, or used as beverages. The analysis on edible parts revealed that major parts of the plants are used as food by the local tribal are fruits with 320 plant taxa, followed by Leaves (294 taxa), underground parts including B, C, Rh, R, T (122 taxa), Seeds (113 taxa), stems (67 taxa), flowers (36 taxa), whole plants (8 taxa) and 19 other parts like Gum, Bark, etc.

   Mukesh Kumar et al. (2013) documented 8 wild edible fruits, 4 flowers, 6 leafy vegetables and one species of grain, 1 tuber and 1 whole plant as edible by the tribal and local inhabitants of Balasore, Bhadrak, Jajpur, Keonjhar and Cuttack districts of Odisha. Sanyasi Rao et al. (2014) reported 55 indigenous food plants consumed by tribal communities of Dumbriguda area of Visakhapatnam district, among them 24 species are used as leafy vegetables, 21 fruits, 6 tubers, 4 tender shoots, 2 each for seeds and flowers. Rajasab and Isaq (2004) have explored 22 species of plants used as edible leaves, stem, flowers, fruits, seeds and roots as part of regular diet by village folks of North Karnataka. Rajasab and Isaq (2004) recorded 51 wild edible plants from the district of Gulbarga of Deccan region of them 27 are edible fruits, 16 leaves/stems, 4 flowers, 3 seeds and the roots of *Decalepis hamiltonii*. Sinha and Lakra (2005) documented 126 wild edible plants of which 50 are edible leaves, 46 fruits, 15 flowers, 14 tubers, 11 seeds and 5 gums are the part of tribal diet in Orissa (Figure 6.1).

FIGURE 6.1 Methodology. (Photos courtesy of Naveena Photo Parlour, Wanaparthy.)

6.4.1 UNDERGROUND PARTS

The underground parts are delicious, provide instant energy and meet food requirements to some extent. The underground parts are washed properly, eaten as raw, burnt, or largely practiced by a complex process of boiled, sliced, cooked to remove the acrid taste and eaten. Some of them are just eaten as raw just after collection and peeling the outer skin owing to its good taste. Some of the tubers were powdered and added for fermentation of local wine (Misra et al., 2013). The tubers of *Brachystelma, Ceropegia* species are edible by the locals in Eastern Ghats (Sadasivaiah and Rao, 2012).

A total of 122 plants belonging to 44 families yield underground edible parts (Figure 6.2). They are categorized as bulbs, corms, root stock, rhizome and tubers. Among 44 families, monocots are dominating than dicots. Zingiberaceae is the dominating family that yields 16 edible underground parts followed by Dioscoreaceae (13 taxa), Araceae (12 taxa) and Asclepiadaceae (11 taxa). The members of Araceae (corms), Dioscoreaceae (tuberous), Liliaceae (bulbous) and Zingiberaceae (rhizomatous) are the major underground parts yielding plants in Monocots and Apiaceae (tuberous), Asclepiadaceae (tuberous) and Cucurbitaceae (root stocks) are in dicots. Tribal people collected 15 species of wild edible tubers during the crucial months and of the 15 tubers 11 are the species of *Dioscorea* (Sinha and Lakra, 2005).

FIGURE 6.2   Edible underground parts/stems/flowers. (Photos courtesy of Naveena Photo Parlour, Wanaparthy.)

The bulbils of *Aponogeton natans* are sweet in taste and are directly eaten as raw as snacks. The fresh tubers of *Dioscorea glabra* and *D. puber* are taken as snacks (Misra and Misra, 2014). The bulbs of *Allium cepa, A. fistulosum* and *A. sativum* are also used as spice. The species of *Amorphophallus, Arisaema tortuosum, Colocasia esculenta* corms are used as vegetable along with other vegetables. Kumar et al. (2013) worked out on the medicinal and food properties of 12 *Dioscorea* species available in the forests of Odisha and stated that, all the species are rich in starch and carbohydrates and also they used to treat diabetes and various skin diseases.

6.4.2 STEMS

The stems of 67 species and pith of 6 species belonging to 37 families are also edible by the tribals of Eastern Ghats. Among the 37 families, Asclepiadaceae, Asteraceae and Poaceae are dominating with 5 species in each family. The stems of *Amorphophallus bulbilfer, Caralluma adscendens, Caralluma umbellata, Cissus quadrangularis,* tender stems of *Achyranthes aspera* and *Curcuma zedoaria* are used as vegetable and the stems of *Saccharum officinarum* used to prepare candy, juice and also eaten as raw. *Saccharum officinarum* is cultivated throughout Eastern Ghats and it is also available in local markets. The pith of stems of the species of *Phoenix* is eaten as raw. 'Sago' is prepared from the pith of *Metroxylon sagu* and *Cycas sphaerica*. Toddy, a local drink is coming from the stems of *Phoenix sylvestris, Borassus flabellifer* and *Caryota urens*. The stems of *Boucerosia* and *Caralluma* are used in chutnies and or directly eaten as raw in Eastern Ghats of Andhra Pradesh.

6.4.3 LEAVES

Good number of leaves is edible by tribal people of Eastern Ghats. Two Hundred and Ninety four taxa of leaves under 77 families are utilized as edible. Most of the edible leaves are used as leafy vegetable (Figure 6.3). Some of the leaves have potential economic value. These leaves have more proteins, vitamins, etc. Some of them are available in local market and most of them are under neglected species. The families like Amaranthaceae (32 plant taxa), Fabaceae (20), Caesalpiniaceae (16), Asteraceae (11) and Euphorbiaceae (10) are the dominant families that yield edible leaves. The leaves of *Alternanthera sessilis, Trianthema portulacastrum, Allmania nodiflora, Digera muricata, Basella alba, Tamarindus indica, Boerhavia diffusa* and some other species of leaves need to be enhanced to market level. The species like stalk of *Allium cepa* and leaves of *Alternanthera sessilis, Amaranthus spinosus, Cassia tora, Celosia argentia, Centella asiatica, Colacasia esculenta, Moringa oleifera, Murraya koenigii, Portulaca oleracea, Trianthema protulacastrum* and *Solanum nigrum* leaves are commonly consumed by adivasi communities all theses leafy vegetable are rich in vitamin-A. There is a good demand for the leaves of *Coriandrum sativum* and *Murraya koenigii* in the market. *Alternanthera philoxeroides* is an exotic cultivated in Eastern Ghats for its edible leaves (Reddy and Raju, 2005).

FIGURE 6.3   Edible leaves. (Photos courtesy of Naveena Photo Parlour, Wanaparthy.)

Kamble et al. (2013) reported that 3,900 plant species are used by tribals as food, out of which 145 species comprise of root and tubers, 521 species of leafy vegetables. Nagalakshmi (2014) surveyed the leafy vegetables of Anantapur district and reported 54 plants under 44 genera of 29 families. Sinha and Lakra (2005) reported 50 wild edible leaves from the forest of Orissa, consumed by *Gond, Sounti, Bhumiz* and *Juang* tribes, among 50 leaves 32 are very popular among all tribes and often consumed in their respective season. All these edible leaves commonly available in forests or found in cultivated fields as weeds. Misra and Misra (2014) analyzed the nutritional values of 27 leafy vegetable of South Odisha, of 27 species *Murraya koenigii* showed the highest total sugar content, followed by *Tamarindus indica* and *Senna tora. Tamarindus indicus* contains the highest vitamin B1 content followed by *Bambusa bambos* while *Moringa oleifera* showed the highest vitamin C content followed by *Cleome viscosa.*

6.4.4 FLOWERS

Flowers of 36 species belonging to 21 families are edible either raw (*Holostemma ada-kodien, Tamarindus indica*) or cooked as vegetable and used in the preparation of curries. Caesalpiniaceae is the dominant family with 8 speices that yields edible flowers. The flower buds of *Syzygium aromaticum* are used as spice. The parts of flowers like stamens and thalamus of *Nelumbo nucifera* are directly eaten by the tribal people living in Eastern Ghats. The flowers of *Holostemma ada-kodien* are sweet in taste and *Tamarindus indica* are sour in taste as said by tribal people of Nallamalais, who live in Andhra Pradesh and Telangana.

Wild edible flowers are consumed seasonally according their availability in the forests by 5 tribal people of Orissa forests, some of them are *Sesbania grandiflora, Cochlospermum religiosum, Tamilnadia uliginosa, Madhuca indica.* Of the 15 species of wild edible flowers 6 are commonly found in summer season, where as 4 species are found in rainy season and very few occur in other season (Sinha and Lakhra, 2005).

6.4.5 FRUITS

Wild fruits play a significant role in human nutrition and they are generally used as raw or processed, which help to compensate the day-to-day requirement of calories. These fruits are sources of carbohydrates, proteins, vitamins, minerals, dietary fiber and enormous medicinal potential (Quebedeaux and Bliss, 1988; Quebedeaux and Eisa, 1990; Craig and Beck, 1999; Wargovich, 2000).

Some fruits like *Annona squamosa, Aegle marmelos, Ziziphus* species are cultivated and also available in market at commercial level, but still the tribal people collected these from forests. *Diospyros melanoxylon, Phoenix sylvestris, Ziziphus oenoplea, Manilkara hexandra* and *Buchanania lanzan* are collected from forests and commonly sold in local markets by the tribals (Reddy, 2012).

Gathering and exploitation of wild edible fruits is a common activity of the indigenous people in Eastern Ghats and Deccan region (Figure 6.4). A total of 320 species of fruits belonging to 70 families were collected from

FIGURE 6.4   Edible fruits. (Photos courtesy of Naveena Photo Parlour, Wanaparthy.)

Eastern Ghats and Deccan. Out of 320 fruits, most of them are found in forests as wild and some of the fruits are available in wild as well as in markets; and few of them are cultivated in home gardens or in brought from local markets. It shows that the indigenous people collect many wild fruits from the area. These wild edible fruits may not be the alternatives for food but they contribute the necessary nutrient requirements of the aboriginal people (Sasi and Rajendran, 2012).

Out of 70 families identified in the widely utilized species belonged to Cucurbitaceae (23 taxa), followed by Solanaceae (20 taxa), Rubiaceae (18), Rutaceae and Moraceae (17) and rest of the families represented 1–14 species. The wild edible species have been meeting the protein, carbohydrates, fat, vitamins and mineral requirements of the local residents to a great extent (Sebastin and Bhadari, 1990; Omo Ohiokpehai, 2003). The fruits also contain antioxidants which protects people against heart disease and certain type of cancers (Saxena, 1999). The dangerous diseases like cancer, cardiovascular disorders, Alzheimer, cataract are reduced due to regular consumption of wild edible fruits (Liu, 2003).

Some of the wild edible fruits were collected by local people and they are selling in nearby urban area for income generation (Sasi and Rajendran, 2012). The species like *Mangifera indica, Murraya koenigii, Pithecellobium dulce, Syzygium cumini, Tamarinus indica* are collected for domestication. Domestication grew out of food gathering almost led to cultivation (FAO, 1999). The fruits of *Trichosanthes cucumerina* are used as vegetable by the tribal people of Simlipal Biosphere Reserve (Tripathy et al., 2014). Sinha and Lakra (2005) documented 46 wild edible fruits, of which 32 are popular to tribal people and they are used as vegetables, used to prepare pickles and some of them are eaten directly. Nayak and Basak (2015) evaluated nutritional properties in 8 wild edible fruits, for example, *Antidesma ghaesembilla, Careya arborea, Dillenia pentagyna, Streblus asper, Carmona retusa, Melastoma malabathricum, Calamus guruba* and *Ficus hispida* reported to be consumed by rural people and tribals of Odisha.

6.4.5   SEEDS

A total of 113 seeds of plant species of 31 families were identified as edible seeds in Eastern Ghats. Fabaceae is the largest family having 27 edible seeds, followed by Poaceae (23 species), Caesalpiniaceae (6 species) and Cucurbitaceae (5 species). Nearly 50% of the seeds are coming from Fabaceae and Poaceae families only (Figure 6.5). These two families are the main source for most of the food grains and pulses. Eleven species of seeds are utilized as spices, such as *Coriandrum sativum, Cuminum cyminum, Piper longum* and *Brassica nigra*. A good number of seeds eaten as raw and some seeds like *Arachis hypogaea, Hibiscus cannabinus, Cucumis sativus, Bauhinia vahlii* roasted and consumed. Some of the seeds used to prepare chutnies, sweets, salads and a few of them used as medicinal. Some of the seeds are boiled and consumed.

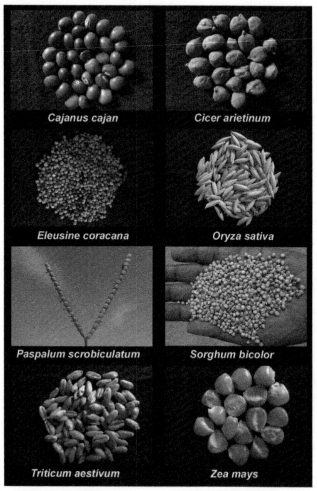

FIGURE 6.5   Edible seeds. (Photos courtesy of Naveena Photo Parlour, Wanaparthy.)

To meet the diet of increasing population, new food resources are increasing, the seeds of wild plants, including the tribal pulses, received more attention in this case, because they are well adapted to adverse environmental conditions, highly resistant to disease and pests, and exhibit good nutritional qualities (Maikhuri et al., 1991). Wild seeds are commonly used in different parts of the world as proteinaceous food (Amubode and Fetuga, 1983). In India different tribal people used some 28 wild legumes as pulses (Arora et al., 1980; Murthy and Pullaiah, 2005). The tribal communities living in Eastern Ghats and adjacent Deccan region collected the seeds of wild legumes randomly in the vicinity of the forests, soak in water and consumed the seed meal after boiling and decanting for four to twelve times (Murthy, 2011).

Sinha and Lakra (2005) recorded 11 edible seeds consumed by the tribal people of Orissa, most of them are coming from trees. Some of the seeds have market value.

6.4.6   OTHERS

The Aerial roots of epiphytic orchids, such as *Pelatantheria insectifera* and *Vanda tessellata* are used as snacks just after burning by children (Misra et al., 2013). The bark of *Cinnamomum verum* is used as spice and it is collected from forests and also brought from markets. The gum of *Anogeissus latifolia* and *Sterculia urens* are eaten as raw and used in the preparation of sweets respectively. The whole plant of *Agaricus bisporus,* a fungi used as vegetable and also as snacks, whereas *Parmelia tinctoria* is a foliage lichen used as spice. The sustainable collection of *Agaricus bisporus* and *Parmelia tinctoria* gives good economic benefit to the tribals in Eastern Ghats. The gum of *Terminalia alata, Terminalia bellirica* and *Ficus benghalensis* are consumed by the local tribal people of Orissa (Sinha and Lakra, 2005).

6.5   USE CATEGORY

Analysis of data on use category showed that the identified edible plants especially wild species provide 5 major edible and allied categories based on local practices, such as food, food supplements, liquor additives, food flavorings (spices and condiments), medicine and sorcery. Every plant has one or more uses. It revealed that majority of taxa have multiple uses that means the same plant serves more than one use category. Some of the plants have some cultural or ritual importance that the tubers of *Asparagus racemosus, Smilax zeylanica, Nymphaea pubescens, Lygodium flexuosum* and the fruits of *Solanum melongena* var. *insanum* are tied to the neck/body of the children/pregnant women to keep away of evil spirits and to maintain god health.

6.6   INCOME GENERATING SPECIES

Besides the household consumption as supplementary food by the tribal people, some wild edible plants are marketable and provide opportunity for earning additional income. According to Misra et al. (2013) 22 edible tubers, Basha et al. (2009) 13 species of wild edible fruits were domesticated and 19 wild edible fruits are having commercial marketing values. Most of the traditional leafy vegetables have commercial value and some of them are already available in local markets nearby forest areas. The commercial usage of the rootstocks of *Decalepis hamiltonii,* the gum of *Sterulia urens,* fruits of *Limonia acidissima* are needed to be enhanced up to national and international levels. Sale of forest food products have potential to increase the purchasing power of the households and therefore contribute indirectly to food products and daily need, such as rice, salt, clothes, etc.

6.7   DISCUSSION

Eastern Ghats are the rich source for a good number of wild edible plants. Most of the wild plants especially leafy vegetables, fruit yielding plants, tuberous and rhizomatous plants can easily grow in the back yards of houses, home gardens, so that it can be used readily. According to Nordeide et al. (1996), Orech et al. (2007), Sundriyal and Sundriyal (2001) the nutritional values of traditional leafy vegetables is higher than common vegetables. Samyadurai et al. (2012) studied 38 wild edible roots and tuberous species of Kolli hills of Eastern Ghats and stated that these species are closely related to socio-economic conditions of tribals of Kolli hills for their day-today requirement. Southern Eastern Ghats of United Andhra Pradesh were explored for wild edible fruits by Basha et al. (2009) and recorded 69 species belonging to 44 genera and 28 families. Of these 69 species, 13 are domesticated by the local communities and 19 species are having commercial value for marketing.

The root stocks of *Cissampelos pareira, Asparagus racemosus, Lygodium flexuosum, Rubia cordifolia, Orthosiphon rubicundus, Elephantophus scaber, Madhuca indica, Clerodendrum serratum* and few others are used as liquor additives (Misra et al., 2013). Some of the edible plants are also used as medicine by the tribal people of Eastern Ghats. Most of the members of wild edible plants of Zingiberace, Dioscoreaceae, Apiaceae, Asclepiadaceae and other family members are also used as medicine by tribal people. This confirms the fact that the food and medicinal plants are closely related particularly in rhizomatous/tuberous species and can lead to the development of pharma-food or nutraceuticals (Etkin and Johns, 1998; Bonet et al., 1999).

6.8 CONCLUSION

The aboriginal people of Eastern Ghats through their traditional knowledge infer what to eat what not to eat. They are thoroughly acquainted with methods of excluding the harmful substances from the edible plants and prepared acceptable recipes. The corms and aerial bulbs of *Dioscorea* eaten as raw, cause a terrible itching sensation in throat, hence for removing itching sensation, they will be peeled, double boiled in tamarind water and smeared with turmeric paste. This is one of the methods devised in the kitchens of the tribal people in Eastern Ghats.

The high diversity of these indigenous species within Eastern Ghats affirmed the importance in sustaining the livelihoods of tribal communities. The hilly undulating terrain with limited cultivable lands, non- availability of sufficient food, poor accessibility and marketability and very low agriculture yield are the main attributes for use of wild edible plants as food. Some of the species like *Dioscorea hamiltonii, Curculigo trichocarpa, Habenaria plantaginea, Vanda tessellata, Amomum dealbatum* consumed by the children and very poor households during normal and difficult times ensured their contribution and further maintenance of indigenous knowledge, however, the transmission of such data and wisdom on edible use of some species in food scarcity, gradually lead to the fading away of traditional knowledge associated with those species. The wild edible fruits are playing a vital role in providing nutritional and economic securities to the poor masses in rural areas, but the commercial and market value of these wild fruits is unknown to them. The commercial and market value of these wild edible plants should be tapped timely and seasonally.

Tribal women are well experienced and they play major role in utilization of wild edible plants. The oral transmission of traditional knowledge is declining day by day, so there is an urgent need to document the traditional knowledge on wild edible plants as well as the preparation modes of food. There is much scope for improving the growth forms of wild edible plants by using modern agronomy techniques. For all such endeavor, thorough field-work in various tribal areas and critical ethno botanical observation on wild edible plants are the basic requirements.

As a supplementary source of income, non-timber forest produce is important to the tribal people. The unsustainable collection, lopping of branches by the forest dwellers and grazing leads to over exploitation of wild resources. Over exploitation of wild edible plants may lead to great threat to certain species. Hence there is a need to find the sustainable methods of collection of wild edible plants for their conservation. The tribal people should be trained in mode of collection and preservation of wild edible plants for better shelf-life period. Forest product conservation and sustainable collection are the two very important factors for the development of economic status of tribal people of Eastern Ghats. All these wild edible plants also have immense potential for the fauna in the forests.

It is evident from many works on wild edible plants collectively that with the effect of several factors, such as migration of tribal inhabitants from core to transitional zone, influence of modern lifestyles on younger generation with change in dietary habits and their impression of forest plants as poor men's food, agriculture encroachment and launching of Government schemes on food security, we have started to lose the indigenous knowledge required to identify, locate, gather and know the consumption pattern of wild edible species. In addition, many of the wild edible plant species are under the pressure of various categories and magnitudes, thus public awareness and community based programs through joint forest management plan need to be encouraged at all levels for *ex-situ* and *in situ* conservation of such species of future potential. Therefore, sustaining the wild edible plants species will be worthwhile only if conservation efforts be made into execution for those potential species to preserve their cultural heritage and to enhance the food security of tribal communities in Eastern Ghats.

TABLE 6.1   List of Food Plants

S. No.	Name of the plant	Edible Part (s)	Mode of preparation	References
1	*Abelmoschus crinitus* Wall.	R, T	Underground fleshy tubers are first boiled and consumed as vegetable	Misra et al. (2013); Misra and Misra (2014)
2	*Abelmoschus esculentus* (L.) Moench	Fr.	Used as vegetable	Singh (2013)
3	*Abrus precatorius* L.	L, Fl.	Occasionally leaves and flowers eaten raw	Reddy et al. (2007); Dhore et al. (2012)
4	*Abutilon indicum* (L.) Sweet	Flb, S, L	Flower buds and seeds are occasionally edible. Leaves used as vegetable	Murthy et al. (2003); Rajasab and Isaq (2004); Reddy et al. (2007)
5	*Acacia concinna* (Willd.) DC.	L	Young leaves are ground to coarse paste along with red chilies and salt to make chutney	Nagalakshmi (2014)
6	*Acacia nilotica* (L.) Delile	S	Rosted with salt and eaten	Murthy et al. (2003)
7	*Acacia pennata* (L.) Willd.	L, Fr.	Cooked and eaten along with boiled rice; fruits edible	Rekka and Senthil Kumar (2014); Sanyasi Rao et al. (2014)
8	*Acacia sinuata* (Lour.) Merr.	L	Tender leaves used as vegetable	Murthy et al. (2003)
9	*Acalypha fruticosa* Forssk.	L	Used as vegetable	Deepa et al. (2014)
10	*Acalypha indica* L.	L	Used as vegetable	Murthy et al. (2003)
11	*Achyranthes aspera* L.	Ts, L	Tender shoots and leaves used as vegetable	Murthy et al. (2003); Reddy et al. (2007); Prabakaran et al. (2013); Misra and Misra (2013); Deepa et al. (2014)
12	*Achyranthes bidentata* Blume	L	Used as vegetable	Deepa et al. (2014)
13	*Acorus calamus* L.	Rh	Used as vegetable	Samydurai et al. (2012)
14	*Acronychia pedunculata* (L.) Miq.	L	Used as vegetable	Murthy et al. (2003)
15	*Adenanthera pavonina* L.	S	Eaten raw	Panda (2014)
16	*Aegle marmelos* (L.) Correa ex Serr.	Fr.	Pulp taken orally. Ripe fruits eaten raw, or prepare juice	Murthy et al. (2003); Reddy et al. (2007); Basha et al. (2009); Singh (2013); Deepa et al. (2014)

TABLE 6.1 *(Continued)*

S. No.	Name of the plant	Edible Part (s)	Mode of preparation	References
17	*Aerva lanata* (L.) Juss.	Ts, L	Tender shoots and leaves used as vegetable	Murthy et al. (2003); Reddy et al. (2007); Misra and Misra (2013)
18	*Aeschynomeme aspera* L.	Ts, L	Tender shoots and leaves used as vegetable	Reddy et al. (2007)
19	*Agave americana* L.	In., St	Tender shoots and inflorescence edible	Murthy et al. (2003)
20	*Aglaia elaeagnoidea* (Juss.) Benth.	Fr.	Fruits eaten raw	Murthy et al. (2003)
21	*Aglaia lawii* (Wight) Sald.	Fr.	Aril eaten raw	Hebber et al., 201
22	*Alangium salvifolium* (L.f.) Wangerin	Fr.	Ripe fruits eaten as raw	Murthy et al. (2003); Reddy et al. (2007); Basha et al. (2009); Prabakaran et al. (2013); Rekka and Senthil Kumar (2014); Deepa et al. (2014)
23	*Albizia lebbeck* (L.) Willd.	S	Seeds roasted and eaten	Murthy et al. (2003)
24	*Allium cepa* L.	B, L	Used as spices. Leaves used as vegetable	Misra and Misra (2013); Singh (2013)
25	*Allium sativum* L.	B	Used as spice	Pandravada et al. (2007); Singh (2014)
26	*Allmania longepedunculata* (Trimen) Gamble	L	Used as vegetable	Deepa et al. (2014)
27	*Allmania nodiflora* (L.) R. Br. ex Wight	L	Used as vegetable	Reddy et al. (2007); Prabakaran et al. (2013); Deepa et al. (2014);
28	*Allmania nodiflora* (L.) R.Br. ex Wight var. *angustifolia* Hook. f.	L	Used as vegetable	Sadasivaiah (2009)
29	*Allmania nodiflora* (L.) R.Br. ex Wight var. *aspera* (Heyne ex Roth) Hook. f.	L	Used tas vegetable	Sadasivaiah (2009)
30	*Allmania nodiflora* (L.) R.Br. ex Wight var. *dichotoma* Hook. f.	L	Used as vegetable	Sadasivaiah (2009)

TABLE 6.1 *(Continued)*

S. No.	Name of the plant	Edible Part (s)	Mode of preparation	References
31	*Allmania nodiflora* (L.) R.Br. ex Wight var. *procumbens* Hook. f.	L	Used as vegetable	Sadasivaiah (2009)
32	*Allmania nodiflora* (L.) R.Br. ex Wight var. *roxburghii* Wight	L	Used as vegetable	Sadasivaiah (2009)
33	*Allophyllus cobbe* (L.) Raeuschet	Fr	eaten as raw	Murthy et al. (2003)
34	*Alocasia fornicata* (Roxb.) Schott	Rh	Boiled with fruit or leaves of jajo (tamarind) and cooked along with other vegetables	Misra et al. (2013)
35	*Alocasia macror-rhiza* (L.) G.Don	Rh	Sliced, successively boiled and cooked with pulses	Misra et al. (2013)
36	*Aloe vera* (L.) Burm.f.	Fl.	Cooked as vegetable	Murthy et al. (2003)
37	*Alpinia galanga* (L.) Willd.	Rh	Rhizomes used as vegetable	Samydurai et al. (2012)
38	*Alternanthera amoena* (Lemaire) Voss	L	Leaves used as vegetable	Sinha and Lakra (2005)
39	*Alternanthera paronychioides* St. Hil.	L	Used as leafy vegetable	Reddy et al. (2007)
40	*Alternanthera philoxeroides* Mart.	L	Leaf and leafy shoots are cooked as vegetable	Panda (2014)
41	*Alternanthera pungens* Kunth	L	Used as leafy vegetable	Deepa et al. (2014)
42a	*Alternanthera sessilis* (L.) R. Br. ex DC.	L	Leaves boiled and squeezed, then add groundnut powder and eaten along with boiled rice	Murthy et al. (2003); Reddy et al. (2007); Prabakaran et al. (2013); Misra and Misra (2013); Rekka and Senthil Kumar (2014); Deepa et al. (2014)
43	*Alternanthera tenella* Colla	L	Leaves boiled and squeezed, then add groundnut powder and eaten along with boiled rice	Rao (2014)
44	*Alysicarpus rugosus* DC.	L	Leaves used as vegetable	Murthy et al. (2003)

TABLE 6.1 *(Continued)*

S. No.	Name of the plant	Edible Part (s)	Mode of preparation	References
45	*Amaranthus caudatus* L.	L	Used as vegetable	Sanyasi Rao (2014)
46	*Amaranthus cruentus* L.	L	Used as vegatable	Rao (2014)
47	*Amaranthus dubius* Mart. ex Thell.	L	Used as vegetable	Singh (2013)
48	*Amaranthus graecizans* L.	L	Used as vegetable	Rao (2014); Nagalakshmi (2014)
49	*Amaranthus hybridus* L.	L	Used as vegetable	Rao (2014); Murthy et al. (2003)
50	*Amaranthus polygamus* L.	L	Used as vegetable	Dhore et al. (2012)
51	*Amaranthus roxburghianus* Nevski	L	Used as vegetable	Rao (2014)
52	*Amaranthus spinosus* L.	L	Used as vegetable	Murthy et al. (2003); Reddy et al. (2007); Misra and Misra (2013); Singh (2013); Rekka and Senthil Kumar (2014); Deepa et al. (2014)
53	*Amaranthus tenuifolius* Willd.	L	Used as vegetable	Singh (2013)
54	*Amaranthus tricolor* L.	L	Used as vegetable	Reddy et al. (2007); Misra and Misra (2013); Deepa et al. (2014); Murthy et al. (2003);
55	*Amaranthus viridis* L.	L	Used to prepare Dall and curry	Murthy et al. (2003); Reddy et al. (2007); Misra and Misra (2013); Singh (2013); Deepa et al. (2014)
56	*Ammannia baccifera* L.	L	Used as vegetable	Reddy (2012)
57	*Amomum dealbatum* Roxb.	Rh	Peeled, burnt or boiled and eaten as chutney by children to stimulate appetite	Misra et al. (2013)
58	*Amorphophallus bulbilfer* (Roxb) Bl.	St, L, Rh, Fl.	All the parts are used as vegetable	Misra et al. (2013)
59	*Amorphophallus campanulatus* (Roxb.) Blume ex Decne	C	Corm used as vegetable	Singh (2014)

TABLE 6.1 *(Continued)*

S. No.	Name of the plant	Edible Part (s)	Mode of preparation	References
60	*Amorphophallus paeoniifolius* (Dennst.) Nicolson	T, L	Petiole/ bulb as vegetable; chutney; underground fleshy tuber and corms are first boiled with rice husk and kept overnight then sliced and cooked along with boiled potato or other vegetables are made into chutney	Murthy et al. (2003); Reddy et al. (2007); Samydurai et al. (2012); Singh (2013); Misra et al. (2013); Misra and Misra (2014); Mukesh Kumar et al. (2014)
61	*Amorphophallus sylvaticus* (Roxb.) Kunth	C, L	Petiole/bulb as vegetable; Chutny; Underground fleshy tuber and corms are first boiled with rice husk and kept overnight then sliced and cooked along with boiled potato or other vegetables are made into chutney	
62	*Ampelocissus latifolia* (Roxb.) Planchon	Fr., L	Ripe fruits edible; tender leaves used as vegetable	Murthy et al. (2003); Rout (2007)
63	*Ampelocissus tomentosa* (Heyne ex Roth) Planchon	Fr.	Ripe fruits edible	Murthy et al. (2003)
64	*Anacardium occidentale* L.	Fr.	Ripe fruits eaten raw	Murthy et al. (2003); Prabakaran et al. (2013); Rekka and Senthil Kumar (2014); Deepa et al. (2014)
65	*Ananas comosus* (L.) Merr.	Fr.	Ripe fruits eaten raw	Naidu and Khasim (2010); Rekka and Senthil Kumar (2014); Deepa et al. (2014)
66	*Anethum graveolens* L.	L	Leaves and young shoots are roasted then eaten.	Misra and Misra (2013)
67	*Anisochilus carnosus* (L.f.) Benth.	L	Used as leafy vegetable	Murthy et al. (2003); Reddy et al. (2007); Misra and Misra (2013)
68	*Annona cherimola* Miller	Fr.	Ripe fruits eaten raw	Rao (2014)
69	*Annona muricata* L.	Fr.	Ripe fruits eaten raw	Rao (2014)

TABLE 6.1   *(Continued)*

S. No.	Name of the plant	Edible Part (s)	Mode of preparation	References
70	*Annona reticulata* L.	Fr.	Ripe fruits eaten raw	Rao (2014)
71	*Annona squamosa* L.	Fr.	Ripe fruits eaten raw	Murthy et al. (2003); Basha et al. (2009); Deepa et al. (2014); Singh (2013)
72	*Anogeissus latifolia* (Roxb.ex DC.) Wall. ex Guill. & Perr.	G	Dried gum eaten raw, used in sweet preparation	Murthy et al. (2003)
73	*Anthocephalus chinensis* (Lam.) A.Rich. ex Walp.	Fr	Eaten raw	Murthy et al. (2003)
74	*Antidesma acidum* Retz.	Fr., L	Ripe fruits edible; leaves used as vegetable	Reddy et al. (2007); Basha et al. (2009); Murthy et al. (2003)
75	*Antidesma bunius* (L.) Spreng.	Fr.	Eaten raw	Murthy et al. (2003)
76	*Antidesma diandrum* Heyne ex Roth	L	Leaves edible	Sinha and Lakra (2005)
77	*Antidesma ghaesambilla* Gaertner	Fr.	Eaten raw	Basha et al. (2009); Sinha and Lakra (2005); Murthy et al. (2003)
78	*Aponogeton natans* (L.) Engl.	R, T	Starchy bulbils and tuberous roots are eaten as raw	Murthy et al. (2003); Reddy et al. (2007); Misra and Misra (2014)
79	*Aponogeton undulatus* Roxb.	R, T	Used as vegetable; starchy bulbils in winter are consumed after cooking.	Misra and Misra (2014)
80	*Arachis hypogaea* L.	S	Edible oil extracted from seeds, seeds eaten raw and roasted, seeds used in various sweets, chutnies, etc.	Singh (2013, 2014)
81	*Ardisia solanacea* Roxb.	L, Fr.	Young leaves and fruits are fried eaten	Murthy et al. (2003); Reddy et al. (2007); Misra and Misra (2013)
82	*Argemone mexicana* L.	L	Very young leaves cooked along with red gram	Nagalakshmi (2014)
83	*Argyria nervosa* (Burm.f.) Boj.	L	Young leaves are fried and eaten	Misra and Misra (2013); Murthy et al. (2003)
84	*Arisaema tortuosum* (Wall.) Schott	C	Corm sliced, double boiled with tamarind and cooked as vegetable	Murthy et al. (2003); Misra et al. (2013)

TABLE 6.1 *(Continued)*

S. No.	Name of the plant	Edible Part (s)	Mode of preparation	References
85	*Arisaema le-schenaultii* Blume	T	Boiled and eaten	Samydurai et al. (2012)
86	*Arthrocnemum indicum* (Willd.) Moq.	Wp	Whole plant used to prepare pickle.	Murthy et al. (2003)
87	*Artocarpus hetero-phyllus* Lam.	Fr.	Ripe fruits edible and seeds used as vegetable	Murthy et al. (2003); Reddy et al. (2007); Prabakaran et al. (2013); Rekka and Senthil Kumar (2014); Deepa et al. (2014)
88	*Artocarpus hirsutus* Lam.	Fr.	Raw fruits cooked and eaten	Rekka and Senthil Kumar (2014)
89	*Artocarpus lacucha* Buch.-Ham.	Fr.	Fruits edible.	Murthy et al. (2003)
90	*Asparagus racemo-sus* Willd.	R, T	Root tubers roasted, cooked and eaten. Sliced tuber pieces are dried for a week and the tuber powder along with sugar made into pudding (Khiri) during festivals and ceremonies. Tubers also eaten as raw	Murthy et al. (2003); Samydurai et al. (2012); Misra et al. (2013); Misra and Misra (2014)
91	*Aspidopteris indica* (Roxb.) Hochr.	T	Tubers are edible.	Murthy et al. (2003)
92	*Asteracantha longi-folia* L.	L	Leaves cooked as vegetable	Panda (2014); Sinha and Lakra (2005)
93	*Atalantia mono-phylla* (L.) Correa	Fr.	Green fruits used to make pickles	Basha et al. (2009); Prabakaran et al. (2013); Rekka and Senthil Kumar (2014); Deepa et al. (2014)
94	*Atylosia scarabae-oides* Benth.	L, Fr., S	Leaves and green pods used as vegetables. Seeds eaten as raw	Reddy et al. (2007); Misra and Misra (2013); Rajasab and Isaq (2004);
95	*Averrhoa caram-bola* L.	Fr.	Ripe fruits edible	Mukesh Kumar et al. (2014)
96	*Azadirachta indica* A. Juss.	Fr.	Ripe fruits edible	Murthy et al. (2003); Basha et al. (2009)
97	*Azima tetracantha* Lam.	Fr.	Fruits edible	Murthy et al. (2003); Reddy et al. (2007)

TABLE 6.1 *(Continued)*

S. No.	Name of the plant	Edible Part (s)	Mode of preparation	References
98	*Bacopa monnieri* (L.) Wettest.	St, L	Leaves and young shoots cooked and eaten	Rajasab and Isaq (2004); Misra and Misra (2013)
99	*Balanites aegyptiaca* (L.) Delile	Fr.	Dried fruit pulp is edible.	Rajasab and Isaq (2004); Murthy et al. (2003)
100	*Baliospermum-montanum* (Willd.) Muel.-Arg.	L	Leaves used as vegetable	Murthy et al. (2003)
101	*Bambusa arundinacea* (Retz.) Roxb.	St., L	Young shoots cut into small pieces cooked with salt and chilly and eaten; used as vegetable	Reddy et al. (2007); Misra and Misra (2013); Sanyasi Rao et al. (2014); Murthy et al. (2003);
102	*Bambusa bambos* (L.) Voss	S, St	Seeds made into flour and are used in cakes.Young shoots used as vegetables	Panda (2014); Misra and Misra (2014)
103	*Benkara malabarica* (Lam.) Tirveng.	Fr.	Ripe fruits edible	Reddy et al. (2007)
104	*Barringtonia acutangula* (L.) Gaertn.	L	Tender leaves edible	Murthy et al. (2003); Reddy et al. (2007)
105	*Basella alba* L.	St, L	Young shoots and leaves used as vegetable	Murthy et al. (2003); Rajasab and Isaq (2004); Misra and Misra (2013); Deepa et al. (2014)
106	*Bauhinia divaricata* L.	Fl.	Flowers are used as vegetable.	Panda (2014)
107	*Bauhinia malabarica* Roxb.	St, L	Young shoots and leaves are eaten raw	Murthy et al. (2003)
108	*Bauhinia purpurea* L.	L, Fl., S	Leaves, flowers, seeds used as vegetable	Murthy et al. (2003); Reddy et al. (2007); Rekka and Senthil Kumar (2014); Dhore et al. (2012)
109	*Bauhinia racemosa* Lam.	L	Leaves, flowers, seeds as Vegetable and prepared chutneys	Naidu and Khasim (2010); Nagalakshmi (2014); Dhore et al. (2012); Murthy et al. (2003)
110	*Bauhinia semla* Wunderlin	L	Young leaves are cooked as curry and taken	Misra and Misra (2013)
111	*Bauhinia vahlii* Wight & Arn.	L, S	Leaves used as vegetable. Seeds used to prepare chutney and roasted and eaten	Murthy et al. (2003); Misra and Misra (2013); Sinha and Lakra (2005)

TABLE 6.1   *(Continued)*

S. No.	Name of the plant	Edible Part (s)	Mode of preparation	References
112	*Bauhinia variegata* L.	L, Fl, S	Leaves and young flowers used as vegetables. Seeds used to prepare chutney	Murthy et al. (2003); Misra and Misra (2013);
113	*Begonia picta* Sm.	L	Leaves used as vegetable.	Murthy et al. (2003); Misra and Misra (2013)
114	*Benincasa hispida* (Thunb.) Cogn.	Fr.	Used as vegetable; prepare sweets	Murthy et al. (2003)
115	*Bidens bipinnata* L.	St	Fleshy shoots used as vegetable	Murthy et al. (2003)
116	*Boerhavia chinensis* (L.) Rottb.	R, L	Tender leaves and leafy shoots as vegetable	Misra and Misra (2013)
117	*Boerhavia diffusa* L.	R, L	Roots and leaves used as vegetable	Murthy et al. (2003); Reddy et al. (2007); Samydurai et al. (2012); Misra and Misra (2013); Deepa et al. (2014)
118	*Boerhavia erecta* L.	L	Cut into small pieces and cooked with salt, chilly and with oil Prepared Fry	Deepa et al. (2014)
119	*Bombax ceiba* L.	L, Fl.	Tender leaves used as vegetable	Reddy et al. (2007)
120	*Borassus flabellifer* L.	Fr., R	Ripe fruits are eaten as raw. Baked young roots are edible	Murthy et al. (2003); Reddy et al. (2007); Basha et al. (2009); Prabakaran et al. (2013); Singh (2013); Deepa et al. (2014)
121	*Boswellia serrata* Roxb. ex Coleber.	Fr., Fl	Fruits used to prepare pickle. Flowers eaten as raw	Murthy et al. (2003); Dhore et al. (2012)
122	*Brachiaria ramosa* (L.) Stapf	S	Used as food grains	Singh (2013)
123	*Brachystelma glabrum* Hook. f.	T	Eaten as raw	Sadasivaiah and Ravi Prasad Rao et al. (2012)
124	*Brachystelma volubile* Hook.f.	T	Eaten as raw	Sadasivaiah and Ravi Prasad Rao (2012)
125	*Brachystelma pullaiahii* Ravi Prasad et al.	T	Eaten as raw	Rao et al. (2011)
126	*Brassica compestris* L.	S	Used as spice	Singh (2013)
127	*Brassica juncea* Czern. & Coss.	S	Used as spice	Rao (2014)

TABLE 6.1   *(Continued)*

S. No.	Name of the plant	Edible Part (s)	Mode of preparation	References
128	*Brassica napus* L. var. *glauca* (Roxb.) Schulz	L	Leaves and stems dried in shade, stored in an earthen container	Misra and Misra (2013)
129	*Brassica nigra* (L.) Czerniak.	S	Used as spices	Sadasivaiah (2009)
130	*Brassica oleracea* L. var. *botrytis* L.	In.	Used as vegetable	Misra and Misra (2013)
131	*Brassica oleracea* L. var. *capitata* L.	L	Used as vegetable – prepare curries	Rao (2014)
132	*Brassica rapa* L.	R, T	Root boiled, cooked, roasted and then consumed as vegetable. Cultivated.	Misra and Misra (2014)
133	*Breynia vitis-idaea* (Burm.f.) Fischer	L	Young leaves used as vegetable	Reddy et al. (2007); Misra andMisra (2013); Murthy et al. (2003)
134	*Bridelia cinerescens* Gaertn.	Fr.	Fruits are edible	Murthy et al. (2003); Naidu and Khasim (2010)
135	*Bridelia monoica* (Lour.) Merr.	Fr.	Fruits are edible	Murthy et al. (2003)
136	*Bridelia montana* (Roxb.) Willd.	Fr.	Fruits are edible	Murthy et al. (2003); Reddy et al. (2007)
137	*Bridelia retusa* (L.) Spreng.	Fr.	Ripe fruits eaten raw	Murthy et al. (2003); Sinha and Lakra (2005)
138	*Bridelia stipularis* (L.) Bl.	Fr.	Fruits edible	Rout (2007)
139	*Bridelia tomentosa* Blume	Fr.	Ripe fruits eaten raw	Nayak and Basak (2015)
140	*Bruguiera gymnorrhiza* (L.) Savi	R	Vegetable	Murthy et al. (2003)
141	*Buchanania axillaris* (Desr.) Ramamoorthy	Fr.	Ripe fruits eaten raw	Murthy et al. (2003); Reddy et al. (2007); Basha et al. (2009); Prabakaran et al. (2013); Rekka and Senthil Kumar (2014)
142	*Buchanania lanzan* Sprengel	Fr., S	Ripe fruits and seeds eaten raw	Murthy et al. (2003); Sinha and Lakra (2005); Reddy et al. (2007); Rout (2007); Basha et al. (2009)
143	*Bupleurum ramosissimum* Wight & Arn.	Wp	Plant used as vegetable, fresh plant as flavoring agent	Girach (2001)

TABLE 6.1 *(Continued)*

S. No.	Name of the plant	Edible Part (s)	Mode of preparation	References
144	*Butea monosperma* (Lam.) Taub.	R, L	Flour prepared from young roots to make bread.	Reddy et al. (2007)
145	*Butea superba* Roxb.	Sap	Watery sap oozing out from the stem on cutting is used for drinking purpose by the tribal people.	Misra and Misra (2013)
146	*Cadaba fruticosa* (L.) Druce	L	Leaves and young parts cooked as curry	Nagalakshmi (2014)
147	*Caesalpinia pulcherima* (L.) Sw.	Fl., S	Flowers and young seeds are edible occasionally.	Rajasab and Isaq (2004)
148	*Caesulia axillaris* Roxb.	L	Young leaves are mixed with rice and cooked then eaten	Misra and Misra (2013); Murthy et al. (2003)
149	*Cajanus albicans* (Wight & Arn.) van der Maesan	S	Seeds used as pulse	Murthy (2011)
150	*Cajanus cajan* (L.) Millsp.	S	Seeds as pulse	Naidu and Khasim (2010); Singh (2013)
151	*Cajanus cajanifolius* (Haines) Maesen	S	Seeds used as pulse	Girach (2001); Reddy et al. (2006)
152	*Calamus guruba* Buch. Ham.	Fr.	Ripe fruits are edible	Nayak and Basak (2015)
153	*Calamus rotang* L.	Fr.	Ripe fruits eaten as raw	Panda (2014)
154	*Callicarpa arborea* Roxb.	St	Young shoots used as vegetable	Murthy et al. (2003)
155	*Callicarpa macrophylla* Vahl	Fr.	Fruits used as raw	Murthy et al. (2003)
156	*Canavalia gladiata* (Jacq.) DC.	Fr.	Green fruits used as vegetable	Rao (2014)
157	*Canavalia ensiformis* (L.) DC.	Fr.	Green fruits used in curries	Rekka and Senthil Kumar (2014); Deepa et al. (2014)
158	*Canna indica* L.	Rh, St.	Rhizome made into pieces and used to prepare curries	Mohan and Kalidass (2010)
159	*Cansjera rheedi* J.F.Gmel.	L, Fr.	Leaves used as vegetable, fruits are edible	Murthy et al. (2003); Hebber et al. (2010); Prabakaran et al. (2013)
160	*Canthium dicoccum* (Gaertn.) Teijsm. & Benn.	Fr.	Ripe fruits are edible	Prabakaran et al. (2013); Hebber et al. (2010)

TABLE 6.1 *(Continued)*

S. No.	Name of the plant	Edible Part (s)	Mode of preparation	References
161	*Canthium parviflorum* Lam.	Fr., L	Ripe fruits are edible; Young leaves are cooked with fermented rice water and eaten	Murthy et al. (2003); Reddy et al. (2007); Basha et al. (2009); Misra and Misra (2013); Deepa et al. (2014)
162	*Capparis brevispina* DC.	Fr.	Fruits used to prepare pickle.	Murthy et al. (2003)
163	*Capparis decidua* (Frossk.) Edgew.	Fr.	Fruits are edible as raw and made into pickle.	Murthy et al. (2003)
164	*Capparis divaricata* Lam.	Fr.	Tender fruits used as vegetable	Murthy et al. (2003); Deepa et al. (2014);
165	*Capparis grandis* L.f.	Fr.	Tender fruits used as vegetable	Murthy et al. (2003);
166	*Capparis roxburghii* DC.	Fr.	Used as vegetable	Murthy et al. (2003);
167	*Capparis zeylanica* L.	Fr.	Unripe fruits used as vegetables, ripe fruits eaten	Murthy et al. (2003); Reddy et al. (2007); Dhore et al. (2012); Rout (2007); Deepa et al. (2014); Panda (2014)
168	*Capsicum annuum* L.	Fr.	Used as spices	Singh (2013)
169	*Caralluma adscendens* R. Br. var. *attenuata* (Wight) Grav. & Mayuranathan	St	Used as raw and also prepared chutny	Murthy et al. (2003); Reddy et al. (2007); Sadasivaiah and Ravi Prasad Rao (2012); Misra and Misra (2013); Deepa et al. (2014);
170	*Caralluma indica* (Wight & Arn.) Plowes	St	Shoots are edible	Sadasivaiah and Rao (2012)
171	*Caralluma pauciflora* (Wight) N.B. Br.	St	Used as raw and also prepared chutney	
172	*Caralluma stalagmifera* Fischer	St	Used as raw and also prepared chutny	Sadasivaiah (2009)
173	*Caralluma umbellata* Haw.	St	Used as raw and also prepared chutney	Murthy et al. (2003); Sadasivaiah and Rao (2012)
174	*Cardamine hirsuta* L.	L	Leaves used as vegetable	Murthy et al. (2003)

TABLE 6.1 *(Continued)*

S. No.	Name of the plant	Edible Part (s)	Mode of preparation	References
175	*Cardiospermum halicacabum* L.	L	Used as vegetable	Reddy et al. (2007); Deepa et al. (2014)
176	*Careya arborea* Roxb.	Fr., L, Fl	Ripe fruits are eaten; Tender leaves are roasted then eaten. Flowers used as vegetable.	Murthy et al. (2003); Reddy et al. (2007); Basha et al. (2009); Dhore et al. (2012); Misra and Misra (2013)
177	*Carica papaya* L.	Fr.	Ripe fruits eaten raw	Naidu and Khasim (2010)
178	*Carissa carandas* L.	Fr.	Chutnies prepared from green fruits and ripe fruits eaten as raw.	Murthy et al. (2003); Basha et al. (2009); Prabakaran et al. (2013); Singh (2013); Deepa et al. (2014)
179	*Carissa inermis* Vahl	Fr.	Chutnies prepared from Green fruits and Ripe fruits eaten as raw	Rao (2014)
180	*Carissa paucinervia* A. DC.	Fr.	Ripe fruits eaten raw	Basha et al. (2009)
181	*Carissa spinarum* L.	Fr.	Chutnies prepared from Green fruits and Ripe fruits eaten as raw	Murthy et al. (2003); Basha et al. (2009); Deepa et al. (2014); Nayak and Basak (2015)
182	*Carmona retusa* (Vahl) Masam	Fr.	Fruits are edible	Nayak and Basak (2015); Murthy et al. (2003)
183	*Carthamus tinctorius* L.	S	Used to oil seeds	Singh (2013)
184	*Caryota urens* L.	Fr.	Stem pith is used as vegetable. Ripe fruits edible; toddy from cut inflorescence stalk drunk	Girach (2001); Murthy et al. (2003); Naidu and Khasim (2010); Sanyasi Rao et al. (2014)
185	*Casearia esculenta* Roxb.	Fr.	Ripe fruits eaten as raw	Murthy et al. (2003); Reddy et al. (2007)
186	*Casearia graveolens* Dalz.	Fr.	Ripe fruits eaten as raw	Murthy et al. (2003); Sinha and Lakra (2005); Panda (2014)
187	*Casearia tomentosa* Roxb.	Fr.	Fruits used to prepare pickle	Murthy et al. (2003)
188	*Cassia auriculata* L.	Fl.	Flowers are edible, young flower buds powdered and used as tea powder.	Rajasab and Isaq (2004); Reddy et al. (2007); Deepa et al. (2014)

TABLE 6.1   *(Continued)*

S. No.	Name of the plant	Edible Part (s)	Mode of preparation	References
189	*Cassia fistula* L.	Fl., L	Leaves, flowers used as vegetable.	Murthy et al. (2003); Reddy et al. (2007); Dhore et al. (2012)
190	*Cassia italica* (Mill.) Andr.	L	Cooked as vegetable	Reddy et al. (2007);
191	*Cassia obtusifolia* L.	L	Cooked as vegetable	Murthy et al. (2003)
192	*Cassia serecea* L.	L	Used as vegetable	Rajasab and Isaq (2004)
193	*Cassia tora* L.	L	Cooked as vegetable	Murthy et al. (2003); Prabakaran et al. (2013); Misra and Misra (2013); Deepa et al. (2014)
194	*Cassine glauca* (Rottb.) Kuntze	Gum	Gum roasted and eaten	Murthy et al. (2003)
195	*Catunaregam spinosa* (Thunb.) Tirvengadum	Fr.	Ripe fruits are eaten after roasting	Murthy et al. (2003)
196	*Cayratia auriculata* (Roxb.) Gamble	St, L	Tender leaves and shoots used as vegetable	Murthy et al. (2003)
197	*Cayratia trifolia* (L.) Domin.	L	Leaves used as vegetable.	Murthy et al. (2003); Reddy et al. (2007);
198	*Celastrus paniculatus* Willd.	Fr.	Unripe fruits boiled and cooked as vegetable, fruits edible.	Murthy et al. (2003); Sinha and Lakra (2005)
199	*Celosia argentea* L.	L	Used to prepare dal and curry	Murthy et al. (2003); Reddy et al. (2007); Sadasivaiah (2009); Misra andMisra (2013); Deepa et al. (2014)
200	*Celosia argentea* L. var. *cristata* Kuntze	L	Leaves cooked as curry	Murthy et al. (2003)
201	*Centella asiatica* (L.) Urban	L	Used to prepare chutnies and make it into powder and eat with boiled rice	Murthy et al. (2003); Reddy et al. (2007); Sadasivaiah (2009); Misra and Misra (2013); Deepa et al. (2014)
202	*Ceriscoides turgida* (Roxb.) Triveng.	Fr.	Ripe fruits are eaten	Murthy et al. (2003); Reddy et al. (2007)
203	*Ceropegia bulbosa* Roxb.	T, L	Tubers and leaves are edible as raw	Murthy et al. (2003); Reddy et al. (2007); Sadasivaiah andRao (2012)

TABLE 6.1 *(Continued)*

S. No.	Name of the plant	Edible Part (s)	Mode of preparation	References
204	*Ceropegia candelabrum* L.	T	Consumed raw in empty stomach to check hyperacidity. Tubers cooked as vegetable	Samydurai et al. (2012); Sadasivaiah and Rao (2012); Misra et al. (2013)
205	*Ceropegia hirsuta* Wight & Arn.	T	Edible as raw	Murthy et al. (2003); Reddy et al. (2007); Sadasivaiah and Rao (2012)
206	*Ceropegia juncea* Roxb.	WP	Edible as raw	Sadasivaiah and Rao (2012)
207	*Ceropegia spiralis* Wight	T	Eaten as raw	Murthy et al. (2003); Sadasivaiah and Rao (2012)
208	*Cheilocostus speciosus* (J.Koenig) C.D.Specht	Rh	Rhizomes used as vegetable. Rhizome is boiled, cooked with pulses and tamarind/tomato, then consumed as curry.	Murthy et al. (2003); Reddy et al. (2007); Samydurai et al. (2012); Misra et al. (2013); Misra and Misra (2014)
209	*Chenopodium album* L.	L	Leaves used as vegetable	Murthy et al. (2003); Sinha and Lakra (2005); Rao (2014)
210	*Chlorophytum arundinaceum* Baker	T	Tubers are edible.	Reddy et al. (2007)
211	*Chlorophytum tuberosum* (Roxb.) Baker	T	Tubers cooked as vegetables	Murthy et al. (2003); Reddy et al. (2007); Samydurai et al. (2012); Misra et al. (2013)
212	*Chrysophyllum roxburghii* G. Don	Fr., Fl	Flowers and fruits roasted and eaten	Murthy et al. (2003)
213	*Cicer arietinum* L.	S, L	Seeds used as pulse food. Leaves used as vegetale.	Dhore et al. (2012); Singh (2013)
214	*Cinnamomum tamala* (Buch.-Ham.) Nees & Eberm.	L	Used as spices	Rao (2014)
215	*Cinnamomum verum* J. Presl	Br.	Used as spices	Rao (2014)
216	*Cissus quadrangularis* L.	L, St	Prepared curries, used to prepare chutnies, cut into small pieces and cooked with salt, chilly and with oil	Murthy et al. (2003); Reddy et al. (2007); Naidu and Khasim (2010); Deepa et al. (2014); Rekka and Senthil Kumar (2014)

TABLE 6.1 *(Continued)*

S. No.	Name of the plant	Edible Part (s)	Mode of preparation	References
217	*Cissus repanda* L.	L, Fr.	Leaves and fruits used as vegetable	Murthy et al. (2003)
218	*Cissus vitiginea* L.	T	Used as vegetable	Murthy et al. (2003)
219	*Citrullus colocynthis* (L.) Schrader	Fr.	Eaten as raw and vegetable	Deepa et al. (2014)
220	*Citrullus lanatus* (Thunb.) Matsum. & Nakai	Fr., S	Ripe fruits. Seeds roasted and eaten	Singh (2013)
221	*Citrus aurantiifolia* (Christm.) Swingle	Fr.	Ripe fruits are edible	Singh (2013, 2014)
222	*Citrus aurantium* L.	Fr.	Ripe fruits are edible	Rao (2014); Singh (2014)
223	*Citrus limon* (L.) Burm.f.	Fr.	Fruit juice is used to prepare salads, sharbaths and various rice items	Rao (2014); Singh (2014)
224	*Citrus medica* L.	Fr.	Pickles prepared with green fruits	Rao (2014); Singh (2014)
225	*Citrus reticulata* Blanco	Fr.	Ripe fruits edible	Rao (2014); Singh (2014)
226	*Citrus sinensis* (L.) Osbeck	Fr.	Ripe fruits edible	Rao (2014); Singh (2014)
227	*Citrus maxima* (Burm.) Merr.	Fr.	Ripe fruits edible	Singh (2013, 2014); Rekka and Senthil Kumar (2014)
228	*Clausena dentata* (Willd.) Roem.	Fr.	Ripe fruits edible	Rekka and Senthil Kumar (2014); Deepa et al. (2014)
229	*Clausena excavata* Burm.	Fr.	Fruits edible	Rout (2007)
230	*Clausena heptophylla* Wight & Arn.	L	Leaves chewed with betel	Murthy et al., 200
231	*Cleistanthus collinus* (Roxb.) Benth.	S	Seed pulp eaten as raw	Murthy et al. (2003)
232	*Cleome gynandra* L.	L, St	Leaves and young shoots roasted, then eaten	Murthy et al. (2003); Sinha and Lakra (2005); Reddy et al. (2007); Misra and Misra (2013);
233	*Cleome isonandra* L.	L	Vegetable	Sinha and Lakra (2005)
234	*Cleome monophylla* L.	L	Cooked with salt, chillies and oil	Misra and Misra (2013); Murthy et al. (2003)

TABLE 6.1 *(Continued)*

S. No.	Name of the plant	Edible Part (s)	Mode of preparation	References
235	*Cleome viscosa* L.	L	Used as vegetable	Misra and Misra (2013, 2014)
236	*Cleome chelidonii* L.f. var. *pallai* V.S. Raju & C.S. Reddy	S	Seeds used as condiment	Reddy et al. (2006)
237	*Clerodendrum serratum* (L.) Moon	L	Leaves used as vegetable.	Murthy et al. (2003); Reddy et al. (2007); Misra et al. (2013);
238	*Coccinia grandis* (L.) Voigt	Fr., L	Unripe fruits as vegetable and pickle, ripe fruits eaten raw. Leaves and leafy shoots, green fruits cooked with salt, chillies and with oil	Murthy et al. (2003); Rout (2007); Misra and Misra (2013, 2014); Deepa et al. (2014)
239	*Cocculus hirsutus* (L.) Diels	L	Tender leaves are allowed to coagulate and after adding sugar, then eaten.	Reddy et al. (2007); Prabakaran et al. (2013); Misra and Misra (2013); Deepa et al. (2014);
240	*Cochlospermum religiosum* (L.) Alston	L, Fl	Used as vegetable	Sinha and Lakra (2005);
241	*Cocos nucifera* L.	Fr.	The liquid endosperm is used to drink, solid endosperm is used to prepare chutnies and also used in masala curries. Sweets are prepared with sugars and candy.	Rao (2014)
242	*Coffea arabica* L.	S	Seeds are made in powder, this powder used to make a Coffee.	Rekka and Senthil Kumar (2014)
243	*Coix lacryma-jobi* L.	S	Used as food grains	
244	*Coleus aromaticus* Benth.	T	Tubers cooked as vegetables to cure heavy cold and asthma	
245	*Coleus forskohlii* (Poir.) Briq.	T	Tubers cooked as vegetables to cure fever, cold and cough	Samydurai et al. (2012)

TABLE 6.1 *(Continued)*

S. No.	Name of the plant	Edible Part (s)	Mode of preparation	References
246	*Colocasia esculenta* (L.) Schott	C, L	Tubers cooked as vegetables to stimulant and indigestion. Used as vegetable and cooked as curries, chutney. Corm pieces smeared with rice flour are fried as cake	Murthy et al. (2003); Reddy et al. (2007); Naidu and Khasim (2010); Samydurai et al. (2012); Misra et al. (2013); Prabakaran et al. (2013); Misra and Misra (2013, 2014); Rekka and Senthil Kumar (2014)
247	*Colocasia gigantea* Hook.f.	L	Used as vegetable	Murthy et al. (2003)
248	*Combretum roxburghii* Spreng.	Sap	Sap used as drink	Misra and Misra (2013)
249	*Commelina benghalensis* L.	L, St	Used as vegetable	Naidu and Khasim (2010); Samydurai et al. (2012); Prabakaran et al. (2013); Misra and Misra (2013); Deepa et al. (2014); Murthy et al. (2003)
250	*Commelina communis* L.	L	Used as vegetable	Murthy et al. (2003)
251	*Commelina ensifolia* R. Br.	L	Used as vegetable	Murthy et al. (2003)
252	*Commelina obliqua* Buch.-Ham.	L	Used as vegetable	Murthy et al. (2003)
253	*Commiphora caudata* (Wight & Arn.) Engler	Fr.	Unripe and ripe fruits eaten raw	Murthy et al. (2003); Singh (2013)
254	*Corchorus aestuans* L.	L, St	Tender leaves and young shoots fried and eaten	Misra and Misra (2013, 2014); Singh (2013)
255	*Corchorus capsularis* L.	L, St	Tender leaves and young shoots fried and eaten	Misra and Misra (2013); Deepa et al. (2014)
256	*Corchorus fascicularis* Lam.	L, St	Tender leaves and young shoots fried and eaten	Misra and Misra (2013); Murthy et al. (2003)
257	*Cordia dichotoma* Forst. f.	Fr.	Ripe fruits edible	Murthy et al. (2003); Rajasab and Isaq (2004); Reddy et al. (2007); Basha et al. (2009); Prabakaran et al. (2013)
258	*Cordia domestica* Roth.	Fr.	Ripe fruits edible	Basha et al. (2009)

TABLE 6.1 *(Continued)*

S. No.	Name of the plant	Edible Part (s)	Mode of preparation	References
259	*Cordia evolutior* (C.B. Clarke) Gamble	Fr.	Ripe fruits edible	Basha et al. (2009)
260	*Cordia gharaf* (Forsskal) Ehrenb.	Fr.	Ripe fruits edible	Basha et al. (2009)
261	*Cordia macleodii* Hook. f. & Thoms.	Fr.	Ripe fruits edible directly	Murthy et al. (2003); Basha et al. (2009)
262	*Cordia monoica* Roxb.	Fr.	Ripe fruits edible	Murthy et al. (2003)
263	*Coriandrum sativum* L.	L, S	Used as spice	Singh (2013)
264	*Corallocarpus epigaeus* Benth. & Hook.f.	T	Used as vegetable	Samydurai et al. (2012)
265	*Cosmostigma racemosa* Wight	Fl	Flowers eaten as raw	Murthy et al. (2003)
266	*Crataeva adansonii* DC.	L	Used as vegetable	Murthy et al. (2003)
267	*Crateva magna* (Lour.) DC.	L	Used as vegetable	Murthy et al. (2003)
268	*Crotalaria pulchra* Andr.	L	Used as vegetable	Murthy et al. (2003)
269	*Cucumis melo* L.	Fr., S, L	Used as vegetable. Ripe fruits eaten as raw. Seeds roasted and eaten.Tender leaves are cooked and eaten.	Misra and Misra (2013) Singh (2013); Rajasab and Isaq (2004)
270	*Cucumis melo* L. var. *agrestis* Naud.	Fr.	Used as vegetable	Singh (2013)
271	*Cucumis sativus* L.	Fr., S	Used as vegetable. Ripe fruits eaten raw. Seeds roasted and eaten.	Singh (2013)
272	*Cucurbita maxima* Duchesne ex Lam.	Fr., S, L	Fruits are used as vegetable curry. Seeds roasted and eaten. Leaves and young stems roasted then eaten	Naidu and Khasim (2010); Misra and Misra (2013)
273	*Cucurbita pepo* L.	Fr., S	Used as vegetable. Ripe fruits eaten as raw. Seeds roasted and eaten	Singh (2013)
274	*Cuminum cyminum* L.	S	Used as spice	Rao (2014)

TABLE 6.1   *(Continued)*

S. No.	Name of the plant	Edible Part (s)	Mode of preparation	References
275	*Curculigo orchioides* Gaertn.	T	Used as vegetable. Burnt in embers and consumed as snacks by teenage girls before or during monthly periods to regularize menstruation; Tubers are used for preparation of local drinks.	Samydurai et al. (2012); Misra et al. (2013); Deepa et al. (2014); Murthy et al. (2003)
276	*Curculigo trichocarpa* (Wight) Bennet et Raizada	T	Burnt and consumed with salt as nut by children and old men	Misra et al. (2013)
277	*Curcuma aeruginosa* Roxb.	Rh	Used as vegetable	Misra and Misra (2014)
278	*Curcuma amada* Roxb.	Rh	Sliced and put in stale rice to enhance flavor and taste and consumed in morning. Ground and added with a pinch of sugar and salt and made in to sherbet frequently during summer. It is also pickled and used as condiment in fresh/stale foods.	Misra et al. (2013); Singh (2013)
279	*Curcuma angustifolia* Roxb.	Rh	Rhizome is rubbed on stone, dissolved in sufficient water, filtered and allowed to evaporate. The starch powder obtained after sundrying is cooked into pudding (khiri) along with sugar.	Reddy et al. (2007); Misra et al. (2013); Misra and Misra (2014)
280	*Curcuma aromatica* Salisb.	Rh	Rhizome is ground or rubbed on the stone and dissolved in water, filtered and allowed to evoperate. The starch powder obtained after drying is cooked along with sugar and made into Khiri (pudding) or cake. Powder is made into sherbet with sugar and taken during summer	Misra et al. (2013); Murthy et al. (2003)
281	*Curcuma caesia* Roxb.	Rh	Edible	Murthy et al. (2003)
282	*Curcuma harita* Mangaly & Sabu	Rh	Edible	Murthy et al. (2003)

TABLE 6.1 *(Continued)*

S. No.	Name of the plant	Edible Part (s)	Mode of preparation	References
283	*Curcuma longa* L.	Rh	Used as spice	Naidu and Khasim (2010); Singh (2013)
284	*Curcuma montana* Roxb.	Rh	Burnt and consumed as snacks. Sliced and used as substitute of spice	Misra et al. (2013)
285	*Curcuma pseudo-montana* Salisb.	St, L	Tender shoots and leaves used as vegetable	Murthy et al. (2003)
286	*Curcuma zedoaria* Rosc.	Rh, St	Ground into a paste and added to curry to enhance taste and digestion. Cut into pieces and made pickles	Murthy et al. (2003); Misra et al. (2013)
287	*Cyamopsis tetragonoloba* L.	Fr.	Used as vegetable	Rao (2014)
288	*Cyanotis tuberosa* Roxb.	Rh	Used as vegetable. Tuberous roots cooked and eaten	Samydurai et al. (2012)
289	*Cycas beddomei* Dyer	P, L	Cut the plant and take out the pith and make it into pieces and used in their diet in case of debility. Tender leaves used as vegetable.	Murthy et al. (2003); Reddy et al. (2006)
290	*Cycas sphaerica* Roxb.	P	Pith pieces are used to make 'sago' flour.	Reddy et al. (2006)
291	*Cynodon dactylon* (L.) Pers.	St, L	Used to prepare chutney	Reddy et al. (2007)
292	*Cyperus bulbosus* Vahl	T	Roasted tubers are eaten	Murthy et al. (2003)
293	*Cyperus esculentus* L.	T	Tubers used as vegetable	Murthy et al. (2003)
294	*Cyperus rotundus* L.	T	Tubers eaten after boiling. Burnt or roasted and consumed as snacks like nuts. Also consumed raw	Samydurai et al. (2012); Misra et al. (2013)
295	*Cyphomandra beta-cea* (Cav.) Sendtn.	Fr.	Used in curries	Rekka and Senthil Kumar (2014)
296	*Cyphostemma auriculatum* (Roxb.) P. Singh & B.V. Shetty	L, St	Cooked and eaten	Misra and Misra (2013)
297	*Cyphostemma setosum* (Roxb.) Alston	T, L	Tubers and leaves used as vegetable	Murthy et al. (2003)

TABLE 6.1 *(Continued)*

S. No.	Name of the plant	Edible Part (s)	Mode of preparation	References
298	*Dactyloctenium ae-gyptium* (L.) Beau.	S	Seeds used as grains	Singh (2013)
299	*Datura metel* L.	Wp	Taken as food to treat cough	Ramakrishna et al. (2014)
300	*Daucas carota* L.	T	Roots boiled, cooked, roasted and consumed as vegetable. Tubers also taken as raw.	Misra and Misra (2014)
301	*Debregeasia longifolia* (Burm.f.) Widd.	L	Used as vegetable	Murthy et al. (2003)
302	*Decalepis hamilto-nii* Wight & Arn.	R	Root pieces pickled. Rhizomes are boiled and eaten; roots boiled in preparation of cool drinks in summer season	Murthy et al. (2003); Reddy et al. (2007);Samydurai et al. (2012); Rekka and Senthil Kumar (2014); Deepa et al. (2014);
303	*Decaschistia cud-dapahensis*Paul & Nayar	L	Used as vegetable	Singh (2013)
304	*Delonix regia* Hook.f.	Fl.	Edible	Rajasab and Isaq (2004)
305	*Dendrocalamus strictus* (Roxb.) Nees	Rh, St	Tender culms, rhizome used as vegetable	Murthy et al. (2003); Reddy et al. (2007); Misra and Misra (2013, 2014)
306	*Dentella repens* (L.) J.R. Forst. & G. Forst.	L, Fr.	Ripe fruits edible; young leaves used as vegetable	Murthy et al. (2003)
307	*Derris scandens* (Roxb.) Benth	L	Leaves used as vegetable	Murthy et al. (2003)
308	*Digera muricata* (L.) Mart.	L	Leaves boiled and squeezed, then add groundnut powder and eaten along with boiled rice or roties	Murthy et al. (2003); Reddy et al. (2007); Sadasiv-aiah (2009); Prabakaran et al. (2013); Singh (2013); Deepa et al. (2014)
309	*Dillenia aurea* Sm.	Fl, Fr.	Edible	Murthy et al. (2003); Sinha and Lakra (2005);
310	*Dillenia indica* L.	Fl, Fr.	Fruits edible	Murthy et al. (2003); Sinha and Lakra (2005); Rout (2007)
311	*Dillenia pentagyna* Roxb.	Fl, Fr.	Used as vegetable, raw fruits edible	Reddy et al. (2007); Rout (2007); Sinha and Lakra (2005); Murthy et al. (2003)

TABLE 6.1   *(Continued)*

S. No.	Name of the plant	Edible Part (s)	Mode of preparation	References
312	*Dimocarpus longan* Lour.	Fr.	Fruits are edible	Hebber et al. (2010);
313	*Dioscorea alata* L.	T	Tuber edible, raw or boiled. Used as vegetable	Samydurai et al. (2012); Misra and Misra (2014); Rekka and Senthil Kumar (2014)
314	*Dioscorea be-lophylla* (Prain) Haines	R, T	Eaten as raw or cooked as vegetable with pulses.	Murthy et al. (2003); Misra et al. (2013); Misra and Misra (2014)
315	*Dioscorea bulbifera* L. var. *vera* Prain & Burkill	T	Used as vegetable.Tuber sliced, boiled and kept over-night in running tap water, further boiled to remove bitterness, cooked then taken as curry	Murthy et al. (2003); Reddy et al. (2007): Naidu and Khasim (2010); Samy-durai et al. (2012); Singh (2013); Misra et al. (2013); Misra and Misra (2014)
316	*Dioscorea esculen-ta* (Lour.) Burkill	T	Tubers eaten	Samydurai et al. (2012)
317	*Dioscorea glabra* Roxb.	T	Tubers eaten as raw or burnt as snacks or cooked with other vegetables	Misra et al. (2013); Misra and Misra (2014); Panda (2014)
318	*Dioscorea hamilto-nii* Hook. f.	T	Fresh tuber is slimy and tasty and eaten raw by children	Murthy et al. (2003); Misra et al. (2013)
319	*Dioscorea hispida* Dennst.	T	Tuber sliced soaked in run-ning water and boiled suc-cessively with the leaves of Tamarind. The excess water is filtered out further cooked as curry and eaten as such	Murthy et al. (2003); Samydurai et al. (2012); Misra et al. (2013); Misra and Misra (2014)
320	*Dioscorea oppositi-folia* L.	T	Tuber peeled and eaten raw or sliced cooked with other vegetable and onion and consumed as curry	Murthy et al. (2003); Reddy et al. (2007); Samydurai et al. (2012); Misra et al. (2013); Misra and Misra (2014); Deepa et al. (2014); Rekka and Senthil Kumar (2014)
321	*Dioscorea pen-taphylla* L.	T	Tubers eaten after boiling. Tuber and bulbils are thor-oughly washed then boiled, sliced and cooked with onion and spice and eaten	Murthy et al. (2003); Reddy et al. (2007); Nai-du and Khasim (2010); Samydurai et al. (2012); Misra et al. (2013); Deepa et al. (2014); Misra and Misra (2014); Rekka and Senthil Kumar (2014)

TABLE 6.1 *(Continued)*

S. No.	Name of the plant	Edible Part (s)	Mode of preparation	References
322	*Dioscorea puber* Blume	T	Tubers and bulbils cooked as curry by frying with oil and spice. It is boiled with salt and taken as chutney	Murthy et al. (2003); Misra et al. (2013); Misra and Misra (2014)
323	*Dioscorea tomentosa* Koen. ex Spreng.	T	Tuberous roots eaten after boiling. Tubers are successively boiled and cooked as vegetable curry	Murthy et al. (2003); Reddy et al. (2007); Samydurai et al. (2012); Misra et al. (2013)
324	*Dioscorea wallichii* Hook.f.	T	Cooked, boiled and used as vegetable. Rhizome, tubers and bulbils are sliced and cooked as curry after successive boiling to remove the acrid principle. Tuber dried, powdered and made into sherbet with sugar.	Sinha and Lakra (2005); Misra et al. (2013); Misra and Misra (2014);
325	*Dioscorea dodecaneura* Vell.	T	Roasted and eaten	Panda (2014)
326	*Diospyros chloroxylon* Roxb.	Fr.	Ripe fruits eaten raw	Murthy et al. (2003); Reddy et al. (2007)
327	*Diospyros ebenum* Koen.	Fr.	Ripe fruits eaten raw	Murthy et al. (2003); Prabakaran et al. (2013)
328	*Diospyros exsculpta* Buch.-Ham.	Fr.	Ripe fruits eaten raw	Murthy et al. (2003)
329	*Diospyros malabarica* (Desr.) Kostel	Fr.	Fruits edible	Rout (2007)
330	*Diospyros melanoxylon* Roxb.	Fr., S	Ripe fruits eaten raw. Seeds are edible	Murthy et al. (2003); Sinha and Lakra (2005); Reddy et al. (2007); Basha et al. (2009)
331	*Diospyros montana* Roxb.	Fr.	Ripe fruits eaten raw.	Murthy et al. (2003)
332	*Diospyros ovalifolia* Wight	Fr.	Ripe fruits eaten raw.	Murthy et al. (2003)
333	*Diospyros peregrina* (Gaertn.) Guerke	Fr., S	Ripe fruits eaten raw. Seeds are edible	Murthy et al. (2003); Sinha and Lakra (2005); Reddy et al. (2007)
334	*Diospyros sylvatica* Roxb.	Fr.	Ripe fruits eaten raw.	Murthy et al. (2003)
335	*Diospyros vera* (Lour.) A.Chev.	Fr.	Ripe fruits eaten raw	Murthy et al. (2003); Basha et al. (2009); Prabakaran et al. (2013)

TABLE 6.1 *(Continued)*

S. No.	Name of the plant	Edible Part (s)	Mode of preparation	References
336	*Diplazium esculentum* (Retz.) Sw.	L	Tender folded leaves are cooked as vegetable	Sanyasi Rao et al. (2014)
337	*Diplocyclos palmatus* (L.) Jeffrey	L, Fr.	Leaves and unripe fruits are used as vegetable.	Murthy et al. (2003); Rajasab and Isaq (2004)
338	*Disporum cantoniense* (Lour.) Merr.	R	Sliced, well boiled, cooked as vegetable	Misra et al. (2013)
339	*Dodonaea viscosa* L.	S	Seeds eaten	Murthy et al. (2003)
340	*Dolichos trilobus* L.	L, T	Used as vegetable	
341	*Drymaria cordata* (L.) Willd. ex Roem. & Schult.	L	Leaves used as vegetable	Murthy et al. (2003)
342	*Drynaria quercifolia* (L.) J. Sm.	Rh	Rhizomes soup drunk to get relief from rheumatic complaints. Boiled and made into soup	Samydurai et al. (2012); Deepa et al. (2014); Rekka and Senthil Kumar (2014)
343	*Drypetes sepiaria* (Wight & Arn.) Pax & Haffm.	Fr., S	Ripe fruits and seeds eaten raw	Murthy et al. (2003)
344	*Echinochloa colonum* (L.) Link	S	Grains used as food	Murthy et al. (2003); Naidu and Khasim (2010); Singh (2013)
345	*Echinochloa crusgalli* (L.) Beauv.	S	Grains used as food	Murthy et al. (2003); Singh (2013, 2014)
346	*Echinochloa frumentacea* (Roxb.) Link.	S	Used as food grains	Singh (2013)
347	*Eclipta prostrata* (L.) L.	L, R	Prepared chutney and eaten along with boiled rice.	Murthy et al. (2003); Samydurai et al. (2012); Rekka and Senthil Kumar (2014); Deepa et al. (2014)
348	*Ehretia acuminata* R. Br.	Fr.	Fruits eaten as raw	Murthy et al. (2003);
349	*Ehretia canarensis* (Cl.) Gamble	Fr.	Ripe fruits edible	Reddy et al. (2007); Misra and Misra (2013)
350	*Ehretia laevis* Roxb.	Fr., Br.	Ripe fruits edible; bark of the stem cooked and eaten	Murthy et al. (2003); Reddy et al. (2007); Misra and Misra (2013)
351	*Ehretia microphylla* Lam.	L	Leaves used as vegetable	Naidu and Khasim (2010)

TABLE 6.1 *(Continued)*

S. No.	Name of the plant	Edible Part (s)	Mode of preparation	References
352	*Eleocharis dulcis* Trin.	T	Tubers used raw	Murthy et al. (2003)
353	*Eleusine coracana* (L.) Gaertn.	S	Make powder and prepare soup and balls. Nutritious syrup is called "*Ambali*" prepared from powder of grains	Naidu and Khasim (2010); Singh (2013)
354	*Eleusine indica* Steudel	S	Used as food grains	Singh (2013)
355	*Embelia ribes* Burm. f.	S	Eaten as raw	Murthy et al. (2003)
356	*Emilia sonchifolia* (L.) DC.	L, St	Prepared chutney and eaten along with boiled rice	Murthy et al. (2003); Misra and Misra (2013)
357	*Entada rheedii* Spreng.	S	Seeds as pulse	Panda (2014)
358	*Erioglossum rubiginosum* (Roxb.) Blume	St	Tender shoots used as vegetable	Murthy et al. (2003)
359	*Erycibe paniculata* Roxb.	Fr.	Ripe fruits are edible	Sinha and Lakra (2005); Panda (2014)
360	*Eryngium foetidum* L.	L	Leaves used as flavoring agent	Murthy et al. (2003); Misra and Misra (2013)
361	*Erythrina suberosa* Roxb.	L, S	Tender leaves used as vegetable. Seeds are edible	Murthy et al. (2003)
362	*Erythrina variegata* L.	L, fr, St	Tender leaves and shoots used as vegetable. Fruits edible	Reddy et al. (2007)
363	*Erythroxylon monogynum* Roxb.	L, Fr.	Leaves boiled along with dal and fried with oil and ground nut powder and eaten. Ripe fruits eaten as raw	Murthy et al. (2003); Reddy et al. (2007); Basha et al. (2009); Prabakaran et al. (2013); Deepa et al. (2014)
364	*Eugenia roxburghii* DC.	Fr.	Fruits edible	Murthy et al. (2003); Nayak and Basak (2015)
365	*Euphorbia caducifolia* Haines	L	Leaves used as vegetable	Murthy et al. (2003)
366	*Euphorbia heterophylla* L.	L	Leaves used as vegetable	Reddy et al. (2007)
367	*Euphorbia hirta* L.	L	Whole plant is used as vegetable, leaf and curry is eaten as galactogogue	Murthy et al. (2003); Reddy et al. (2007); Naidu and Khasim (2010); Deepa et al. (2014)

TABLE 6.1 *(Continued)*

S. No.	Name of the plant	Edible Part (s)	Mode of preparation	References
368	*Euphorbia thymifolia* L.	St, L	Tender leaves and shoots are cooked then eaten	Murthy et al. (2003); Misra and Misra (2013);
369	*Ficus auriculata* Lour.	Fr.	Unripe fruits used as vegetable	Murthy et al. (2003)
370	*Ficus benghalensis* L.	Fr., Gum, L	Ripe fruits edible.Tender leaves used as vegetable	Murthy et al. (2003); Basha et al. (2009); Deepa et al. (2014)
371	*Ficus carica* L.	Fr.	Ripe fruits edible	Sinha and Lakra (2005)
372	*Ficus hispida* L.f.	Fr.	Ripe fruits edible	Murthy et al. (2003); Reddy et al. (2007); Basha et al. (2009); Panda (2014)
373	*Ficus microcarpa* L.f.	Fr.	Ripe fruits edible	Deepa et al. (2014)
374	*Ficus palmata* Forssk.	Fr.	Ripe fruits edible	Murthy et al. (2003)
375	*Ficus racemosa* L.	Fr.	Ripe fruits edible	Murthy et al. (2003); Reddy et al. (2007); Basha et al. (2009); Prabakaran et al. (2013); Deepa et al. (2014)
376	*Ficus religiosa* L.	Fr., L	Ripe fruits edible. Young leaves and shoots are used as vegetable	Murthy et al. (2003); Basha et al. (2009); Dhore et al. (2012); Deepa et al. (2014)
377	*Ficus rumphii* Bl.	Fr.	Ripe fruits edible	Murthy et al. (2003)
378	*Ficus semicordata* Buch.-Ham. ex J.E. Smith	Fr.	Ripe fruits edible	Murthy et al. (2003)
379	*Ficus virens* Ait.	Fr., L	Ripe fruits edible. Young leaves used as vegetable	Murthy et al. (2003); Deepa et al. (2014)
380	*Flacourtia indica* (Burm. f.) Merr.	Fr.	Ripe fruits eaten raw; fruits are used to make pickle	Murthy et al. (2003); Reddy et al. (2007); Basha et al. (2009); Prabakaran et al. (2013); Deepa et al. (2014)
381	*Flacourtia jangomas* (Lour.) Raeusch.		Ripe fruits edible	Murthy et al. (2003); Sinha and Lakra (2005)
382	*Foeniculum vulgare* Miller	S	Used as spice	Rao (2014)
383	*Garcinia cowa* Roxb. ex Choisy	Fr.	Ripe fruits edible	Sinha and Lakra (2005); Panda (2014)

TABLE 6.1   *(Continued)*

S. No.	Name of the plant	Edible Part (s)	Mode of preparation	References
384	*Garcinia spicata* (Wight & Arn.) Hook.f.	Fr.	Fruits edible	Murthy et al. (2003)
385	*Garcinia xanthochymus* T. And.	Fr.	Fruits edible	Murthy et al. (2003)
386	*Gardenia gummifera* L. f.	Fr.	Unripe and ripe fruits edible	Murthy et al. (2003); Basha et al. (2009); Deepa et al. (2014)
387	*Gardenia latifolia* Ait.	Fr.	Ripe fruits edible	Murthy et al. (2003); Reddy et al. (2007); Basha et al. (2009)
388	*Gardenia resinifera* Roth	Fr.	Ripe fruits edible	Murthy et al. (2003); Reddy et al. (2007)
389	*Garuga pinnata* Roxb.	Fr.	Ripe fruits edible	Murthy et al. (2003); Sinha and Lakra (2005); Reddy et al. (2007); Basha et al. (2009)
390	*Gisekia pharnaceoides* L.	L	Used as vegetable	Murthy et al. (2003)
391	*Givotia moluccana* (L.) Sree.	S	Eaten as raw	Basha et al. (2009)
392	*Glinus lotoides* L.	L	Tender shoots and leaves used as vegetable	Murthy et al. (2003)
393	*Glinus oppositifolius* (L.) DC.	L	Used as vegetable	Misra and Misra (2013, 2014)
394	*Globba marantina* L.	T	Rhizome used as vegetable	Misra and Misra (2014)
395	*Glycosmis mauritiana* (Lam.) Tanaka	Fr	Edible raw	Murthy et al. (2003)
396	*Glycosmis pentaphylla* (Retz.) DC.	Fr.	Edible raw	Hebber et al. (2010); Nayak and Basak (2015)
397	*Glycyrrhiza glabra* L.	R	Cooked as vegetable	Samydurai et al. (2012)
398	*Gmelina arborea* Roxb.	Fr.	Ripe fruits edible	Panda (2014)
399	*Gnaphalium polycaulon* Pers.	L	Used as vegetable	Reddy et al. (2007)
400	*Gouania leptostachya* DC.	L	Used as vegetable	Murthy et al. (2003)

TABLE 6.1   *(Continued)*

S. No.	Name of the plant	Edible Part (s)	Mode of preparation	References
401	*Grewia abutilifolia* Juss.	Fr.	Ripe fruits edible	Murthy et al. (2003)
402	*Grewia asiatica* L.	Fr.	Ripe fruits edible	Murthy et al. (2003); Rekka and Senthil Kumar (2014)
403	*Grewia damine* Gaertn.	Fr.	Ripe fruits edible	Murthy et al. (2003); Basha et al. (2009)
404	*Grewia elastica* Royle	Fr.	Ripe fruits edible	Reddy et al. (2007)
405	*Grewia flavescens* Juss.	Fr.	Ripe fruits edible	Murthy et al. (2003); Reddy et al. (2007); Basha et al. (2009)
406	*Grewia hirsuta* Vahl	Fr.	Ripe fruits edible and also used as vegetable	Murthy et al. (2003); Reddy et al. (2007); Basha et al. (2009); Deepa et al. (2014)
407	*Grewia nervosa* (Lour.) Panigrahi	Fr.	Fruits edible	Hebber et al. (2010)
408	*Grewia rhamnifolia* Roth	Fr.	Ripe fruits edible	Murthy et al. (2003)
409	*Grewia rothii* DC.	Fr.	Ripe fruits edible	Murthy et al. (2003)
410	*Grewia tenax* (Frossk.) Fiori	Fr.	Ripe fruits edible	Murthy et al. (2003); Basha et al. (2009)
411	*Grewia tiliifolia* Vahl	Fr.	Ripe fruits are edible and also used as vegetable	Murthy et al. (2003); Reddy et al. (2007); Basha et al. (2009); Deepa et al. (2014)
412	*Grewia villosa* Willd.	Fr.	Ripe fruits edible	Murthy et al. (2003); Reddy et al. (2007); Basha et al. (2009)
413	*Guazuma ulmifolia* Lam.	Fr.	Fruits edible	Reddy et al. (2007)
414	*Guizotia abyssinica* (L.f.) Cass.	S	Seeds for edible oil	Singh (2014)
415	*Gymnema sylvestre* R. Br.	R, L	Roots and leaves cooked as vegetable to cure Diabetes	Murthy et al. (2003); Samydurai et al. (2012); Deepa et al. (2014)
416	*Habenaria plantaginea* Lindl.	T	Burnt and consumed as snacks; preferred by children due to its sweet taste	Misra et al. (2013)

TABLE 6.1 *(Continued)*

S. No.	Name of the plant	Edible Part (s)	Mode of preparation	References
417	*Hedychium coccineum* Buch.-Ham. ex Sm.	Rh	Sliced into pieces, cooked with other vegetables and made in to a curry and eaten	Misra et al. (2013)
418	*Hedychium coronarium* Koen.	Rh	Sliced, peeled and made into curry. Also boiled and consumed during food scarcity	Misra et al. (2013); Misra and Misra (2014); Murthy et al. (2003)
419	*Hedyotis auriculata* L.	L	Leaves used as vegetable	Murthy et al. (2003)
420	*Helianthus annuus* L.	Fr.	Raw or roasted and eaten; roasted and made into powder with chillies	Singh (2013)
421	*Hemidesmus indicus* (L.) R. Br.	R	Roots are roasted, cooked, then consumed. Roots boiled several times and make it into a jelly, it is used in preparation of cool drinks in summer season. Powder of the roots used in coffee, etc.	Rajasab and Isaq (2004); Reddy et al. (2007); Samydurai et al. (2012); Deepa et al. (2014); Misra and Misra (2014)
422	*Heritiera littoralis* Aiton	S	Seeds eaten	Murthy et al. (2003)
423	*Hibiscus aculeatus* Walt.	L	Used as leafy vegetable	Singh (2013)
424	*Hibiscus cannabinus* L.	L, S	Leaves used as vegetable and dried seeds roasted and eaten	Singh (2013)
425	*Hibiscus rosa-sinensis* L.	Fl.	Flowers used to prepare chutney	Reddy et al. (2007)
426	*Hibiscus sabdariffa* L.	L	Tender leaves are cooked with tamarind pulp or chutney is prepared from raw leaves and the eaten	Misra and Misra (2013)
427	*Hibiscus surattensis* L.	Fr.	Used in curries to get sour taste	Murthy et al. (2003); Sadasivaiah (2009)
428	*Holarrhena pubescens* Wall.	Fl.	Used as vegetable	Sinha and Lakra (2005); Panda (2014)
429	*Holoptelia integrifolia* Planch.	Fl.	Flowers are used as vegetable	Murthy et al. (2003)
430	*Holostemma adakodien* Schultes	Rh, Fl., Fr., L	Cooked as vegetable; flowers eaten raw; fruits edible. Leaves used as vegetable	Murthy et al. (2003); Reddy et al. (2007); Samydurai et al. (2012)

TABLE 6.1 *(Continued)*

S. No.	Name of the plant	Edible Part (s)	Mode of preparation	References
431	*Homonoia riparia* Lour.	L	Tender leaves used as vegetable	Murthy et al. (2003); Reddy et al. (2007)
432	*Hordeum vulgare* L.	S	Cereal	Sahu et al. (2013)
433	*Hugonia mystax* L.	Fr.	Fruits edible	Murthy et al. (2003)
434	*Hydrolea zeylanica* (L.) Vahl	WP	Used as vegetable	Reddy et al. (2007); Misra and Misra (2013)
435	*Hygrophila auriculata* (Schum.) Heine	L	Used as vegetable	Murthy et al. (2003); Reddy et al. (2007); Misra and Misra (2013)
436	*Hygrophila salicifolia* Nees	L	Used as vegetable	Murthy et al. (2003)
437	*Hyptis suaveolens* (L.f.) Wall.	L	Used as vegetable	Murthy et al. (2003)
438	*Impatiens balsamina* L.	L, S	Leaves and seeds eaten raw	Murthy et al. (2003)
439	*Indigofera glabra* Roxb.	L	Used as vegetable.	Murthy et al. (2003)
440	*Indigofera glandulosa* Roxb. ex Willd.	L	Used as vegetable	Rajasab and Isaq (2004)
441	*Indigofera pulchella* Roxb.	Fl.	Flowers edible	Sinha and Lakra (2005)
442	*Ipomoea aquatica* Forssk.	L, St	Leaves and tender shoots are cooked and eaten	Murthy et al. (2003); Sinha and Lakra (2005); Reddy et al. (2007); Misra and Misra (2013)
443	*Ipomoea batatas* (L.) Lam.	T	Used as vegetable and also eaten raw	Samydurai et al. (2012); Dhore et al. (2012); Misra and Misra (2014)
444	*Ipomoea cairica* (L.) Sweet	T	Underground fleshy tubers are first boiled and then consumed as vegetable	Misra and Misra (2014)
445	*Ipomoea eriocarpa* R.Br.	L, St	Leaves and young shoots used as vegetable	Murthy et al. (2003)
446	*Ipomoea staphylina* Roem. & Schult.	R	Eaten as raw	Mohan and Kalidass (2010)
447	*Ixora arborea* Smith	Fr.	Ripe fruits edible	Reddy et al. (2007)
448	*Ixora undulata* Roxb.	Fr.	Ripe fruits edible	Panda (2014)
449	*Jasminum auriculatum* Vahl	L, St	Used as vegetable	Murthy et al. (2003)

TABLE 6.1 *(Continued)*

S. No.	Name of the plant	Edible Part (s)	Mode of preparation	References
450	*Jatropha heynei* Balakr.	L	Tender leaves used as vegetable	Murthy et al. (2003)
451	*Justicia betonica* L.	L	Tender leaves used as vegetable	Misra and Misra (2013)
452	*Justicia glauca* Rottl.	L	Tender leaves used as vegetable	Murthy et al. (2003)
453	*Justicia tranquebariensis* L.f.	L	Tender leaves used as vegetable	Deepa et al. (2014)
454	*Kalanchoe pinnata* (Lam.) Pers.	L	Leaves are used to prepare sauce along with other ingredients.	Rajasab and Isaq (2004)
455	*Kedrostis foetidissima* (Jacq.) Cogn.	T, L	Tubers used as vegetable, leaves eaten raw	Murthy et al. (2003)
456	*Lablab purpureus* (L.) Sweet	Fr.	Used as vegetable	Singh (2013)
457	*Lactuca runcinata* DC.	Wp	Used as vegetable	Murthy et al. (2003)
458	*Lactuca scariola* L.	L	Leaves used as vegetable, eaten raw (salad) or cooked.	Rajasab and Isaq (2004)
459	*Lagenaria siceraria* (Molina) Standly	Fr.	Tender fruits used as vegetable	Naidu and Khasim (2010); Singh (2013)
460	*Lannea coromandelica* (Houtt.) Merr.	Fr., L	Ripe fruits eaten raw. Leaves used as vegetable	Murthy et al. (2003); Reddy et al. (2007); Basha et al. (2009)
461	*Lantana camara* L. var. *aculeata* (L.) Moldenke	Fr.	Ripe fruits are edible	Murthy et al. (2003); Rajasab and Isaq (2004); Basha et al. (2009)
462	*Lantana montevidensis* (Spreng.) Briq.	Fr.	Ripe fruits are edible	Rekka and Senthil Kumar (2014)
463	*Lasia spinosa* (L.) Thwaites	T, Rh, L	Rhizome properly washed, spines are peeled, cut into small pieces, fried with tamarind, salt and chilly then consumed. Leaves are roasted and taken as food	Murthy et al. (2003); Reddy et al. (2007); Misra et al. (2013); Misra and Misra (2013); Misra and Misra (2014)
464	*Lathyrus odoratus* L.	S	Used to prepare curry	Singh (2013)
465	*Launaea procumbens* (Roxb.) Ramayya & Rajagopal	L	Leaves cooked as curry	Rajasab and Rajshekhar (2012)

TABLE 6.1 *(Continued)*

S. No.	Name of the plant	Edible Part (s)	Mode of preparation	References
466	*Leea asiatica* (L.) Ridsdale	L, Fr.	Leaves and fruits edible	Murthy et al. (2003)
467	*Leea indica* (Burm.) Merr.	L, Fr.	Leaves and fruits edible	Murthy et al. (2003)
468	*Leea macrophylla* Roxb. ex Hornem	L, Fr.	Leaves and fruits edible	Murthy et al. (2003)
469	*Lens culinaris* Medikus	S	Used as vegetable	Singh (2013)
470	*Lepisanthes tetraphylla* (Vahl) Radlk.	Fr.	Fruits edible	Murthy et al. (2003)
471	*Leptadenia reticulata* (Retz.)Wight & Arn.	L	Leaves used as vegetable	Nagalakshmi (2014)
472	*Leucas aspera* (L.) Link.	L, St	Leaves and young shoots are roasted and taken as food	Murthy et al. (2003); Reddy et al. (2007); Misra and Misra (2013, 2014)
473	*Leucas cephalotes* Spreng.	L	Leaves used as vegetable.	Sinha and Lakra (2005)
474	*Leucas decemdentata* (Willd.) Sm.	L, St	Leaves and young shoots are roasted and taken as food	Misra and Misra (2013)
475	*Limnophila indica* (L.) Druce	St, L	Leaves and young shoots are cooked and eaten	Murthy et al. (2003); Misra and Misra (2013);
476	*Limonia acidissima* Groff	L, Fr.	Leaves used to make pickles. Ripe fruits are edible with sugar or candy	Basha et al. (2009); Prabakaran et al. (2013); Singh (2013); Rekka and Senthil Kumar (2014); Deepa et al. (2014)
477	*Lippia nodiflora* Roxb.	L	Leaves edible	Murthy et al. (2003)
478	*Litsea glutinosa* C.B. Robins.	Fr.	Fruits edible	Murthy et al. (2003)
479	*Litsea monopetala* (Roxb.) Pers.	L	Leaves edible	Murthy et al. (2003)
480	*Lobelia alsinoides* Lam.	L	Leaves as vegetable	Murthy et al. (2003)
481	*Ludwigia adscendens* (L.) Hara	L, St	Young leaves and tender shoots as vegetable	Murthy et al. (2003)
482	*Luffa acutangula* (L.) Roxb.	Fr., L	Fruits as vegetable; tender leaves are mixed with fish and cooked as curry	Misra and Misra (2013)

TABLE 6.1 *(Continued)*

S. No.	Name of the plant	Edible Part (s)	Mode of preparation	References
483	*Luffa cylindrica* (L.) M. Roemer	Fr.	Used as vegetable	Singh (2013)
484	*Luffa hermaprodita* Singh & Bhandari	Fr.	Used as vegetable	Pandravada et al. (2014)
485	*Luffa tuberosa* Roxb.	Fr.	Used as vegetable	Murthy et al. (2003)
486	*Lycopersicon esculentum* Mill.	Fr.	Used as vegetable	Sadasivaiah (2009)
487	*Maclura cochinchinensis* (Lour.) Corner	L	Young leaves used as vegetable	Murthy et al. (2003)
488	*Macrotyloma uniflorum* (Lam.) Verdc.	S	Seeds are cooked and eaten	Naidu and Khasim (2010); Singh (2013)
489	*Madhuca indica* Gmel.	Fr., Fl, S	Ripe fruits and flowers used to prepare arrack. Seeds are edible.	Sinha and Lakra (2005); Reddy et al. (2007); Basha et al. (2009); Prabakaran et al. (2013)
490	*Madhuca longifolia* (Koen.) Macbr.	Fr., L	Ripe fruits edible. Leaves are roasted and taken as food. Flowers used as vegetable. Seeds edible.	Sinha and Lakra (2005); Misra and Misra (2013)
491	*Malva sylvestris* L.	L	Leaves used as vegetable	Murthy et al. (2003)
492	*Malvastrum coromandelianum* (L.) Garcke	L	Tender leaves used as vegetable	Reddy et al. (2007)
493	*Mangifera indica* L.	Fr.	Green fruits used to prepare pickles, chutnies, dal and ripe fruits edible	Murthy et al. (2003); Basha et al. (2009); Prabakaran et al. (2013); Deepa et al. (2014); Singh (2013, 2014)
494	*Manihot esculenta* Crantz	T	Fleshy swollen/tuberous roots are consumed raw or after boiling	Samydurai et al. (2012); Misra and Misra (2014); Rekka and Senthil Kumar (2014)
495	*Manilkara hexandra* (Roxb.) Dubard.	Fr.	Ripe fruits edible	Murthy et al. (2003); Reddy et al. (2007); Basha et al. (2009)
496	*Manilkara roxburghii* (Wight) Dubard.	Fr.	Ripe fruits edible	Basha et al. (2009)

TABLE 6.1 *(Continued)*

S. No.	Name of the plant	Edible Part (s)	Mode of preparation	References
497	*Maranta arundinacea* L.	Rh	Cooked as vegetable	Samydurai et al. (2012)
498	*Marsilea polycarpa* Hook. & Grev.	L	Used as vegetable	Misra and Misra (2013)
499	*Marsilea quadrifolia* L.	L	Used as vegetable	Murthy et al. (2003); Sinha and Lakra (2005); Misra and Misra (2013); Deepa et al. (2014)
500	*Maytenus emarginatus* (Willd.) Ding Hou	Fr.	Ripe fruits edible	Murthy et al. (2003); Mukesh Kumar et al. (2013)
501	*Melastoma malabathricum* L.	L, Fr.	Tender leaves used as vegetable. Fruits edible	Murthy et al. (2003); Reddy et al. (2007); Nayak and Basak (2015)
502	*Melia azedarach* L.	Fr.	Ripe fruits edible	Prabakaran et al. (2013)
503	*Melicope lunuankenda* (Gaertn.) T.G. Hartley	L	Leaves used as flavoring agent	Murthy et al. (2003)
504	*Melochia corchorifolia* L.	L	Used as vegetable	Murthy et al. (2003); Ramakrishna et al. (2014)
505	*Memecylon edule* Roxb.	Fr.	Edible	Murthy et al. (2003); Basha et al. (2009)
506	*Memecylon umbellatum* Burm.f.	Fr.	Edible	Murthy et al. (2003); Basha et al. (2009)
507	*Mentha arvensis* L.	L	Used as leafy vegetable and spice	Rao (2014)
508	*Mentha spicata* L.	L	Used as vegetable and spice and used to prepare chutney	Sanyasi Rao et al. (2014)
509	*Menya laxiflora* Robyns	L	Used vegetable	Sinha and Lakra (2005)
510	*Merremia emarginata* (Burm.f.) Hallier f.	L, St	Leaves and tender shoots cooked and eaten	Murthy et al. (2003); Misra and Misra (2013)
511	*Merremia gangetica* (L.) Cuf.	L	Used as vegetable	Rajasab and Isaq (2004)
512	*Merremia vitifolia* (Burm.f.) Hall.f.	L	Tender leaves used as vegetable	Murthy et al. (2003)
513	*Mesua ferrea* L.	Fr.	Ripe fruits are edible	Panda (2014)
514	*Metroxylon sagu* Rottb.	P	Prepare juice, kheer	Rao (2014)

TABLE 6.1 *(Continued)*

S. No.	Name of the plant	Edible Part (s)	Mode of preparation	References
515	*Miliusa tomentosa* (Roxb.) Sinclair	Fr.	Ripe fruits eaten as raw	Murthy et al. (2003); Reddy et al. (2007)
516	*Miliusa velutina* (Dunal) Hook.f. & Thoms.	Fr.	Ripe fruits eaten as raw	Murthy et al. (2003)
517	*Mimosa intsia* L.	Fr.	Fruits are edible	Prabakaran et al. (2013)
518	*Mimusops elengi* L.	Fr.	Ripe fruits are edible	Murthy et al. (2003); Nayak and Basak (2015)
519	*Mollugo cerviana* (L.) Ser.	St, L	Tender shoots and leaves used as vegetable	Murthy et al. (2003)
520	*Mollugo pentaphylla* L.	L	Cut into small pieces and cooked with salt, chilly and with oil	Murthy et al. (2003); Misra and Misra (2013)
521	*Momordica charantia* L.	Fr., L	Fruits used as vegetable; Leaves eaten after frying	Misra and Misra (2013); Singh (2013)
522	*Momordica cymbalaria* Hook.f.	Fr.	Used as vegetable.	Rajasab and Isaq (2004)
523	*Momordica dioica* Roxb. ex Willd.	T, Fr.	Used as vegetable; Tender fruit curry is utilized as blood purifer	Murthy et al. (2003); Sinha and Lakra (2005); Reddy et al. (2007); Naidu and Khasim (2010)
524	*Monochoria hastata* (L.) Solms	L	Leaves are eaten after roasting	Misra and Misra (2013)
525	*Morchella esculenta*	Wp	Whole plant is edible (Mushroom)	Vasundhra (2009)
526	*Morinda citrifolia* L.	Fr.	Fruits edible	Nayak and Basak (2015)
527	*Morinda pubescens* J.E. Smith	Fr.	Ripe fruits edible	Murthy et al. (2003); Reddy et al. (2007); Deepa et al. (2014)
528	*Morinda umbellata* L.	L	Used as vegetable	Murthy et al. (2003)
529	*Moringa concanensis* Nimmo ex Gibs.	L, Fr.	Fruits and leaves used as vegetables	Murthy et al. (2003); Basha et al. (2009); Deepa et al. (2014)
530	*Moringa oleifera* Bedd.	L, Fr.,	Fruits and leaves used as vegetables	Murthy et al. (2003); Reddy et al. (2007); Naidu and Khasim (2010); Misra and Misra (2013, 2014)
531	*Morus alba* L.	Fr.	Ripe fruits edible	Rao (2014)

TABLE 6.1 *(Continued)*

S. No.	Name of the plant	Edible Part (s)	Mode of preparation	References
532	*Mucuna atropurpurea* DC.	S	Eaten raw	Singh (2013)
533	*Mucuna pruriens* (L.) DC.	Fr., S	Unripe fruits roasted and consumed	Murthy et al. (2003); Reddy et al. (2007)
534	*Mukia maderaspatana* (L.) Roemer	L, Fr.	Leaves and seeds used to prepare curries. Fruits are edible	Murthy et al. (2003); Rajasab and Isaq (2004); Deepa et al. (2014)
535	*Muntigia calabura* L.	Fr.	Ripe fruits eaten as raw	Rao (2014)
536	*Murraya koenigii* (L.) Spreng.	L, Fr.	Leaves used in curries and made into powder; ripe fruits are edible	Murthy et al. (2003); Reddy et al. (2007); Basha et al. (2009); Misra and Misra (2013, 2014); Singh (2013)
537	*Musa X paradisiaca* L.	Rh, Fr.	Fruits edible. Rhizome boiled and cooked then eaten as vegetable	Naidu and Khasim (2010); Misra and Misra (2014); Singh (2014)
538	*Naravelia zeylanica* (L.) DC.	T	Used as vegetable	Murthy et al. (2003)
539	*Naringi crenulata* (Roxb.)	Fr.	Ripe fruits are edible	Prabakaran et al. (2013)
540	*Nelsonia canescens* (Lam.) Spreng.	L	Used as vegetable	Murthy et al. (2003)
541	*Nelumbo nucifera* Gaertn.	Rh, Fl, S	Rhizome boiled and cooked then eaten. Stamens and thalamus eaten raw. Rhizomes used as vegetable	Murthy et al. (2003); Reddy et al. (2007); Samydurai et al. (2012); Misra and Misra (2014)
542	*Neptunia oleracea* Lour.	L	Tender leaves and young pods used as vegetable	Murthy et al. (2003); Reddy et al. (2007); Misra and Misra (2013)
543	*Nervilia aragoana* Gaud.	T	Burnt and taken as snacks, also boiled and taken with salt	Misra et al. (2013)
544	*Nervilia discolor* (Bl.) Schltr.	T	Tuber burnt and eaten	Misra et al. (2013)
545	*Nothopegia heyneana* (Hook.f.) Gamble	Fr.	Ripe fruits edible.	Murthy et al. (2003)
546	*Nothosaerva brachiata* (L.) Wight & Arn.	L	Used as vegetable	Murthy et al. (2003); Reddy et al. (2007)

TABLE 6.1 *(Continued)*

S. No.	Name of the plant	Edible Part (s)	Mode of preparation	References
547	*Nymphaea nouchali* Burm. f.	Rh, Fr.	Rhizome is eaten after boiling. Fruits edible raw	Murthy et al. (2003); Misra and Misra (2014);
548	*Nymphaea pubescens* Willd.	Rh, S	Sliced and cooked as curry with potato and onions. Rhizome locally called madhi is eaten after boiling.	Murthy et al. (2003); Basha et al. (2009); Misra et al. (2013); Misra and Misra (2014)
549	*Ocimum americanum* L.	Wp	Used as vegetable	Murthy et al. (2003)
550	*Olax imbricata* Roxb.	Fr.	Aril is edible	Hebber et al. (2010)
551	*Olax scandens* Roxb.	St, L	Tender stems used as vegetable; Leaves are roasted then eaten	Murthy et al. (2003); Sinha and Lakra (2005); Reddy et al. (2007); Misra and Misra (2013)
552	*Oldenlandia trinervia* Retz.	L	Leaves used as vegetable	Misra and Misra (2013)
553	*Operculina turpethum* (L.) Silva Manso	L	Leaves with rice is made into cake and eaten	Murthy et al. (2003); Misra and Misra (2013)
554	*Opuntia dillenii* (Ker-Gawl.) Haw.	Fr.	Ripe fruits eaten raw and used to prepare jelly and is used as coloring agent for cool drinks	Murthy et al. (2003); Rajasab and Isaq (2004); Reddy et al. (2007); Basha et al. (2009); Deepa et al. (2014)
555	*Opuntia elatior* Mill.	Fr.	Ripe fruits edible	Rekka and Senthil Kumar (2014)
556	*Opuntia vulgaris* Mill.	Fr.	Ripe fruits edible	Basha et al. (2009)
557	*Oroxylum indicum* (L.) Vent.	Fl.	Used as vegetable	Murthy et al. (2003); Reddy et al. (2007)
558	*Orthosiphon rubicundus* (D.Don) Benth.	R	Sliced pieces are ground along with broken rice, made flour and fried into cakes.	Murthy et al. (2003); Misra et al. (2013)
559	*Oryza jeyporensis* Govindasw. & Chandrasekh.	S	Used as food grains	Reddy et al. (2006)
560	*Oryza rufipogon* Griff.	S	Used as food grains	
561	*Oryza sativa* L.	S	Food grains	Singh (2014)
562	*Osbeckia stellata* Buch.-Ham. ex Ker Gawl	L	Used as vegetable	Murthy et al. (2003)

TABLE 6.1 *(Continued)*

S. No.	Name of the plant	Edible Part (s)	Mode of preparation	References
563	*Ottelia alismoides* (L.) Pers.	Fr.	Eaten raw	Murthy et al. (2003); Sadasivaiah (2009)
564	*Oxalis corniculata* L.	L	Cut into small pieces and cooked with salt, chilly and with oil and eaten along with boiled rice	Murthy et al. (2003); Reddy et al. (2007); Misra and Misra (2013); Rekka and Senthil Kumar (2014); Deepa et al. (2014)
565	*Oxalis latifolia* HBK	L	Used as vegetable	Rekka and Senthil Kumar (2014)
566	*Oxystelma esculentum* R.Br.	L	Used as vegetable	Murthy et al. (2003)
567	*Paederia foetida* L.	L	Leaves used as vegetable	Reddy et al. (2007)
568	*Pandanus tectorius* Parkinson ex Du Roi	St	Tender shoots used as vegetable	Murthy et al. (2003)
569	*Panicum miliaceum* L.	S	Used as food grains	Sadasivaiah (2009); Singh (2013)
570	*Panicum sumatrense* Roem. & Schult.	S	Food grains	Murthy et al. (2003)
571	*Papaver somniferum* L.	S	Used as spice	Rao (2014)
572	*Paracalyx scariosus* (Roxb.) Ali	S, T	Seeds used as pulse	Murthy et al. (2003); Murthy and Sambasiva Rao (2009)
573	*Parkinsonia aculeata* L.	S	Seeds eaten raw and roasted and eaten	Murthy et al. (2003)
574	*Paspalum scrobiculatum* L.	S	Used as food grains	Sadasivaiah (2009); Singh (2013)
575	*Passiflora edulis* Sims	Fr.	Ripe fruits are eaten. Juice prepared from ripe fruits	Rekka and Senthil Kumar (2014)
576	*Passiflora foetida* L.	Fr.	Ripe fruits edible directly	Murthy et al. (2003)
577	*Pavetta indica* L.	Fl., Fr.	Flowers used to prepare curry; tender unripe fruits are edible	Murthy et al. (2003); Reddy et al. (2007)
578	*Pavonia odorata* Willd.	L	Eaten raw	Murthy et al. (2003)
579	*Pelatantheria insectifera* (Reichb. f.) Ridley	AR	Stripped off the skin, burnt and taken as snacks	Misra et al. (2013)
580	*Pennisetum glaucum* (L.) R. Br.	S	Food grains	Naidu and Khasim (2010); Singh (2013, 2014);

TABLE 6.1 *(Continued)*

S. No.	Name of the plant	Edible Part (s)	Mode of preparation	References
581	*Pergularia daemia* (Frossk.) Chiov.	L	As vegetable	Nagalakshmi (2014); Murthy et al. (2003)
582	*Persea americana* Mill.	Fr.	Juice prepared from ripe fruits	Rekka and Senthil Kumar (2014)
583	*Persicaria barbata* (L.) H. Hara	L, St	Tender leaves and shoots are cooked then eaten	Misra and Misra (2013);
584	*Persicaria glabra* (Willd.) M.Gomez	Rh, St	Rhizome consumed as vegetable.	Murthy et al. (2003); Misra and Misra (2014)
585	*Phoenix acaulis* Buch.-Ham. ex Roxb.	R, Fr.,	Fruits are edible; peeled and eaten raw as snacks by children	Reddy et al. (2007); Misra andMisra (2013); Misra et al. (2013)
586	*Phoenix loureiroi* Kunth.	Fr., P	Ripe fruits eaten raw; pith also eaten raw	Reddy et al. (2007); Basha et al. (2009); Rekka and Senthil Kumar (2014); Deepa et al. (2014)
587	*Phoenix pusilla* Gaertner	Fr., P	Ripe fruits eaten raw; pith eaten raw	Rao (2014)
588	*Phoenix sylvestris* (L.) Roxb.	Fr., P	Ripe fruits eaten raw; pith eaten raw	Murthy et al. (2003); Reddy et al. (2007); Basha et al. (2009); Naidu and Khasim (2010); Singh (2013)
589	*Phyla nodiflora* Lour.	L	Leaves used as vegetable	Misra and Misra (2013);
590	*Phyllanthus acidus* (L.) Skeels	Fr.	Unripe fruits used in preparation of pickle and ripe fruits edible	Rao (2014); Nayak and Basak (2015)
591	*Phyllanthus emblica* L.	Fr.	Unripe fruits used in preparation of pickle and ripe fruits eat directly	Murthy et al. (2003); Reddy et al. (2007); Basha et al. (2009); Prabakaran et al. (2013); Singh (2013); Deepa et al. (2014)
592	*Phyllanthus indofischeri* Bennet	Fr.	Unripe fruits used in preparation of pickle and ripe fruits eaten raw	Reddy et al. (2006)
593	*Phyllanthus reticulatus* Poir.	Fr.	Ripe fruits edible	Murthy et al. (2003); Deepa et al. (2014)
594	*Phyllanthus virgatus* Forst.f.	Fr.	Ripe fruits edible	Murthy et al. (2003)
595	*Physalis angulata* L.	Fr.	Ripe fruits edible	Reddy et al. (2007)

TABLE 6.1 *(Continued)*

S. No.	Name of the plant	Edible Part (s)	Mode of preparation	References
596	*Physalis minima* L.	L, Fr.	Ripe fruits edible, leaves used as vegetable	Murthy et al. (2003); Reddy et al. (2007); Basha et al. (2009); Deepa et al. (2014)
597	*Physalis minima* L. var. *indica* Clarke	Fr.	Ripe fruits edible	Reddy et al. (2007)
598	*Pilea melastomoides* Bl.	L	Leaves used as vegetable	Murthy et al. (2003)
599	*Pimpinella heyneana* (DC.) Benth.	S, L	Used as spices	Murthy et al. (2003)
600	*Piper longum* L.	S	Used as spice	Naidu and Khasim (2010)
601	*Piper nigrum* L.	Fr.	Used as condiment	Deepa et al. (2014); Sanyasi Rao et al. (2014)
602	*Pistia stratiotes* L.	L	Used as vegetable	Nagalakshmi (2014)
603	*Pisum sativum* L.	S	Unripe seeds edible, ripe seeds as pulse	Rao (2014)
604	*Pithecellobium dulce* (Roxb.) Benth.	Fr.	Aril edible	Murthy et al. (2003);Reddy et al. (2007); Basha et al. (2009); Prabakaran et al. (2013); Deepa et al. (2014)
605	*Plantago asiatica* L.	L	Pot herb	Murthy et al. (2003)
606	*Plectranthus barbatus* Andr.	T	Used as vegetable	Murthy et al. (2003)
607	*Plectronia didyma* Kurz.	Fr.	Ripe fruits eaten as raw	Rekka and Senthil Kumar (2014)
608	*Plumbago zeylanica* L.	R	Cooked as vegetable	Samydurai et al. (2012)
609	*Polyalthia cerasoides* (Roxb.) Bedd.	Fr.	Ripe fruits edible	Reddy et al. (2007); Basha et al. (2009); Prabakaran et al. (2013); Murthy et al. (2003)
610	*Polyalthia suberosa* (Roxb.) Thw.	Fr.	Ripe fruits eaten raw	Nayak and Basak (2015)
611	*Polygala arvensis* Willd.	L	Used as vegetable	Misra and Misra (2013)
612	*Polygonum chinense* L.	L	Used as vegetable	Murthy et al. (2003)
613	*Polygonum hydropiper* Willd.	L	Used as vegetable	Murthy et al. (2003)

TABLE 6.1   *(Continued)*

S. No.	Name of the plant	Edible Part (s)	Mode of preparation	References
614	*Polygonum plebeium* R. Br.	L, Fl	Used as vegetable	Murthy et al. (2003); Sinha and Lakra (2005); Panda (2014)
615	*Polygonum pulchrum* Bl.	L	Used as salads	Murthy et al. (2003)
616	*Portulaca oleracea* L.	L, St	As leafy vegetable, tender leaves and shoots are roasted and eaten	Murthy et al. (2003); Reddy et al. (2007); Misra and Misra (2013); Deepa et al. (2014)
617	*Portulaca quadrifida* L.	L, S	Boiled, squeezed and mixed with Ground nut powder and eat with boiled rice	Murthy et al. (2003);Rajasab andIsaq (2004); Reddy et al. (2007)
618	*Portulaca tuberosa* Roxb.	L	Used as vegetable	Murthy et al. (2003)
619	*Pouteria sapota* (Jacq.) H.E.Moore & Stearn	Fr.	Ripe fruits are edible	Rekka and Senthil Kumar (2014)
620	*Pouzolzia zeylanica* (L.) Benn.	Rh	Rhizome consumed as vegetable	Reddy et al. (2007); Misra and Misra (2014)
621	*Premna latifolia* Roxb.	L	Leaves used to prepare curries	Murthy et al. (2003); Reddy et al. (2007); Misra and Misra (2014)
622	*Premna mollissima* Roth	L	*Ambila* (liquid curry) is prepared from leaves with tamarind pulp	Misra and Misra (2013)
623	*Premna tomentosa* Willd.	Fr., L	Ripe fruits are edible; tender leaves used as vegetable	Murthy et al. (2003); Reddy et al. (2007); Prabakaran et al. (2013); Deepa et al. (2014)
624	*Protium serratum* (Colebr.) Engl.	Fr.	Fruits eaten raw	Murthy et al. (2003)
625	*Prunus jenkinsii* Hook.f.	Fr.	Fruits eaten raw	Murthy et al. (2003)
626	*Pseudarthria viscida* (L.) Wight & Arn.	Fr.	Fruits eaten raw	Murthy et al. (2003)
627	*Psidium guajava* L.	Fr.	Ripe fruits edible	Deepa et al. (2014)
628	*Pterolobium hexapetalum* (Roth) Santapu & Wagh	L, Fr.	Tender leaves used as vegetable, fruits are eaten raw	Prabakaran et al. (2013); Deepa et al. (2014)

TABLE 6.1 *(Continued)*

S. No.	Name of the plant	Edible Part (s)	Mode of preparation	References
629	*Pueraria tuberosa* (Roxb. ex Willd.) DC.	T	Dried powder is taken with water. Also consumed raw or burnt as snacks	Reddy et al. (2007); Misra et al. (2013); Singh (2013)
630	*Punica grantum* L.	Fr.	Ripe fruits edible directly	Murthy et al. (2003); Rao (2014); Singh (2014)
631	*Pupalia lappacea* (L.) Juss.	L	Tender leaves are cooked then eaten	Misra and Misra (2013); Murthy et al. (2003)
632	*Pyrus communis* L.	Fr.	Ripe fruits are edible	Rekka and Senthil Kumar (2014)
633	*Radermachera xylocarpa* (Roxb.) K. Schum.	Fr.	Tender fruits used as vegetable	Murthy et al. (2003)
634	*Raphanus raphanistrum* L. ssp. *sativus* (L.) Domin	Rh, L	Rhizomes sliced, roasted or cooked with spice then consumed; leaves fried along with other vegetable and eaten	Misra and Misra (2013, 2014)
635	*Remusatia vivipara* (Roxb.) Schott	C	Sliced, boiled successively till irritation disapperas and cooked with spice	Misra et al., 201
636	*Rhaphidophora pertusa* (Roxb.) Schott.	Fr.	Fruits eaten raw	Murthy et al. (2003)
637	*Rhizophora mucronata* Lam.	St, Fl,	Used as vegetable	Murthy et al. (2003)
638	*Rhus mysorensis* G. Don	Fr.	Ripe fruits eaten raw	Rao (2014); Murthy et al. (2003)
639	*Rhynchosia bracteata* Benth.	S	Seeds used as pulses	Murthy and Emmanuel (2011)
640	*Rhynchosia cana* DC.	S	Eaten raw	
641	*Rhynchosia filipes* Benth.	S	Eaten Raw	
642	*Rhynchosia heynei* Wight & Arn.	Fr.	Eaten Raw	Murthy et al. (2003);
643	*Rhynchosia rufescens* (Willd.) DC.	S	Eaten Raw	
644	*Rhynchosia suaveolens* (L.f.) DC.	S	Eaten Raw	
645	*Ricinus communis* L.	S	Seeds used in masala curries	Singh (2013)

TABLE 6.1   *(Continued)*

S. No.	Name of the plant	Edible Part (s)	Mode of preparation	References
646	*Rivea hypocrateri-formis* (Desr.) Choisy	L, Fr.	Tender leaves and fruits are used as vegetable or raw.	Murthy et al. (2003)
647	*Rivea ornata* Choisy	L	Cooked as vegetable	Reddy et al. (2007)
648	*Rothia indica* (L.) Druce	L	Leaves used as vegetable	
649	*Rubus ellipticus* Sm.	Fr.	Ripe fruits edible	Murthy et al. (2003)
650	*Rumex vesicarius* L.	L	Used as vegetable	Rao (2014)
651	*Rungia pectinata* (L.) Nees	L	Tender leaves cooked and eaten	Misra and Misra (2013)
652	*Saccharum officinarum* L.	St	Edible raw, sugar and jaggery edible, candy is prepared from juice	Singh (2013, 2014)
653	*Salacia chinensis* L.	Fr.	Fruits edible	Reddy et al. (2007)
654	*Salvadora persica* L.	L	Used as vegetable	Murthy et al. (2003)
655	*Santalum album* L.	Fr.	Ripe fruits edible	Rajasab and Isaq (2004)
656	*Sapindus emar-ginatus* Vahl	S	Seeds roasted and eaten raw	
657	*Schleichera oleosa* (Lour.)Oken	Fr.	Edible	Murthy et al. (2003); Reddy et al. (2007); Prabakaran et al. (2013)
658	*Schrebera swi-etenioides* Roxb.	Fr.	Fruits are edible	Murthy et al. (2003); Reddy et al. (2007)
659	*Scolopia crenata* Clos.	Fr.	Fruits are edible	Murthy et al. (2003)
660	*Scutia myrtina* (Burm.f.) Kurz.	Fr.	Ripe fruits edible	Murthy et al. (2003); Reddy et al. (2007); Basha et al. (2009); Prabakaran et al. (2013); Deepa et al. (2014)
661	*Sechium edule* (Jacq.) Sw.	Fr.	Used as vegetable	Rao (2014)
662	*Securinega leu-copyrus* (Willd.) Muell.-Arg.	Fr.	Ripe fruits edible	Murthy et al. (2003); Reddy et al. (2007); Basha et al. (2009); Rekka and Senthil Kumar (2014)

TABLE 6.1 *(Continued)*

S. No.	Name of the plant	Edible Part (s)	Mode of preparation	References
663	*Semecarpus anacardium* L.f.	Fr., S	Ripe fruits eaten raw. Seeds also edible.	Murthy et al. (2003); Sinha and Lakra (2005); Reddy et al. (2007); Basha et al. (2009); Naidu and Khasim (2010); Prabakaran et al. (2013); Rekka and Senthil Kumar (2014)
664	*Senna alexandrina* Mill.	L	Tender leaves used as vegetable	Deepa et al. (2014)
665	*Senna occidentalis* (L.) Link	Fr., L, St	Fruits used as vegetable; leaves and young shoots fried then eaten	Misra and Misra (2013)
666	*Senna sophera* (L.) Roxb.	L	Leaves are cooked and then taken as food	Misra and Misra (2013);
667	*Sesamum indicum* L.	S	Used as oil seed, raw seeds eaten	Sadasivaiah (2009); Singh (2013)
668	*Sesbania grandiflora* (L.) Poir.	L, S, Fl.	Leaves, flowers and seeds used as vegetables	Misra and Misra (2013); Rajasab and Isaq (2004)
669	*Sesbania sesban* (L.) Poir.	L	Leaves used as vegetable	Murthy et al. (2003)
670	*Sesuvium portulacastrum* L.	L	Leaves used as vegetable	Murthy et al. (2003)
671	*Setaria italica* (L.) P. Beauv.	S	Used as food grains	Naidu and Khasim (2010); Singh (2013)
672	*Setaria verticillata* (L.) P. Beauv.	S	Used as food grains	Singh (2013)
673	*Shorea robusta* Gaertn.	S, Fr.	Raw fruit and seed are used as vegetable	Murthy et al. (2003); Sinha and Lakra (2005); Panda (2014)
674	*Sida cordata* Burm.f	L	Leaves used as vegetable	Panda (2014)
675	*Smilax zeylanica* L.	R, Fr., S	Roots cooked with other vegetable. Ripe fruits edible. Young seedlings used as vegetable	Misra et al. (2013); Misra and Misra (2014); Murthy et al. (2003)
676	*Solanum americanum* Mill.	L	Cooked with redgram and eaten	Nagalakshmi (2014)
677	*Solanum anguivi* Lam.	Fr.	Fruits cooked and eaten	Murthy et al. (2003)
678	*Solanum erianthum* D. Don	Fr.	Unripe fruits used as vegetable	Murthy et al. (2003)

TABLE 6.1   *(Continued)*

S. No.	Name of the plant	Edible Part (s)	Mode of preparation	References
679	*Solanum indicum* L.	Fr.	Unripe fruits used as vegetable	Murthy et al. (2003)
680	*Solanum melongena* L.	Fr.	Used as vegetable	Singh (2013)
681	*Solanum melongena* L. var. *incanum* (L.) Kuntze	Fr.	Used as vegetable	Sadasivaiah (2009)
682	*Solanum melongena* L. var. *insanum* (L.) Prain	Fr.	Unripe fruits as vegetable	Sadasivaiah (2009)
683	*Solanum nigrum* L.	Fr., L	Ripe fruits eaten raw; leaves used as vegetable; unripe fruits as vegetable	Murthy et al. (2003); Reddy et al. (2007); Basha et al. (2009); Deepa et al. (2014); Ramakrishna et al. (2014)
684	*Solanum pubescens* Willd.	Fr.	Unripe fruits used as vegetable and ripe fruits eaten directly	Murthy et al. (2003); Basha et al. (2009); Deepa et al. (2014)
685	*Solanum surattense* Burm. f.	Fr.	Used as vegetable and cooked as curries	Murthy et al. (2003)
686	*Solanum torvum* Sw.	Fr., L, St	Leaves and young shoots are cooked with salt and chilly and eaten. Green fruits salted, dried, roasted in oil and eaten	Murthy et al. (2003); Misra and Misra (2013); Rekka and Senthil Kumar (2014); Nayak and Basak (2015)
687	*Solanum trilobatum* L.	L, T, Fr.	Underground fleshy tubers and fruits sliced, boiled, cooked as curry then consumed as vegetable. Leaves also used as vegetable	Deepa et al. (2014)
688	*Solanum tuberosum* L.	T	Underground fleshy tubers sliced, boiled, cooked as curry then consumed as vegetable.	Misra and Misra (2014)
689	*Solanum viarum* Dunal	Fr.	Fruit used as vegetable.	Panda (2014)
690	*Solanum virginianum* L.	Fr.	Fruits used as vegetable	Reddy et al. (2007); Mukesh Kumar et al. (2013)
691	*Solena amplexicaulis* (Lam.) Gandhi	T, L, Fr.	Tubers eaten raw or burnt; cooked as curry with other vegetables. Leaves are roasted and taken as food	Murthy et al. (2003); Sinha and Lakra (2005); Hebber et al. (2010); Misra et al. (2013); Misra and Misra (2013, 2014)

TABLE 6.1 *(Continued)*

S. No.	Name of the plant	Edible Part (s)	Mode of preparation	References
692	*Sorghum bicolor* (L.) Moench.	S	Used as food grains	Naidu and Khasim (2010); Singh (2013, 2014)
693	*Sorghum halepense* Pers.	S	Seeds used to prepare flour, which is mixed with flour of sorghum	Rajasab and Isaq (2004)
694	*Soymida febrifuga* A. Juss.	Fr.	Fruits edible	Dhore et al. (2012)
695	*Spermacoce articularis* L.f.	L	Leaves used as vegetable	Murthy et al. (2003)
696	*Spermacoce hispida* L.	L	Young leaves are fried, then eaten	Murthy et al. (2003); Misra and Misra (2013)
697	*Sphenoclea zeylanica* Gaertn.	St	Younger shoots used as vegetable	Murthy et al. (2003)
698	*Sphaeranthus indicus* L.	L, St, S	Seedlings, leaves and tender shoots used as vegetable	Murthy et al. (2003); Misra and Misra (2013);
699	*Spinacea oleracea* L.	L	Used as vegetable	Murthy et al. (2003)
700	*Spondias pinnata* (L.f) Kurz.	Fr.	Ripe fruits are edible, made into pickle. Leaves used to prepare chutney.	Rao (2014)
701	*Stemona tuberosa* Lour.	T	Tubers eaten after cooking	Samydurai et al. (2012)
702	*Sterculia guttata* Roxb. ex DC.	S	Roasted and eaten	Hebber et al. (2010)
703	*Sterculia urens* Roxb.	G, S	Gum used in preparation of sweet. Seeds and gum consumed directly	Murthy et al. (2003)
704	*Sterculia villosa* Roxb.	S	Seeds roasted and eaten	Murthy et al. (2003)
705	*Streblus asper* Lour.	Fl, Fr.	Flowers are used as vegetable. Ripe fruit is edible.	Murthy et al. (2003); Sinha and Lakra (2005); Panda (2014); Nayak and Basak (2015)
706	*Streblus taxoides* (Roth) Kurz	Fr.	Ripe fruit is edible.	Nayak and Basak (2015);
707	*Strychnos potato-rum* L.f.	Fr., S	Fruit pulp eaten raw	Reddy et al. (2007); Prabakaran et al. (2013); Murthy et al. (2003)
708	*Suaeda maritima* (L.) Dunn.	L	Leaves used as vegetable	Murthy et al. (2003)

TABLE 6.1 *(Continued)*

S. No.	Name of the plant	Edible Part (s)	Mode of preparation	References
709	*Suaeda monoica* Forsk. ex Gmel.	L	Leaves used as vegetable	Murthy et al. (2003)
710	*Suregada multiflora* (A. Juss.) Baill.	Fr.	Fruits edible	Sinha and Lakra (2005)
711	*Synedrella nodiflora* Gaertn.	L	Leaves eaten raw	Murthy et al. (2003)
712	*Syzygium alternifolium* (Wight) Walp.	Fr.	Ripe fruits edible	Murthy et al. (2003); Basha et al. (2009); Singh (2013)
713	*Syzygium aromaticum* (L.) Merr. & Perry	Flb	Used as spice	Singh (2013)
714	*Syzygium cumini* (L.) Skeels	Fr.	Ripe fruits edible	Murthy et al. (2003); KN. Reddy et al. (2007); Basha et al. (2009); Prabakaran et al. (2013); Singh (2013); Deepa et al. (2014)
715	*Syzygium jambos* (L.) Alston	Fr.	Ripe fruits edible	Singh (2013)
716	*Syzygium nervosum* A. Cunn ex DC.	Fr.	Ripe fruits edible	Murthy et al. (2003); Sinha and Lakra (2005)
717	*Syzygium salicifolium* (Wight) J. Graham	Fr.	Ripe fruits edible	Murthy et al. (2003)
718	*Syzygium samarangense* (Bl.) Merr. & Perry.	Fr.	Ripe fruits edible	Murthy et al. (2003)
719	*Tacca leontopetaloides* (L.) O. Kuntze	T	Tubers consumed as vegetable	Reddy et al. (2007)
720	*Tali minor* (Gaertn.) Almeida	Fr.	Aril is edible	Hebber et al. (2010)
721	*Talinum portulacifolium* (Forsk.) Asch.& Schweinf	L	Leaves used as vegetable	Murthy et al. (2003)
722	*Tamarindus indica* L.	L, Fl., Fr., S	Tender leaves and flowers, tender fruits used as vegetable. Ripe fruits yield tamarind which is used in most of the curries. Seeds roasted and soaked in water for two days and eaten; green fruits are used in pickles	Murthy et al. (2003); Basha et al. (2009); Prabakaran et al. (2013); Misra and Misra (2013); Deepa et al. (2014)

TABLE 6.1 *(Continued)*

S. No.	Name of the plant	Edible Part (s)	Mode of preparation	References
723	*Tamilnadia uliginosa* (Retz.) Tirveng.	Fr.	Fruits used in preparation of curry. Unripe fruits roasted and eaten.	Murthy et al. (2003); Reddy et al. (2007); Hebber et al. (2010)
724	*Tarenna asiatica* (L.) Kunize ex Schumann	Fr.	Tender unripe fruits edible	Murthy et al. (2003); Prabakaran et al. (2013); Deepa et al. (2014)
725	*Tephrosia purpurea* (L.) Pers.	L	Leaves used as vegetable	Misra and Misra (2013)
726	*Teramnus labialis* (L. f.) Spreng	S, L	Eaten as raw and Vegetable	Murthy et al. (2003)
727	*Terminalia alata* Heyne ex Roth	Gum	Gum is edible	Sinha and Lakra (2005)
728	*Terminalia bellirica* (Gaertn.) Roxb.	Fr., S	Fruits edible; Seeds and kernal eaten raw	Murthy et al. (2003); Prabakaran et al. (2013); Deepa et al. (2014); Sanyasi Rao et al. (2014)
729	*Terminalia catappa* L.	S	Seeds eaten raw	Rao (2014)
730	*Terminalia chebula* Retz.	S, Fr.	Seeds and kernal eaten as raw. Pickle is made from fruits	Deepa et al. (2014)
731	*Terminalia citrina* Roxb. ex Fleming	Fr.	Fruits edible	Nayak and Basak (2015)
732	*Tetrastigma lanceolarium* Planch.	L, Fr.	Leaves and fruits used as raw	Murthy et al. (2003)
733	*Theobroma cacao* L.	Fr.	Ripe fruits edible	Prabakaran et al. (2013)
734	*Theriophonum indicum* (Dalz.) Engler	T	Tubers are edible	Sinha and Lakra (2005)
735	*Tinospora cordifolia* (Willd.) Hook.f. & Thoms.	L	Leaves used as vegetable	Reddy et al. (2007)
736	*Toddalia asiática* (L.) Lam.	L, Fr.	Leaves used as vegetable. Ripe fruits are edible	Murthy et al. (2003);Reddy et al. (2007); Misra and Misra (2013); Deepa et al. (2014); Nayak and Basak (2015)
737	*Toona ciliata* Roem.	L	Tender leaves used as vegetable	Murthy et al. (2003)

TABLE 6.1 *(Continued)*

S. No.	Name of the plant	Edible Part (s)	Mode of preparation	References
738	*Trachyspermum ammi* (L.) Sprague	S	Used as spice	Rao (2014)
739	*Trema orientalis* (L.) Bl.	Fr.	Fruits consumed directlly	Murthy et al. (2003)
740	*Trianthema decandra* L.	L	Used as vegetable – Leaves used in preparation of Dall	Murthy et al. (2003); Sinha and Lakra (2005); Reddy et al. (2007)
741	*Trianthema portucastrum* L.	L, St	Leaves and young shoots used as vegetable	Murthy et al. (2003); Reddy et al. (2007); Misra and Misra (2013); Deepa et al. (2014)
742	*Tribulus terrestris* L.	L, Ts	Leaves and tender shoots are boiled and squeezed, add groundnut powder and eat with boiled rice	Reddy et al. (2007); Deepa et al. (2014); Rajasab and Isaq (2004)
743	*Trichodesma indicum* (L.) R. Br. ex Lehm.	L	Leaves used as vegetable	Murthy et al. (2003); Misra and Misra (2013)
744	*Trichosanthes anguina* L.	Fr.	Used as vegetable	Rao (2014)
745	*Trichosanthes cucumerina* L.	Fr.	Fruits used as vegetable	Prakash Kumar et al. (2014)
746	*Tridax procumbens* (L.) L.	L, St	Leaves and tender shoots used as vegetable	Misra and Misra (2013, 2014)
747	*Trigonella foenumgraecum* L.	L, S	Leaves used as leafy vegetable and seeds as condiment	Misra and Misra (2013)
748	*Triticum aestivum* L.	S	Used as food grains	Singh (2013, 2014)
749	*Triticum dicoccum* Schubl.	S	Used as food grains	Singh (2013)
750	*Typha angustata* Bory & Chaub.	R, St, In	Flour extracted from roots is used with Sorghum powder. Young male spike is a delicious vegetable. Young shoots eaten raw or boiled	Rajasab and Isaq (2004)
751	*Vallisneria spiralis* L.	L	Young leaves used as vegetable	Murthy et al. (2003)
752	*Vanda tessellata* (Roxb.) Hook. ex G.Don	AR	Burnt and consumed by children as snacks	Misra et al. (2013)
753	*Vernonia cinerea* (L.) Less.	L	Leaves eaten raw	Murthy et al. (2003)

TABLE 6.1 *(Continued)*

S. No.	Name of the plant	Edible Part (s)	Mode of preparation	References
754	*Vicoa indica* (L.) DC.	L	Leaves used vegetable	Murthy et al. (2003)
755	*Vigna aconitifolia* (Jacq.) Marechal	S	Seeds used as pulse	Naidu and Khasim (2010)
756	*Vigna mungo* (L.) Hepper	S	Seeds used as pulse	Naidu and Khasim (2010)
757	*Vigna radiata* (L.) Wilczek	S, L, St	Seeds used as pulse; leaves and young shoots are cut into small pieces cooked with salt and chilly and eaten	Naidu and Khasim (2010); Misra and Misra (2013); Singh (2013)
758	*Vigna radiata* (L.) Wilczek var. *subolata* (Roxb.) Verdc.	Fr., S	Fruits as vegetable and seeds as pulse	Singh (2013)
759	*Vigna trilobata* (L.) Verdc.	Fr., S	Fruits as vegetable and seeds as pulse	Murthy et al. (2003)
760	*Vigna unguiculata* (L.) Walp.	Fr., S	Fruits as vegetable and seeds as pulse	Naidu and Khasim (2010)
761	*Vitex glabrata* R.Br.	Fr.	Fruits edible	Sinha and Lakra (2005)
762	*Vitex leucoxylon* L.f.	Fr.	Fruits edible	Murthy et al. (2003)
763	*Vitis heyneana* Roem.et Schult.	T	Boiled, cooked and consumed	Misra et al. (2013);
764	*Vitis vinifera* L.	Fr.	Ripe fruits edible	Singh (2013, 2014);
765	*Wattakaka volubilis* (L.f.) Stapf	L	Used as vegetable	Murthy et al. (2003); Deepa et al. (2014); Panda (2014)
766	*Withania somnifera* (L.) Dunal	Fr.	Fruits eaten raw	Murthy et al. (2003)
767	*Woodfordia fruticosa* (L.) Kurz.	St	Tender shoots as vegetable	Murthy et al. (2003)
768	*Wrightia arborea* (Dennest.) Mabb.	Fr.	Tender fruits are edible	Murthy et al. (2003)
769	*Xanthium strumarium* L.	St	Young shoots are used as vegetable	Murthy et al. (2003)
770	*Xantolis tomentosa* (Roxb.) Raf.	Fr.	Fruits used as vegetable	Murthy et al. (2003)
771	*Ximenia americana* L.	Fr.	Ripe fruits are edible	Reddy et al. (2007); Sanyasi Rao et al. (2014); Murthy et al. (2003)

TABLE 6.1 *(Continued)*

S. No.	Name of the plant	Edible Part (s)	Mode of preparation	References
772	*Xylia xylocarpa* (Roxb.) Taub.	S	Seeds used as vegetable	Sinha and Lakra (2005); Murthy et al. (2003)
773	*Xyris pauciflora* Willd.	B	Bulbs eaten raw	Murthy et al. (2003)
774	*Zanthoxylum rhetsa* (Roxb.) DC	L	Used as vegetable	Murthy et al. (2003)
775	*Zea mays* L.	S	Used as food grains	Singh (2013)
776	*Zingiber officinale* Rosc.	Rh	Used as spice; Rhizome used in preparation of masala Tea	Singh (2013); Deshpande and Kulakarni (2013)
777	*Zingiber purpureum* Rosc.	Rh	Sliced rhizome or dry powder is put in to curry to enhance flavour	Misra et al. (2013)
778	*Ziziphus glabrata* Roth	Fr.	Unripe and ripe fruits edible	Rekka and Senthil Kumar (2014)
779	*Ziziphus mauritiana* Lam.	Fr., S	Unripe and ripe fruits edible	Murthy et al. (2003); Sinha and Lakra (2005); Reddy et al. (2007); Basha et al. (2009); Prabakaran et al. (2013); Singh (2013); Deepa et al. (2014); Rekka and Senthil Kumar (2014)
780	*Ziziphus mauritiana* Lam. var. *fruticosa* (Haines) Seb. & Henry	Fr.	Unripe and ripe fruits edible, seeds edible	Reddy et al. (2007); Basha et al. (2009); Prabakaran et al. (2013); Singh (2013); Rekka and Senthil Kumar (2014); Deepa et al. (2014)
781	*Ziziphus nummularia* Wight & Arn.	Fr.	Unripe and ripe fruits edible	Rajasab and Isaq (2004)
782	*Ziziphus oenoplea* (L.) Mill.	Fr.	Unripe and ripe fruits are edible	Murthy et al. (2003); Reddy et al. (2007); Basha et al. (2009); Prabakaran et al. (2013); Deepa et al. (2014); Rekka and Senthil Kumar (2014)
783	*Ziziphus rotundifolia* Lam.	Fr.	Ripe fruits edible	Panda (2014)
784	*Ziziphus rugosa* Lam.	Fr.	Unripe and ripe fruits edible	Murthy et al. (2003); Nayak and Basak (2015)
785	*Ziziphus xylopyrus* (Retz.) Willd.	Fr., S	Unripe and ripe fruits edible	Murthy et al. (2003)

ACKNOWLEDGEMENTS

Authors are thankful to Mr. Ramesh, Naveena Photo Studio, Wanaparthy for providing tribal photos.

KEYWORDS

- **Edible Plants**
- **Ethnic Food Plants**
- **Ethnic Food Preparation**
- **Fruits**
- **Vegetables**

REFERENCES

Alagesaboopathi, C., Balu, S. & Dwarakan, P. (1996). Edible fruit yielding plants of Shevaroy hills in Tamil Nadu. *Ancient Science of Life 16*(2), 148–151.

Amubode, F.A. & Fetuga, B.L. (1983). Proximate composition and chemical assay of methionine, lysine, tryptophan in some Nigerian forest trees. *Food Chem.*, *12,* 67–72

Anonymous (1970–1988). Wealth of India: Raw Materials. Council of Scientific and Industrial Research, Delhi, 1–12 (reprinted).

Anonymous (1996). Report on the State of the World's Plant Genetic Resources for Food and Agriculture, Rome, Italy. pp. 511.

Ansari, A.A., Diwakar, P.G. & Dwarakan, P. (1993). Less known edible plants of Shevoroy and Kolli hills. *J. Econ. Taxon. Bot. 17,* 245.

Appalanaidu, P. (2013). Life and Livelihood strategies among the Chenchu: Forest related tribal group (FRTG) in Andhra Pradesh. *Abhinav*, *2*(7)*,* 41–48.

Arora, R.K. (1991). Conservation and Management concept and Approach in Plant Genetic resources, R.S. Paroda & R.K. Arora (Eds.). IBPGR, Regional Office South and Southeast Asia, New Delhi, p.25.

Arora, R.K., Chandel, K.P.S., Joshi, B.S. & Pant, K.C. (1980). Rice bean: Tribal pulse of Eastern India. *Economic Bot. 34,* 260–263.

Bagul, R.M. (2013). Some ethnomedicinal plant species of Satpuda forest region of east Khandesh Jalgain district, Maharashtra. *J. New Biological Rep. 2*(3), 264–271.

Balemie, K. & Kebebew, F. (2006). Ethnobotanical study of wild edible plants in Derashe and Kucha Districts, South Ethiopia. *J. Ethnobiol Ethnomed., 2,* 53–61.

Basha, S. (2009). Diversity, quantification and conservation of tree resources of Nallamalais, Andhra Pradesh. PhD thesis, Sri Krishnadevaraya University, Anantapur.

Basha, S., Sadasivaiah, B. & Ravi Prasad Rao, B. (2009). Wild edible fruit resources in southern Eastern Ghats of Andhra Pradesh. *Int. J. For Usuf. Mngt., 10*(2), 20–25.

Behera, K.K., Mishra, N.M., Dhal, N.K. & Rout, N.C. (2008). Wild Edible plants Mayurbhanj district, Orissa, India. *J. Econ. Taxon. Bot. 32(suppl.)*, 305–314.

Bharucha, Z. & Pretty, J. (2010). The roles and values of wild foods in agricultural systems. *Phil. Trans. Royal Soc. B., 365,* 2913–2926.

Bonet, M.A., Parada, M., Selga, A. & Valles, J. (1999). Studies on pharmaceutical ethno botany in the regions of L' Alt Emporada and Les Guilleries (Catalonia, Iberian Peninsula). *J. Ethnophramacol, 68*(1–3), 145–168.

Census of India (2001). Provisional population statistics The Registrar General & Census Commissioner, Government of India.

Chauhan, K.P.S. (1998). Framework for conservation and sustainable use of biological diversity: Action plan for the Eastern Ghats region. In Proceedings of Seminar on Conservation of Eastern Ghats. pp. 345–357.

Craig, W. & Beck, L. (1999). Phytochemicals: Health Protective Effects. *Can. J. Diet. Pract. Res. 60,* 78–84.

Deepa, P., Murugesh, S., Sowndhararajan, K. & Manikandan, P. (2014). Ethnobotanical Studies on wild edible plants used by Malayali tribals of Melur, Bodha Hills, Southern Eastern Ghats, Namakkal District, Tamil Nadu, India. *World J. Pharmaceut. Res., 3*(7), 621–633.

Deshpande, S. & Kulkarni, D.K. (2013). *Theriophonum indicum* (Dalz.) Engler (Araceae) –Leafy vegetable of Gondia tribe, Vidarbha region, Maharashtra. *Indian J. Fundam. Appl. Life Sci., 3*(4), 35–38.

Devaraj, M., Ganapathy, M.S. & Mahadeva, M.M. (2006). Extraction and marketing of non-timber forest products. *My Forest, 42*(3), 239–249.

Dhole, J.A., Dhole, N.A. & Bodke, S.S. (2009). Ethnomedicinal studies of some weeds in crop fields of Marathwada region, India. *Ethnobotanical Leaflets, 13,* 1443–1452.

Dhore M.M., Lachure, P.S., Bharsakale, D.B. & Dabhadkar, D.K. (2012). Exploration of some wild edible plants of Digras Tahsil, Dist. Yavatmal, Maharashtra, India. *International J. Scientif. Res. Publications*, 2(5), 1–5.

Etkin, N.L. & Johns, T. (1998). 'Pharmafoods' and 'nutraceuticals': paradigm shifts in bio-therapeutics. In: H.D.V. Prendergast, N.L. Etkin & D.R.P.J. Harris (eds.). Plants for Food and Medicine. Royal Botanic Gardens, Kew. pp. 3–16.

FAO (1999). Use and Potential of Wild Plants. Information Division, Food and Agricultural Organization of the United Nations, Rome, Italy.

FAO (2004). Annual Report: The state of food insecurity in the world, monitoring the progress towards the world food summit and millennium development goals. Rome.

FAO, WFP & IFAD (2012). The State of Food Insecurity in the World 2012. Economic growth is necessary but not sufficient to accelerate reduction of hunger and malnutrition. Rome, FAO.

Ganesan, R. & Setty, R.S. (2004). Regeneration of amla, an important non-timber forest product from Southern India. *Conservation Society, 2*(2), 365–375.

Gayatri, K. & Srividya, N. (2008). Ethnomedicinal knowledge of traditionally used edible leaves, seeds flowers among women—A transgenerational study. International Seminar on Medicinal Plants and Herbal products. 7th–9th March 2008, p. 60.

Girach, R.D., Brahmam, M. & Misra, M.K. (1997). Some less known plant foods from Bhadrak district of Orissa. *J. Econ. Taxon. Bot. 21*(1), 107–111.

Girach, R.D. & Aminuddin (1992). Some little known edible plants from Orissa. *J. Econ. Taxon. Bot., 16*(1), 61–68.

Girach, R.D., Aminuddin & Ahmed, I. (1992). Observation on wild edible plants from tribal pockets of Orissa. *J. Econ. Taxon. Bot. 16*(3), 589–594.

Goud, S.P. & Pullaiah, T. (1996). Ethno-botany of Kurnool District: some wild plants used as food. *J. Econ. Taxon. Bot. Addl. Ser. No. 12*, 224–227.

Haimendorf, C.V.F. (1943). The Chenchus. London: MacMillan & Co.

Haimendorf, C.V.F. (1945). The Reddis of Bison Hills. London: MacMillan & Co.

Haimendorf, C.V.F. (1979). The Gonds of Andhra Pradesh. New Delhi: Vikas.

Hebbar, S.S., Harsha, V.H., Shripahi, V. & Hegde, G.R. (2003). Wild edible fruits of Dharwad, Karnataka. *J. Econ. Taxon. Bot. 27*(4), 982–988.

Jain, S.K. (1981). Glimpses of Indian Ethnobotany. New Delhi, India: Oxford & IBH Publishing Co.

Jyothi Arun, B., Venkatesh, K., Chakrapani, P. & Roja Rani, A. (2011). Phytochemical and pharmacological potential of *Annona cherimola*—A Review. *Intern. J. Phytomedicine 3*, 439–447.

Kamble, S.V. & Jadhav, V.D. (2013). Traditional Leafy vegetables: A future herbal medicine. *International J. Agricultural and Food Science*, *3*(2), 56–58.

Karuppusamy, S. & Pullaiah, T. (2005). Selected medicinal plant species of Sirumalai hills, south India, used by natives and antibacterial screening of plants. *J. Trop. Med. Plants,* *6*(1), 99–109.

Krishnamachari, K.S. (1900). *Erythroxylum monogynum* leaves and *Aloe* roots as food. *Indian Forester 26*, 619–620.

Kumar S., Jena, P.K. & Tripathy, P.K. (2012). Study of wild edible plants among tribal groups of Simlipal Biosphere Reserve Forest, Odisha, India: with reference to *Dioscorea* species. *Int. J. Biol. Tech.*, *3*(1), 11–19.

Kumar, S., Parida, A.K. & Jena, P.K. (2013). Ethno-Medico-Biology of Bān-Aālu (*Dioscorea* species): A neglected tuber crop of Odisha, India. *Int. J. Pharm., Life Sci.*, *4*(12), 3143–3150.

Liu, R.H. (2003). Health benefits of fruit and vegetables are from additive and synergistic combination of phytochemicals. *Amer. J. Clinical Nutrition, 78*(3), 517–520.

Mahapatra, A.K. & Panda, P.C. (2009). Wild Edible Fruits of Eastern Ghats. Regional Plant Resource Centre, Bhubaneswar, India.

Mahapatra, A.K., Mishra, S., Basak, U.C. & Panda, P.C. (2012). Nutrient analysis of some selected wild edible fruits of deciduous forests of India: an explorative study towards non-conventional bio-nutrition. *Advance J. Food Sc. Tech. 4*(1), 15–21.

Maikhuri, R.K., Nautiyal, M.C. & Khali, M.P. (1991). Lesser-known crops of foods value in Garhwal Himalaya and a strategy to conserve them. *FAO/IBPGR Plant Genrt.Res. Newslett. 86*, 33–36.

Mamatha, N., Pavan, Murthy, K.R.K. & Venkatesh, D.A. (2006). Data on 100 medicinal plants used by Soligas of Biligirirangan hills of Mysore district, Karnataka. *My Forest, 42*(2), 212–139.

Mathew, K.W. (1983). Flora of Tamil Nadu Carnatic. Tiruchirapalli, India: The Rapinat Herbarium.

Misra, M.K. (2013). Biodiversity and traditional Knowledge and Village Ecosystem Sustainability. *The Eco Scan*; *Special Issue, 3*, 235–240.

Misra, M.K., Panda, A. & Sahu, D. (2012). Survey of useful plants of South Odisha, India. *Indian J. Trad. Know. 11*, 658–666.

Misra, R.C., Sahoo, H.K., Pani, D.R. & Bhandari, D.C. (2013). Genetic resources of wild tuberous food plants traditionally used in Similipal Biosphere Reserve, Odisha, India. *Genet Resour Crop Evol*, doi: 10.1007/s10722-013-9971-6.

Misra, S. (2009). Farming System in Jeypore tract of Orissa, India. *Asian Agri-History, 13*(4), 271–292.

Misra, S. & Misra, M.K. (2013). Leafy Vegetable plants of South Odisha, India. *Intern. J. Agric. Food Sci., 3*(4), 131–137.

Misra, S. & Misra, M.K. (2014). Ethno-botanical study of plants with edible underground parts of South Odisha, India. *Intern. J. Agric. Food Sci., 4*(2), 51–58.

Misra, S. & Misra, M.K. (2014a). Nutritional evaluation of some leafy vegetable used by the tribal and rural people of south Odisha, India. *J. Nat. Prod. Plant Resour., 4*(1), 23–28.

MoEF (2013). Annual Report 2012–13. Ministry of Environment and Forests, Government of India.

Mohan, V.R. & Kalidass, C. (2010). Nutritional and antinutritional evaluation of some unconventional and wild edinle plants. *Tropical and subtropical Agroecosystems 12 (3)*, 495–506.

Murali, K.S., Uma Shaanker, R., Ganeshaiah, K.N. & Bawa, K.S. (1998). Exraction of non–timber forest products in the forests of Biligirirangan hills, India. 2. Impact of NTFP extraction on regeneration, population structure and species composition. *Economic Bot., 50*(3), 252–269.

Mukesh Kumar, Husaini, S.A., Qamar Uddin, Aminuddin, Kumar, K. & Samiulla. L. (2013). Ethnobotanical study of the wild edible plants from Odisha, India. *Life Sciences Leaflets, 7,* 13–20.

Murthy, K.S.R. (2011). Nutritional potential and biochemical compounds in *Cajanus albicans* (Wight & Arn.) van der Maesan for food and agriculture. *J. Agric. Tech., 7(1),* 161–171.

Murthy, K.S.R. & Emmanuel, S. (2011). Nutritional and antinutritional properties of the underexploited wild legume *Rhynchosia bracteata* Benth. *Bangladesh J. Sci. Industr. Res. 46*(2), 141–146.

Murthy, K.S.R. & Pullaiah, T. (2005). Wild relatives and related species of cultivated crop plants of Eastern Ghats, India. Recent trends in plant Sciences. Pullaiah et al. (Eds.) pp. 96–103; Regency Publications, New Delhi, India.

Murthy, K.S.R. & Sambasiva Rao, K.R.S. (2009). Chemical composition and nutritional evaluation of *Paracalyx scariosus* (Roxb.) Ali – a wild relative of *Cajanus* from Southern Peninsular India. *Tropical and Subtropical Agroecosystems, 10,* 121–127.

Murthy, K.S.R., Sandhya Rani, S. & Pullaiah, T. (2003). Wild edible plants of Andhra Pradesh, India. *J. Econ.Taxon. Bot., 27*(3), 613–630.

Nagalakshmi, N.V.N. (2014). Diversity of wild greens knowledge from the rural households of Anantapur district, A.P. *Intern. J. Res. Appl. Nat. Social Sci., 2*(5), 157–160.

Naidu, K.A. & Khasim, S.M. (2010). Contribution to the floristic diversity and Ethno botany of Eastern Ghats in Andhra Pradesh, India. *Ethnobotanical Leaflets, 14,* 20–41.

Nandini, N. & Shiddamallayya, N. (2014). Wild edible plants of old Mysore district, Karnataka, India. *Plant Sciences Feed, 4*(4), 28–32.

Nayak, J. & Basak, U.C. (2015). Analysis of some nutritional properties in eight wild edible fruits of Odisha, India. *Int. J. Curr. Sci., 14,* 55–62.

Nordeide, M.B., Hatloy, A., Folling, M., Lied, E. & Oshoug, A. (1996). Nutrient composition and nutritional importance of green leaves and wild foods in an agricultural district, Koutiala, in Southern Mali. *Int. J. Food Sci. Nutr. 47*(6), 455–468.

Orech, F.O., Aagaard-Hansen, J. & Friis, H. (2007). Ethnoecology of traditional leafy vegetables of the Luo people of Bondo district, Western Kenya. *Int. J. Food. Sci. Nutr. 58*(7), 522–530.

Pal, D.C. & Banerjee, D.K. (1971). Some less known plant foods among the tribals of Andhra Pradesh and Orissa states. *Bull. Bot. Surv. India 13,* 221–223.

Panda, T. (2014). Traditional knowledge on wild edible plants as livelihood food in Odisha, India. *J. Biol. Earth Sci., 4*(2), B144–B159.

Pandravada, S.R., Sivaraj, N., Kamala, V., Sunil, N., Sarath Babu & Varaprasada, K.S (2007). Agri-Biodiversity of Eastern Ghats- exploration, collection, and conservation of crop genetic resources. Proc. National Seminar on Conservation of Eastern Ghats, pp. 19–27.

Pandravada, S.R., Sivaraj, N., Jairam, R., Sunil, N., Begum, H., Thirupathi Reddy, M., Chakrabarty, S.K., Bisht, I.S. & Bansal, K.C. (2014). *Luffa hermaphrodita*: First Report of its distribution and cultivation in Adilabad, Andhra Pradesh, South India. *Asian Agri-History, 18*(2), 123–132.

Prabakaran, R., Senthil Kumar, T. & Rao, M.V. (2013). Role of Non Timber Forest Products in the livelihood of *Malayali* Tribe of Chitteri Hills of Southern Eastern Ghats, Tamil Nadu, India. *J. Appl. Pharmaceut. Sci., 3*(5), 56–60.

Prasad, V.K., Rajagopal, T., Kanit, Y. & Badrinath, K.V.S. (1999). Food plants of Konda Reddis of Rampa Agency. East Godavari district, Andhra Pradesh—A case study. *Ethnobotany 11,* 92–96.

Pushpangadan, P. (1994). Ethnobiology in India. A status report, Ministry of Environment and Forest, GOI, New Delhi.

Quebedeaux, B. & Bliss, F.A. (1988). Horticulture and human health: Contributions of fruits and vegetables. Proc. 1st Intl. Symp. Hort. and Human Health. Prentice Hall, Englewood NJ.

Quebedeaux, B. & Eisa, H.M. (1990). Horticulture and Human Health: Contributions of Fruits and Vegetables. Proc. 2nd Intl. symp. Hort. and Human Health. *Hort. Science 25,* 1473–1532.

Rajasab, A.H. & Mahamad Isaq (2004). Documentation of folk knowledge on edible wild plants of North Karnataka. *Indian J. Trad.Know., 3*(4), 419–429.

Rajasab, A.H. & Rajshekhar, S.B. (2012). *Launea procumbens* – a wild edible plant of north Karnataka, India. *Life Sci. Leaflets, 7,* 84–87.

Ramakrishna, N., Saidulu, Ch. & Hindumathi, A. (2014). Ethnomedicinal uses of some plant species by tribal healers in Adilabad distrct of Telangana state, India. *World J. Pharmaceut. Res., 3*(8), 545–561.

Rao, B.R.P., Prasad, K., Sadasivaiah, B., Khadar Basha, K., Suresh Babu, M.V. & Prasanna, P.V. (2011). A New Species of *Brachystelma* R. Br. (Apocynaceae: Asclepiadoideae – Ceropegieae) from India. *Taiwania, 56*(3), 223–226.

Reddy, C.S. & Raju, V.S. (2005). Invasion of Alligator weed (*Alternanthera philoxeroides*) in Andaman Islands. *J. Bombay Nat. Hist. Soc., 102*(1), 133.

Reddy, C.S., Reddy, K.N., Pattanaik, C. & Raju, V.S. (2006). Ethnobotaniocal observations on some endemic plants of Eastern Ghats, India. *Ethnobotanical Leaflets, 10,* 82–91.

Reddy, K.N., Pattanaik, C., Reddy, C.S. & Raju, V.S. (2007). Traditional knowledge on wild food plants in Andhra Pradesh. *Indian J. Trad. Know. 5,* 368–372.

Reddy, M. (2012). Wild edible plants of Chandrapur district, Maharashtra, India. *Indian J. Natural Products and Resources, 3*(1), 110–117.

Rekka, R. & Senthil Kumar, S. (2014). Ethnobotanical notes on wild edible plants used by Malayali tribals of Yercaud Hills, Eastern Ghats, Salem District, Tamil Nadu. *Intern.J. Herbal Med., 2*(1), 39–42.

Rout, S.D. (2007). Ethnobotany of Diversified wild edible fruit plants in Similipal Biosphere Reserve, Orissa. *Ethnobotany 19,* 137–139.

Sadasivaiah, B. (2009). Diversity, quantification and conservation of Herbaceous plant resources of Nallamalais, Andhra Pradesh. PhD thesis, Sri Krishnadevaraya University, Anantapur.

Sadasivaiah, B. & Rao, B.R.P. (2012). Tribe: Ceropegieae (Apocynaceae, Asclepidoideae) in Eastern Ghats of Andhra Pradesh, India. In: G.G. Maiti & S.K. Mukherjee (Eds.). Multidisciplinary Approaches in Angiosperm Systematics. Publication Cell, University of Kalayni, West Bengal, India. Vol. 1. pp. 86–94.

Sahu, C.R., Nayak, R.K. & Dhal, N.K. (2013). The plant wealth of Boudh district of Odisha, India with reference to Ethnobotany. *Int. J. Curr. Biotechnol. 1*(6), 4–10.

Samant, S.S. & Dhar, U. (1997). Diversty, Endemism and Economic potential of Wild Edible plants of Indian Himalaya. *Inter. J. Sustain. Develop. World Ecol., 4,* 179–191.

Samydurai, P., Thangapandian, V. & Aravinthan, V. (2012). Wild habits of Kolli Hills being staple food of inhabitant tribes of Eastern Ghats, Tamil Nadu, India. *Intern. J. Nat. Prod. Resources, 3*(3), 432–437

Sanyasi Rao, M.L., Yesudas, S. & Kiran, S. (2014). Indigenous plant foods which are commonly consumed by the tribal communities in Dumbriguda area of Visakhapatnam district, Andhra Pradesh, India. *Biolife, 2*(3), 866–875.

Sasi, R. & Rajendran, A. (2012). Diversity of Wild fruits in Nilgiri Hills of the Southern Western Ghats-Ethnobotanical aspects. *Intern. J. Appl. Biol. Pharmaceut. Tech., 3*(1), 82–87.

Sasi, R., Rajendran, A. & Maharaj, M. (2011). Wild Edible plant diversity of Kotagiri Hills- a part of Nilgiri Biosphere Reserve, Southern India. *J. Res.Biol., 2,* 80–87.

Satyavathi, K., Sandhya, D., Deepika & Pada, S.B. (2014). Ethnomedicinal plants used by the *Bagata* Tribes of Paderu Forest Division, Andhra Pradesh, India. *Int. J. Adv. Res. Sci. Technol., 3*(2), 36–39.

Saunders, C.F. (1920). Useful Wild Plants of the United States and Canada. Robert M. McBride & Co., New York.

Saxena, R. (1999). How green is your diet? *Nutrition 33(3),* 9.

Sebastian, M.K. & Bhandari, M.M. (1990). Edible wild plants of the forest areas of Rajasthan. *J. Econ. Taxon. Bot., 14*(3), 689–694.

Seema Mundoli (2011). Impacts of government policies on sustenance of tribal people in the Eastern Ghats. Report Submitted to Dhaatri Resource Centre for Women and Children & Samata.

Sharma, J.R., Mudgal, V. & Hajra, P.K. (1997). Floristic diversity-review, scope and perspective. In: V. Mudgal & P.K. Hajra (Eds.) Floristic Diversity and Conservation Strategies in India, vol. 1. Botanical Survey of India, Calcutta, pp. 1–45.

Singh, A.K. (2013). Probable Agricultural Biodiversity Heritage Sites in India: XVII. The South-Central Region of Eastern Ghats. *Asian Agri-History, 17*(3), 199–220.

Singh, A.K. (2014). Probable Agricultural Biodiversity Heritage Sites in India: XIX. The North-western Deccan Plateau Region, the Leeward Side of the Western Ghats. *Asian Agri-History, 18*(2), 101–122.

Singh, H.B. & Arora, R.K. (1978). Wild edible plants of India. 1st edition. New Delhi, ICAR Publication

Sinha, R. & Lakra, V. (2005). Wild tribal food plants of Orissa. *Indian J. Trad. Know., 4*(3), 246–252.

Srivastava, R.P. & Ali, M. (2004). Nutritional quality of common pulses: Indian Institute of pulses Research, Kanpur. pp. 14–22.

Subbaiah, M., Singaram, R. & Arunachalam, S. (2012). Plants used for non-medicinal purposes by Malayali tribals in Jawadhu hills of Tamil Nadu, India. *Global J Res. Med. Plants & Indigen. Med.*, *1*(12), 663–669.

Subramanyam, V. & Rama Mohan, K.R. (2001). Tribal Ecology and Food Security: A Study in Visakha Agency area of Andhra Pradesh. *J. Hum. Ecol.*, *12*(5), 351–356.

Subramanyam, Veerabhadrudu, B. (2013). Environment and sustainable development: A study among the tribes of Eastern Ghats in Andhra Pradesh. *Nature Environment and Pollution Technology*, *12*(3), 425–434.

Sudhakar, A. & Vedavathy, S. (1999). Wild edible plants used by the tribals of Chittoor District (Andhra Pradesh), India. *J. Econ. Taxon. Bot.*, *23*(2), 321–329.

Tripathy, P.K., Sanjeet Kumar, S. & Jena, P.K. (2014). Assessment of food, ethnobotanical and antibacterial activity of *Trichosanthes cucumerina* L. *IJPSR*, *5*(7), 2919–2926.

Uma, J. & Singh, V.A. (1987). Census of edible species of *Diospyros* L. in India. *J. Econ. Taxon. Bot.*, *10*(2), 416–419.

Umashankar, K.S., Murali, Umashaanker, R., Ganeshaiah, K.N. & Bawa, K.S. (1996). Extraction of non-timber forest products in the forests of Biligirirangan hills, India. 3. Productivity, extraction and prospects of sustainable harvest of *Phyllanthus emblica* (Euphorbiaceae). *Economic Bot.*, *50*(3), 270–279.

Vadivel, V., Doss, A. & Pugalenthi, M. (2010). Evaluation of nutritional value and protein quality of raw and differentially processed sword bean (*Canavalia gladiata* (Jacq.) DC) seeds. *African J. Food, Agriculture, Nutrition and Development*, *10*(7), 2850–2865.

Vaidyanathan, D., Salai, M.S., Senthilkumar & Ghouse Basha, M. (2013). Studies on ethnomedicinal plants used by Malayali tribals in Kolli hills of Eastern Ghats, Tamilnadu, India. *Asian J. Plant Sci. Res.*, *3*(6), 29–45.

Vasundhara (2009). Report: Biodiversity Assessment in some selected hill forests of south Orissa. Vasundhara, Orissa.

Verma, R.C. (1995). Indian Tribes: Through the Ages. Director, Publication Division, Ministry of Information and Broadcasting, Government of India, New Delhi.

Wargovich, M.J. (2000). Anticancer properties of fruits and vegetables. *Hort. Science, 35,* 573–575.

Wilson, E.O. (1992). The Diversity of Life. Penguin, London, UK. pp. 432.

Xavier, T.F., Fred Rose, A. & Dhivyaa, M. (2011). Ethnomedicinal survey of Malayali tribes in Kolli Hills of Eastern Ghats of Tamilnadu, India. *Indian J. Trad. Knowl.*, *10*(3), 559–562.

CHAPTER 7

ETHNOMEDICINAL PLANTS OF EASTERN GHATS AND ADJACENT DECCAN REGION

S. KARUPPUSAMY[1] and T. PULLAIAH[2]

[1]Department of Botany, The Madura College (Autonomous), Madurai–625011, Tamilnadu, India, E-mail: ksamytaxonomy@gmail.com

[2]Department of Botany, Sri Krishnadevaraya University, Anantapur–515003, Andhra Pradesh, India, E-mail: pullaiah.thammineni@gmail.com

CONTENTS

ABSTRACT

Ethnomedicinal plants of Eastern Ghats and adjacent Deccan region of India was surveyed with available literature in Google search engine and various published sources. The study revealed that about 54 different ethnic tribal communities used 782 ethnomedicinal plants for their primary health care. These plants are belonging to 132 families and 384 genera. Leguminosae is a largest ethnomedicinal plant yielding family and it is contributing about 67 species of medicinal plants. The plant families like Apocynaceae, Orchidaceae, Solanaceae and Rubiaceae contribute more than 20 species are of local medicinal values. The predominant ethnomedicinal plant genera of Eastern Ghats are *Cassia* and *Solanum* (12 spp.), *Acacia* and *Euphorbia* (11 spp. each), *Ficus* (8 spp.), *Curcuma* (7), *Andrographis* (6), *Bauhinia*, *Habenaria* and *Terminalia* (5 spp. each). The habit-wise distribution of ethnomedicinal plants of Eastern Ghats showed herbs (41%), trees (24%), shrubs (22%) and climbers (13%). Some 75 plant species are endemic to this region. These plants played vital role in health care and local livelihood among the native communities of Eastern Ghats and some these plants are also contributing in codified Indian medicinal systems like Siddha, Ayurvedha and Unani. The present study aimed to compile the tribal medicinal plant wealth of Eastern Ghats for further promotion of utility, screening for clinical potential and enhancing conservation measures.

7.1 INTRODUCTION

The Eastern Ghats are discontinuous mountain ranges along the east coast of Indian Peninsula running parallel to the Bay of Bengal about 1750 km from Odisha through Andhra Pradesh to Tamilnadu and also passing in some parts of Karnataka. Some perennial rivers originated from Western Ghats and cut through Eastern Ghats ranges especially Mahanadhi, Godavari, Krishna and Cauveri, finally they flow into Bay of Bengal. The rocks of Eastern Ghats are considered Gondwana origin and made up of charnokites, granites gneiss, khondalites, metamorphic gneiss and quartzite. Several hill ranges and hillocks of Eastern Ghats locally divided into three parts viz, northern Eastern Ghats (mostly in Odisha and northern Andhra Pradesh), middle Eastern Ghats (Andhra Pradesh) and southern Eastern Ghats (Tamilnadu). The vegetation of Eastern Ghats characteristically comprised with deciduous and scrub forests interspersed with grassy open canopies.

The Eastern Ghats and adjacent Deccan region is inhabited by about 54 tribal communities, which constitute about 30% of tribal population of India. They depend mostly on various forest resources available locally for their livelihood. In Odisha, major part of the northern Eastern Ghats, many numbers of tribal communities namely Bathudi, Bhottoda, Bhumia, Bhumiji, Dharua, Gadaba, Gond, Khanda, Kandha, Gouda, Kolha, Koya, Munda, Paroja, Omanatya, Santal and Saora (Nayak and Sahoo, 2002) are found. Among them many native communities belong to primitive scheduled tribes like Birhor, Bonda, Didayi, Dogaria Kondh, Hill Kharia, Juang, Kutia Kondh, Lanjia Saora, Mankirdia, Paudi Bhuyan and Saora. The hill tribes of Andhra Pradesh in Eastern Ghats area are Chechus, Gadabas, Savaras, Konda Reddis, Koyas, Khonds, Kolamis, Nayakpods, Valmikis, Bhagatas, Jatayus, Yanadis and Yerukalas and they constitute 6.3% of total population of Andhra Pradesh (Verma, 1998). The typical aboriginal tribes are Yanadis, Chenchus, Koyas and Savaras still live with indigenous habit in Eastern Ghats (Solomon Raju and Jonathan, 2006). Tamilnadu part of Eastern Ghats is inhabited by tribal communities which include Malayalis in Chitteri hills, Kolli hills, Shervarayan hills, Kalrayan hills, Javadhu hills and Jinji hills (Xavier et al., 2011). Paliyans, a small group of scheduled tribe lives in Sirumalai hills of southernmost tip of Eastern Ghats (Karuppusamy and Pullaiah, 2006) (for more details see Chapter 2 of this volume).

During the last few decades, there has been an increasing interest in the study of medicinal plants and their traditional use in different tribal communities of India. According to World Health Organization (WHO) as many as 80% of the world's population depend on traditional medicines and in India about 65% of the population in rural areas uses traditional medicines for their primary health care. Ethnobotany deals with the relationship between human societies and plants. It has been recognized recently as a multidisciplinary science comprising of many interesting and useful aspects of plant science, chemistry, environment, anthropology, history, culture, pharmacology and literature. It gives varied economic uses of plants among the primitive human societies, which are equally beneficial to modern man. The results of the ethnobotanical researches brought out numerous little known and unknown uses of plants (Jain, 1981).

The botanical study of folk drugs in recent years has led to the discovery of large number of new medicinal compounds having potent therapeutic activities. During the last few decades, a succession of so called "wonder drugs", such as reserpine, ajmalicine, quinine, ephedrine, conine, cocaine, emetine, colchicine, digoxin, berberine and artimisine have been discovered from plants with rich ethnobotanical role in primitive human societies. Currently about 720 active principles isolated from higher plants are utilized in allopathy system after being validated by ethnopharmacology. About 70% of the plant derived allopathy drugs find the same therapeutic applications in original traditional medicine. Unfortunately, traditional knowledge of primitive human societies has only oral or verbal traditions without any written documents. Due to the fast changing life style by urbanization, modernization and intrusion of modern civilization among tribal communities, the traditional knowledge on useful plants acquired and accumulated through generation is gradually getting lost. Hence, documentation of traditional knowledge of wild plants has become imperative lest the vital clues they hold for the quality life of modern man would be lost forever.

7.2   METHODS

In recent years, many of the methods used by ethnobiologists have been compiled into filed manuals, most notably the series titled "People and Plants Conservation Manuals" developed by the World Wildlife Fund, UNESCO and Kew Royal Botanical Gardens as part of the People and Plants Initiative (Cunningham, 2000). Prior to mid-1950s, researcher in ethnobotany was primarily descriptive. A large amount of data was collected regarding traditional names and uses of plants and animals for a number of sociolinguistic groups (for more details see Chapter 3 of this volume). Within ethnobotany, researchers were increasingly becoming concerned with understanding emic perceptions of people and plants. A detailed account of this fascinating history is provided by D'Andrade (1995). Ethnobotanists have been at the forefront of participatory methods, developing innovative strategies for training indigenous collaborators and conservation of local biological resources. Modern studies in ethnobotany are distinguishable from earlier studies of useful plants, in that modern ethnobotanical studies tend to include more information about the ethnic cultural groups that use the plants medicinally or other purposes. It includes formation about cultural beliefs surrounding illness, treatment and healing methods; human ecological relationships and the role of medicinal plants in larger societies; rituals, ceremonies and other uses of medicinal plants; the role of traditional healers, shamans or other ritual specialists who has medicinal plants to treat patients.

Usual data collection of ethnobotanical information and practice within any culture vary by geographical origin, residence, ethnicity, religion, age and gender. After an exhaustive search on ethnobotanical and ethnographical works available online (Google search engine; www.google.com) and various published literature, we compiled the plant list and their ethnomedicinal uses relevant to the Eastern Ghats ranges and Deccan region. It also includes about 782 ethnomedicinal plant species enumerated from about 150 published literature of various sources. All the information in this work therefore refers to wild plants used in medicinal purposes at least during last 50–100 years.

7.3   DIVERSITY OF ETHNOMEDICINAL PLANTS IN EASTERN GHATS AND ADJACENT DECCAN REGION

India has a rich tradition of plant-based knowledge on healthcare since time immemorial. A large number of plants, plant products, decoctions, pastes are equally used by tribals and folklore traditions in India for treatment of several ailments. The present review thus attempt to analyze the ethnomedicinal knowledge base for treatment of all kinds of ailments, methods employed by tribal and folklore practices previously in Eastern Ghats of Peninsular India. Out of 7,500 species of medicinal plants estimated in India, about 1,800 species are known to occur in Eastern Ghats region. At least 800 medicinal and 40 aromatic plants are concentrated in this area which are used in various medicinal systems including codified and folklore. From the present review and a case study enumerated about 782 plants species includes 50 Pteridophytes and two Gymnosperm species (Table 7.1).

These 782 ethnomedicinal plants species belonging to 132 families and 384 genera were recorded from various sources and published literature. Leguminosae (67 spp.), Apocynaceae (29 spp.), Malvaceae (26 spp.), Euphorbiaceae (25 spp.), Orchidaceae (22 spp.), Solanaceae and Rubiaceae (16 spp. each), Asteraceae (15 spp.), Acanthaceae, Asteraceae and Lamiaceae (14 spp. each), Cucurbitaceae and Zingiberaceae (13 spp. each), Rutaceae (12 spp.) and Araceae (10 spp.) were the dominant families of ethnomedicinal plants. *Cassia* and *Solanum* (12 spp.), *Acacia* and *Euphorbia* (11 spp. each), *Ficus* (8 spp.), *Curcuma* (7), *Andrographis* (6), *Bauhinia*, *Habenaria* and *Terminalia* (5 spp. each), *Albizia* and *Dioscorea* (4 spp. each) were the dominant medicinal plant genera (Table 7.1).

TABLE 7.1   Ethnomedicinal Plants of Eastern Ghats and Deccan

S. No.	Name of plant Species	Useful Parts	Medicinal uses against	References
1.	*Abelmoschus esculentus*	Leaves	Dysentery	Padal et al. (2013a)
2.	*Abelmoschus moschatus*	Seed	Fever	Padal et al. (2013b)
3.	*Abrus precatorius*	Leaves, seeds, root	Leucorrhoea, bronchitis, cold, cough, fever, eczema, urinary disorders, hepatitis, snake bite, swelling, skin diseases, menstrual pains, for abortion, contraceptive, poisonous bite	Raja Reddy et al. (1989); Vijayakumar and Pullaiah (1998); Rao and Pullaiah (2001); Chaudhari and Hutke (2002); Pullaiah et al. (2003); Sen and Behera (2003); Basha et al. (2011); Murthy (2012); Lingaiah and Rao (2013); Padal et al. (2013b); Sahu et al. (2013c); Vaidyanathan et al. (2013); Kannan and Kumar (2014); Manikandan and Lakshmanan (2014); Satyavathi et al. (2014); Ramakrishna and Saidulu (2014b); Shanmukha Rao et al. (2014)
4.	*Abrus pulchellus*	Whole plant	Aphrodisiac	Rekha and Senthilkumar (2014)
5.	*Abutilon crispum*	Leaves	Dysentery, jaundice and piles	Manikandan and Lakshmanan (2014)
6.	*Abutilon hirtum*	Leaves	Bronchitis	Reddy et al. (2006)
7.	*Abutilon indicum*	Leaves, root	Jaundice, dental problems, skin problems, epileptic fits, menstrual disorders, scorpion sting, diabetes	Panda (2007); Xavier et al. (2011); Venkata Subbiah and Savithramma (2012); Sahu et al. (2013c); Ramakrishna and Saidulu (2014); Prasanthi et al. (2014); Satyavathi et al. (2014)
8.	*Acacia auriculiformis*	Flowers	Joint pains, rheumatism	Kumar and Pullaiah (1999)
9.	*Acacia caesia*	Stem, bark, leaves	Respiratory troubles, cough, skin diseases, to heal wounds	Raja Reddy et al. (1989); Vijayakumar and Pullaiah (1998); Xavier et al. (2011); Murthy (2012); Venkata Subbiah and Savithramma (2012)
10.	*Acacia concinna*	Pods	Purgative, relieves biliousness	Murthy (2012)
11.	*Acacia farnesiana*	Stem, bark	Diarrhea dysentery, cough, dog bite	Pullaiah et al. (2003); Murthy (2012); Padal et al. (2013b)
12.	*Acacia leucophloea*	Root, stem bark	Abortifacient, arthritis, wounds	Prusti (2007); Padal et al. (2013b); Satyavathi et al. (2014)

TABLE 7.1   *(Continued)*

S. No.	Name of plant Species	Useful Parts	Medicinal uses against	References
13.	*Acacia mangia*	Stem bark	Paralysis	Padal et al. (2013b)
14.	*Acacia nilotica (A.arabica auct non)*	Tender leaves, stem bark, flowers	Leucorrhoea, syphilitic ulcers, dysentery, diarrhea, scabies; snake bite, diabetes	Raja Reddy et al. (1989); Sen and Behera (2003); Parthipan et al. (2011); Xavier et al. (2011); Murthy (2012); Padal et al. (2013b); Satyavathi et al. (2014); Prasanthi et al. (2014)
15.	*Acacia pennata*	Leaves, stem	Asthma, febrifuge	Reddy et al. (2006); Murthy (2012); Senthilkumar et al. (2013); Vaidyanathan et al. (2013)
16.	*Acacia polyacantha*	Root bark	Abortifacient	Prusti (2007)
17.	*Acacia rugata*	Fruit	Leucoderma	Padal et al. (2013a) Padal and Sandhyasri (2013)
18.	*Acacia sinuata*	Fruits	Dandruff	Parthipan et al. (2011)
19.	*Acalypha indica*	Leaves, whole plant	Insanity, skin diseases; get rid of intestinal worms, general tonic, STDs, jaundice	Sen and Behera (2003); Venkataratnam and Raju (2004); Jeevan Ram et al. (2007); Parthipan et al. (2011); Xavier et al. (2011); Murthy (2012); Savithramma et al. (2012); Lingaiah and Rao (2013); Padal et al. (2013b); Shanmukha Rao et al. (2014)
20.	*Acampe carinata*	Leaves	Ear problems	Padal et al. (2013c)
21.	*Acampe praemorsa*	Leaves	Skin disease	Padal et al. (2013c)
22.	*Acanthospermum hispidum*	Leaves	Cuts and wounds	Padal et al. (2013b)
23.	*Achyranthes aspera*	Roots, whole plant, leaves	Diuresis, ringworm, eczema, asthma, delivery, snake bite, scorpion sting, tooth ache, cuts and wounds, jaundice, for bone setting,	Raja Reddy et al. (1989); Sen and Behera (2003); Reddy et al. (2006); Prusti (2007); Parthipan et al. (2011); Murthy (2012); Lingaiah and Rao (2013); Padal et al. (2013b); Kannan and Kumar (2014); Ramakrishna and Saidulu (2014); Koteswara Rao et al. (2014)
24.	*Achyranthes bidentata*	Leaves	Cholera, testis pain and swellings	Dhatchanamoorthy et al. (2013)
25.	*Acorus calamus*	Rhizome	Fever, cough	Padal et al. (2013a, b)
26.	*Acrocephalus indicus*	Root	Leucorrhoea	Venkataratnam and Raju (2004, 2005)

TABLE 7.1 *(Continued)*

S. No.	Name of plant Species	Useful Parts	Medicinal uses against	References
27.	*Actiniopteris radiata*	Root	Snake bite, To promote fertility	Vijayakumar and Pullaiah (1998); Padal et al. (2013) b
28.	*Adansonia digitata*	Leaves	Dysentery	Padal et al. (2013b)
29.	*Adhatoda vasica (*Syn.: *A. zeylanica)*	Leaves	Tuberculosis, cough	Raja Reddy et al. (1989); Murthy (2012); Padal et al. (2013) b
30.	*Adiantum lunulatum*	Rhizome	Snake bite	Prusti (2007)
31.	*Adiantum philippense*	Leaves, root	Eczema, cough	Padal and Sandhyasri (2013); Padal et al. (2013b)
32.	*Adiantum raddianum*	Leaves	Stomach problems	Karuppusamy et al. (2009)
33.	*Aegle marmelos*	Fruit, leaves, bark	Piles, dysentery, asthma, jaundice, diabetes, diarrhea, cuts and wounds, scabies, pimples, itches, skin rashes	Vijayakumar and Pullaiah (1998); Girach (2001); Pullaiah et al. (2003); Reddy et al. (2003); Sen and Behera (2003); Venkataratnam and Raju (2004); C.S. Reddy et al. (2006); Murthy (2012); Padal et al. (2013a, b); Vaidyanathan et al. (2013); Satyavathi et al. (2014); Koteswara Rao et al. (2014); Shanmukha Rao et al. (2014)
34.	*Aerva lanata*	Whole plant, flowers, leaves, root	Kidney stones, urinary disorders, stomach problems, fever, diabetes, snake bite	Goud and Pullaiah (1996); Rao and Pullaiah (2001); Prusti (2007); Parthipan et al. (2011); Senthilkumar et al. (2013); Kannan and Kumar (2014); Prasanthi et al. (2014); Shanmukha Rao et al. (2014)
35.	*Aganosma cymosa*	Root	Cobra bite	Raja Reddy et al. (1989)
36.	*Agave americana*	Leaves	Rheumatism	Rekha and Senthilkumar (2014)
37.	*Agave cantula*	Leaf	Leucoderma	Padal et al. (2013a)
38.	*Agave sisalana*	Leaf	Ear drop for ear ache	Xavier et al. (2011)
39.	*Ageratum conyzoides*	Leaves	Skin infections, itching; cuts and wounds	Sen and Behera (2003); Padal et al. (2013a, b); Anandakumar et al. (2014); Koteswara Rao et al. (2014)
40.	*Ageratum racemosus*	Root	Snake bite	Kannan and Kumar (2014);

TABLE 7.1   *(Continued)*

S. No.	Name of plant Species	Useful Parts	Medicinal uses against	References
41.	*Ailanthes excelsa*	Stem bark, root, leaves	Skin infections; leucorrhoea and menorrhoea, snake bite, abscess, cough, swellings	Jeevan Ram and Raju (2001); Reddy et al. (2003); Venkata Subbiah and Savithramma (2012); Padal et al. (2013a, b); Lingaiah and Rao (2013); Ramakrishna and Saidulu (2014a, b)
42.	*Ajuga macrosperma*	Stem	Bodyache	Girach (2001)
43.	*Alangium hexapetalum*	Stem bark, leaves	Improve sexual potency	Rekha and Senthilkumar (2014)
44.	*Alangium salvifolium*	Root, leaves, seeds, stem bark, cotyledons	Snake bites, rabies, bone fracture, fever, paralysis, skin diseases, diarrhea, menorrhoea, diabetes, to remove poison, aphrodisiac	Goud and Pullaiah (1996); Vijayakumar and Pullaiah (1998); Jeevan Ram and Raju (2001); Jeevan Ram et al. (2007); Prusti (2007); Karuppusamy et al. (2009); Xavier et al. (2011); Murthy (2012); Padal et al. (2013)
45.	*Albizia amara*	Stem bark. Leaves, roots	Wound, Inflammation, snake bite, scorpion sting, skin diseases	Vijayakumar and Pullaiah (1998); Reddy et al. (2003); Senthilkumar et al. (2013); Vaidyanathan et al. (2013)
46.	*Albizia chinensis*	Leaves, roots	Cuts and wounds	Kumar and Pullaiah (1999);
47.	*Albizia lebbeck*	Stem bark, leaves	Neck pains, diarrhea tooth ache, snake bite, scorpion sting, eye injury, stomachache	Raja Reddy et al. (1989); Pullaiah et al. (2003); Jeevan Ram et al. (2007); Prusti (2007); Vaidyanathan et al. (2013); Manikandan and Lakshmanan (2014); Ramakrishna and Saidulu (2014)
48.	*Albizia odoratissima*	Stem bark	Skin diseases, cough, diabetes	Reddy et al. (2006); Murthy (2012)
49.	*Albizia thompsonii*	Leaves, bark, stem	Ulcers, skin diseases	C.S. Reddy et al. (2006); Vaidyanathan et al. (2013)
50.	*Allium cepa*	Bulb	Stomach problems	Pullaiah et al. (2003); Xavier et al. (2011)
51.	*Allium sativum*	Bulb	Wound healing, swelling, snake bite	Sen and Behera (2003); Karuppusamy et al. (2009); Kannan and Kumar (2014); Ramakrishna and Saidulu (2014b); Koteswara Rao et al. (2014)

TABLE 7.1 *(Continued)*

S. No.	Name of plant Species	Useful Parts	Medicinal uses against	References
52.	*Aloe barbadensis*	Stem	Skin allergy, leucorrhoea	Lingaiah and Rao (2013)
53.	*Aloe vera* (Syn.: *Aloe barbadensis*)	Leaves	Conjunctivitis, antifertility, skin diseases; cuts and wounds	Raja Reddy et al. (1989); Karuppusamy et al. (2009); Murthy (2012); Koteswara Rao et al. (2014)
54.	*Alpinia calcarata*	Rhizome	Stomach problems	Karuppusamy et al. (2009)
55.	*Alpinia galanga*	Rhizome	Indigestion, rheumatism	Xavier et al. (2011); Padal et al. (2013b)
56.	*Alseodaphne semecarpifolia*	Stem bark	Poisonous bites	Karuppusamy et al. (2009)
57.	*Alstonia scholaris*	Stem bark, whole plant	Lice, fever, rheumatism	Reddy et al. (2011); Sahu et al. (2013c)
58.	*Alstonia venenata*	Stem bark	Snake bite	Karuppusamy et al. (2009)
59.	*Alternanthera pungens*	Leaves	Kidney stones	Parthipan et al. (2011)
60.	*Alternanthera sessilis*	Leaves, root	For lactation, leucorrhoea	Raja Reddy et al. (1989); Lingaiah and Rao (2013)
61.	*Alysicarpus vaginalis*	Leaves	Cough	Reddy et al. (2006)
62.	*Ammania baccifera*	Leafy shoots	Poisonous bites	Parthipan et al. (2011)
63.	*Ammi majus*	Leaves	Dysentery	Padal et al. (2013b)
64.	*Amorphophallus campanulatus*	Corm	Piles, abdominal pain, asthma and tumors	Ramanathan et al. (2014)
65.	*Amorphophallus peonifolius*	Corm	Piles, asthma and bronchitis	Ramanathan et al. (2014)
66.	*Amorphophallus sylvaticus*	Corm	Piles	Anandakumar et al. (2014)
67.	*Ampelocissus latifolia*	Leaves, root	Dental troubles, dysentery	Murthy (2012)
68.	*Ampelocissus tomentosa*	Root tuber, bark	Aphrodisiac, fever, check bleeding	Vijayakumar and Pullaiah (1998); Prusti (2007)
69.	*Andrographis echioides*	Whole plant	Liver diseases	Sabjan et al. (2014)

TABLE 7.1 *(Continued)*

S. No.	Name of plant Species	Useful Parts	Medicinal uses against	References
70.	*Andrographis elongata*	Whole plant	Snake bite, scorpion sting, diabetes, jaundice and constipation	Alagesaboopathi (2012)
71.	*Andrographis lineata*	Leaves	Snake bite	Xavier et al. (2011)
72.	*Andrographis nallamalayana*	Root	Leucorrhea	Venkataratnam and Raju (2004, 2005)
73.	*Andrographis ovata*	Leaves	Scabies	Girach (2001)
74.	*Andrographis paniculata*	Leaves, whole plant, roots	Snake bite, cold, cough, diabetes, fevers, malaria, scabies, warts, snake bite	Raja Reddy et al. (1989); Vijaya-kumar and Pullaiah (1998); Girach (2001); Jeevan Ram and Raju (2001); Xavier et al. (2011); Murthy (2012); Sahu et al. (2013c); Kannan and Kumar (2014); Shanmukha Rao et al. (2014)
75.	*Andrographis serpyllifolia*	Leaves	Epiepsy,	Raja Reddy et al. (1989)
76.	*Angiopteris evecta*	Leaves	Diarrhea	Panda et al. (2012)
77.	*Anisochilus carnosus*	Leaves, shoots	Sores, bronchitis	Raja Reddy et al. (1989); C.S. Reddy et al. (2006)
78.	*Anisomeles malabarica*	Leaves	Fever, wound healing	Jeevan Ram and Raju (2001); Parthipan et al. (2011); Xavier et al. (2011)
79.	*Annona squamosa*	Seed paste, root, bark leaf	To kill lice, dys-entery, inflamma-tion, carbuncle, scorpion sting	Rao and Pullaiah (2001); Pulla-iah et al. (2003); Sen and Behera (2003); Parthipan et al. (2011); Vaidyanathan et al. (2013)
80.	*Anogeissus latifolia*	Stem bark, gum, leaves	Scabies, asthma, stomachache, can-cer, wounds, skin infections, cough	Jeevan Ram and Raju (2001); P.R. Reddy et al. (2003); Venkataratnam and Raju (2004); C.S. Reddy et al. (2006); Murthy (2012); Vaidyana-than et al. (2013)
81.	*Ardisia solanacea*	Stem bark	Toothache, cough, skin diseases	Girach (2001); Reddy et al. (2006); Karuppusamy et al. (2009)
82.	*Areca catechu*	Seeds	Chronic wounds	Karuppusamy et al. (2009)
83.	*Argemone mexicana*	Shoot, seeds, leaf, root, latex	Skin allergies, leu-corrhea; eczema, leucoderma, syphilis, scorpion sting	Goud and Pullaiah (1996); Pulla-iah et al. (2003); Sarangi and Sahu (2004); Venkata Ratnam and Raju (2005); Basha et al. (2011); Kumar et al. (2012); Vaidyanathan et al. (2013)

TABLE 7.1 *(Continued)*

S. No.	Name of plant Species	Useful Parts	Medicinal uses against	References
84.	*Argyreia cymosa*	Root, leaves	Micturition, Wound healing	Girach (2001); Karuppusamy et al. (2009)
85.	*Argyreia kleiniana*	Leaves	Tonsillitis	Vijayakumar and Pullaiah (1998);
86.	*Argyreia nervosa*	Stem, root	Syphilis, rheumatism, filaria, wound healing, paralysis	Sarangi and Sahu (2004); Panda (2007); Padal et al. (2013b); Sahu et al. (2013c)
87.	*Arisaema tortuosum*	Tuber	Kidney problems; curing headache	Padal et al. (2013c); Satyavathi et al. (2014)
88.	*Aristida funiculata*	Whole plant	Dantruff	Padal et al. (2013c)
89.	*Aristolochia bracteolata*	Leaves, while plant, root	Poisonous bites, nackache, colic, wounds, menstrual pains, dysmenorrhoea, chest pain, diabetes, abortifacient	Raja Reddy et al. (1989); Kumar and Pullaiah (1999); Rao and Pullaiah (2001); Sarangi and Sahu (2004); Xavier et al. (2011); Murthy (2012)
90.	*Aristolochia indica*	Roots, leaves	Snake bite, diarrhea, rash, cholera, leucorrhea; skin infection, abortifacient, impetigo, rash	Goud and Pullaiah (1996); Kumar and Pullaiah (1999); Jeevan Ram and Raju (2001); Pullaiah et al. (2003); Venkata Ratnam and Raju (2005); Jeevan Ram et al. (2007); Venkata Subbiah and Savithramma (2012); Murthy (2012); Lingaiah and Rao (2013); Kannan and Kumar (2014)
91.	*Artabotrys odoratissimus*	Leaves	To treat antifertility	Ranganathan et al. (2012)
92.	*Artemisia nilagirica*	Leaves	Respiratory disorder	Karuppusamy et al. (2009)
93.	*Artocarpus heterophyllus*	Root, latex	Skin diseases, ulcer and asthma, delayed delivery	Prusti (2007); Manikandan and Lakshmanan (2014)
94.	*Artocarpus hirsutus*	Fruits	Stomach problems	Karuppusamy et al. (2007)
95.	*Artocarpus integrifolia*	Stem bark	Head ache by a nasal drops	Das and Misra (1988)
96.	*Asclepias curassavica*	Latex	Inflammation	Senthilkumar et al. (2013)

TABLE 7.1 *(Continued)*

S. No.	Name of plant Species	Useful Parts	Medicinal uses against	References
97.	*Asparagus racemosus*	Root tubers	Diuresis, sun stroke, somach-ache, urinay prob-lems, migraine, dyspepsia, itching, aphrodisiac, galac-togogue, nervine tonic, for fertility, lactation	Raja Reddy et al. (1989); Panda (2007); Prusti (2007); Girach (2001); Rao and Pullaiah (2001); Chaudhari and Hutke (2002); Sarangi and Sahu (2004); Murthy (2012); Manikandan and Laksh-manan (2014); Shanmukha Rao et al. (2014)
98.	*Aspidoptrys cordata*	Root	Cough	Vijayakumar and Pullaiah (1998)
99.	*Asplenium aethiopicum*	Leaves	Venereal diseases	Karuppusamy et al. (2009)
100.	*Atalantia monophylla*	Leaves, fruits, stem bark, root	Allergic and skin diseases, cough, phlegm, rheuma-tism, post natal complaints	Jeevan Ram and Raju (2001); Jeevan Ram et al. (2007); Murthy (2012); Ranganathan et al. (2012); Vaidyanathan et al. (2013)
101.	*Azadirachta indica*	Leaves, root bark	small pox, scabies, itching, dysentery, menorrhea, snake bite, stomach pain	Raja Reddy et al. (1989); Pullaiah et al. (2003); Venkata Ratnam and Raju (2005); Lingaiah and Rao (2013); Vaidyanathan et al. (2013); Kannan and Kumar (2014)
102.	*Bacopa monnieri*	Leaves	Dysentery	Rao and Pullaiah (2001)
103.	*Balanites roxburghii*	Roots	Asthma, leprosy, eye infection, her-nia, tonic, to avoid conception	Satyavathi et al. (2014); Dinesh and Balaji (2015)
104.	*Bambusa arundinacea*	Stem, tender shoots, scrapings of stem	Pimples, cuts and wounds, abortive, indigestion	Raja Reddy et al. (1989); Vijaya-kumar and Pullaiah (1998); P.R. Reddy et al. (2003); Anandakumar et al. (2014); Senthilkumar et al. (2013); Sahu et al. (2013)
105.	*Barleria cristata*	Leaves	Asthma	Reddy et al. (2006)
106.	*Barleria longiflora*	Roots	Dropsy, cystitis	Jeevan Ram et al. (2007);
107.	*Barleria prionitis*	Leaves, fruits	Earache, asthma, toothache, cold, fever, whooping cough, for fertility	Raja Reddy et al. (1989); Sarangi and Sahu (2004); Padal and Sand-hyasri (2013); Sahu et al. (2013c); Satyavathi et al. (2014)
108.	*Barleria strigosa*	Root	Blood purifier	Prusti (2007)
109.	*Barringtonia acutangula*	Stem bark	Eczema, body pain	Jeevan Ram and Raju (2001); Reddy et al. (2007)

TABLE 7.1   *(Continued)*

S. No.	Name of plant Species	Useful Parts	Medicinal uses against	References
110.	*Basella rubra*	Leaves	Dyspepsia, piles	Padal et al. (2013a, b)
111.	*Bauhinia purpurea*	Seeds, bark	Lice, dysentery, leucorrhoea	Sahu et al. (2013); Padal et al. (2013b) Satyavathi et al. (2014)
112.	*Bauhinia racemosa*	Stem bark, leaves, fruits	Menorrhea, spermatorrhoea, diorrhoea, dysentery, malaria, to kill intestinal worms	Vijayakumar and Pullaiah (1998); Pullaiah et al. (2003); Venkata Ratnam and Raju (2004, 2005); Murthy (2012)
113.	*Bauhinia tomentosa*	Flowers	Diarrhoea, dysentery	Ranganathan et al. (2012)
114.	*Bauhinia vahlii*	Seeds, stem bark, leaves	Ulcers on tongue, blood dysentery, arthritis	Raja Reddy et al. (1989); Pullaiah et al. (2003); Panda et al. (2012); Murthy (2012); Padal et al. (2013a, b); Satyavathi et al. (2014)
115.	*Bauhinia variegata*	Leaves, stem bark, flowers	Stomach disorders, diarrhea, dysentery, asthma	Panda (2007); Pullaiah et al. (2003); Reddy et al. (2006)
116.	*Begonia malabarica*	Leaves	Wound healing	Senthilkumar et al. (2013)
117.	*Benincasa hispida*	Stem, fruit	Skin diseases, reduce stomach ache and ulcer pain	Sen and Behera (2003); Padal et al. (2013a, b)
118.	*Benkara malabarica* (Syn.: *Xeromphis malabarica)*	Root	Snake bite	Raja Reddy et al. (1989)
119.	*Beta vulgaris*	Tuber	Extract is given blood purifier	Ramanathan et al. (2014)
120.	*Benkara malabarica*	Stem bark	Asthma	Reddy et al. (2006)
121.	*Bidens pilosa*	Leaves	Whitlow	Padal et al. (2013b)
122.	*Biophytum candolleanum*	Leaves	Scorpion sting	Vaidyanathan et al. (2013)
123.	*Biophytum sensitivum*	Plant, root	Eczema, gonorrhea, poisonous bite	Vaidyanathan et al. (2013); Rekka and Senthil Kumar (2014); Ramakrishna and Saidulu (2014b)
124.	*Bixa orellana*	Roots, seeds, leaves	Fever, sprain, jaundice	Chaudhari and Hutke (2002); Padal et al. (2013b); Satyavathi et al. (2014)

TABLE 7.1 *(Continued)*

S. No.	Name of plant Species	Useful Parts	Medicinal uses against	References
125.	*Blepharis maderaspatensis*	Whole plant	Poisonous bites, jaundice and malaria	Prabu and Kumuthakalavalli (2012)
126.	*Blepharis repens*	Leaves	Syphilis, dysentery	Vijayakumar and Pullaiah (1998); Jeevan Ram et al. (2007)
127.	*Blepharispermum subsessile*	Roots	Leucorrhoea, rheumatism	Venkata Ratnam and Raju (2004); Prusti (2007)
128.	*Blumea eriantha*	Root	Rabies	Goud and Pullaiah (1996)
129.	*Boerhavia chinensis*	Root	For strength and vigor	Goud and Pullaiah (1996)
130.	*Boerhavia diffusa*	Root	Anti-emetic, pimples, jaundice, skin diseases, jaundice	Raja Reddy et al. (1989); Sen and Behera (2003); Padma Rao et al. (2007); Xavier et al. (2011); Padal et al. (2013b); Shanmukha Rao et al. (2014)
131.	*Bombax ceiba*	Root, fruits, stem bark	Menorrhagia, pimples, urinary problems, leucor-rhoea, spermator-rhoea, diarrhea, diabetes, fertility	Raja Reddy et al. (1989); Sen and Behera (2003); Sarangi and Sahu (2004); Murthy (2012); Padal et al. (2013a, b)
132.	*Borassus flabellifer*	Ramen-tum, stem sap	Cuts, skin diseases and ulcer	Raja Reddy et al. (1989); Manikan-dan and Lakshmanan (2014)
133.	*Borreria articularis*	Root	Delayed delivery	Prusti (2007)
134.	*Borreria hispida*	Stems, seeds	Gingivitis	Raja Reddy et al. (1989)
135.	*Borreria verticillata*	Root	Leucorrhea	Manikandan and Lakshmanan (2014)
136.	*Boswellia ovalifoliolata*	Resin	Scorpion sting	C.S. Reddy et al. (2006)
137.	*Boswellia serrata*	Resin, stem bark, leaves	Skin eruptions, fever, dysentery, diarrhea, skin allergies, scorpion sting, cuts and wounds	Jeevan Ram and Raju (2001); Rao and Pullaiah (2001); Pullaiah et al. (2003); Padal et al. (2013b); Vaidy-anathan et al. (2013); Ramakrishna and Saidulu (2014); Koteswara Rao et al. (2014)
138.	*Boucerosia umbellata* (Syn: *Caralluma umbellata*)	Stem	Diabetes	Anandakumar et al. (2014)

TABLE 7.1   *(Continued)*

S. No.	Name of plant Species	Useful Parts	Medicinal uses against	References
139.	*Bougainvillea spectabilis*	Leaves	Diabetes	Prasanthi et al. (2014)
140.	*Brassica juncea*	Seeds, leaves	Diarrhoea, snake bite, ear wound	Padal et al. (2013b); Vaidyanathan et al. (2013); Kannan and Kumar (2014)
141.	*Brassica nigra*	Seed oil	Rheumatoid arthritis	Padal et al. (2013a)
142.	*Breynia retusa*	Root	Nervous trouble	Padal et al. (2013d)
143.	*Breynia rhamnoides*	Root bark	For eye problem	Das and Misra (1988)
144.	*Bridelia crenulata*	Stem bark	Cuts, wounds	Prabakaran et al. (2013)
145.	*Bridelia montana*	Stem bark	Skin diseases, jaundice	Padal and Sandhyasri (2013); Padal et al. (2013b)
146.	*Bridelia retusa*	Stem bark	Asthma, chest pain	Reddy et al. (2006); Satyavathi et al. (2014)
147.	*Bryophyllum calycinum*	Leaves	Dysentery	Panda et al. (2012)
148.	*Bryophyllum pinnatum*	Leaf	Abdominal pain	Prusti (2007)
149.	*Buchanania axillaris*	Stem bark	Bone fracture	Rekka and Senthil Kumar (2014)
150.	*Buchanania cochinchinensis*	Stem bark	Diarrhea	Satyavathi et al. (2014)
151.	*Buchanania lanzan*	Leaves, gum, stem bark	Heart pain, diarrhea, indigestion	Vijayakumar and Pullaiah (1998); Girach (2001); Jeevan Ram and Raju (2001); Murthy (2012); Panda et al. (2012)
152.	*Buddleja asiatica*	Leaves	Wound healing	Karuppusamy et al. (2009)
153.	*Bulbophyllum neilgherrense*	Pseudo-bulb	Aphrodisiac	Karuppusamy et al. (2009)
154.	*Bupleurum mucronatum*	Seeds	Indigestion	Karuppusamy (2007)
155.	*Butea monosperma*	Resin, seeds, leaves, stem bark	Jaundice, cough, ringworm, as contraceptive, pain, wounds, itches. Leucorrhoea,	Vijayakumar and Pullaiah (1998); Jeevan Ram and Raju (2001); Venkata Ratnam and Raju (2004); Sarangi and Sahu (2004); Reddy et al. (2006); Padma Rao et al. (2007); Lingaiah and Rao (2013); Padal et al. (2013a, b); Manikandan and Lakshmanan (2014)

TABLE 7.1   *(Continued)*

S. No.	Name of plant Species	Useful Parts	Medicinal uses against	References
156.	*Butea superba*	Root, shoots, seeds	Diarrhea, arthritis, piles, anthelmintic	Murthy (2012); Panda et al. (2012); Satyavathi et al. (2014)
157.	*Byttneria herbacea*	Rhizome, whole plant	Ulcers, nervous disorders	Sen and Behera (2003); Savithramma et al. (2012)
158.	*Cadaba fruticosa*	Leaves, fruit	Leucoderma, eczema, itching, bone fracture	Jeevan Ram and Raju (2001); Pullaiah et al. (2003); Padal et al. (2013b); Vaidyanathan et al. (2013)
159.	*Caesalpinia bonduc*	Leaves, seeds	Hydrocele, asthma, boils	Raja Reddy et al. (1989); Vijayakumar and Pullaiah (1998); Sen and Behera (2003); Reddy et al. (2006); Jeevan Ram et al. (2007)
160.	*Caesalpinia pulcherrima*	Flower	Fever	Padal et al. (2013b)
161.	*Cajanus cajan*	Leaves	Toothache	Raja Reddy et al. (1989)
162.	*Cajanus scarabaeoides*	Whole plant	Skin itching	Pullaiah et al. (2003)
163.	*Caladium bicolor*	Tuber	Snake bite	Padal et al. (2013a, b)
164.	*Calamus rotang*	Seeds, leaf	Cold and cough	Manikandan and Lakshmanan (2014); Satyavathi et al. (2014)
165.	*Calophyllum inophyllum*	Seeds	Scabies	Manikandan and Lakshmanan (2014)
166.	*Calotropis gigantea*	Leaves, roots, flower, latex	Earche, rheumatism, tonsilites, snake bite, cramps, pain, arthritis, to induce abortion, purulous skin affections, syphilitic ulcers	Raja Reddy et al. (1989); Vijayakumar and Pullaiah (1998); P.R. Reddy et al. (2003); Sen and Behera (2003); Parthipan et al. (2011); Lingaiah and Rao (2013); Kannan and Kumar (2014)
167.	*Calotropis procera*	Root, Latex, shoots	Scorpion sting, mumps, skin diseases, migraine	Raja Reddy et al. (1989); Venkata Ratnam and Raju (2004); Jeevan Ram et al. (2007); Senthilkumar et al. (2013)
168.	*Calycopteris floribunda*	Leaves, Stem bark	Wound healing	P.R. Reddy et al. (2003); Padal et al. (2013b)
169.	*Canarium strictum*	Bark	Rheumatic pains, skin diseases	Vaidyanathan et al. (2013)
170.	*Canavalia gladiata*	Leaves, root	Antifertility, anthelmintic, liver enlargement	Venkataratnam and Raju (2004); Rao and Pullaiah (2001); Sabjan et al. (2014)

TABLE 7.1 *(Continued)*

S. No.	Name of plant Species	Useful Parts	Medicinal uses against	References
171.	*Canavalia virosa*	Flower	Asthma	Vaidyanathan et al. (2013)
172.	*Canna edulis*	Tuber	Throat pain	Padal et al. (2013b)
173.	*Canna indica*	Roots, tuber	Cuts and wounds, ringworm	Padal et al. (2013b); Koteswara Rao et al. (2014)
174.	*Cannabis sativa*	Leaves	Diarrhoea, cuts and wounds	Padal et al. (2013b); Koteswara Rao et al. (2014)
175.	*Canscora decussata*	Whole plant	Mouth ulcers	Rao and Pullaiah (2001)
176.	*Canthium dicoccum*	Leaves	Stomach pain	Anandakumar et al. (2014)
177.	*Canthium parviflorum*	Leaves, root	Fever, boils, anthelmintic, ringworm	Goud and Pullaiah (1996); P.R. Reddy et al. (2003); Venkata Subbiah and Savithramma (2012)
178.	*Capparis divaricata*	Root, stem bark	Asthma, bronchitis, jaundice; chest pain	Goud and Pullaiah (1996); Anandakumar et al. (2014); Rekka and Senthil Kumar (2014)
179.	*Capparis grandis*	Leaves	Skin eruptions, diarrhea, dysentery	Jeevan Ram and Raju (2001); Pullaiah et al. (2003)
180.	*Capparis sepiaria*	Root, leaves, root bark	Cooling, skin problems, post delivery pains, lactation	Raja Reddy et al. (1989); Vijayakumar and Pullaiah (1998); Jeevan Ram et al. (2007); Padal et al. (2013b)
181.	*Capparis zeylanica*	Leaves, stem bark, root bark, root, fruits	Piles, haematuria, jaundice, syphilis, urinary calculi, earache, furits, improves immunity	Raja Reddy et al. (1989); Girach (2001); Sarangi and Sahu (2004); Jeevan Ram et al. (2007); Murthy (2012); Satyavathi et al. (2014)
182.	*Capsicum annuum*	Ripe fruits	Cuts	P.R. Reddy et al. (2003)
183.	*Caralluma adscendens*	Stem	Abdominal pains, cough	Goud and Pullaiah (1996); Reddy et al. (2006)
184.	*Caralluma attenuata*	Stem	Bone fracture	Dhatchanamoorthy et al. (2013)
185.	*Cardiospermum halicacabum*	Leaves	Ear ache, for treating fits	Raja Reddy et al. (1989); Xavier et al. (2011); Vaidyanathan et al. (2013)
186.	*Cardiospermum luridum*	Leaves, whole plant	Laxative, diuretic, dandruff, dysentery	Ranganathan et al. (2012)

TABLE 7.1 *(Continued)*

S. No.	Name of plant Species	Useful Parts	Medicinal uses against	References
187.	*Careya arborea*	Stem bark, leaves	Cough, skin diseases	Pullaiah et al. (2003); Reddy et al. (2006); Murthy (2012)
188.	*Carica papaya*	Leaves, young fruits, fruit, latex	Jaundice, indigestion, cuts and wounds, galactogogue, used for abortion;	Raja Reddy et al. (1989); Kumar and Pullaiah (1999); Sen and Behera (2003); Parthipan et al. (2011); Padal et al. (2013b); Koteswara Rao et al. (2014)
189.	*Carmona retusa*	Leaves	Wounds	Jeevan Ram et al. (2007);
190.	*Caryota urens*	Seed, toddy	Dandruff, urinary problems	Prusti (2007); Murthy (2012); Satyavathi et al. (2014)
191.	*Cascabela thevetia*	Latex, leaves	Muscle pains, ringworm and scabies, skin diseases, boils	Rao and Pullaiah (2001); P.R. Reddy et al. (2003); Sen and Behera (2003); Padal et al. (2013b)
192.	*Casearia elliptica*	Stem bark	To heal wounds	Murthy (2012); Koteswara Rao et al. (2014)
193.	*Cassia absus*	Leaves	Stomachache, itches, fever, conjunctivitis	Raja Reddy et al. (1989); Jeevan Ram and Raju (2001), Jeevan Ram et al. (2007); Padal et al. (2013b)
194.	*Cassia alata*	Leaves	Bronchitis, eczema, snake bite	Padal et al. (2013a, b); Vaidyanathan et al. (2013)
195.	*Cassia auriculata*	Leaves, flowers, flower buds, seeds	Snake bite, scorpion sting, menorrhoea, leucorrhoea, diabetes, dysentery	Raja Reddy et al. (1989); Venkataratnam and Raju (2004, 2005); Padal et al. (2013b); Vaidyanathan et al. (2013); Ramakrishna and Saidulu (2014b); Prasanthi et al. (2014)
196.	*Cassia fistula*	Fruit, leaves, bark, shoots, seeds, flowers	Intestinal worms, cataract, labor pain, wounds, burns, mad dog bite, indigestion, menstrual disorders, skin diseases, tonic, jaundice, dog bite, diarrhea, heart pain, leprosy, fever	Vijayakumar and Pullaiah (1998); Jeevan Ram and Raju (2001); P.R. Reddy et al. (2003); Venkataratnam and Raju (2004); Panda (2007); Parthipan et al. (2011); Murthy (2012); Kumar et al. (2012); Padal et al. (2013a, b); Vaidyanathan et al. (2013); Sahu et al. (2013c)
197.	*Cassia italica* (Syn.: *C. obtusa*)	Leaves	Constipation, jaundice	Raja Reddy et al. (1989); Jeevan Ram et al. (2007);
198.	*Cassia montana*	Stem bark	Leucorrhea	Venkata Ratnam and Raju (2005)

TABLE 7.1   *(Continued)*

S. No.	Name of plant Species	Useful Parts	Medicinal uses against	References
199.	*Cassia obtusifolia*	Leaves	Scorpion sting	Lingaiah and Rao (2013)
200.	*Cassia occidentalis*	Root, fruit, leaves	Bone setting, cough, itching, diabetes, pains; cuts and wounds, stomachache	Raja Reddy et al. (1989); Pullaiah et al. (2003); Reddy et al. (2006); Lingaiah and Rao (2013); Vaidyanathan et al. (2013); Koteswara Rao et al. (2014)
201.	*Cassia pumila*	Root	Eczema	Pullaiah et al. (2003)
202.	*Cassia senna*	Leaves	Urticaria	Sen and Behera (2003)
203.	*Cassia siamea*	Whole plant	Laxative, snake bite	Vaidyanathan et al. (2013)
204.	*Cassia tora*	Leaves, root	Asthma, cough, skin diseases; stomach pain, diabetes	Raja Reddy et al. (1989); C.S. Reddy et al. (2006); Venkata Subbiah and Savithramma (2012); Anandakumar et al. (2014); Prasanthi et al. (2014)
205.	*Cassytha filiformis*	Plant	Body pains, flatulence	Rao and Pullaiah (2001); Rekka and Senthil Kumar (2014)
206.	*Catharanthus pusillus*	Leaves, whole plant	Dandruff, lice, sores on the head, ulcer	Raja Reddy et al. (1989); Dhatchanamoorthy et al. (2013)
207.	*Catharanthus roseus*	Whole plant, flowers	Blood pressure, cuts and wounds, diabetes	Murthy (2012); Koteswara Rao et al. (2014); Prasanthi et al. (2014)
208.	*Catunaregum spinosa*	Fruit	Diarrhoea	Pullaiah et al. (2003)
209.	*Cayratia pedata*	Leaves, fruit	Hydrocele, lice, dandruff, asthma, scabies, ulcer, uterine reflexes	Raja Reddy et al. (1989); Gritto et al. (2012); Murthy (2012); Vaidyanathan et al. (2013)
210.	*Ceiba pentandra*	Leaves; stem bark, root	Boils, skin disease, impotency, scorpion sting	Padal et al. (2013b); Vaidyanathan et al. (2013); Ramakrishna and Saidulu (2014b)
211.	*Celastrus paniculatus*	Seeds, root, stem bark	Body pains, snake bite, syphilis	Rao and Pullaiah (2001); Sarangi and Sahu (2004); Prusti (2007)
212.	*Celosia argentea*	Seeds, leaves	Cough, stomach ulcer	Reddy et al. (2006); Padal et al. (2013b)

TABLE 7.1 *(Continued)*

S. No.	Name of plant Species	Useful Parts	Medicinal uses against	References
213.	*Centella asiatica*	Leaves, roots	To improve memory, as tonic, hemiphlegia, dysentery, headache, skin diseases, gastric ulcer, nervous weakness, snake bite	Raja Reddy et al. (1989); Rao and Pullaiah (2001); Panda (2007); Prusti (2007); Kumar and Pullaiah (1999); Basha et al. (2011); Xavier et al. (2011); Kannan and Kumar (2014); Shanmukha Rao et al. (2014)
214.	*Ceratopteris thallictroides*	Whole plant	Cough	Kumar and Pullaiah (1999)
215.	*Ceriscoides turgida*	Fruit, stem bark	Insomnia, leucorrhoea	Vijayakumar and Pullaiah (1998); Ramakrishna and Saidulu (2014b)
216.	*Ceropegia bulbosa*	Leaves	Dysentery, diarrhea	Basha et al. (2011)
217.	*Ceropegia juncea*	Stem	Intestinal ulcer	Karuppusamy et al. (2009)
218.	*Ceropegia spiralis*	Root tuber	Indigestion	C.S. Reddy et al. (2006)
219.	*Chenopodium ambrosioides*	Whole plant	Anthelmintic	Karuppusamy et al. (2009)
220.	*Chlorophytum arundinaceum*	Root tuber	Galactagogue	Padal et al. (2013a)
221.	*Chlorophytum borivilianum*	Roots	Cough, tonic,	Chaudhari and Hutke (2002)
222.	*Chlorophytum tuberosum*	Root tubers	Scorpion sting	Ramana naidu et al. (2012)
223.	*Chloroxylon swietenia*	Gum, stem bark, leaves	Urinary disorders, itches, asthma, snake bite, scorpion sting, fever, diabetes, ulcers	Rao and Pullaiah (2001); Jeevan Ram and Raju (2001); Sen and Behera (2003); Venkataratnam and Raju (2004); Reddy et al. (2006); Prasanthi et al. (2014)
224.	*Christella dentata*	Leaves	External skin infection	Kumar et al. (2012)
225.	*Chromolaena odorata*	Leaves	Cuts and wounds	Koteswara Rao et al. (2014)
226.	*Chrysopogon zizanioides*	Root	Dandruff, skin diseases	Ranganathan et al. (2012)
227.	*Cipadessa baccifera*	Stem bark, leaves	Toothache, chickenpox, diarrhea	Girach (2001); Padal et al. (2013a, b); Vaidyanathan et al. (2013)

TABLE 7.1   *(Continued)*

S. No.	Name of plant Species	Useful Parts	Medicinal uses against	References
228.	*Cissampelos periara*	Root, leaves, whole plant	Bronchitis, stomachache, epilepsy, fever, itching, snake bite, diarrhea	Raja Reddy et al. (1989); C.S. Reddy et al. (2006); Jeevan Ram et al. (2007); Prusti (2007); Basha et al. (2011); Padal et al. (2013b); Vaidyanathan et al. (2013)
229.	*Cissus quadrangularis*	Stem, roots, leaves	Bone fracture, body pains, gas trouble, burns, wounds, cough, dyspepsia, paralysis, asthma	Kumar and Pullaiah (1999); Rao and Pullaiah (2001); Jeevan Ram et al. (2007); Parthipan et al. (2011); Murthy (2012); Padal et al. (2013b); Vaidyanathan et al. (2013)
230.	*Cissus setosa*	Leaves	Worms	Vaidyanathan et al. (2013);
231.	*Cissus vitiginea*	Fruits	Asthma	Reddy et al. (2006)
232.	*Citrullus colocynthis*	Root, fruit mesocarp	Epilepsy, cough, hysteria	Rajareddy et al. (1989); Reddy et al. (2006); Jeevan Ram et al. (2007);
233.	*Citrus aurantifolia*	Fruit	Wounds	P.R. Reddy et al. (2003)
234.	*Citrus limon*	Fruits	Stomach ache, diarrhea, dandruff, hair fall, snake bite	Lingaiah and Rao (2013); Padal et al. (2013b); Kannan and Kumar (2014)
235.	*Citrus medica*	Fruits	Dysentery	Padal et al. (2013b)
236.	*Clausena dentata*	Leaves	Chronic wounds	Karuppusamy et al. (2009)
237.	*Clausena excavata*	Root	Diarrhea	Panda et al. (2012)
238.	*Cleistanthus collinus*	Seeds	To commit suicide	Ramana naidu et al. (2012)
239.	*Clematis gouriana*	Leaves, stem	To remove body odor, to kill lice	Karuppusamy and Pullaiah (2005); Basha et al. (2011)
240.	*Clematis roylei*	Leaves	Skin diseases	Rao and Pullaiah (2007)
241.	*Clematis wightiana*	Leaves	Rheumatism	Rao and Pullaiah (2007)
242.	*Cleome gynandra*	Leaves	Earache, asthma, skin allergies, ringworm, paralysis, cold	Raja Reddy et al. (1989); Pullaiah et al. (2003); Sen and Behera (2003); Reddy et al. (2006); Padal et al. (2013a); Vaidyanathan et al. (2013)
243.	*Cleome monophylla*	Leaves	Ear ache	Senthilkumar et al. (2013)

TABLE 7.1   *(Continued)*

S. No.	Name of plant Species	Useful Parts	Medicinal uses against	References
244.	*Cleome viscosa*	Root, leaves	Tooth pain, menstrual problems; head ache and tooth ache	Prusti (2007); Parthipan et al. (2011); Dhatchanamoorthy et al. (2013)
245.	*Clerodendrum inerme*	Leaves	Diabetes	Raja Reddy et al. (1989);
246.	*Clerodendrum phlomidis*	Leaves	Urinary disorder	Xavier et al. (2011)
247.	*Clerodendrum serratum*	Stem bark, young shoots, roots, leaves	Asthma and bronchitis; head ache, dropsy, wounds, nervous disorders, malaria, burns	Jeevan Ram and Raju (2001); Reddy et al. (2006); Jeevan Ram et al. (2007); Prusti (2007); Padal et al. (2013a); Sahu et al. (2013c)
248.	*Clitoria ternatea*	Leaves, roots, white flowers, whole plant	Thorn cracks, cough, snake bite, memory enhancer, pimples, swelling of legs	Chaudhari and Hutke (2002); Sen and Behera (2003); Dhatchanamoorthy et al. (2013); Vaidyanathan et al. (2013); Kannan and Kumar (2014)
249.	*Coccinia grandis*	Leaves, whole plant	Mouth ulcer, diabetes	Manikandan and Lakshmanan (2014); Prasanthi et al. (2014)
250.	*Cocculus hirsutus*	Leaves, root, entire plant, leaves	Stomach pain, eczema, fever, menorrhoea, leucorrhoea, menstrual pain, gonorrhea, diabetes, post natal pains, boils, blood purifier, leukemia, cooling tonic	Raja Reddy et al. (1989); Rao and Pullaiah (2001); P.R. Reddy et al. (2003); Venkataratnam and Raju (2004); Jeevan Ram et al. (2007); Basha et al. (2011); Parthipan et al. (2011); Murthy (2012); Padal et al. (2013b); Vaidyanathan et al. (2013); Ramakrishna and Saidulu (2014b)
251.	*Cochlospermum religiosum*	Resin, stem bark	Itches, bone fracture	Jeevan Ram and Raju (2001); Rao and Pullaiah (2001); Vaidyanathan et al. (2013)
252.	*Cocos nucifera*	Flowers, root	Diabetes, palpitation	Prusti (2007); Prasanthi et al. (2014)
253.	*Colocasia esculenta*	Rhizome	Haemorrhage	Murthy (2012)
254.	*Corchorus capsularis*	Seeds	Stomach ache	Padal et al. (2013b)

TABLE 7.1 *(Continued)*

S. No.	Name of plant Species	Useful Parts	Medicinal uses against	References
255.	*Colocasia esculenta*	Tuber	Hair tonic	Ramanathan et al. (2014)
256.	*Coldenia procumbens*	Leaves, whole plant	Scorpion poison, Hydrocele, scabies, itching, anemia, rheumatism	Rajareddy et al. (1989); Vijayakumar and Pullaiah (1998); Pullaiah et al. (2003); Padal and Sandhyasri (2013); Padal et al. (2013b)
257.	*Colebrookia oppositifolia*	Leaves	Headache, wound healing	Girach (2001); Karuppusamy et al. (2009)
258.	*Coleus aromaticus*	Leaves	Cough	Xavier et al. (2011)
259.	*Coleus forskohlii*	Tuber	Cold, cough and fever	Ramanathan et al. (2014)
260.	*Combretum albidum*	Leaves, flowers	Skin allergies, dysentery, diarrhea	Pullaiah et al. (2003); Karuppusamy et al. (2009)
261.	*Combretum ovalifolium*	Bark	Jaundice	Vaidyanathan et al. (2013)
262.	*Combretum roxburghii*	Leaves	Diarrhoea	Girach (2001)
263.	*Commelina benghalensis*	Whole plant, stem	Herpes, wound healing; diuretic, febrifuge	Sen and Behera (2003); Parthipan et al. (2011); Ranganathan et al. (2012)
264.	*Commiphora caudata*	Leaves, fruits	Skin diseases, sex stimulant in males, stomach ache, to improve digestion	Vijayakumar and Pullaiah (1998); Rao and Pullaiah (2001); Karuppusamy et al. (2009); Vaidyanathan et al. (2013)
265.	*Commiphora wightii*	Latex	Ulcers	Sen and Behera (2003)
266.	*Convolvulus sepiaria*	Root	Infertility	Murthy (2012)
267.	*Corallocarpus epigaeus*	Tuber, leaves, root	Snake bite, asthma; allergic and dermatitis, epilepsy, rheumatism	Kumar and Pullaiah (1999); Reddy et al. (2006); Basha et al. (2011); Dhayapriya and Senthilkumar (2014)
268.	*Corchorus aestuans*	Root	Asthma	Rao and Pullaiah (2001)
269.	*Corchorus olitorius*	Seed	Ear pain	Padal et al. (2013b)
270.	*Cordia dichotoma*	Leaves	Jaundice	Padal et al. (2013b)

TABLE 7.1 *(Continued)*

S. No.	Name of plant Species	Useful Parts	Medicinal uses against	References
271.	*Coriandrum sativum*	Seeds	Sinusitis	Dhayapriya and Senthilkumar (2014)
272.	*Costus speciosus*	Rhizome	Skin diseases, ringworm, rheumatism, snake bite, anthelmintic, chicken pox, galactogogue, tonic	Jeevan Ram and Raju (2001); Chaudhari and Hutke (2002); Prusti (2007); Murthy (2012); Venkata Subbiah and Savithramma (2012); Padal et al. (2013b); Shanmukha Rao et al. (2014)
273.	*Crinum asiaticum*	Bulb	To eject stritching thorns, anthelmintic, aphrodisiac	Karuppusamy et al. (2009); Murthy (2012)
274.	*Crotalaria laburnifolia*	Root	Snake bite	Padal et al. (2013b)
275.	*Crotalaria pallida*	Seeds	Narcotic	Padal et al. (2013b)
276.	*Crotalaria paniculata*	Leaves	Eczema	C.S. Reddy et al. (2006)
277.	*Crotalaria ramosissima*	Leaves	Diarrhoea	Pullaiah et al. (2003)
278.	*Crotalaria spectabilis*	Root	Dysentery	Panda et al. (2012)
279.	*Croton bonplandianus* (Syn: *C. sparsiflorus*)	Leaves, root	Healing wounds, to relieve pain, treat dysentery	Vijayakumar and Pullaiah (1998); Panda et al. (2012)
280.	*Croton roxburghii*	Stem	Wounds	Sen and Behera (2003)
281.	*Croton scabiosus*	Stem bark	Leucorrhea	Venkata Ratnam and Raju (2005)
282.	*Cryptolepis buchanani*	Root	Whitlow	Sen and Behera (2003)
283.	*Cucumis sativus*	Fruit	Urinary trouble	Padal et al. (2013b)
284.	*Curculigo orchioides*	Rhizome, leaves, root tuber	Diabetes, itches, scabies, pimples, whitlow, cough, dysentery, as aphrodisiac, impotency; cuts and wounds, irregular menstruation, headache, snake bite	Vijayakumar and Pullaiah (1998); Pullaiah et al. (2003); Jeevaan Ram and Raju (2001); Chaudhari and Hutke (2002); Sen and Behera (2003); Prusti (2007); Murthy (2012); Ranganathan et al. (2012); Padal et al. (2013b); Ramakrishna and Saidulu (2014b); Kannan and Kumar (2014); Koteswara Rao et al. (2014); Shanmukha Rao et al. (2014)

TABLE 7.1   *(Continued)*

S. No.	Name of plant Species	Useful Parts	Medicinal uses against	References
285.	*Curcuma angustifolia*	Rhizome	Cuts, wounds, body pain and bone fracture, galactogogue	Prabu and Kumuthakalavalli (2012); Padal et al. (2013b)
286.	*Curcuma aromatica*	Rhizome	Skin diseases	Padal et al. (2013b)
287.	*Curcuma caesia*	Rhizome	Cuts, wounds	Padal et al. (2013c)
288.	*Curcuma inodora*	Rhizome	Chronic wounds; swellings	C.S. Reddy et al. (2006); Patanaik et al. (2009)
289.	*Curcuma longa*	Rhizome	Scabies, cough and cold, diabetes	Prabu and Kumuthakalavalli (2012); Prasanthi et al. (2014)
290.	*Curcuma pseudomontana*	Rhizome	Antiseptic	Patanaik et al. (2009)
291.	*Cuscuta reflexa*	Whole plant	Urticaria, aphrodisiac, piles, pruritus	Sen and Behera (2003); Karuppusamy et al. (2007); Prusti (2007); Padal et al. (2013b)
292.	*Cyanotis axillaris*	Tuberous root	Skin diseases	Rajareddy et al. (1989)
293.	*Cyanotis tuberosa*	Root tubers	For strength, blood motions, cough	Goud and Pullaiah (1996); Jeevan Ram et al. (2007); Murthy (2012)
294.	*Cycas beddomei*	Underground rhizome	Aphrodisiac	Patanaik et al. (2009)
295.	*Cycas circinalis*	Young cone	Snake bite; venereal disease	Gritto et al. (2012); Karuppusamy et al. (2009)
296.	*Cyclea peltata*	Leaves, tubers	Boils and blisters, stomachache	Rao and Pullaiah (2007); Murthy (2012)
297.	*Cymbidium aloifolium*	Pseudobulb, leaves	Wound healing, ear problems	Karuppusamy et al. (2009); Xavier et al. (2011)
298.	*Cynanchum callialatum*	Latex	Ulcer	Karuppusamy et al. (2009)
299.	*Cynodon dactylon*	Whole plant, leaves, root	Skin diseases, ear problems, snake bite	Raja Reddy et al. (1989); Sen and Behera (2003); Prusti (2007); Kannan and Kumar (2014)
300.	*Cyperus rotundus*	Root, tubers	Haematuria, skin diseases, blood purifier, jaundice, stomach problems, asthma, skin diseases, scorpion sting	Sen and Behera (2003); Ramanathan et al. (2014); Karuppusamy et al. (2007) (2009); Ramakrishna and Saidulu (2014)

TABLE 7.1 *(Continued)*

S. No.	Name of plant Species	Useful Parts	Medicinal uses against	References
301.	*Dalbergia latifolia*	Root, bark	Menorrhagia, fever	Vaidyanathan et al. (2013); Satyavathi et al. (2014)
302.	*Dalbergia paniculata*	Stem bark	Baldness	Murthy (2012)
303.	*Dalbergia sissoo*	Stem bark, root, leaves	Urinary infection, heat of body, diarrhea	Padal et al. (2013b); Vaidyanathan et al. (2013); Satyavathi et al. (2014)
304.	*Datura fastuosa*	Leaves	Boils	P.R. Reddy et al. (2003)
305.	*Datura innoxia*	Fruit, leaves	Dog bite, headache, diabetes, eczema, scabies, skin diseases, itching	Vijayakumar and Pullaiah (1998); Kumar and Pullaiah (1999); Pullaiah et al. (2003); Dhatchanamoorthy et al. (2013); Padal et al. (2013b)
306.	*Datura metel*	Fruit, leaves, roots, stem	Whitlow, asthma; eye disorders, pains, sprains, cuts and wounds, scabies, paralysis, for lactation	Sen and Behera (2003); Reddy et al. (2006); Jeevan Ram et al. (2007); Panda (2007); Xavier et al. (2011); Lingaiah and Rao (2013); Sahu et al. (2013c); Koteswara Rao et al. (2014); Shanmukha Rao et al. (2014)
307.	*Decalepis hamiltonii*	Fruits, roots	For easy delivery, cooling, tonic,	Raja Reddy et al. (1989); Vijayakumar and Pullaiah (1998); C.S. Reddy et al. (2006); Basha et al. (2011)
308.	*Decaschistia caddapahensis*	Roots	Aphrodisiac	C.S. Reddy et al. (2006)
309.	*Delonix elata*	Leaves	Constipation, rheumatism	Raja Reddy et al. (1989); Padal et al. (2013b)
310.	*Delonix regia*	Root	Asthma	Satyavathi et al. (2014)
311.	*Dendrobium herbaceum*	Stem	Drops used for eye trouble	Reddy et al. (2005)
312.	*Dendrobium macrostachyum*	Leaves	Ear ache	Patanaik et al. (2009)
313.	*Dendropthoe falcata*	Stem bark	Asthma	Reddy et al. (2006)
314.	*Derris indica*	Root	Snake bite	Padal et al. (2013b)
315.	*Derris scandens*	Roots	Stomachache	Goud and Pullaiah (1996)
316.	*Desmodium biarticulatum*	Whole plant	Asthma	Reddy et al. (2006)

TABLE 7.1 *(Continued)*

S. No.	Name of plant Species	Useful Parts	Medicinal uses against	References
317.	*Desmodium gangeticum*	Leaves, root	Boils, asthma, fever, dysentery, rheumatism	Sen and Behera (2003); Reddy et al. (2006); Panda (2007); Padal and Sandhyasri (2013); Padal et al. (2013b)
318.	*Desmodium gyrans*	Plant	Kidney stone	Rekha and Senthil Kumar (2014)
319.	*Desmodium pulchellum*	Leaves	Wounds	Padal et al. (2013b)
320.	*Desmodium triflorum*	Leaves	Dysentery, diarrhea, convulsions	Murthy (2012)
321.	*Dichrostachys cinerea*	Leaves, root, bark	Menstrual disorders, leucorrhoea, menorrhoea, rheumatism, diarrhea, joint pains	Rao and Pullaiah (2001); Jeevan Ram et al. (2007); Karuppusamy et al. (2009); Murthy (2012); Padal et al. (2013b)
322.	*Didymocarpus gambleanus*	Leaves	Wound healing	Karuppusamy et al. (2009)
323.	*Dillenia indica*	Fruit, stem bark	Abdominal pain, stomach ache	Basha et al. (2011); Padal et al. (2013a, b)
324.	*Dillenia pentagyna*	Stem bark	Stomach ache	Murthy (2012)
325.	*Dioscorea alata*	Tuber	For indigestion	Ramanathan et al. (2014)
326.	*Dioscorea bulbifera*	Tubers	Wounds; boils, cough, cold, piles, sterility	Vijayakumar and Pullaiah (1998); Jeevan Ram and Raju (2001); Murthy (2012); Shanmukha Rao et al. (2014)
327.	*Dioscorea esculenta*	Tuber	Rheumatoid arthritis	Ramanathan et al. (2014)
328.	*Dioscorea oppositifolia*	Tuber	Leucorrhoea, kidney trouble, snake bite, scorpion sting	Goud and Pullaiah (1996); Venkataratnam and Raju (2004); Basha et al. (2011); Ramanathan et al. (2014)
329.	*Diospyros chloroxylon*	Leaves, stem bark	Wounds	P.R. Reddy et al. (2003); Murthy (2012)
330.	*Diospyros ebenum*	Stem bark	Wounds, ulcer	Ramana Naidu et al. (2012)
331.	*Diopyros malabarica*	Stem bark	Dysentery	Panda et al. (2012)
332.	*Diospyros melanoxylon*	Root, stem bark, fruit	Malaria, anemia in pregnant women, post natal pains, cough	Rao and Pullaian (2001); Venkataratnam and Raju (2004); Prusti (2007); Sahu et al. (2013a)

TABLE 7.1 *(Continued)*

S. No.	Name of plant Species	Useful Parts	Medicinal uses against	References
333.	*Diospyros ovalifolia*	Stem bark	Burning wounds, swellings	Venkataratnam and Raju (2004); Anandakumar et al. (2014)
334.	*Diplocyclos palmatus*	Root, seeds	Snake bite, tooth decay, improve fertility	Rajareddy et al. (1989); Padal et al. (2013a, b)
335.	*Dodonaea angustifolia*	Leaves, stem, roots	Sprains, bone fracture, fever, diarrhea, rheumatism,	Raja Reddy et al. (1989); Vaidyanathan et al. (2013)
336.	*Dodonea viscosa*	Leaves	Intestinal worms, swellings, fits	Prabu and Kumuthakalavalli (2012); Satyavathi et al. (2014)
337.	*Dolichos lablab*	Leaf	Sprain	Prusti (2007);
338.	*Drimia indica* (Syn.: *Scilla indica*)	Bulb	Diarrhoea	Pullaiah et al. (2003)
339.	*Drosera burmannii*	Whole plant	Blood dysentery	Vaidyanathan et al. (2013)
340.	*Drynaria quercifolia*	Rhizome	Rheumatism	Ramanathan et al. (2014)
341.	*Dysophylla quadrifolia*	Leaves	Chickenpox	Padal et al. (2013b)
342.	*Echinops echinatus*	Root	Bronchitis, wounds	Raja Reddy et al. (1989); Kumar and Pullaiah (1999);
343.	*Eclipta prostrata* (Syn.: *E. alba*)	Whole plant, leaves	Fever, filariasis, jaundice, bleeding, dysentery, skin allergy, hair fall, dandruff	Raja Reddy et al. (1989); Padal et al. (2013); Lingaiah and Rao (2013); Sahu et al. (2013c); Satyavathi et al. (2014); Shanmukha Rao et al. (2014)
344.	*Elephantopus scaber*	Leaves, root	Tongue dryness; diarrhea, tooth decay, scabies, skin diseases	Jeevan Ram and Raju (2001); Sen and Behera (2003); Padal et al. (2013a, b); Satyavathi et al. (2014)
345.	*Eleutherine bulbosa*	Bulb	Dysentery	Panda et al. (2012)
346.	*Elytraria acaulis*	Leaves	Post delivery pains, antiseptic, as tonic, scabies, ringworm, abscess of mammary glands	Vijayakumar and Pullaiah (1998); Pullaiah et al. (2003); Jeevan Ram et al. (2007); Padal et al. (2013b)

TABLE 7.1   *(Continued)*

S. No.	Name of plant Species	Useful Parts	Medicinal uses against	References
347.	*Embelia basaal*	Fruits	To prevent pregnancy	Karuppusamy et al. (2009b)
348.	*Embelia ribes*	Seeds	Asthma	Patanaik et al. (2009)
349.	*Emilia sonchifolia*	Leaves	Galactagogue, fits	Padal et al. (2013a, b)
350.	*Enhydra fluctuans*	Leaves	Ulcers	Sen and Behera (2003)
351.	*Enicostemma axillare*	Leaves, whole plant	Leucorrhoea, snake bite	Raja Reddy et al. (1989); Goud and Pullaiah (1996)
352.	*Enicostemma littorale*	Whole plant	Diabetes	Prasanthi et al. (2014)
353.	*Entada pursaetha*	Seeds, leaves	Ulcer, inflammation, rheumatism, to induce sterility	Vijayakumar and Pullaiah (1998); Patanaik et al. (2009); Vaidyanathan et al. (2013); Shanmukha Rao et al. (2014)
354.	*Entada rheedii*	Seeds	Boils, rheumatism	Sen and Behera (2003); Satyavathi et al. (2014)
355.	*Enicostemma axillare*	Shoot	Leucorrhea	Venkata Ratnam and Raju (2005)
356.	*Eryngium foetidum*	Root	Stomachache	Padal et al. (2013b)
357.	*Erythrina stricta*	Stem bark	Dysentery	Pullaiah et al. (2003);
358.	*Erythrina variegata*	Gum, leaves, bark	Jaundice, joint pains, backache	Padma Rao et al. (2007); Ramana Naidu et al. (2012); Padal et al. (2013b)
359.	*Erythroxylum monogynum*	Oil from wood, Root, stem bark	Skin diseases, allergy, rheumatism, itches, for easy delivery	Rajareddy et al. (1989); Rao and Pullaiah (2001); Jeevan Ram and Raju (2001); Venkataratnam and Raju (2004); Savithramma et al. (2012); Anandakumar et al. (2014);
360.	*Eucalyptus globulus*	Leaves	Body pain	Parthipan et al. (2011)
361.	*Eulophia epidendraea*	Pseudo-bulb	Venereal diseases	Reddy et al. (2005)
362.	*Eulophia graminea*	Tubers	Ear and eye infections	Karuppusamy et al. (2009)
363.	*Euphorbia antiquorum*	Stem juice, latex	Deafness, bone fracture, sprain, cancer, diabetes	Lingaiah and Rao (2013); Dhayapriya and Senthilkumar (2014)
364.	*Euphorbia cyathophora* (Syn.: *E. heterophylla*)	Leaves	Diuretic, increasing lactation	Kumar and Pullaiah (1999); Xavier et al. (2011)

TABLE 7.1 *(Continued)*

S. No.	Name of plant Species	Useful Parts	Medicinal uses against	References
365.	*Euphorbia heyneana*	Whole plant	Leucorrhea	Venkata Ratnam and Raju (2005)
366.	*Euphorbia hirta*	Latex, plant	Wound healing, hemorrhage, leucorrhoea, dysentery, cold, fever	Venkataratnam and Raju (2004); Prusti (2007); Parthipan et al. (2011); Xavier et al. (2011); Padal et al. (2013b); Shanmukha Rao et al. (2014)
367.	*Euphorbia hypericifolia*	Leaf	Herpes, eczema	Goud and Pullaiah (1996)
368.	*Euphorbia indica*	Leaves	Skin itching	Pullaiah et al. (2003)
369.	*Euphorbia ligularia*	Latex	Backache	Padal et al. (2013b)
370.	*Euphorbia nivulia*	Latex	Cuts, cure boils and blisters	Padal et al. (2013b); Anandakumar et al. (2014)
371.	*Euphorbia prostrata*	Leaves, whole plant	Scabies, menorrhea	Raja Reddy et al. (1989); Venkataratnam and Raju (2005)
372.	*Euphorbia thymifolia*	Latex	Burn wound	Sen and Behera (2003)
373.	*Euphorbia tirucalli*	Latex, stem bark	Cough, skin allergy, galactogogue, rheumatoid arthritis	Reddy et al. (2006); Jeevan Ram et al. (2007); Padal et al. (2013b); Ramakrishna and Saidulu (2014b)
374.	*Evolvulus alsinoides*	Leaves, flowers, whole plant, roots	Poisonous bites, dysentery, nervous disorder, asthma, for conception, cuts	Rajareddy et al. (1989); Pullaiah et al. (2003); P.R. Reddy et al. (2003); Jeevan Ram et al. (2007); Padal et al. (2013b); Manikandan and Lakshmanan (2014)
375.	*Exacum pedunculatum*	Leaves	Cold, cough and fever	Karuppusamy et al. (2009)
376.	*Ficus benghalensis*	Tender prop roots, stem, latex, leaves, bark	Menorrhagia, dysentery, rheumatism, impotency, skin allergy, wounds	Raja Reddy et al. (1989); Rao and Pullaiah (2001); P.R. Reddy et al. (2003); Sen and Behera (2003); Parthipan et al. (2011); Padal et al. (2013b); Ramakrishna and Saidulu (2014b)
377.	*Ficus glomerata*	Fruits, stem bark	Stomach problems, snake bite	Karuppusamy et al. (2007); Ramakrishna and Saidulu (2014)

TABLE 7.1 *(Continued)*

S. No.	Name of plant Species	Useful Parts	Medicinal uses against	References
378.	*Ficus hispida*	Latex, stem bark, leaves	Diarrhoea, to heal crack, ringworm	Prusti (2007); Xavier et al. (2011); Padal et al. (2013b)
379.	*Ficus microcarpa*	Leaves, bark	Stomach problems, dysentery	Padal et al. (2013b); Senthilkumar et al. (2013)
380.	*Ficus mollis*	Leaves	Skin allergies, earache,	Pullaiah et al. (2003); Venkataratnam and Raju (2004);
381.	*Ficus racemosa*	Bark, fruit, leaves	Cuts, wounds, ulcers, cooling agent, diabetes, gonorrhea, spermatorrhoea, leucorrhoea	Sen and Behera (2003); Sarangi and Sahu (2004); Senthilkumar et al. (2013); Prasanthi et al. (2014)
382.	*Ficus religiosa*	Stem bark, fruits	Cuts, dysentery, rheumatism, for immunity, impotency, skin allergy, hepatitis, STDs	Raja Reddy et al. (1989); Pullaiah et al. (2003); Parthipan et al. (2011); Murthy (2012); Lingaiah and Rao,2013; Ramakrishna and Saidulu (2014b)
383.	*Ficus tinctoria*	Leaves	Snake bite	Ramakrishna and Saidulu (2014)
384.	*Flacourtia indica*	Root, stem bark	Poisonous bites, cholera	Raja Reddy et al. (1989); Rao and Pullaiah (2001)
385.	*Flemingia nana*	Root	Diarrhea, dysentery	Panda et al. (2012)
386.	*Flemingia semialata*	Roots	External application	Murthy (2012)
387.	*Flemingia wightiana*	Root	Fever	Girach (2001)
388.	*Garcinia xanthochymus*	Stem bark	Rheumatism	Rao and Pullaiah (2007)
389.	*Gardenia gummifera*	Resin, gum	Ringworm, intestinal worms	Jeevan Ram and Raju (2001); Chaudhari and Hutke (2002)
390.	*Gardenia latifolia*	Gum, stem bark	Constipation, cuts and wounds	Rao and Pullaiah (2001); Koteswara Rao et al. (2014)
391.	*Gardenia resinifera*	Leaves, gum, resin	Liver disorders, anthelmintic, diabetes	Kumar and Pullaiah (1999); Karuppusamy et al. (2009)
392.	*Garuga pinnata*	Leaves, roots, fruits, stem bark	Asthma, pulmonary infections, vermifuge, anthelmintic	Reddy et al. (2006); Murthy (2012); Vaidyanathan et al. (2013)

TABLE 7.1   *(Continued)*

S. No.	Name of plant Species	Useful Parts	Medicinal uses against	References
393.	*Geniosporum tenuiflorum*	Leaves	Dog bite	Sahu et al. (2013)
394.	*Givotia rottleriformis*	Stem bark	Psoriasis, leucorrhea	Venkata Ratnam and Raju (2004, 2005)
395.	*Glinus lotoides*	Leaves	Urinary tract infections	Raja Reddy et al. (1989)
396.	*Globba bulbifera*	Rhizome	Leucoderma, antiseptic	Jeevan Ram and Raju (2001); Ramana Naidu et al. (2012)
397.	*Globba oriexensis*	Rhizome	Scorpion sting	Das and Misra (1988)
398.	*Glochidion tomentosum*	Stem bark	Wound healing	C.S. Reddy et al. (2006); Patanaik et al. (2009)
399.	*Gloriosa superba*	Rhizome, leaves	Boils, blister, typhoid, back ache, malaria, asthma, skin eruptions, headache, lice, rheumatism, for abortion	Raja Reddy et al. (1989); Goud and Pullaiah (1996); Rao and Pullaiah (2001); Jeevan Ram and Raju (2001); Chaudhari and Hutke (2002); Patanaik et al. (2009); Padal et al. (2013b); Shanmukha Rao et al. (2014)
400.	*Glossocardia bosvallea*	Whole plant	Dysmenorrhoea	Raja Reddy et al. (1989)
401.	*Glycosmis cochinchinensis*	Leaves	Poisonous bites	Raja Reddy et al. (1989)
402.	*Glycosmis pentaphylla*	Stem bark, leaves	Bronchitis; stomach problems, wounds	Reddy et al. (2006); Karuppusamy et al. (2009); Padal et al. (2013b)
403.	*Glycyrrhiza glabra*	Root	Cold and cough	Dhatchanamoorthy et al. (2013); Padal et al. (2013b)
404.	*Gmelina arborea*	Root bark, stem bark, leaves	Lactogogue, impotency, fever, headache	Venkataratnam and Raju (2004); Murthy (2012); Padal et al. (2013b); Manikandan and Lakshmanan (2014)
405.	*Gmelina asiatica*	Root, root bark, fruit pulp	Meningitis, tooth ache, dandruff	Raja Reddy et al. (1989); Vijayakumar and Pullaiah (1998); Karuppusamy and Pullaiah (2005); Jeevan Ram et al. (2007)
406.	*Gomphrena decumbens*	Root	Cough	Ramana Naidu et al. (2012)
407.	*Gossypium arboreum*	Leaves	Conjunctivitis	Raja Reddy et al. (1989)

TABLE 7.1 *(Continued)*

S. No.	Name of plant Species	Useful Parts	Medicinal uses against	References
408.	*Gossypium barbadense*	Leaf extract	Menstrual disorder	Anandakumar et al. (2014)
409.	*Gossypium herbaceum*	Seeds	Diabetes	Prasanthi et al. (2014)
410.	*Grewia gamblei*	Leaves, root, bark	Scorpion sting	Vaidyanathan et al. (2013)
411.	*Grewia hirsuta*	Leaves, roots	Bruises, dysentery, diarrhea, snake bite	Sen and Behera (2003); Venkataratnam and Raju (2004); Jeevan Ram et al. (2007); Panda et al. (2012)
412.	*Grewia rothii*	Root	Gonorrhoea	Rao and Pullaiah (2001)
413.	*Grewia tiliifolia*	Stem bark, root	Cough; bone fracture, cuts and wounds	K.N. Reddy et al. (2006); Satyavathi et al. (2014); Koteswara Rao et al. (2014)
414.	*Gymnema sylvestre*	Leaves, root	Eye injuries, poisonous bites, cataract, diabetes, jaundice, scorpion sting, diabetes, cardiac disorder, biliousness	Raja Reddy et al. (1989); Vijayakumar and Pullaiah (1998); Venkataratnam and Raju (2004); Jeevan Ram et al. (2007); Patanaik et al. (2009); Parthipan et al. (2011); Sahu et al. (2013c); Ramakrishna and Saidulu (2014); Prasanthi et al. (2014); Shanmukha Rao et al. (2014)
415.	*Gymnosporia montana*	Young leaves	Skin allergy	Ramakrishna and Saidulu (2014b)
416.	*Gyrocarpus americanus*	Leaves, stem bark	Boils, chest pain	P.R. Reddy et al. (2003); Rao and Pullaiah (2001)
417.	*Habenaria longicornu*	Root tuber	Nervous disorder	Karuppusamy et al. (2009)
418.	*Habenaria fusifera*	Root tuber	Rheumatism	K.N. Reddy et al. (2005)
419.	*Habenaria longicorniculata*	Root tuber	General tonic	K.N. Reddy et al. (2005)
420.	*Habenaria plantaginea*	Whole plant	Antiseptic	K.N. Reddy et al. (2005)
421.	*Habenaria roxburghii*	Root tubers	Wound healing, snake bite	C.S. Reddy et al. (2006); Patanaik et al. (2009)
422.	*Hackelochloa granularis*	Plant	Hepatomegaly	Raja Reddy et al. (1989)
423.	*Haldinia cordifolia* (Syn.:*Adina cordifolia*)	Stem bark	Dysentery, leucorrhoea	Vijayakumar and Pullaiah (1998); Venkataratnam and Raju (2004); Padal et al. (2013b)

TABLE 7.1 *(Continued)*

S. No.	Name of plant Species	Useful Parts	Medicinal uses against	References
424.	*Hardwickia binata*	Stem bark	Dog bite	Venkataratnam and Raju (2004)
425.	*Hedyotis corymbosa*	Whole plant	Jaundice	Sahu et al. (2013)
426.	*Hedyotis pubeula*	Leaves	Burns	Dhatchanamoorthy et al. (2013)
427.	*Helicteres isora*	Fruits, root	Scabies, bronchitis, diarrhea, dysentery, fever, post natal pains	Jeevan Ram and Raju (2001); Pullaiah et al. (2003); Venkataratnam and Raju (2004); K.N. Reddy et al. (2006); Jeevan Ram et al. (2007); Padal et al. (2013b); Vaidyanathan et al. (2013); Koteswara Rao et al. (2014)
428.	*Heliotropium indicum*	Root, leaves	Diabetes, dog bite	Prusti (2007); Reddy et al. (2011); Padal et al. (2013b)
429.	*Heliotropium scabrum*	Leaves, whole plant	Dandruff, easy delivery	Raja Reddy et al. (1989)
430.	*Heterostemma deccanense*	Stem bark	Indigestion; stomach ache	C.S. Reddy et al. (2006); Patanaik et al. (2009)
431.	*Hemidesmus indicus*	Roots	Cooling agent, rheumatism, diarrhea, tooth ache, syphilis, spermatorrhoea, tonic, kidney stones, snake bite, migraine, fever, skin disorders, menstrual disorders	Raja Reddy et al. (1989); Chaudhari and Hutke (2002); Sarangi and Sahu (2004); Parthipan et al. (2011); Murthy (2012); Lingaiah and Rao (2013); Sahu et al. (2013c); Kannan and Kumar (2014); Satyavathi et al. (2014); Shanmukha Rao et al. (2014)
432.	*Hibiscus ovalifolius*	Leaf	Leucorrhea	Venkata Ratnam and Raju (2005)
433.	*Hibiscus rosa-sinensis*	Leaves, flowers	Boils, nervous disorder; hair tonic	Sen and Behera (2003); Parthipan et al. (2011); Anandakumar et al. (2014)
434.	*Hibiscus vitifolius*	Leaves, root	Asthma, tumor	Padal et al. (2013b); Anandakumar et al. (2014)
435.	*Hildegardia populifolia*	Stem bark	Malaria	C.S. Reddy et al. (2006); Patanaik et al. (2009)
436.	*Hiptage benghalensis*	Leaves	Scabies	Girach (2001)

TABLE 7.1   *(Continued)*

S. No.	Name of plant Species	Useful Parts	Medicinal uses against	References
437.	*Holarhaena pubescens* (Syn.: *Holarrhena antidysenterica*)	Stem bark, root	Asthma, cough, diarrhea, leuco-derma, dysentery, arthritis, worms, stomach problems, dental caries	Vijayakumar and Pullaiah (1998); Jeevan Ram and Raju (2001); Pul-laiah et al. (2003); Venkataratnam and Raju (2004); Prusti (2007); Rekka and Senthil Kumar (2014); Shanmukha Rao et al. (2014)
438.	*Holoptela integrifolia*	Leaves, bark	Migraine, wounds, boils, dental prob-lems; rheumatism	Raja Reddy et al. (1989); P.R. Reddy et al. (2003); Sahu et al. (2013); Anandakumar et al. (2014)
439.	*Holostemma ada-kodien*	Leaves	Galactogogue	Venkataratnam and Raju (2004);
440.	*Hoya pendula*	Root	Jaundice	Padal et al. (2013b)
441.	*Hugonia mystax*	Root bark	Snake bite	Ramana Naidu et al. (2012)
442.	*Hybanthus enneaspermus*	Leaves, seeds, tubers, entire plant	Snake bite, aph-rodisiac, diuretic, fever, demulcent, impotency, for immunity	Raja Reddy et al. (1989); Goud and Pullaiah (1996); Murthy (2012); Prabu and Kumuthakalavalli (2012); Vaidyanathan et al. (2013); Shanmukha Rao et al. (2014)
443.	*Hygrophila auriculata*	Leaves, Plant ash, entire plant	Heart failure, piles, leucorrhoea, urticaria	Raja Reddy et al. (1989); Sen and Behera (2003); Karuppusamy et al. (2007); Murthy (2012)
444.	*Hymenodictyon orixense*	Sten	Cuts and wounds	Koteswara Rao et al. (2014)
445.	*Hypericum gaitii*	Leaves	Skin eruption	C.S. Reddy et al. (2006)
446.	*Hyptis suaveolens*	Leaves	Keintra	Sen and Behera (2003);
447.	*Ichnocarpus frutescens*	Stem, roots	Tooth brush, ulcer, epilepsy	Prusti (2007); Karuppusamy et al. (2009); Shanmukha Rao et al. (2014)
448.	*Impatiens balsamina*	Leaves, seeds, stems	Snake bite, skin diseases	Vaidyanathan et al. (2013)
449.	*Imperata cylindrica*	Roots	Piles	Ramana Naidu et al. (2012)
450.	*Indigofera aspalathoides*	Whole plant	Diarrhoea	Vaidyanathan et al. (2013)
451.	*Indigofera cassioides*	Roots	Dysentery	Panda et al. (2012)
452.	*Indigofera linnaei*	Leaves	Asthma	Padal et al. (2013b)

TABLE 7.1 *(Continued)*

S. No.	Name of plant Species	Useful Parts	Medicinal uses against	References
453.	*Indigofera tinctoria*	Roots	Asthma, dry cough and respiratory disorder	Rao and Pullaiah (2001); Manikandan and Lakshmanan (2014)
454.	*Indoneesiella echioides*	Stem	Cuts and wound	Koteswara Rao et al. (2014)
455.	*Ipomoea batatas*	Leaf	Snake bite	Prusti (2007)
456.	*Ipomoea mauritiana*	Fruits	Abdominal tumors, weakness in children	Sahu et al. (2013c)
457.	*Ipomoea pestigridis*	Leaves	Carbuncle	Sen and Behera (2003);
458.	*Ipomoea staphylina*	Latex	Swellings	Anandakumar et al. (2014)
459.	*Ixora pavetta*	Root bark	Abdominal pain	Ramana naidu et al. (2012)
460.	*Jasminum angustifolium*	Root	Antidote for poison	Kumar and Pullaiah (1999)
461.	*Jasminum auriculatum*	Root	Snake bite	Kannan and Kumar (2014)
462.	*Jasminum grandiflorum*	Flowers	Skin diseases	Senthilkumar et al. (2013)
463.	*Jatropha curcus*	Leaves, root, latex, seeds	Diarrhoea, skin diseases; purgative, wounds	Pullaiah et al. (2003); Parthipan et al. (2011); Murthy (2012); Padal et al. (2013b); Manikandan and Lakshmanan (2014)
464.	*Jatropha glandulifera*	Latex	Asthma, mouth ulcer	Vijayakumar and Pullaiah (1998); Anandakumar et al. (2014)
465.	*Jatropha gossypifolia*	Leaves, latex, root	Rheumatism, skin diseases, syphilis, toothache, leucorrhoea, wounds, anemic condition	Raja Reddy et al. (1989); Goud and Pullaiah (1996); Rao and Pullaiah (2001); P.R. Reddy et al. (2003); Sarangi and Sahu (2004); Padal and Sandhyasri (2013)
466.	*Justicia adhatoda*	Leaves, stem, root	Nasal bleeding, asthma, fever, cough; cuts and wounds	Girach (2001); Parthipan et al. (2011); Lingaiah and Rao (2013); Sahu et al. (2013c); Koteswara Rao et al. (2014); Shanmukha Rao et al. (2014)
467.	*Justicia gendarussa*	Root, leaves	Leucorrhea	Venkata Ratnam and Raju (2005); Ramakrishna and Saidulu (2014b)
468.	*Justicia glauca*	Leaves	Cuts and wounds, backache	Padal et al. (2013b); Koteswara Rao et al. (2014)

TABLE 7.1   *(Continued)*

S. No.	Name of plant Species	Useful Parts	Medicinal uses against	References
469.	*Justicia tranqubariensis*	Leaves	Common cold and fever	Parthipan et al. (2011)
470.	*Kalanchoe pinnata*	Leaves	Bone fracture; rheumatism, kidney stones, dysentery, scorpion sting	Padal et al. (2013); Prabu and Ku-muthakalavalli (2012); Panda et al. (2012); Vaidyanathan et al. (2013)
471.	*Kedrostis foetidissima*	Leaves	Infant cold	Karuppusamy et al. (2009)
472.	*Kleinia grandiflora*	Leaves	Venereal disease	Karuppusamy et al. (2009)
473.	*Kydia calycina*	Leaves	Abscess	Jeevan Ram and Raju (2001)
474.	*Lagenaria siceraria*	Leaves	Burns	Padal et al. (2013)
475.	*Lagerstroemia reginae*	Leaves	Constipation	Panda (2007)
476.	*Lagerstroemia parviflora*	Stem bark	Leucorrhoea	Murthy (2012)
477.	*Lannaea coromandelica*	Stem bark, fruits, latex, leaves	For bone setting, to treat wounds, bronchitis, dys-tentery, cuts and wounds, body pain, inflammation	Vijayakumar and Pullaiah (1998); Jeevan Ram and Raju (2001); P.R. Reddy et al. (2003); C.S. Reddy et al. (2006); Das and Misra (1988); Vaidyanathan et al. (2013); Koteswara Rao et al. (2014)
478.	*Lantana camara*	Leaves	Ringworm	Venkata Subbiah and Savithramma (2012)
479.	*Lawsonia inermis*	Leaves	Leucorrhoea, headache, hair tonic, eczema	Raja Reddy et al. (1989); Rao and Pullaiah (2001); Sen and Behera (2003); Padal et al. (2013b); Sent-hilkumar et al. (2013)
480.	*Leonotis nepetifolia*	Leaves, flowers	Rheumatic pains, cuts	Padal et al. (2013a, b)
481.	*Lepidagathis cristata*	Plant	Eczema, dandruff, skin diseases	Goud and Pullaiah (1996)
482.	*Lepisanthes tetraphylla*	Leaves	Cough	Vaidyanathan et al. (2013)
483.	*Leptadenia reticulata*	Latex, leaves	Nasal stimulant, cold, asthma, cough	Raja Reddy et al. (1989); Jeevan Ram et al. (2007); Ramana naidu et al. (2012)

TABLE 7.1 *(Continued)*

S. No.	Name of plant Species	Useful Parts	Medicinal uses against	References
484.	*Leucas aspera*	Leaves, whole plant	Scorpion sting, eczema, leucorrhea, head ache, boils, snake bite	Rao and Pullaiah (2001); Pullaiah et al. (2003); Sen and Behera (2003); Venkata Ratnam and Raju (2005); Padma Rao et al. (2007); Parthipan et al. (2011); Kannan and Kumar (2014)
485.	*Leucas cephalotes*	Leaves	Headache	Padal et al. (2013b)
486.	*Leucas lavandulifolia*	Leaves	Galactogogue	Prusti (2007)
487.	*Limonia acidissima*	Fruit pulp, bark, root	Dysentery, diarrhea, wounds, snake bite	Raja Reddy et al. (1989); Vijayakumar and Pullaiah (1998); Pullaiah et al. (2003); Sen and Behera (2003); Padal et al. (2013b); Vaidyanathan et al. (2013)
488.	*Linocera zeylanica*	Leaves	Leucorrhea	Venkata Ratnam and Raju (2005)
489.	*Lippia nodiflora*	Leaves, root	Antibacterial	Basha et al. (2011)
490.	*Litsea deccanensis*	Bark	Cuts and wounds	Koteswara Rao et al. (2014)
491.	*Litsea glutinosa*	Stem bark	Snake bite, blood dysentery, muscular bleeding, chest pain, heal wounds	Chaudhari and Hutke (2002); Patanaik et al. (2009); Murthy (2012)
492.	*Lobelia nicotianifolia*	Leaves	Common cold, snake bite	Karuppusamy et al. (2009); Kannan and Kumar (2014)
493.	*Lonicera ramiflora*	Inflorescence	Asthma	Ramana Naidu et al. (2012)
494.	*Luffa aegyptiaca*	Leaves, root	Leprotic wound	Sen and Behera (2003);
495.	*Lycopersicon esculentum*	Leaves	Fever	Prusti (2007)
496.	*Lygodium flexuosum*	Leaves, rhizome	Wounds, convulsions, dysentery	Goud and Pullaiah (1996); Prusti (2007); Panda et al. (2012)
497.	*Macaranga peltata*	Stem bark	External application	Murthy (2012)
498.	*Madhuca indica*	Stem bark, flowers, fruits, leaves; seed oil	Fever, rabies, stomachache, fever, cuts and wounds, hair care, diabetes	Kumar and Pullaiah (1999); Rao and Pullaiah (2001); Prusti (2007); Padal et al. (2013a, b); Koteswara Rao et al. (2014); Prasanthi et al. (2014)

TABLE 7.1   *(Continued)*

S. No.	Name of plant Species	Useful Parts	Medicinal uses against	References
499.	*Maducha longifolia*	Stem bark	Skin disease	Manikandan and Lakshmanan (2014)
500.	*Maerua apetala*	Root	Leucoderma	Rao and Pullaiah (2001)
501.	*Maerua oblongifolia*	Root	Head ache	Padal et al. (2013b)
502.	*Malaxis acuminata*	Whole plant	Wound healing	K.N. Reddy et al. (2005)
503.	*Malaxis rheedei*	Whole plant	Cuts, wounds	K.N. Reddy et al. (2005)
504.	*Mallotus philippensis*	Leaves, seeds, stem bark	Skin diseases; scabies, dysmenorrhoea	Kumar and Pullaiah (1999); Jeevan Ram and Raju (2001); Pullaiah et al. (2003); Prusti (2007)
505.	*Mangifera indica*	Latex, bark, seeds	Skin diseases, diarrhea, dysentery, diabetes	Sen and Behera (2003); Prusti (2007); Vaidyanathan et al. (2013)
506.	*Manilkara hexandra*	Stem bark	Galactagogue	Padal et al. (2013a)
507.	*Manilkara zapota* (Syn.: *Achras sapota)*	Latex	Purgative	Kumar and Pullaiah (1999)
508.	*Marsilea quadrifolia*	Leaves	Skin diseases	Kumar et al. (2012); Padal et al. (2013b)
509.	*Martynia annua*	Fruit	Scorpion sting	Padal et al. (2013b)
510.	*Maytenus emarginata*	Bark, leaves	Gingivitis, cuts and wounds, fertility in women	Rajareddy et al. (1989); Jeevan Ram and Raju (2001); Venkataratnam and Raju (2004); Koteswara Rao et al. (2014)
511.	*Melastoma malabathricum*	Leaves	Cuts, wounds	Ramana Naidu et al. (2012)
512.	*Melia azedarach*	Leaves, seed oil, fruits	Dysmenorrhoea, antiseptic; menstrual disorder, skin and bone TB	Raja Reddy et al. (1989); Manikandan and Lakshmanan (2014); Padal et al. (2013); Sahu et al. (2013c)
513.	*Melia dubia*	Stem bark, leaves seeds	Small pox, rheumatism, promote general health	Senthilkumar et al. (2013); Vaidyanathan et al. (2013)
514.	*Melicope lunuankenda*	Root	Asthma, bronchitis	Rao and Pullaiah (2007)
515.	*Memecylon umbellatum*	Leaves	Wound scar	Karuppusamy et al. (2009)

TABLE 7.1   *(Continued)*

S. No.	Name of plant Species	Useful Parts	Medicinal uses against	References
516.	*Merremia emarginata*	Leaves	Hiccups, peptic ulcers	Rajareddy et al. (1989)
517.	*Merremia gangetica*	Root	Eye diseases	Padal et al. (2013b)
518.	*Merremia tridentata*	Whole plant	Bone fracture	Ramana Naidu et al. (2012)
519.	*Mesua ferrea*	Leaves	Dysentery	Panda et al. (2012)
520.	*Michelia champaca*	Leaves	Scorpion sting	Vaidyanathan et al. (2013)
521.	*Miliusa montana*	Leaves	Fever	Rao and Pullaiah (2007)
522.	*Miliusa tomentosa*	Leaves	Head application	Murthy (2012)
523.	*Millettia extensa*	Root	Vomiting	Prusti (2007)
524.	*Mimosa hamata*	Fruit, stem, seed	Skin itching, snake bite, sexual weakness in male	Pullaiah et al. (2003); Vaidyanathan et al. (2013)
525.	*Mimosa prainiana*	Root	After child birth	Goud and Pullaiah (1996)
526.	*Mimosa pudica*	Root, leaves, entire plant	Skin diseases, piles, head ache, fever, snake bite, filarial, blood pressure, piles, menstrual bleeding	Sen and Behera (2003); Jeevan Ram et al. (2007); Panda (2007); Prusti (2007); Parthipan et al. (2011); Murthy (2012); Lingaiah and Rao (2013); Vaidyanathan et al. (2013); Kannan and Kumar (2014); Ramakrishna and Saidulu (2014)
527.	*Mimusops elengi*	Leaves, stem bark	Diabetes, tuberculosis, leprosy wound, dental problems	Raja Reddy et al. (1989); Sen and Behera (2003); Jeevan Ram et al. (2007); Manikandan and Lakshmanan (2014)
528.	*Mirabilis jalapa*	Root	Reduce obesity	Anandakumar et al. (2014)
529.	*Mitragyna parvifolia*	Leaves, fruits, root	Eye diseases; dysentery, snake bite	Rao and Pullaiah (2001); Venkataratnam and Raju (2004); Padal et al. (2013b); Ramana Naidu et al. (2012)
530.	*Mollugo disticha*	Leaves	Puerperal sepsis	Raja Reddy et al. (1989)
531.	*Mollugo nudicaulis*	Leaves	Polydipsia	Raja Reddy et al. (1989);
532.	*Mollugo pentaphylla*	Whole plant	For smooth delivery	Rao and Pullaiah (2001)
533.	*Momordica charantia*	Leaves, root	Ulcers, ringworm, jaundice, diabetes, snake bite	Sen and Behera (2003); Venkata Subbiah and Savithramma (2012); Lingaiah and Rao (2013); Kannan and Kumar (2014)

TABLE 7.1 *(Continued)*

S. No.	Name of plant Species	Useful Parts	Medicinal uses against	References
534.	*Momordica dioica*	Fruits, root	Anthelmintic, snake bite	Karuppusamy et al. (2009); Kannan and Kumar (2014)
535.	*Morinda pubescens*	Root, leaf	Skin diseases, diarrhea, dysentery	Pullaiah et al. (2003); Padal et al. (2013b)
536.	*Morinda tomentosa*	Stem bark, roots	Boils, rabies, wounds, leucorrhoea	Jeevan Ram and Raju (2001); P.R. Reddy et al. (2003); Sarangi and Sahu (2004); Jeevan Ram et al. (2007)
537.	*Moringa concanensis*	Resin, leaves	Liver diseases, anemia, jaundice	Gritto et al. (2012); Murthy (2012)
538.	*Moringa pterygosperma* (Syn.: *M. oleifera*)	Leaves, stem bark, root	Boils, blisters, poisonous bites, skin diseases, to increase sperm count in man	Raja Reddy et al. (1989); Sen and Behera (2003); Xavier et al. (2011); Lingaiah and Rao (2013)
539.	*Mucuna atropurpurea*	Seed	Bone setting	Vaidyanathan et al. (2013)
540.	*Mucuna gigantea*	Root	Stomach upset	Vaidyanathan et al. (2013);
541.	*Mucuna pruriens*	Seeds, plant, fruit, root	Male sterility; snake bite, reduce swellings, tooth ache, dysentery	Murthy (2012); Lingaiah and Rao (2013); Padal et al. (2013a, b); Sahu et al. (2013c); Manikandan and Lakshmanan (2014);
542.	*Mukia maderaspatana*	Leaves	Bronchitis	Reddy et al. (2006)
543.	*Murraya koenigii*	Leaves	Vomiting, diabetes	Padal et al. (2013b); Prasanthi et al. (2014)
544.	*Murraya paniculata*	Leaves, bark	Wounds, body pain, fever, skin problems, snake bite	Vaidyanathan et al. (2013); Kannan and Kumar (2014); Manikandan and Lakshmanan (2014)
545.	*Musa paradisiaca*	Root, pith, flowers, tuber	Boils, blister, fever, diabetes, dysentery, snake bite	Raja Reddy et al. (1989); Prusti (2007); Padal et al. (2013b); Kannan and Kumar (2014); Prasanthi et al. (2014)
546.	*Musa sapientum*	Stem	Black spots	Sen and Behera (2003)
547.	*Myristica fragrans*	Seeds and aril	Infant cough	Karuppusamy et al. (2009)
548.	*Naravelia zeylanica*	Leaves	Skin diseases	Vaidyanathan et al. (2013)
549.	*Naringi alata*	Stem bark	Post natal complaints	Jeevan Ram et al. (2007)

TABLE 7.1   *(Continued)*

S. No.	Name of plant Species	Useful Parts	Medicinal uses against	References
550.	*Naringi crenulata*	Root	Snake bite	Vaidyanathan et al. (2013)
551.	*Nelumbo nucifera*	Rhizome, seeds	Diarrhoea, dysentery, infertility	Rao and Pullaiah (2001); Murthy (2012); Padal et al. (2013b)
552.	*Nerium oleander*	Leaves	Heart problems, skin diseases	Prabu and Kumuthakalavalli (2012)
553.	*Nerium indicum*	Root	Skin diseases	Padal et al. (2013b)
554.	*Nervilia aragoana*	Leaves	Skin diseases	Patanaik et al. (2009)
555.	*Nervilia plicata*	Leaves	Snake bite	Reddy et al. (2005)
556.	*Nicandra physaloides*	Leaves	Septic wounds	Karuppusamy and Pullaiah (2005)
557.	*Nyctanthes arbor-tristis*	Leaves, bark	Cough, piles, fever	Reddy et al. (2006); Padal et al. (2013a); Sahu et al. (2013c)
558.	*Nymphaea nouchali*	Tuber, root	Diabetes, tumors, urinary problems in children	Sahu et al. (2013c); Vaidyanathan et al. (2013); Ramanathan et al. (2014)
559.	*Nymphaea pubescens*	Root	Dysentery; goiter	Padal et al. (2013b); Sahu et al. (2013)
560.	*Oberonia wightiana*	Leaves	Wound healing	Reddy et al. (2005)
561.	*Ochna obtusata*	Stem bark	Snake bite	Ramana Naidu et al. (2012)
562.	*Ocimum americanum*	Leaves	Head ache, fever, repel mosquitos	Raja Reddy et al. (1989); Dhatchanamoorthy et al. (2013)
563.	*Ocimum basilicum*	Leaves	Ear ache, leucorrhoea	Raja Reddy et al. (1989); Padal et al. (2013a)
564.	*Ocimum tenuiflorum* (Syn: *O. sanctum*)	Leaves, seeds	Tuberculosis, cough, to avoid night blindness, skin allergy, cuts and wounds, ulcers	Kumar and Pullaiah (1999); Sen and Behera (2003); Parthipan et al. (2011); Murthy (2012); Padal et al. (2013a); Lingaiah and Rao (2013); Koteswara Rao et al. (2014)
565.	*Olax scandens*	Leaves	Psoriasis	Jeevan Ram and Raju (2001)
566.	*Operculina turpethum*	Leaves	Ringworm	Sen and Behera (2003)
567.	*Ophiorrhiza mungos*	Roots	Snake bite	Karuppusamy et al. (2009b)
568.	*Opuntia stricta* (Syn. *Opuntia dillenii*)	Stem tissue	Joint pains, blisters	Rao and Pullaiah (2001); Anandakumar et al. (2014)
569.	*Opuntia vulgare*	Flowers	Boils	P.R. Reddy et al. (2003);

TABLE 7.1   *(Continued)*

S. No.	Name of plant Species	Useful Parts	Medicinal uses against	References
570.	*Oroxylum indicum*	Stem bark, leaves, root bark	Rheumatic pain, mumps, pain, cough, jaundice; asthma, pain, prurutis, menstrual pain, leucoderma, snake bite, abdominal pains	Girach (2001); Chaudhari and Hutke (2002); Sen and Behera (2003); Panda (2007); Lingaiah and Rao (2013); Padal et al. (2013a); Padal and Sandhyasri (2013); Sahu et al. (2013c); Ramakrishna and Saidulu (2014b)
571.	*Ottelia alismoides*	Leaves	Fever	Ramana Naidu et al. (2012)
572.	*Oxystelma esculentum*	Leaves	Scabies	Sahu et al. (2013)
573.	*Oxalis corniculata*	Leaves	Dysentery, nervous weakness	Padal et al. (2013a); Panda et al. (2012); Vaidyanathan et al. (2013)
574.	*Paederia foetida*	Leaves	Stomach problems	Panda (2007)
575.	*Pandanus fascicularis*	Leaves, flowers	Leprosy, headache	Kumar and Pullaiah (1999);
576.	*Parthenium hysterophorus*	Leaves	Ear and teeth problems	Goud and Pullaiah (1996); Prusti (2007)
577.	*Passiflora foetida*	Leaves, root	Diarrhoea, giddiness	Raja Reddy et al. (1989); Padal et al. (2013a)
578.	*Pavetta indica*	Leaves, roots	Boils, itches	Venkataratnam and Raju (2004)
579.	*Pavonia odorata*	Leaves	Gonorrhoea	Ramakrishna and Saidulu (2014b)
580.	*Pavonia zeylanica*	Root	Diarrhoea	Padal et al. (2013a)
581.	*Pedalium murex*	Leaves	Leucorrhoea, cystitis, boils, menorrhoea, hydrocele	Raja Reddy et al. (1989); P.R. Reddy et al. (2003); Jeevan Ram et al. (2007); Sahu et al. (2013)
582.	*Pentanema indicum*	Root	For termination of ovum	Murthy (2012)
583.	*Peperomia tetraphylla*	Plant	Kidney stones	Girach (2001)
584.	*Pergularia daemia*	Leaves, root, latex	Funiculitis, throat infection, fever, urinary problems, fits, amenorrhoea, leprotic wounds, antifertility, bone fracture, swellings, snake bite, scorpion sting, headache; for abortion	Raja Reddy et al. (1989); Goud and Pullaiah (1996); Rao and Pullaiah (2001); Sen and Behera (2003); Sarangi and Sahu (2004); Venkataratnam and Raju (2004); Jeevan Ram et al. (2007); Murthy (2012); Lingaiah and Rao (2013); Padal et al. (2013a); Kannan and Kumar (2014); Manikandan and Lakshmanan (2014)

TABLE 7.1 *(Continued)*

S. No.	Name of plant Species	Useful Parts	Medicinal uses against	References
585.	*Peristylus lawii*	Whole plant	Antiseptic	Reddy et al. (2005)
586.	*Peuraria tuberosa*	Root tuber	Women related diseases	Karuppusamy et al. (2007)
587.	*Phoenix humilis* var. *pedunculata*	Fruits	Laxative, coolant	Kumar and Pullaiah (1999);
588.	*Phoenix sylvestris*	Leaves	Ulcerous tongue	Raja Reddy et al. (1989)
589.	*Pholidota imbricata*	Pseudo-bulb	Aphrodisiac	Reddy et al. (2005)
590.	*Phyllanthus amarus*	Leaves, whole plant, root, fruit	Jaundice, ring worm, urinary diseases, fever, jaundice	Raja Reddy et al. (1989); Kumar and Pullaiah (1999); Padma Rao et al. (2007); Parthipan et al. (2011); Lingaiah and Rao (2013); Shanmukha Rao et al. (2014)
591.	*Phyllanthus emblica*	Leaves, fruits, roots,	Burn wound, diarrhea, diabetes, blood pressure, dysentery, skin diseases, STDs, stomachache, snake bite	Sen and Behera (2003); Prusti (2007); Murthy (2012); Lingaiah and Rao (2013); Padal et al. (2013a); Kannan and Kumar (2014); Manikandan and Lakshmanan (2014); Prasanthi et al. (2014)
592.	*Phyllanthus fraternus*	Roots, leaves, seeds	Malaria, conjunctivitis, leucorrhoea	Sarangi and Sahu (2004); Sahu et al. (2013c)
593.	*Phyllanthus maderaspatensis*	Leaves	Jaundice	Raja Reddy et al. (1989)
594.	*Phyllanthus pinnatus*	Leaves	Intestinal worms	Vijayakumar and Pullaiah (1998);
595.	*Phyllanthus reticulatus* (Syn.: *Kirganelia reticulata*)	Leaves	Skin diseases, psoriasis	Raja Reddy et al. (1989); Jeevan Ram and Raju (2001)
596.	*Phyllanthus virgatus*	Plant	Dysuria	Rekka and Senthil Kumar (2014)
597.	*Physalis minima*	Leaves, fruits	Lymphogranuloma, kidney problems, for liver protection	Raja Reddy et al. (1989); Xavier et al. (2011); Sabjan et al. (2014)
598.	*Physalis peruviana*	Whole plant	Labour pain	Anandakumar et al. (2014)

TABLE 7.1 *(Continued)*

S. No.	Name of plant Species	Useful Parts	Medicinal uses against	References
599.	*Pimpinella tirupatiensis*	Tuberous root	Stomachache, scorpion sting	Raja Reddy et al. (1989); C.S. Reddy et al. (2006)
600.	*Piper betel*	Leaves	Snake bite, digestive	Kannan and Kumar (2014); Manikandan and Lakshmanan (2014)
601.	*Piper hymenophyllum*	Leaves	Mouth ulcers, indigestion	Venkataratnam and Raju (2004)
602.	*Piper longum*	Seeds	Chest pain and dry cough	Venkata Subbiah and Savithramma (2012)
603.	*Piper nigrum*	Seeds	Cough, cold, snake bite	Parthipan et al. (2011); Kannan and Kumar (2014)
604.	*Pithecellobium dulce*	Seeds	Diabetes	Padal et al. (2013b)
605.	*Plectranthus amboinicus*	Leaves	Running nose and cough	Dhatchanamoorthy et al. (2013)
606.	*Plumbago indica*	Whole plant, root	Ringworm, abortifacient, gynecolgical, rheumatism	Sen and Behera (2003); Patanaik et al. (2009); Sahu et al. (2013c)
607.	*Plumbago rosea*	Root bark	Piles	Murthy (2012)
608.	*Plumbago zeylanica*	Leaves, root, stem	Conjunctivitis, body pain, rheumatism, paralysis, skin diseases, cancer, tumors, scabies, leucoderma, eczema, ring worm, stomach disorders, piles, diarrhea, ulcers, abortifacient	Raja Reddy et al. (1989); Vijayakumar and Pullaiah (1998); Jeevan Ram and Raaju (2001); Chaudhari and Hutke (2002); Pullaiah et al. (2003); Venkataratnam and Raju (2004); Jeevan Ram et al. (2007); Panda (2007); Prusti (2007); Basha et al. (2011); Parthipan et al. (2011)
609.	*Plumeria rubra*	Root	Bruises, stomach ache	Sen and Behera (2003); Sahu et al. (2013)
610.	*Polycarpaea aurea*	Stem, leaf	Cold, meningitis	Rajareddy et al. (1989)
611.	*Polygala erioptera*	Root	Increase lactation in women	Karuppusamy et al. (2007)
612.	*Polyalthia longifolia*	Stem bark	Rheumatism	Padal et al. (2013b)
613.	*Pongamia pinnata*	Stem bark, seeds, root bark, stem, leaves	Meningitis, scabies, tooth brush, leprosy, leucorrhea, blood pressure, paralysis, pains	Raja Reddy et al. (1989); Vijayakumar and Pullaiah (1998); Pullaiah et al. (2003); Venkata Ratnam and Raju (2005); Prusti (2007); Lingaiah and Rao (2013)

TABLE 7.1 *(Continued)*

S. No.	Name of plant Species	Useful Parts	Medicinal uses against	References
614.	*Portulaca quadrifida*	Whole plant	Cough	Manikandan and Lakshmanan (2014)
615.	*Premna calycina*	Stem bark	Stomach ache	C.S. Reddy et al. (2006); Patanaik et al. (2009)
616.	*Premna tomentosa*	Leaves	Stomach pain, scabiess, skin rashes, to expel placenta, easy delivery, biliousness, headache	Goud and Pullaiah (1996); Vijaya-kumar and Pullaiah (1998); Girach (2001); Rao and Pullaiah (2001); Venkataratnam and Raju (2004); Jeevan Ram et al. (2007)
617.	*Priva cordifolia*	Leaves	To increase sexual desire in women	Vijayakumar and Pullaiah (1998)
618.	*Prosopis cineraria*	Leaf, root bark	Leucorrhea; skin diseases	Venkata Ratnam and Raju (2005); Padal et al. (2013b); Ramakrishna and Saidulu (2014b)
619.	*Psidium guajava*	Leaves, fruits	Diabetes, itches, dysentery, mouth ulcers	Kumar and Pullaiah (1999); Sen and Behera (2003); Parthipan et al. (2011); Lingaiah and Rao (2013)
620.	*Pseudarthria viscida*	Whole plant	Rheumatisim	Gritto et al. (2012)
621.	*Psoralia corylifolia*	Leaves, roots fruits	Rheumatoid arthritis, tetanus, worm infection	Sahu et al. (2013c)
622.	*Pteris vittata*	Herb	Skin diseases	Kumar and Pullaiah (1999)
623.	*Pterocarpus marsupium*	Stem bark, wood oil, whole plant	Boils, fever, tooth ache, cough, diarrhea, dysentery, jaundice, diabetes	Jeevan Ram and Raju (2001); Murthy (2012); Padal et al. (2013b); Sahu et al. (2013c); Manikandan and Lakshmanan (2014); Prasanthi et al. (2014)
624.	*Pterocarpus santalinus*	Wood, stem	Diabetes	Raja Reddy et al. (1989); C.S. Reddy et al. (2006)
625.	*Pterospermum acerifolium*	Stem bark	Dysentery	Panda et al. (2012)
626.	*Pterospermum suberifolium*	Whole plant	Bruises	Vaidyanathan et al. (2013)
627.	*Pueraria tuberosa*	Tubers	Galactogogue, nervine tonic, aphrodisiac, cuts and wounds, peptic ulcers, rheumatoid arthritis	Vijayakumar and Pullaiah (1998); Murthy (2012); Koteswara Rao et al. (2014); Shanmukha Rao et al. (2014)

TABLE 7.1 *(Continued)*

S. No.	Name of plant Species	Useful Parts	Medicinal uses against	References
628.	*Pulicaria wightiana*	Leaves	Head ache	Ramana Naidu et al. (2012)
629.	*Punica granatum*	Fruits	Dystentery and diarrhea	Manikandan and Lakshmanan (2014)
630.	*Putranjiva roxburghii*	seeds	Impotency	Murthy (2012)
631.	*Randia dumetorum*	Root bark	Dandruff	Karuppusamy et al. (2007)
632.	*Randia uliginosa*	Fruits	Dysentery	Panda et al. (2012)
633.	*Rauvolfia serpentina*	Apical bud, leaves, roots	Wounds, rheumatism, toothache, snake bite, expel threadworms, gonorrhea, reduce blood pressure,	Chaudhari and Hutke (2002); Sarangi and Sahu (2004); Panda (2007); Prusti (2007); Prabu and Kumuthakalavalli (2012); Sahu et al. (2013c); Kannan and Kumar (2014); Shanmukha Rao et al. (2014)
634.	*Rauvolfia tetraphylla*	Root	Snake bite, blood pressure	Prabakaran et al. (2013); Shanmukha Rao et al. (2014)
635.	*Raphanus sativus*	Tuber	Dysentery	Padal et al. (2013b)
636.	*Rhinacanthus nasutus*	Root, leaves, roots	Eczema, ringworm, cuts, wounds, snake bite	Jeevan Ram et al. (2007); Prabu and Kumuthakalavalli (2012); Kannan and Kumar (2014)
637.	*Rhus mysorensis*	Seeds	Antifertility	Jeevan Ram et al. (2007);
638.	*Rhynchosia suaveolens*	Leaves	Diarrhoea, dysentery	Pullaiah et al. (2003);
639.	*Rhynchosia viscosa*	Seeds, leaves	Body strength, to cure fever	Kumar and Pullaiah (1999);
640.	*Ricinus communis*	Leaves, twigs, seed oil	Body pain, bone fracture, head ache, pains, jaundice, cut wounds, boils, blisters	Raja Reddy et al. (1989); P.R. Reddy et al. (2003); Sen and Behera (2003); Prusti (2007); Parthipan et al. (2011); Lingaiah and Rao (2013)
641.	*Rivea hypocrateriformis*	Roots, leaves	Purgative, get relief from labor pain	Venkata Ratnam and Raju (2004); Ramana Naidu et al. (2012)
642.	*Rubia cordifolia*	Root	Fever, jaundice	Prusti (2007); Padal et al. (2013a)
643.	*Rubus niveus*	Root	Cuts, wounds	Senthilkumar et al. (2013)
644.	*Ruellia tuberosa*	Leaves	For bone setting	Padal et al. (2013a)
645.	*Ruta graveolens*	Leaves, whole plant	Dysentery, rheumatic pains	Senthilkumar et al. (2013); Vaidyanathan et al. (2013)

TABLE 7.1   *(Continued)*

S. No.	Name of plant Species	Useful Parts	Medicinal uses against	References
646.	*Rynchosia beddomei*	Leaves	Abortifacient	C.S. Reddy et al. (2006)
647.	*Salvadora persica*	Stem bark	Peptic ulcers	Rao and Pullaiah (2001)
648.	*Salvia splendens*	Leaves	Cold	Padal and Sandhyasri (2013)
649.	*Samanea saman*	Bark, leaves	Colds, sore throat, headache, diarrhea	Vaidyanathan et al. (2013)
650.	*Sansevieria roxburghiana*	Leaves, rhizome	Ear pain, snake bite	Raja Reddy et al. (1989); Vijaya-kumar and Pullaiah (1998); Rao and Pullaiah (2001); Reddy et al. (2011); Kannan and Kumar (2014)
651.	*Santalum album*	Bark, root, stem, stem bark	Gonorrhoea, reducing blood sugar, skin diseases	Jeevan Ram and Raju (2001); Basha et al. (2011); Manikandan and Lakshmanan (2014)
652.	*Sapindus emarginatus*	Endo-sperm, fruit rind, leaves	Sinusitis, jaundice	Venkata Ratnam and Raju (2004); Padma Rao et al. (2007); Reddy et al. (2011)
653.	*Saraca asoca*	Stem bark	Bodyache, gyneco-logical disorders, bone fracture, skin diseases	Girach (2001); Sahu et al. (2013c); Manikandan and Lakshmanan (2014)
654.	*Sarcostemma viminale* (Syn.: *S. acidum*)	Stem, Latex	Boils, cough, ulcer	Rao and Pullaiah (2001); K.N. Reddy et al. (2006); Karuppusamy et al. (2009)
655.	*Senecio candicans*	Stem bark	Diarrhoea	Girach (2001)
656.	*Schefflera racemosa*	Stem bark	Stomach problems	Karuppusamy et al. (2009)
657.	*Schefflera stellata*	Stem	Keep away from evil spirit	Anandakumar et al. (2014)
658.	*Schleichera oleosa*	Stem bark	Dropsy, scabies, dysentery, pain	Jeevan Ram et al. (2007); Prusti (2007); Pullaiah et al. (2003); Panda et al. (2012)
659.	*Schleichera racemosa*	Seeds	Leg swelling	Murthy (2012)
660.	*Scindapsus officinalis*	Root	Bone fracture	Padal et al. (2013)
661.	*Scoparia dulcis*	Leaves, whole plant	Convulsions, kidney disorder, diabetes,	Prusti (2007); Senthilkumar et al. (2013); Prasanthi et al. (2014)

TABLE 7.1   *(Continued)*

S. No.	Name of plant Species	Useful Parts	Medicinal uses against	References
662.	*Sebastiania chamealea*	Leaves	Constipation, indigestion	Goud and Pullaiah (1996);
663.	*Securinega leucopyrus*	Leaves, bark, fruit	Dysentery, dandruff, intestinal worms	Raja Reddy et al. (1989); Goud and Pullaiah (1996); Vijayakumar and Pullaiah (1998)
664.	*Selaginella rupestris*	Whole plant	Leucorrhoea	Sarangi and Sahu (2004)
665.	*Semecarpus anacardium*	Twigs, resin, gum, seeds	Fracture, neck pain, cuts and wounds, skin diseases	Jeevan Ram and Raju (2001); Sen and Behera (2003); Prusti (2007); Sahu et al. (2013); Koteswara Rao et al. (2014)
666.	*Senna sophora*	Leaves	Ringworm and scabies	Manikandan and Lakshmanan (2014)
667.	*Sesbania grandiflora*	Leaves, flowers	Old ulcers, cough; diarrhea	Sen and Behera (2003); K.N. Reddy et al. (2006); Padal et al. (2013b)
668.	*Shorea robusta*	Latex	Skin diseases	Sen and Behera (2003)
669.	*Shorea roxburghii*	Stem bark, root	Dysentery, diarrhea, cholera	Panda et al. (2012); Vaidyanathan et al. (2013)
670.	*Shorea tumbuggaia*	Leaves	Ear ache in children	C.S. Reddy et al. (2006)
671.	*Shuteria densiflora*	Stem	Bodyache	Girach (2001)
672.	*Sida acuta*	Leaves, roots	Skin diseases, headache, boils, blisters, cut wound, swellings, indigestion, rheumatoid arthritis; hemorrhoids, impotence	Raja Reddy et al. (1989); Rao and Pullaiah (2001); P.R. Reddy et al. (2003); Sen and Behera (2003); Prusti (2007); Murthy (2012); Sahu et al. (2013c); Senthilkumar et al. (2013); Vaidyanathan et al. (2013); Anandakumar et al. (2014)
673.	*Sida cordifolia*	Leaves, seeds, roots	Bronchitis, asthma, piles, phthisis, insanity, dysentery	Vaidyanathan et al. (2013); Manikandan and Lakshmanan (2014)
674.	*Sida rhombifolia*	Leaves	Dental problems	Reddy et al. (2011)
675.	*Smilax zeylanica*	Root, tubers	Dysentery, diabetes, gynecological diseases, tonic	Pullaiah et al. (2003); Sarangi and Sahu (2004); Prusti (2007); Karuppusamy et al. (2007); Murthy (2012)
676.	*Solanum anguivi*	Fruits	Anthelmintic	Ramana Naidu et al. (2012)

TABLE 7.1 *(Continued)*

S. No.	Name of plant Species	Useful Parts	Medicinal uses against	References
677.	*Solanum americanum*	Leaves	Cough; ulcers and stomach pain	Reddy et al. (2006); Dhatchanamoorthy et al. (2013)
678.	*Solanum incanum*	Root	Wounds	P.R. Reddy et al. (2003)
679.	*Solanum macranthum*	Fruit	Toothache	Prusti (2007)
680.	*Solanum melongena*	Whole plant	Syphilitic ulcers	Sen and Behera (2003)
681.	*Solanum nigrum*	Fruits, leaves, whole plant, roots	Jaundice, piles, indigestion, colic, piles, leucorrhoea, diarrhea, mouth ulcer, iron deficiency, cuts, wounds, gonorrhea; snake bite	Goud and Pullaiah (1996); Kumar and Pullaiah (1999); Pullaiah et al. (2003); Jeevan Ram et al. (2007); Karuppusamy et al. (2009); Parthipan et al. (2011); Murthy (2012); Kannan and Kumar (2014); Koteswara Rao et al. (2014); Shanmukha Rao et al. (2014)
682.	*Solanum pubescens*	Leaves	Whooping cough	Reddy et al. (2006)
683.	*Solanum surattense*	Seeds, root	Toothache, asthma, urinary problems, leprosy	Raja Reddy et al. (1989); C.S. Reddy et al. (2006); Jeevan Ram et al. (2007); Reddy et al. (2011)
684.	*Solanum torvum*	Fruits	Anthelmintic, colic	Xavier et al. (2011); Anandakumar et al. (2014)
685.	*Solanum trilobatum*	Leaves	Cold and cough	Xavier et al. (2011)
686.	*Solanum verbascifolium*	Leaves	Increase food intake	Anandakumar et al. (2014)
687.	*Solanum virginianum*	Unripe fruit	Cough	Manikandan and Lakshmanan (2014)
688.	*Sonchus wightianus*	Leaves	Bleeding	Girach (2001)
689.	*Soymida febrifuga*	Stem bark	Leucorrhoea, diarrhea, body pains, dysentery, fever, snake bite	Goud and Pullaiah (1996); Rao and Pullaiah (2001); Pullaiah et al. (2003); Prusti (2007); Padal et al. (2013b); Ramakrishna and Saidulu (2014)
690.	*Spermacoce hispida*	Whole plant	Cut wounds	Senthilkumar et al. (2013)
691.	*Sphaeranthus indicus*	Leaves, root, inflorescence	Elephantiasis, fever, migraine, rheumatism, anthelmintic	Goud and Pullaiah (1996); Das and Misra (1998); Girach (2001); Prusti (2007); Shanmukha Rao et al. (2014)

TABLE 7.1 *(Continued)*

S. No.	Name of plant Species	Useful Parts	Medicinal uses against	References
692.	*Spilanthes calva*	Flower head	Tooth problems	Karuppusamy et al. (2009)
693.	*Spondias pinnata*	Bark	Tetanus	Sahu et al. (2013c)
694.	*Stachytarpheta jamaicensis*	Leaves	Purgative, cuts and wounds	Raja Reddy et al. (1989); Shanmukha Rao et al. (2014)
695.	*Stemona tuberosa*	Root, tuber	Stomachache, gynacological disorder	Raja Reddy et al. (1989); Patanaik et al. (2009)
696.	*Sterculia foetida*	Bark, fruit, seed	Gonorrhoea, skin disease, abortifacient	Vaidyanathan et al. (2013)
697.	*Sterculia urens*	Gum, stem bark, leaves	Leucorrhea, oligospermia, spermatorrhoea, dysentery, skin diseases, wounds, cracked skin, leukemia, for free motions	Vijayakumar and Pullaiah (1998); Jeevan Ram and Raju (2001); Venkata Ratnam and Raju (2004, 2005); Prusti (2007); Murthy (2012); Vaidyanathan et al. (2013); Reddy et al. (2014)
698.	*Sterculia villosa*	Stem bark, gum	Menorrhea, body pains	Venkata Ratnam and Raju (2005); Murthy (2012)
699.	*Stereospermum chelonoides*	Bark	Burn wounds	Sen and Behera (2003)
700.	*Streblus asper*	Tender shoots, stem bark, leaves	Fever, menstrual pain, to brush teeth, eczema	Girach (2001); Sen and Behera (2003); Murthy (2012); Ramakrishna and Saidulu (2014b)
701.	*Strychnos nux-vomica*	Root paste, root bark, seeds; stem bark, leaves	Snake bite, paralysis, hypertension, diabetes, leucoderma, scabies, psoriasis	Raja Reddy et al. (1989); Goud and Pullaiah (1996); Rao and Pullaiah (2001); Sen and Behera (2003); Venkata Ratnam and Raju (2004); Jeevan Ram et al. (2007); Murthy (2012); Padal and Sandhyasri (2013); Prasanthi et al. (2014); Shanmukha Rao et al. (2014)
702.	*Strychnos potatorum*	Seeds, root	Scorpion sting, diarrhea, dysentery, conjunctivities, diabetes, kidney problems, spermatorrhoea, impotency, blood pressure, skin diseases	Goud and Pullaiah (1996); Pullaiah et al. (2003); Venkata Ratnam and Raju (2004); Jeevan Ram et al. (2007); Murthy (2012); Sahu et al. (2013); Padal et al. (2013b); Ramakrishna and Saidulu (2014a, b); Prasanthi et al. (2014); Shanmukha Rao et al. (2014)

TABLE 7.1 *(Continued)*

S. No.	Name of plant Species	Useful Parts	Medicinal uses against	References
703.	*Swertia angustifolia*	Whole plant	Stomach problems	Karuppusamy et al. (2009)
704.	*Symphorema involucratum*	Seeds	Snake bite	Goud and Pullaiah (1996)
705.	*Symphorema polyandrum*	Root	Skin diseases	Sen and Behera (2003)
706.	*Symplocos racemosa*	Stem bark	White discharge in women	Karuppusmy et al. (2007)
707.	*Synedrella nodiflora*	Leaves	Diarrhoea	Senthilkumar et al. (2013)
708.	*Syzygium alternifolium*	Fruits, seeds	Diabetes	Raja Reddy et al. (1989); C.S. Reddy et al. (2006); Jeevan Ram et al. (2007); Patanaik et al. (2009)
709.	*Syzygium cumini*	Seeds, bark	Dysentery, diabetes, urinary problems	Murthy (2012); Manikandan and Lakshmanan (2014); Prasanthi et al. (2014)
710.	*Tacca leontopetaloides*	Fruits	Body pain	Patanaik et al. (2009)
711.	*Tagetes erecta*	Leaves	Ringworm	Sen and Behera (2003)
712.	*Tamarindus indica*	Stem bark, leaves, fruit	Diarrhoea, body pain, swelling, burn wound, eye infection, female contraceptive	Pullaiah et al. (2003); Sen and Behera (2003); Prusti (2007); Senthilkumar et al. (2013); Vaidyanathan et al. (2013)
713.	*Tamarix ericoides*	Root	Skin diseases	Sen and Behera (2003)
714.	*Tarenna asiatica*	Leaves	Poisonous bites; for eat problems	Senthilkumar et al. (2013); Karuppusamy et al. (2009)
715.	*Tectona grandis*	Bark	Body heat	Murthy (2012)
716.	*Tephrosia procumbens*	Root	Stomachache	Padal et al. (2013b)
717.	*Tephrosia purpurea*	Leaves, roots, shoot, seeds	Sprain, stomach ache, menorrhea, respiratory diseases, leucoderma, scorpion sting, diabetes, urinary problems,	Raja Reddy et al. (1989); Vijayakumar and Pullaiah (1998); Kumar and Pullaiah (1999); Sen and Behera (2003); Venkata Ratnam and Raju (2005); Lingaiah and Rao (2013); Ramakrishna and Saidulu (2014); Prasanthi et al. (2014)
718.	*Tephrosia villosa*	Root	Diabetes	Prasanthi et al. (2014)
719.	*Teramnus labialis*	Root	Fever	Padal et al. (2013b)

TABLE 7.1   *(Continued)*

S. No.	Name of plant Species	Useful Parts	Medicinal uses against	References
720.	*Terminalia arjuna*	Stem bark, leaves	Skin diseases, cuts, wounds, menstrual problems, dysentery, earache	Sen and Behera (2003); Gritto et al. (2012); Vaidyanathan et al. (2013); Koteswara Rao et al. (2014)
721.	*Terminalia bellirica*	Fruit, whole plant	Jaundice, menorrhea, nervine tonic, asthma, stomach problems	Venkata Ratnam and Raju (2004, 2005); Murthy (2012); Vaidyanathan et al. (2013); Shanmukha Rao et al. (2014)
722.	*Terminalia catappa*	Gum	Cough and dysentery	Manikandan and Lakshmanan (2014)
723.	*Terminalia chebula*	Leaf galls, fruit, seed	For bone-setting, cough, diabetes, piles, snake bite, digestive, antiseptic, diuretic	Goud and Pullaiah (1996); K.N. Reddy et al. (2006); Jeevan Ram et al. (2007); Lingaiah and Rao (2013); Vaidyanathan et al. (2013); Kannan and Kumar (2014); Shanmukha Rao et al. (2014)
724.	*Terminalia pallida*	Fruit	Diuresis, fever, dysentery	Rajareddy et al. (1989); C.S. Reddy et al. (2006)
725.	*Thalictrum foliolosum*	Root	Rheumatism	Padal et al. (2013b)
726.	*Theriophonum fischeri*	Tuber	Rheumatic pain	Karuppusamy (2007)
727.	*Thespesia lampas*	Stem bark	Skin allergies	Pullaiah et al. (2003);
728.	*Thespesia populnea*	Stem bark, young fruits	Dandruff, skin diseases	Raja Reddy et al. (1989); Parthipan et al. (2011)
729.	*Thunbergia laevis*	Root	Insect bite	Prusti (2007)
730.	*Thunbergia fragrans*	Root	Redness of eyes	Girach (2001)
731.	*Tiliacora acuminata*	Leaves	Stomachache, snake bite	Rao and Pullaiah (2001); Vaidyanathan et al. (2013); Ramakrishna and Saidulu (2014)
732.	*Tinospora cordifolia*	Stem, leaves, stem bark, root, fruit, whole plant	Malarial fever, diuresis, pimples, diabetes, STDs, ulcers, scabies, rheumatism, diarrhea, dysentery, health tonic, gas trouble	Raja Reddy et al. (1989); Sen and Behera (2003); Panda (2007); Basha et al. (2011); Parthipan et al. (2011); Murthy (2012); Lingaiah and Rao (2013); Vaidyanathan et al. (2013); Shanmukha Rao et al. (2014)

TABLE 7.1 *(Continued)*

S. No.	Name of plant Species	Useful Parts	Medicinal uses against	References
733.	*Tinospora sinensis*	Leaves, stem	Snake bite, scorpion sting, rheumatic arthritis	Rao and Pullaiah (2007); Vaidyanathan et al. (2013)
734.	*Toddalia asiatica*	Fruit, leaves, whole plant	Dysentery, diarrhea, insect bites, fever, fits, cough	Rao and Pullaiah (2001); Jeevan Ram and Raju (2001); Vaidyanathan et al. (2013); Anandakumar et al. (2014)
735.	*Torenia indica*	Leaves	Ear ache	C.S. Reddy et al. (2006); Patanaik et al. (2009)
736.	*Trachyspermum ammi*	Leaves	Bronchitis	K.N. Reddy et al. (2006)
737.	*Tragia involucrata*	Leaves	Local swellings	Prabu and Kumuthakalavalli (2012)
738.	*Trapa natans*	Leaves	Blood dysentery	Das and Misra (1988)
739.	*Trema orientalis*	Leaves	Boils	Jeevan Ram and Raju (2001)
740.	*Trianthema decandra*	Leaves, roots	Meningites, Asthma, fertility	Raja Reddy et al. (1989); Jeevan Ram et al. (2007)
741.	*Trianthema portulacastrum*	Root tuber, leaves	Jaundice, stomach problems, kidney diseases, increase urination	Padma Rao et al. (2007); Parthipan et al. (2011); Lingaiah and Rao (2013); Anandakumar et al. (2014)
742.	*Tribulus terrestris*	Whole plant, fruits, leaves	Leucorrhoea, gonorrhea, scabies, kidney troubles, impotency, asthma, jaundice	Sarangi and Sahu (2004); Jeevan Ram et al. (2007); Parthipan et al. (2011); Xavier et al. (2011); Lingaiah and Rao (2013); Ramakrishna and Saidulu (2014b); Shanmukha Rao et al. (2014)
743.	*Trichosanthes cucumerina*	Fruits	Promote antifertility and abortion	Karuppusamy et al. (2009)
744.	*Trichosanthes lobata*	Leaves	Fever	Karuppusamy et al. (2009)
745.	*Trichosanthes tricuspidata*	Tuber	Cough, fever, stomach pain, epileptic pain	Sahu et al. (2013c)
746.	*Trichurus monsoniae*	Whole plant	Whooping cough	Ramana Naidu et al. (2012)
747.	*Tridax procumbens*	Leaves	Toothache, wound healing, antiseptic, jaundice	Raja Reddy et al. (1989); Vijayakumar and Pullaiah (1998); P.R. Reddy et al. (2003); Parthipan et al. (2011); Lingaiah and Rao (2013); Koteswara Rao et al. (2014); Shanmukha Rao et al. (2014)

TABLE 7.1 *(Continued)*

S. No.	Name of plant Species	Useful Parts	Medicinal uses against	References
748.	*Trigonella foenum-graecum*	Leaves	Skin diseases, diabetes	Sen and Behera (2003); Lingaiah and Rao (2013); Prasanthi et al. (2014)
749.	*Triumfetta pentandra*	Leaves	Diabetes	Goud and Pullaiah (1996);
750.	*Triumfetta rhomboidea*	Root, leaves	Ulcers, wounds, abdominal colic	Murthy (2012); Padal et al. (2013b); Vaidyanathan et al. (2013)
751.	*Tylophora fasciculata*	Leaves	Rabies, snake bite	Goud and Pullaiah (1996);
752.	*Tylophora indica*	Leaves	Asthma, bronchitis, wounds, snake bite; dysentery	Raja Reddy et al. (1989); Murthy (2012); Lingaiah and Rao (2013); Sahu et al. (2013c); Ramakrishna and Saidulu (2014); Shanmukha Rao et al. (2014)
753.	*Typha angustata*	Rhizome	For body cooling	Ramana Naidu et al. (2012)
754.	*Urena lobata*	Root, leaves	Stomach ache, cure wounds	Murthy (2012); Padal et al. (2013a, b)
755.	*Urginea indica*	Bulb	Swelling, menstrual disorder, snake bite	Prusti (2007); Padal et al. (2013a); Kannan and Kumar (2014);
756.	*Vanda tesellata*	Leaves	Ear drops for infection	Prusti (2007); K.N. Reddy et al. (2005)
757.	*Vanda testacea*	Leaves	For bone setting	K.N. Reddy et al. (2005)
758.	*Ventilago maderaspatana*	Stem bark, root	Diarrhoea, fever, venereal diseases, stomach disorders	Pullaiah et al. (2003); Karuppusamy et al. (2009); Vaidyanathan et al. (2013)
759.	*Vernonia anthelmintica*	Leaves, seeds	Ringworm; dyspepsia, wounds	Jeevan Ram and Raju (2001); P.R. Reddy et al. (2003); Venkata Subbiah and Savithramma (2012); Padal et al. (2013a)
760.	*Vernonia cinerea*	Leaves, whole plant, root, seed, shoots	Eye problems; wound healing, menstrual cycle, malarial fever, fever, fertility, leucoderma	Jeevan Ram et al. (2007); Parthipan et al. (2011); Murthy (2012); Padal et al. (2013b); Shanmukha Rao et al. (2014)
761.	*Vetiveria zizanioides*	Roots	Dandruff	Padal et al. (2013c)
762.	*Vigna aconitifolia*	Seeds	Fever	Padal et al. (2013b)

TABLE 7.1 *(Continued)*

S. No.	Name of plant Species	Useful Parts	Medicinal uses against	References
763.	*Vigna unguiculata* subsp. *unguiculata* (Syn.: *Dolichos biflorus)*	Leaves	Sores	Raja Reddy et al. (1989)
764.	*Vitex altissima*	Stem bark, root	Galactogogue, stomach problems; snake bite	Jeevan Ram et al. (2007); Anandakumar et al. (2014); Satyavathi et al. (2014)
765.	*Vitex leucoxylon*	Seed	Scabies	Pullaiah et al. (2003);
766.	*Vitex negundo*	Leaves, shoots	Arthritis, cold, coug, headache, bronchitis, pains, jaundice, fertility, wounds	Raja Reddy et al. (1989); C.S. Reddy et al. (2006); Jeevan Ram et al. (2007); Lingaiah and Rao (2013); Shanmukha Rao et al. (2014)
767.	*Vitex trifolia*	Leaves	Diarrhoea, eczema	Prabu and Kumuthakalavalli (2012)
768.	*Waltheria indica*	Leaves, whole plant,	Inflammation, cough, hemorrhages	Ramana Naidu et al. (2012); Vaidyanathan et al. (2013)
769.	*Wattakaka volubilis*	Leaves	Boils and abscesses	Ramana Naidu et al. (2012)
770.	*Wedelia calendulacea*	Leaves	Hair tonic	Karuppusamy et al. (2009)
771.	*Withania somnifera*	Root	Ulcer, paralysis, rheumatism, fertility improvement in men	Lingaiah and Rao (2013); Sahu et al. (2013c); Manikandan and Lakshmanan (2014); Ramanathan et al. (2014)
772.	*Woodfordia fruticosa*	Flower, root, bark	Menstrual disorder, dysentery, diarrhea, jaundice, cuts and wounds	Padal et al. (2013); Panda et al. (2012); Satyavathi et al. (2014); Koteswara Rao et al. (2014); Shanmukha Rao et al. (2014)
773.	*Wrightia arborea*	Root, root bark	Diarrhoea, dysentery, menstrual disorders	Pullaiah et al. (2003); Shanmukha Rao et al. (2014)
774.	*Wrightia tinctoria*	Stem bark, root, leaves, latex,	Cuts, injuries, sprains, snake bite, scorpion sting, skin diseases, asthma, bronchitis, typhoid, obesity, rheumatism	Raja Reddy et al. (1989); Goud and Pullaiah (1996); Vijayakumar and Pullaiah (1998); Rao aand Pullaiah (2001); Venkata Ratnam and Raju (2004); Reddy et al. (2006); Murthy (2012); Shanmukha Rao et al. (2014)
775.	*Xanthium indicum*	Leaf, seeds	Skin allergies, leucorrhea; small pox	Pullaiah et al. (2003); Venkataratnam and Raju (2005); Padal et al. (2013b)

TABLE 7.1   *(Continued)*

S. No.	Name of plant Species	Useful Parts	Medicinal uses against	References
776.	*Xanthium strumarium*	Leaves, root	Menorrhoea, leucorrhoea, boils	Venkata Ratnam and Raju (2004); Shanmukha Rao et al. (2014)
777.	*Ximenia americana*	Stem bark	Eczema	Pullaiah et al. (2003);
778.	*Xylia xylocarpa*	Stem bark	Skin eruptions	Murthy (2012)
779.	*Zaleya decandra*	Root,	Jaundice	Goud and Pullaiah (1996);
780.	*Zanthoxylum armatum*	Stem bark	Toothache, dysentery and vomiting in children	Girach (2001); Padal et al. (2013)
781.	*Zanthoxylum rhesta*	Stem bark	Dysentery	Patanaik et al. (2009)
782.	*Zingiber officinale*	Rhizome	Epilepsy, throat infection; cold, cough, asthma, fever	Raja Reddy et al. (1989); Panda (2007); Parthipan et al. (2011); Lingaiah and Rao (2013); Padal et al. (2013); Shanmukha Rao et al. (2014)
783.	*Zingiber roseum*	Rhizome	Stimulant, to treat arthritis, tumors	Patanaik et al. (2009); Murthy (2012); Satyavathi et al. (2014)
784.	*Zingiber zerumbet*	Rhizome	Septic wounds	Padal et al. (2013c)
785.	*Ziziphus mauritiana* (Syn.: *Z. jujuba)*	Leaves, fruit, bark, flower, root	Diuresis, dental disorder, diarrhea, cold, wounds, ulcer, fever, vomiting diarrhea	Raja Reddy et al. (1989); Padal et al. (2013a); Padal et al. (2013b); Vaidyanathan et al. (2013)
786.	*Ziziphus nummularia*	Leaves	Scabies	Padal et al. (2013b)
787.	*Ziziphus oenoplia*	Leaves, fruits	Bronchitis, dyspepsia	K.N. Reddy et al. (2006); Murthy (2012)
782.	*Ziziphus xylopyrus*	Stem bark, leaves	Cough, asthma, skin eruptions	K.N. Reddy et al. (2006); Murthy (2012); Padal et al. (2013b)

The maximum richness and habit-wise distribution of ethnomedicinal plants of Eastern Ghats showed herbs (41%), trees (24%), shrubs (22%) and climbers (13%). About 75 species of higher plants are thought to be endemic to Eastern Ghats region of India spread across the states of Odisha, Andhra Pradesh, Tamilnadu and Karnataka. In Odisha forest areas confined the following species namely *Cryptocarya amygdalina, Callicarpa vestita, Cerbera manghas, Aglaia cucullata, Bowenia serrulata, Hunteria zeylanica, Lasicocca comberi, Odisha cleistantha* and *Pomatocalpa decipiens.* Similarly in Andhra Pradesh region have *Andrographis nallamalayana, Cajanus cajanifolia, Dimorphocalyx glabellatus, Cycas beddomei, Shorea tumbaggaea, Boswellia ovalifoliolata, Pimpinella tirupatiensis Pterocarpus santalinus, Syzygium alternifolium* and *Terminalia pallida* which are narrow endemic medicinal plants species. Southern Eastern Ghats of Tamilnadu have also many narrow endemic plant species which are used in ethnomedicinal systems, such as *Aristolochia tagala, Curcuma nilagirica, Nervilia aragoana, Crotalaria shevaroyensis,* etc.

7.4 INDIGENOUS UTILIZATION OF MEDICINAL PLANTS

Different parts of the ethnomedicinal plants were used in treatment of ailments/diseases of different body parts, such as bones, bronchitis, ears, eyes, head, kidney, intestine, joints, liver, lungs, muscles, nervous system, sex organs, skin, stomach, teeth, throat, etc. Maximum species were used for stomach problems (157 spp.), followed by skin (122 spp.), eyes (78 spp.), blood (47 spp.) and liver (61 spp.) problems (Table 7.1). Based on the review, the proportion of ethnomedicinal plant parts used by tribal communities showed that leaves are most useful part and higher proportion about 283 plant species followed by roots (133 spp.), stems (130 spp.), whole plant (45 spp.), seeds (42 spp.), fruits (35 spp.), flowers (20 spp.), resins and gums (10 spp.) and latex (10 spp.). Mode of administration of herbal drugs among tribal communities of Eastern Ghats mostly by oral administration (57%) and topical application (40%), remaining plant drugs administered by inhalation (1%) and others uses (2%). All the tribal communities in the Eastern Ghats and adjacent Deccan region are invariably use many medicinal plants for same therapeutic purposes and such plants are *Achyranthes aspera, Acorus calamus, Aegle marmelos, Albizia amara, Allium cepa, Alpinia calcarata, Andrographis paniculata, Aristolochia indica, Asparagus racemosus, Cardiospermum halicacabum, Cassia fistula, Centella asiatica, Costus speciosus, Curculigo orchioides, Curcuma longa, Euphorbia hirta, Gymnema sylvestre, Leucas aspera, Pergularia daemia, Phyllanthus amarus, Piper betel, Plumbago zeylanica, Ricinus communis, Solanum nigrum, Terminalia chebula, Tribulus terrestris, Vitex negundo* and *Zingiber officinale*. This shows the cross-cultural similarities between the different human communities of the region. Some of the plants are already included in the rare, endangered and threatened list (IUCN, 2014), they used frequently by local communities for their health care, such as *Artocarpus heterophyllus, Boswellia ovalifoliolata, Saraca asoca, Rauvolfia serpentina, Pterocarpus santalinus* and *Gardenia resinifera*. In some occasions, alien or invasive species are also noted for ethnobotanical uses especially *Adansonia digitata, Agerataum conyzoides, Lantana camara* and *Wedelia calendulacea*.

Ethnomedicobotany of certain ethnically distinct primitive human societies from the Eastern Ghats explored by many investigators; Yanadis of Andhra Pradesh (Sudarsanam and Siva Prasad, 1995; Sudarsanam and Balaji Rao, 1994; Jeevanram and Venkataraju, 2001; Vedavathy and Mrudula, 1996; Karthikeyani, 2003), Chenchus of Andhra Pradesh (Vijayalakshmi, 1993; Reddy et al., 1988; Rao et al., 1995; Ramachandra Reddy et al., 2003; Padma Rao et al., 2007), Sugalis of Andhra Pradesh (Jeevan Ram, 2002), Gonds of Karimnagar ditrict (Reddy et al., 2003), Konda Reddis of Andhra Pradesh (Raju et al., 2011), Khonds of Visakhapatnam (Rao et al., 2006), Khonds of Karimnagar district (Reddy et al., 2003), Bagata of Visakhapatnam (Sandhya Sri and Seetharami Reddi, 2011), Irulas of Javadhu hills (Dhatchanamoorthy et al., 2013), Didahi tribes of Malakangiri (Pattanaik et al., 2008), Malayali tribes of Yercaud hills (Muruganandam et al., 2012; Senthilkumar et al., 2013; Rekha and Senthilkumar, 2014), Juang, Koha and Munda tribes of Odisha (Satpathy, 2008), Malayali's of North Arcot district of Tamilandu (Viswanathan, 1997), Paliyar tribes of Sirumalai hills (Karuppusamy, 2002; Maruthapandian et al., 2011), Sugali tribes of Yerramalais (Basha et al., 2011), Kond Dora tribes of Araku valley (Padal et al., 2013a), Dongaria Kandha of Odisha (Kumar et al., 2012), Bondo tribes of Odisha (Aminuddin and Girach, 1991), Paudi Bhunya tribes of Odisha (Aminuddin and Girach, 1996), Kandha tribes of Odisha (Girach, 1992; Das and Misra, 1998; Behera and Misra, 2005), Bonda's of Koraput district (Misra, 1992; Prusti, 2007), Kondha and Gond tribes of Odisha (Mohapatra and Sahoo, 2008)and Gadabas of Visakhapatnam (Koteswara Rao et al., 2011) have been studies so far.

Ethnomedicobotanical survey of any specific geographical and political province may have one or more ethnic groups. Such studies were also carried out in Eastern Ghats ranges and adjacent Deccan region are Chittoor district of Andhra Pradesh (Sudarsanam, 1987; Vedavathy et al., 1997), Srikakulam district (Harasreeramulu, 1980; Ramachandra Naidu and Seetharami Reddi, 2005), Adilabad district (Ravishankar and Henry, 1992), Anantapur district (Reddy et al., 1989), Kurnool district (Goud, 1995), Guntur district (Rao and Pullaiah, 2001), Kadapa district (Basi Reddy et al., 1989), Khammam district (Reddy, 2002; Raju and Reddy, 2005), Karimnagar distirct (Kaplan and Kapoor, 1980), Mahabubnagar district (Kumar, 1997), Ranga Reddy district (Padma Rao and Reddy, 1999; Ramachandra Reddy and Padma Rao, 2002), Boudh district of Odisha (Sahu et al., 2013), Bargarh district of Odhisa (Sen and Behera, 2003), Bhadrak district of Odisha (Girach et al., 1997 and 1998), Kandhamal district of Odisha (Behera et al., 2006), Koraput district of Odisha (Das and Misra, 1987, 1988, 1996; Pattanaik et al., 2006; Dhal et al., 2011), Kalahandi district of Odisha (Nayak et al., 2004; Sarangi and Sahu, 2004; Panda, 2007; Panda et al., 2008), Malakangiri district of Odisha (Pattanaik et al., 2008), Mayrubhnaj district of Odisha (Mudgal and Pal, 1980; Saxena et al., 1988; Sarkar et al., 1995; Rout et al., 2009), Phulbani district of Odisha (Subudhi and Chaudhury, 1985; Sahoo and Mudgal, 1995), Sunargarh district of Odisha (Mukherjee and Namhata, 1990; Satpathy, 1992; Satparhy and Panda,1992; Singh et al., 2010)and Nawarangpur district of Odisha (Dhal et al., 2014), Adilabad district (Murthy, 2012; Ramakrishna and Saidulu, 2014)and Kancheepuram district of Tamilandu (Muthu et al., 2006). A few studies have revealed the ethnomedicinal plants of entire state of Odisha (Saxena and Dutta, 1975; Chaudhuri et al., 1975; Das et al., 2003). These studies compiled the information regarding the traditional medicinal plant resources of the particular geographical ranges. Many ethnobotanical studies confined only the small areas of Ghat ranges especially in single Ghat areas of Javadhu hills (David and Sudarsanam, 2011; Ranganathan et al., 2012), Paderu division (Padal et al., 2012), Seshachalam hills (Savithramma and Sudharshanamma, 2006), Nallamalais (Jeevanram and Venkataraju, 2001; Ramana Naidu et al., 2012), Gundlabrahmeswaram Wildlife Sanctuary (Ventakataratnam and Venkataraju, 2004), Mahendragiri hills of Odisha (Giarch, 2001), Pachamalai hills (Kolar and Bhasha, 2013; Anandakumar et al., 2014), Araku valley (Banerjee, 1977; Gupta et al., 1997; Padal et al., 2013a, b, c), Paderu of Visakhapatnam (Rao et al., 2010), Kalrayan hills (Manikandan and Alagu Lakshmanan, 2014), Talakona (Basha et al., 2014), Kolli hills (Dwarakan and Ansari, 1992, 1996; Ranjithakani et al., 1992; Rajendran and Manian, 2011;Ramanathan et al., 2014; Suneela and Jyothi, 2014; Kadirvelmurugan et al., 2014), Pichavaram mangrove areas of Tamilandu (Ravindran et al., 2005), Gandhamardhan hills of Odisha (Misra, 2004) and Simlipal Biosphere Reserve (Pandey and Rout, 2006; Rout and Pandey, 2007), Penchalakona forest (Savithramma et al., 2012) and Warangal north forest division (Suthari et al., 2014).

Many ethnobotanical investigations have been undertaken based on utility in particular ailments. It is for specific purposes, such as skin diseases (Jeevan Ram et al., 1999, 2004; Venkata Subbaiah and Savithramma, 2012; Kumar et al., 2012), antidotes (Sudarsanam and Siva Prasad, 1995; Reddy et al., 1996; Ramakrishna and Saidulu, 2014), antidiabetic medicinal plants (Basha et al., 2011; Elavarasi and Saravanan, 2012), diabetes (Prasanthi et al., 2014), gynecological and abortive medicinal plants of Rayalaseema region of Andhra Pradesh (Nagalakshmi, 2001), gynecological and veneral diseases curing medicinal plants from Odisha (Sarangi and Sahu, 2004), snake bites and scorpion sting (Savithramma et al., 2013), antidiarrheal plants (Panda et al., 2012) and for cuts and wounds (Koteswara Rao et al., 2014). Ethnobotany has also been made with particular plant groups and families as the focus. Ethnomedicinal uses of genus *Phyllanthus* in Eastern Ghats areas have been studied by Lakshmi Narasimhudu (2013). Ethnomedicine of families like Cucurbitaceae of Eastern Ghats (Sri Ramamurthy et al. 2013) have been compiled. Sudhakar Reddy et al. (2006) and Rao and Pullaiah (2007) have made the investigation on ethnobotanical values of rare and endemic ethnomedicinal plants of Eastern Ghats. Naidu et al. (2012) described herbal remedies for rheumatoid arthritis used by the tribes of Vizianagaram district of Andhra Pradesh.

7.5 TRIBES AND MEDICINAL PLANT OF EASTERN GHATS

Santals, Kols and Kharias of Mayurbhanj district of Odisha used about 58 plant species belonging to 34 families and some medicinal information recorded particularly for *Aristolochia indica, Ficus racemosa, Hygrophila auriculata, Morinda citrifolia, Puraria tuberosa, Soymida febrifuga* and *Syzygium cerasoides* are unique and different from other communities of Eastern Ghats (Rout et al., 2009). Similarly Paroja, Saora, Bhumia, Godaba, Dogaria and Kondha communities in Koraput district of Odisha used many plants for their liver complaints, such as *Caryota urens, Curcuma montana, Sansevieria roxburghiana, Sesbania grandiflora* and *Elephantopus scaber*

(Smita et al., 2012). Bonda, Didayi, Koya, Bhatoda and Kondh communities of Malakangiri district of Odisha used 34 plant species ethnomedicinally for various ailments includes *Plumbago zeylanica* for abortifacient, *Barleria prionitis* for cough, *Semecarpus anacardium* for wound healing, *Pterocarpus marsupium* for diabetes, *Bauhinia vahlii* for dysentery, *Ricinus communis* for head ache and *Cassia fistula* for leprosy (Pattanaik et al., 2007). Tribal communities in Similipal Biosphere Reserve employed about 77 plant species for treating diarrhea alone such plants are *Bombax ceiba, Buchanania lanzan, Butea superba, Coccinia grandis, Curculigo orchioides, Eleutherine bulbosa, Flemingia nana, Helicteres isora, Lannea coromandelica, Mesua ferrea* and *Smilax zeylanica* (Rout and Pandey, 2007). Baidyas of Similipal area have several guarded knowledge on medicinal plants and also have some unique medicinal plant uses for treating malaria and other diseases such plants are *Rauvolfia serpentina, Oroxylum indicum, Kalanchoe pinnata, Clausena excavata, Curcuma angustifolia, Spondias pinnata, Shorea robusta, Solanum surattense,* etc. (Mohanta et al., 2006). Eastern Ghats of Odisha has a potential ethnomedicinal resources for treating various human diseases particularly rheumatism for about 62 genera with 78 plant species includes *Acanthus ilicifolius, Thunbergia fragrans, Cerbera odollum, Guizotia abyssinica, Derris scandens, Flacourtia indica, Pandanus fascicularis, Sesamum indicum* and *Stachytarpheta jamaicensis* (Panda et al., 2014).

Tribals of Rayalaseema region of Eastern Ghats are using about 54 plant species belonging to 50 genera and 34 families for treating asthma alone (Anjaneyulu and Sudarsanam, 2013). The tribal areas Rayalaseema have reported about 70 medicinal plants species for gynecological and abortive properties (Nagalakshmi, 2001). Gonds of Andhra Pradesh utilized several medicinal plants for their health care especially *Acacia arabica* for antidote, *Albizia odoratissima* for poisonous bites, *Atalantia monophylla* for treating rheumatism, *Cayratia pedata* for uterine relaxes, *Convolvulus sepiaria* for improving fertility, *Cyanotis tuberosa* for treating cough, *Litsea glutinosa* for wound healing, *Putranjiva roxburghii* for impotency, *Sterculia urens* for treating male sterility and *Xylia xylocarpa* for skin eruptions (Murthy, 2012). Konda doras, Kotias and Konds of Visakhapatnam district have vast knowledge on medicinal plant lore about treating diseases like wounds with *Annona squamosa*, rheumatism by *Polyalthia longifolia*, stomachic by *Cissampelos pariera*, dystentery by *Nelumbo nucifera*, diarrhea by *Brassica juncea*, asthma by *Ziziphus xylopyrus*, eczema by *Pterocarpus marsupium*, etc. (Padal et al., 2013b). Konda doras alone used about 68 medicinal plants for their medicinal purposes (Padal et al., 2013a). Bagata tibes of Paduru forest division in Visakhapatnam district of Andhra Pradesh are using 30 species of plants for their health care.

Chenchu's of Nallamalai hills used many plants for treating liver diseases, such as *Andrographis echioides, Andrographis paniculata, Boerhavia diffusa, Canavalia gladiata, Phyllanthus amarus, Physalis minima* and *Tephrosia purpurea* (Sabjan et al., 2014). Chenchu's are the good source of medicinal plant knowledge since long back and they have been treating about 67 diseases known to cure by using 69 medicinal plants includes *Cocculus hirsutus, Abutilon indicum, Sida acuta, Urena lobata, Melochia corchorifolia, Grewia flavescens, Chloroxylon swietenia, Careya arborea, Lepidagathis cristata, Aerva lanata, Glochidion ellipticum* and *Pouzolzia zeylanica* (Ravi Prasad Rao and Sunita, 2011). Yerukalas and Sugalis of Nallamala region have practiced of about 93 genera with 54 families of plants species as medicinal remedies mostly for gastrointestinal disorder, parasitic infections and external injuries. The plant species like *Acacia caesia, Anogeissus latifolia, Alternanthera sessilis, Bacopa monnieri, Boswellia ovalifoliolata, Carissa spinarum, Cleome viscosa, Eclipta prostrata, Linociera zeylanica, Spermacoce articularis, Trichurus monsoniae* and *Waltheria indica* are used for medicinal purposes (Ramana Naidu et al., 2012). Sugalis of Erramalai's used *Alangium salvifolium, Ammannia baccifera, Cassia occiendentalis, Habenaria roxburghii, Lantana indica, Strychnos nux-vomica* and *Wattakaka volubilis* for treating snake bite (Basha and Sudarsanam, 2012). Stem barks of *Albizia procera, Butea monosperma, Eugenia jombos, Gmelina arborea, Madhuca indica, Soymida febrifuga, Sterculia urens* and *Wrightia tinctoria* are used for various medicinal utilizations by Gondu tribes of Adilabad district (Suman Kumar et al., 2013) (Plates 7.1–7.4).

PLATE 1

a – *Androgaraphis echioides*   b – *Cadaba fruticosa*
c – *Capparis zeylanica*   d – *Buchanania lanzan*

PLATE 2

a – *Calamus rotang*   b – *Cassia occidentalis*
c – *Costus speciosus*   d – *Hemidesmus indicus*

PLATE 3

a – *Cynanchum callialatum*   b – *Ophiorrihza mungos*
c – *Kleinia grandiflora*   d – *Indigofera aspalathoides*

PLATE 4

a – *Sphaeranthus indicus*
c – *Semecarpus anacardium*
e – *Anogeissus latifolia*

b – *Secamone emetica*
d – *Toddalia asiatica*
f – *Citrullus colocynthis*

Irulas of Tamilnadu have been using 57 medicinal plant species for treating the health problems who are residing in Javadhu hills of Eastern Ghats (Mohamed Tariq et al., 2012). Dhatchanamoorthy et al. (2013) have reported that Irulas of Javadhu hills used some unique medicinal plants species for their local ailments, such as *Achyranthes bidentata, Blepharis maderaspatensis, Caralluma attenuata, Cymbopogon citratus, Datura innoxia, Hedyotis puberula, Ocimum americanum, Solanum virginianum* and *Vernonia cinerea*. Malayali's of Jawadhu hills used about 150 plants species for their health maintanence (Senthilkumar et al., 2014a) and the Yelagiri hills of Eastern Ghats reported a total of 175 ethnomedicinal species belonging to 147 genera and 56 families (Senthilkuamr et al., 2014b). Some primitive plant groups are ethnomedicinal importance in Kolli hills of Eastern Ghat ranges, it was reported about 30 species of Pteridophytes used for treating various ailments by Malayali tribes and important Pteridophytes species are *Actiniopteris radiata, Adiantum lunulatum, Ceratopteris thalictroides, Cheilanthes tenuifolia, Hemionities arifolia, Pityrogramma calomelanos, Pteris biaurita, Pteris vittata* and *Vateria elongata* (Karthik et al., 2011). Malayali tribes of Kolli hills used about 250 species of higher plants for their healthcare system include the species like *Naravelia zeylanica* (skin disease), *Michelia champaca* (scorpion sting), *Tinospora sinensis* (rheumatism), *Nymphaea nouchali* (urinary problem), *Helicteres isora* (diarrhea), *Impatiens balsamina* (snake bite), *Toddalia asiatica* (fever), *Cissus setosa* (worms), *Lepisanthes tetraphylla* (cold), *Mimosa pudica* (blood purifier), *Spermacoce hispida* (diarrhea), *Heliotropium indicum* (wound healing), *Solanum eriantum* (dysentery), *Polygonum hydropiper* (stimulant) and *Rumex nepalensis* (purgative) (Vaidyanathan et al., 2013; Xavier et al., 2011). Kolli hills is a rich biodiversity area of Eastern Ghats and there has been reported large number of medicinal plants for poisonous bites include 34 plant species for snake bite, 25 species for scorpion sting, 12 species for dog bites used by local tribes (Prabu and Kumuthakalavalli, 2012). Fifteen species of threatened medicinal plants used for disease like ulcers, diarrhea, indigestion, hemorrhage, stomach problems, skin diseases, heart problems and the plant species like *Acorus calamus, Aegle marmelos, Cayratia pedata, Celastrus paniculatus, Cycas circinalis, Decalepis hamiltonii, Gloriosa superba, Hildegardia populifolia, Maduca longifolia, Piper nigrum, Pseudarthria viscida, Santalum album, Smilax zeylanica* and *Terminalia arjuna* are medicinal plant species used to treat by ethnic communities of Pachamaial hills of Eastern Ghats (Gritto et al., 2012). Kolar and Bhasha (2013) reported 190 medicinal plants spread over 158 genera and

67 families which include *Hiptage benghalensis, Hugonia mystax, Mallotus philippensis, Solanum erianthum, Rubus cordifolia* and *Wrightia tinctoria.*

7.6 CONCLUSION

According to the review, 54 tribal communities mentioned for the areas of Eastern Ghats, ethnobotanical documentation available only for 40 tribal communities and remaining are living in remote areas and safe guarding their knowledge secretly themselves which are the main problems of lacunae in documentation. Studies exist on the documentation of herbal drugs used among various societies of Eastern Ghats (Rama Rao, 2006). However, Eastern Ghats region of Peninsular India mostly occupied by hill tribes namely Bathudi, Bhottoda, Bhumia, Bhumiji, Dharua, Gadaba, Gond, Khanda, Kandha, Gouda, Kolha, Koya, Munda, Paroja, Omanatya, Santal, Saora, Chechus, Savaras, Konda Reddis, Khonds, Kolamis, Nayakpods, Valmikis, Bhagatas, Jatayus, Yanadis, Yerukalas, Irulars, Malayalis and Paliyans. Yet they sustained a good wealth of traditional knowledge on preparation of herbal drugs for their primary healthcare (Nayak and Sahoo, 2002). They are sole custodians of the wild genetic resources of this area, and still practice traditional agriculture and they live close vicinity and utility of the wild vegetation. The documentation of the traditional knowledge on medicinal plants sporadically recorded in the past and there is still ample knowledge with primitive societies living in the hilly tracts that are needed to be documented. Most of these traditional knowledge and practices are only on folklore system (Solomon Raju and Jonathan, 2006).

Native communities of Eastern Ghats depend on plants for their health care and diseases curing system. Maximum numbers of plants were used for curing stomach problems, cold, cough and fever, and skin diseases. It was also found that a single plant species may be used for curing many ailments, such as *Andrographis paniculata* is used against jaundice, worms and to treat snake bite. Majority of tribes believed traditional superstition and they have strong faith in tree-spirit, evil eye and magics. The worship was possibly the earliest and the most prevalent form of religion. Almost all tribes are basically religious hence trees are treated by them as Gods. At the same time the tree fulfill their life requirements, such as medicine. It can be evidenced that most of the tribal communities used sacred plants of genera*Ficus, Acacia, Albizia, Alangium, Ocimum, Vitex*, etc. and all the plants are used for treating various ailments. There was an overlapping of many medicinal plant species in nearly all categories, which reflects the cross-cultural similarities among the tribes of Eastern Ghats.

Many numbers of plants that are used by tribes of Eastern Ghats are mentioned in ancient literature, such as Siddha, Ayurrvedha and Unani. The use of medicinal plant species treating a particular ailment is also fairly common among different communities of tribes. The present review evidenced that medicinal plants continue to play an important role in the health care system of traditional communities.

Due to constant association with the forest environment, ethnic communities have evolved knowledge by trial and error and have developed their own way of diagnosis and treatment for different ailments. The ethnic drug formulations need clinical tests to prove their efficacy and also to develop new herbal drugs for the effective treatment. This data provides basic source for further studies aimed at conservation, cultivation, improvement of ethnic traditional medicine and economic welfare of rural and tribal population of the region. The traditional botanical knowledge will provide secure livelihood to the native tribes that minimize the resource depletion, environmental degradation, cultural disruption and social instability. The medico-botanical survey of the area revealed that the people of the area possessing good knowledge of herbal drugs but as the people are in progressive exposure to modernization, their knowledge of traditional uses of plants may be lost in due course. So it is important to study and record the uses of plants by different tribes and sub-tribes for future study. Such studies may also provide some information to biochemists and pharmacologists in screening of individual species and in rapid assessing of phytoconstituents for the treatment of various diseases.

KEYWORDS

- **Eastern Ghats**
- **Ethnobotany**
- **Ethnomedicine**
- **Medicinal Plants**
- **Tribal Medicine**

REFERENCES

Alagesaboopathi, C. (2011a). Ethnobotanical studies of useful plants of Kanjamalai Hills of Salem District of Tamilnadu, Southern India. *Arch. Appl. Sci. Res. 3*(5), 532–539.

Alagesaboopathi, C. (2011b). Ethnomedicinal plants used as medicine by the Kurumba tribals in Pennagaram region, Dharmapuri of Tamil Nadu, India. *Asian J. Exo. Biol. Sci. 2*(1), 140–142.

Alagesaboopathi, C. (2012a). Ethnomedicinal uses of *Andrographis elongata* T. And. – An endemic medicinal plant of India. *International J. Recent Sci. Res. 3*(4), 231–233.

Alagesaboopathi, C. (2012b). Ethnobotanical studies on useful plants of Sirumalai Hills of Eastern Ghats, Dindigul District of Tamilnadu, Southern India. *Int. J. Biosci. 2*(2), 77–84.

Alagesaboopathi, C., Duvarakan, P. & Babu, S. (1999). Plants used as Medicine by tribals of Shervaray Hills, Tamilnadu. *J. Econ. Taxon. Bot. 23*, 391–393.

Aminuddin & Girach, R.D. (1991). Ethnobotanical studies on Bondo tribe of district Koraput (Orissa), India. *Ethnobotany, 3*, 15–19.

Aminuddin & Girach, R.D. (1993). Observations on ethnobotany of the Bhunjia – A tribe of Sonabara plateau. *Ethnobotany 5*, 83–86.

Aminuddin & Girach, R.D. (1996). Native phytotherapy among the Paudi Bhunya of Bonai hills. *Ethnobotany, 8*, 66–70.

Amutha, P. & Prabhakaran, R. (2010). Ethnobotanical studies on Malayali tribe in Nalamankadai, Chitteri Hills, Eastern Ghats, India. *Ethnobotanical leaflets. 14*, 942–951.

Anand, R.M., Nandakumar, N., Karunakaran, I., Ragunathan, M. & Murugan, V. (2006). A Survey of Medicinal plants in Kollimalai hill tracts, Tamilnadu. *Nat. Product Rad. 5*(2), 139–143.

Anandakumar, D., Rathinakumar, S.S. & Prabakaran, G. (2014). Ethnobotanical survey of Athinadu Pachamalai Hills of Eastern Ghats in Tamilnadu, South India. *Int. J. Adv. Interdisciplinary Res. 1*(4), 7–11.

Anjaneyulu, E. & Sudarsanam, G. (2013). Folk medicinal plants used in the treatment of asthma in Rayalaseema region of Andhra Pradesh, India. *Res. J. Pharmaceutical, Biol. Chem. Sci. 1*(4), 833–839.

Anonymous. (1990). Ethnobiology in India: A Status Report. Ministry of Environment & Forests, Govt. of India, New Delhi, pp. 1–68.

Arunachalam, G., Karunanithi, M., Subramani, N., Ramachandran, V. & Selvamuthukumar, S. (2009). Ethnomedicines of Kolli Hills at Namakkal District in Tamilnadu and its significance in Indian System of Medicine. *J. Pharm. Sci. Res. 1*(1), 1–15.

Ayyangar, M. & Ignacimuthu, S. (2009). Medicinal plants used by tribal inhabitants in the forest areas of Tamil Nadu: A review. In: Trivedi, P.C. (ed.). Indigenous Ethnomedicinal Plants. Jaipur: Aavishkar Publishers, pp. 94–111.

Babu, N.C., Naidu, M.T. & Venkaiah, M. (2011). Ethnomedicinal plants used by the tribes of Vizianagaram district, Andhra Pradesh. *Annals Pharm. Pharmaceut. Sci. 2*, 1–4.

Bal, S., Misra, R.C., Sahu, D. & Dhal, N.K. (2007). Therapeutic uses of some orchids among the tribes of Simlipal Biosphere Reserve, Orissa, India. *Med. Plants, 8*(2), 270–277.

Balaji Rao, N.S., Rajasekha, R.D. & Chengal Raju, D. (1995). Folk medicine of a Rayalaseema region, Andhra Pradesh: II blood purifiers. *Bull Pure Appl Sci, 14A*(2), 69–72.

Balasubramanian, P. & Rajasekaran, A. (1998). Utilization of wild plants by tribal communities in Sathyamangalam forest division, Eastern Ghats. In: *Eastern Ghats – Proc. Natl. Sem. Conserv. Eastern Ghats*. Envis center, EPTRI, Hyderabad, India.

Banerjee, D.K. (1977a). Observations on ethnobotany of Araku valley. *J. Bombay Nat. Hist. Soc. 53*, 153–155.

Banerjee, D.K. (1977b). Observations on ethnobotany of Araku valley, Visakhapatnam district, Andhra Pradesh. *J. Sci. Club 33*, 14–21.

Bapuji, J.L. & Ratnam, S.V. (2009). Traditional uses of some medicinal plants by tribals of Gangaraju Madugula Mandal of Visakhapatnam District, Andhra Pradesh. *Ethnobotanical Leaflets, 3*(2), http://opensiuc.lib.siu.edu/ebl/vol2009/iss3/2.

Basha, S.K. & Sudarsanam, G. (2010). Ethnobotanical studies on medicinal plants used by Sugalis of Yerramalais in Kurnool district, Andhra Pradesh, India. *Intern. J. Phytomedicie 2*(4), 349–353.

Basha, S.K., Sudarsanam, G., Mohammed, M.S. & Niaz Parveen (2011). Investigations on antidiabetic medicinal plants used by Sugali tribal inhabitants of Yerramalais of Kurnool District, Andhra Pradesh, India. *Stamford J. Pharamaceut. Sci. 4*(2), 19–24.

Basha, S.K.M., Uma Shankar, M. & John Paul, M. (2014). Ethnobotanical study of Talakona, Eastern Ghats, Andhra Pradesh. *Photon. 122*, 877–883.

Basha, S.K.M., Umamaheswari, P., Rambabu, M. & Savitramma, N. (2011). Ethnobotanical study of Mamandur forest (Kadapa-Nallamalai Range) in Eastern Ghats, Andhra Pradesh. *J. Phytology. 3*(10), 44–47.

Basi Reddy, M., Bryan, H.H., Lance, C.J. & McMillan, J.R.R.T. (1993). A survey of plant crude drugs of Anantapur District, Andhra Pradesh, India. *Econ. Bot. 47*, 79–88.

Basi Reddy, M., Raja Reddy, K. & Reddy, M.N. (1988). A survey of the medicinal plants of Chenchu tribe of Andhra Pradesh, India. *Int. J. Crude Drug Res. 26*, 189–196.

Basi Reddy, M., Raja Reddy, K. & Reddy, M.N. (1989). Ethnobotany of Cuddapah district. *Int. J. Pharmacognosy. 29*, 1–8.

Behera, S.K. & Misra, M.K. (2005). Indigenous phytotherapy for genito-urinary diseases used by the Kandha tribe of Orissa, India. *J. Ethnopharmacol. 102*, 319.

Behera, K.K., Mishra, N.M. & Rout, G.R. (2008). Potential ethnomedicinal plants at Kaptipada forest range, Orissa, India and their uses. *J. Econ. Taxon. Bot. 32 (Suppl.)*, 194–202.

Behera, S.K., Panda, A., Behera, S.K. & Misra, M.K. (2006). Medicinal plants used by the Kandhas of Kandhamal district of Orissa. *Indian J. Trad. Know. 5*, 519.

Bhakshu, L.M. & Venkata Raju, R.R. (2007a). Ethnomedico-botanical studies of certain Euphorbiaceous medicinal plants from Eastern Ghats of Andhra Pradesh. In: *Proc. Natl. Sem. Conserv. Eastern Ghats.* ENVIS Centre, EPTRI, Hyderabad. pp. 102–106.

Bhakshu, L.M. & Venkata Raju, R.R. (2007b). Ethnomedico-botanical studies of certain threatened medicinal plants from Eastern Ghats of Andhra Pradesh. In: *Proc. Natl. Sem. Conserv. Eastern Ghats.* ENVIS Centre, EPTRI, Hyderabad. pp. 129–136.

Bhaskar, A. & Samant, L.R. (2012). Traditional medication of Pachamalai hills, Tamilnadu, India. *Global J. Pharmacol. 6*(1), 47–51.

Brahmam, M., Dhal, N.K. & Saxena, H.O. (1996). Ethnobotanical studies among the Tanla of Malyagiri hills in Dhenkenal district, Orissa, India. In: S.K. Jain (Ed.). Ethnobiology in Human Welfare. New Delhi: Deep Publications. pp. 393–396.

Brahmam, M. & Saxena, H.O. (1990). Ethnobotany of Gandhamardhan hills – Some noteworthy folk medicinal uses. *Ethnobotany 2*, 71–79.

Chaudhari, U.S. & Hutke, V. (2002). Ethno-medico-botanical information on some plants used by Melghat tribes of Amaravati district, Maharashtra. *Ethnobotany 14*, 100–102.

Chaudhuri, R.N.N., Pal, D.C. & Tarafder, C.R. (1975). Less known uses of some plants from the tribal areas of Orissa. *Bull. Bot. Surv. India 17*, 132.

Cunningham, T. (2000). Applied Ethnobotany: People, Wild Plant Use and Conservation. London: Earthseen.

D'Andrade, R. (1995). The Development of Cognitive Anthropology. Cambridge: Cambridge University Press.

Das, P.K. & Kanth, R. (1998). Ethnobotanical studies of the tribal belt of Koraput (Orissa). *Bull. Medico Ethn. Bot. Res., 9,* 123–128.

Das, P.K. & Misra, M.K. (1987). Some medicinal plants used by the tribals of Deomali and adjacent areas of Koraput district, Orissa. *Ind. J. Forestry 10,* 301–303.

Das, P.K. & Misra, M.K. (1988a). Some ethnomedicinal plants of Koraput. *Ancient Sci. Life 1*(1), 60–67.

Das, P.K. & Misra, M.K. (1988b). Some ethnomedicinal plants among Kandhas around Chandrapur (Koraput). *J. Econ. Taxon. Bot. 12,* 103–107.

Das, S., Dash, S.K. & Padhy, S.N. (2003). Ethnomedicinal information from Orissa state. J. *Human Ecol. 14,* 165–227.

Dash, S.S. & Misra, M.K. (1996). Tribal uses of plants from Narayanapatna region of Koraput district, Orissa. *Ancient Sci. Life, 15,* 230.

David, B.C. & Sudarsanam, G. (2011). Ethnomedicinal plant knowledge and practice of people of Javadhu Hills in Tamilnadu. *Asian Pacific J. Trop. Med.* S79–S81.

Dhal, N.K., Panda, S.S. & Muduli, S.D. (2014). Ethnobotanical studies in Nawarangpur district, Odisha, India. *Ann. J. Phytomed., Clinc. Therapeutics. 2,* 257–276.

Dhal, Y., Sahu, R.K. & Deo, B. (2011). Ethnomedicinal Survey of Koraput District, Odisha. An update. *J. Pharm. Res. 4*(11), 414–415.

Dhatchanamoorthy, N., Ashok Kumar, N. & Karthick, K. (2013). Ethnomedicinal plants used by Irular tribes in Javadhu Hills of Southern Eastern Ghats, Tamilnadu, India. *Intern. J. Curr. Res., Dev. 2*(1), 31–37.

Dhayapriya, R.G. & Senthilkumar, S. (2014). Studies on ethnomedicinal plant of Malayali tribes in Bodamalai hills of southern Eastern Ghats, Tamilnadu, India. *Intern. J. Pharmaceut. Res. Develop. 6*(3), 115–118.

Dinesh, V. & Balaji, K. (2015). Ethnobotanical studies on *Balanites aegyptiaca* (L.) Del. Among the folk peoples of Nizamabad district, Telangana state. *Int. J. Ayu. Pharm. Chem. 2,* 56–61.

Dinesh, V., Bembrekar, S.K. & Sharma, P.P. (2013). Herbal formulations used in treatment of kidney stone by native folklore of Nizamabad district, Andhra Pradesh, India. *Bioscience Discovery, 4.*

Dinesh, V., Bembrekar, S.K. & Sharma, P.P. (2014). Ethnobotanical studies on *Borassus flabellifer* L. among the folk peoples of Nizamabad district, Andhra Pradesh. *Int. J. Universal Pharm. Biosci. 3,* 58–59.

Dinesh, V. & Sharma, P.P. (2010). Traditional uses of plants in indigenous folklore of Nizamabad district, Andhra Pradesh, India. *Ethnobotanical leaflets 14,* 29–45.

Dhayapriya, R.G. & Senthilkumar, S. (2014). Studies on ethnomedicinal plant of Malayali tribes in Bodamalai hills of southern Eastern Ghats, Tamilnadu, India. *Intern. J. Pharmaceutical Res. Development 6*(3), 115–118.

Dwarakan, P. & Ansari, A.A. (1992). Ethnobotanical notes on Vallikattupatti and surroundings of Kollimalais of Salem district, Tamilnadu. *J. Econ. Taxon. Bot. 10,* 495–499.

Dwarakan, P. & Ansari, A.A. (1996). Less known uses of plants of Kollimalai (Salem district, Tamilnadu) in south India. *J. Econ. Taxon. Bot. 12,* 284–286.

Elavarasi, S. & Saravanan, K. (2012). Ethnobotanical study of plants used to treat diabetes by tribal people of Kolli Hills, Namakkal District, Tamilnadu, Southern India. *Int. J. Pharm. Tech. Res. 4*(1), 404–411.

Ganesan, S., Pandi, N.R. & Banumathy, N. (2009). Ethnobotanicl studies on the flora of Alagar hills, Eastern Ghats of Tamil Nadu, India. In: Trivedi, P.C. (ed.). Indigenous ethnomedicinal plants. Jaipur: Aavishkar Publishers, pp. 132–148.

Ghatapanadi, S.R., Johnson, N. & Rajasab, A.H. (2011). Documentation of folk knowledge on medicinal plants of Gulbarga district, Karnataka. *Indian J. Tradit. Knowle. 10*, 349–353.

Girach, R.D. (1992). Medicinal plants used by Kondha tribe of district Phulbani (Orissa) in Eastern India. *Ethnobotany, 4*, 53–66.

Girach, R.D. (2001). Ethnobotanical notes on some plants of Mahendragiri hills, Orissa. *Ethnobotany 13*, 80–83.

Girach, R.D. & Aminuddin, Ahmad, M. (1998). Medicinal ethnobotany of Sundargarh, Orissa. *Indian Pharmaceut. Biol. 36*, 20–24.

Girach, R.D., Aminuddin, Ahmad, M., Brahmam, M. & Mishra, M.K. (1996). Native phytotherapy among rural population of district Bhadrak, Orissa.In: Ethnobiology in Human Welfare. S.K. Jain (Ed.). New Delhi: Deep Publications. pp. 162–164.

Girach, R.D., Aminuddin, Ahmad, M., Brahmam, M. & Mishra, M.K. (1998). Euphorbiaceae in native health practices of district Bhadrak, Orissa, India. *Fitoterapia 49*, 24.

Girach, R.D., Aminuddin, Brahmam, M. & Mishra, M.K. (1997). Observations on ethnomedicinal plants of Bhadrak district, Orissa, India. *Ethnobotany 9*, 44–51.

Goud, P.S.P. (1995). Ethno-medico-botanical studies in Kurnool district, Andhra Pradesh. PhD thesis, Sri Krishnadevaraya University, Anantapur.

Goud, P.S.P. & Pullaiah, T. (1996). Ehtnobotany of Kurnool district, Andhra Pradesh. In: S.K. Jain (Ed.). Ethnobiology in Human Welfare. New Delhi: Deep Publications. pp. 410–412.

Goud, P.S.P., Pullaiah, T. & Sri Rama Murthy, K. (1999). Native phytotherapy for fever and malaria from Kurnool district, Andhra Pradesh. *J. Econ. Tax. Bot. 23*, 15–18.

Gritto, M.J., Aslam, A. & Nandagopalan, V. (2012). Ethnomedicinal survey of threatened plants in Pachamalai hills, Tiruchirappalli district, Tamilnadu. *IJRAP 3*(6), 844–846.

Gunasekaran, M. & Balasubramanian, P. (2012). Ethnomedicinal uses of Sthalavrikshas (temple trees) in Tamil Nadu, Southern India. *Ethnobiol. Res. App. 10*, 253–268.

Gupta, V.C., Hussain, S.J. & Imam, S. (1997). Medico-ethno-botanical survey at Paderu forest of Araku valley, Andhra Pradesh, India. *Fitoterapia 68*, 45–48.

Harasreeramulu, S. (1980). Some useful medicinal plants from Srikakulam District, Andhra Pradesh. *J. Indian Bot. Soc. 59 (Suppl.)* 168.

Hemadri, K. (1981). Rheumatism: tribal medicine. *Anc. Sci. Life 1*, 117–120.

Hemadri, K. (1985). Medicinal plants wealth of Chittoor district, Andhra Pradesh, India. *Indian Medicine 34*, 13–15.

Hemadri, K. (1989). Folk-lore claims of Koraput and Phulbani districts of Orissa state. *Indian Medicine 1*, 11–13; *2*, 4–6; *3*, 10–14.

Hemadri, K. (1990). Contribution to the medicinal flora of Karimnagar and Warangal districts, Andhra Pradesh. *Indian Medicine 2*, 16–28.

Hemadri, K. (1991). Contribution to the medicinal flora of Srikakulam district, Andhra Pradesh. *Indian Medicine 3*, 17–34.

Hemadri, K. (1992). Tribals of Andhra Pradesh – their knowledge in nutritional & medicinal herbs. *Indian Med. 4*, 1–6.

Hemadri, K., Raj, P.V., Rao, S.S. & Sarma, C.R.R. (1980). Folklore claims from Andhra Pradesh. *J. Sci. Res. Plant Med. 1*, 37–49.

Hemadri, K. & Rao, S.S. (1983a). Leucorrhoea and menorrhagia: Tribal medicine. *Anc. Sci. Life. 3,* 40–41.

Hemadri, K. & Rao, S.S. (1983b). Antifertility, abortifacient and fertility promoting drugs from Dandakaranya. *Ancient Sci. Life 3,* 103–107.

Hemadri, K. & Rao, S.S. (1984). Jaundice: Tribal medicine. *Ancient Sci. Life, 4,* 209–212.

Hemadri, K. & Rao, S.S. (1989). Folklore claims of Koraput and Phulbani districts of Orissa state. *Indian Medicine 4,* 11–13.

Hemadri, K., Sarma, C.R.R. & Rao, S.S. (1987a). Medicinal plant wealth of Andhra Pradesh I. *Ancient Sci. Life 6,* 167–186.

Hemadri, K., Sarma, C.R.R. & Rao, S.S. (1987b). Medicinal plant wealth of Andhra Pradesh II. *Ancient Sci. Life 7,* 55–64.

Hosagoudar, V.B. & Henry, A.N. (1996). Ethnobotany of Soligas in Biligiri Rangana Betta, Karnataka, southern India. *J. Econ. Taxon Bot. Addl. Ser. 12,* 228–243.

Jain, S.K. (1991). Dictionary of Indian Folk Medicine and Ethnobotany. Deep Publications, New Delhi.

Jayasree, G. (2002). Impact of development Programs on the analysis of Javadhu Hills, Tamilnadu. PhD Thesis Univ. of Madras.Dept. of Botany.

Jeevan Ram, A., Bhakshu, Md. C. & Venkata Raju, R.R. (2004). *In vitro* antimicrobial activity of certain medicinal plants from Eastern Ghats, India, used for Skin diseases. *J. Ethnopharmacol. 90*(2), 353–357.

Jeevan Ram, A., Reddy, P.A. & Venkata Raju, R.R. (1999). The Medicobotanical studies on crude drugs for skin diseases used by tribals from Eastern Ghats of Andhra Pradesh, India. In: M. Sivadasan & Philip Mathew (Eds.). Biodiversity, Taxonomy and Conservation of Flowering Plants, pp. 337–347.

Jeevan Ram, A., Reddy, R.V., Adharvana Chari, A. & Venkata Raju, R.R. (2007). Rare and little known medicinal plants from Nallamalais of the Eastern Ghats, India. *J. Plant Sci. 2*(1), 113–117.

Jeevan Ram, A., Raja, K., Eswara Reddy, K. & Venkata Raju, R.R. (2002). Medicinal plant lore of Sugalis of Gooty forests, Andhra Pradesh, *Ethnobotany, 14,* 37–42.

Jeevan Ram, A. & Raju, R.R.V. (2001). Certain potential crude drugs used by the tribals of Nallamalais, Andhra Pradesh for Skin disease. *Ethnobotany, 13,* 110–115.

Joshi, P. (1995). Ethnobotany of Primitive Tribes in Rajasthan, Rupa books, Jaipur.

Kadavul, K. & Dixit, A.E. (2009). Ethnomedicinal studies of the woody species of Kalrayan and Shevarayan Hills, Eastern Ghats, Tamilnadu. *Indian J. Tradit. Knowle. 8*(4), 592–597.

Kadirvelmurugan, V., Raju, K., Arumugam, T., Karthik, V. & Ravikumar, S. (2014). Ethnobotany of medi-flora of Kolli hills, Tamilnadu. *Archives Appl. Sc. Res. 6*(1), 159–164.

Kannan, P. & Kumar, P.S. (2014). Antidotes against snake bite from ethnobotanical practices of primitive tribes of Tamilnadu. *J. Sci. Trans. Environ. Technov. 8*(1), 33–39.

Kapoor, S.L. & Kapoor, L.D. (1980). Medicinal plant wealth of Karimnagar district of Andhra Pradesh. *Bull. Med. Ethnobot. Res. 1,* 120–144.

Karuppusamy, S. (2002). Floristic studies with special reference to ethnomedicobotany of Sirumalai hills, Tamilnadu, India. PhD thesis, Gandhigram Rural Institute, Deemed University, Gandhigram, Dindigul, Tamilnadu.

Karuppusamy, S. (2007). Medicinal plants used by Paliyan tribes of Sirumalai hills of southern India. *Nat. Prod. Radiance 6*(5), 436–442.

Karuppusamy, S., Muthuraja, G. & Rajasekaran, K.M. (2009). Lesser known ethnomedicinal plants of Alagar hills, Madurai district of Tamilnadu, India. *Ethnobotanical Leaflet 13,* 1426–1433.

Karuppusamy, S. & Pullaiah, T. (2005). Selected medicinal plant species of Sirumalai hills, south India, used by natives and antibacterial screening of plants. *J. Trop. Med. Plants 6*(1), 99–109.

Karuppusamy, S., Rajasekaran, K.M. & Pullaiah, T. (2009). Plant diversity and conservation in Sirumalai hills of Eastern Ghats, Tamilnadu, south India. In: In T. Pullaiah, (Ed.). Emerging Trends in Biological Sciences. New Delhi: Daya Publishing House, pp. 133–171.

Kathikeyani, T.P. (2003). Ethnobotanical Studies among Yanandis of Sathyavedu Mandal, Chittor District, Andhra Pradesh, *Plant Archive, 3*, 21–27.

Kathirvel, K., Ramya, S., Sudha, T.P.S., Ravi, A.V., Rajasekaran, C., Vanitha Selvi, R. & Jayakumararaj, R. (2010). Ethnomedicinal survey on plants used by tribals in chitteri Hills. *Environ. Int. J. Sci. Tech. 5*, 35–46.

Kolar, A.B. & Bhasha, G.M. (2013). Survey of medicinal plants of Pachamalai hills, a part of Eastern Ghats, Tamilnadu. *Intl. J. Curr. Sci. 5*(2), 3923–3929.

Koteswara Rao, J., Prasanthi, S., Aniel Kumar, O. & Seetharami Reddi, T.V.V. (2013). Ethnomedicinal plants used by the tribal groups of North-coastal Andhra Pradesh for healing bone fractures. *Ethnobotany, 25*, 164–165.

Koteswara Rao, J., Prasanthi, S., Aniel Kumar, O. & Seetharami Reddi, T.V.V. (2014). Ethnomedicine for cuts and wounds by the primitive tribe groups of north coastal Andhra Pradesh. *J. Non-Timber For. Prod. 21*(4), 237–240.

Koteswara Rao, J., Suneetha, J., Seetharami Reddi, T.V.V. & Aniel Kumar, O. (2011). Ethnomedicine of the Gadabas, a primitive tribe of Visakhapatnam Distirct, Andhra Pradesh. *International Multidisciplinary Research Journal, 1*, 10–14.

Kottaimuthu, R. (2008). Ethnobotany of Valaiyans of Karandamalai, Dindigul District, Tamilnadu, India. *Ethnobotanical Leaflets. 12*, 195–203.

Krishnamohan, R. & Bhirava Murthy, P.V. (1992). Plants used in traditional medicine by tribals of Prakasam district, Andhra Pradesh. *Ancient Sci. Life 11*, 176–181.

Kumar, A.K. & Niteswar, K. (1983). An enquiry into the folk lore medicines of Addateegala an agency tract of East Godavari District of Andhra Pradesh. *Indian Medicine 2*, 2–4.

Kumar, S., Jena, P.K., Sabnam, S., Kumari, M. & Tripathy, P.K. (2012). Study of plants used against the skin diseases with special reference to *Cassia fistula* L. among the King (Dongaria Kandha) of Niyamgir: a primitive tribe of Odisha, India. *Intern. J. Drug Dev. Res. 4*(2), 256–264.

Kumar, T.D.C. (1997). Ethnomedico botanical studies in Mahabubnagar district in Andhra Pradesh. PhD thesis, Sri Krishnadevaraya University, Anantapur.

Kumar, T.D.C. & Pullaiah, T. (1999). Ethno-medical uses of some plants of Mahabubnagar district, Andhra Pradesh, India. *J. Econ. Taxon. Bot. 23*, 341–345.

Lakshmi Narasimhudu, C. & Venkata Raju, R.R. (2013). Medicobotanical properties of *Phyllanthus* species (Euphorbiaceae) used by the Aboriginal Adivasis of Eastern Ghats, Andhra Pradesh. *Indian J. Tradit. Knowl. 12*, 326–333.

Lingaiah, M. & Rao, P.N. (2013). An ethnobotanical survey of medicinal plants used by traditional healers of Adilabad district, Andhra Pradesh, India. *Biolife 1*, 17–23.

Madhava Chetty, K. & Rao, K.N. (1989). Ethnobotany of Sarakallu and adjacent areas of Chittoor District, Andhra Pradesh. *Vegetos, 1*(2), 160–163.

Madhava Chetty, K. & Lakshmipathi Chetty. (1998). Ethno-medicobotany of some aquatic angiosperms in Chittoor district of Andhra Pradesh, India. *Fitoterapia 69*(1), 7–12.

Malik, S. (1996). Belief and healthcare: A study among the Saoras of Koraput dist. (Orissa). In: Ethnobiology in Human Welfare. Ed. S.K. Jain. Deep Publications, New Delhi. pp. 151–152.

Manikandan, S. & Lakshmanan, G.M.A. (2014). Ethnobotanical survey of medicinal plants in Kalrayan Hills, Eastern Ghats, Tamilnadu. *International Letters of Natural Sciences* 2(2), 111–121.

Maruthapandian, A., Mohan, V.R. & Kottaimuthu, R. (2011). Ethnomedicinal plants used for the treatment of diabetes and jaundice by Palliyar tribals in Sirumalai hills, Western Ghats, Tamil Nadu, India. *Indian J. Nat. Prod. Resour. 2*, 493–497.

Mishra, R.C. (1992). Medicinal plants among the tribal of upper Bonda region, Koraput (Orissa). *J. Econ. Taxon. Bot. Addl. Ser. 10*, 275–279.

Mishra, R.C. (2004a). Therapeutic uses of some seeds among the tribals of Gandhamardhan hill range, Orissa. *Indian J. Trad.Knowl. 3*, 105–115.

Misra, M.K. (2004b). Ethno-medico-botany of Orissa – A review. In: P.C. Trivedi & N.K. Sharma (Eds.). Ethnomedicinal Plants. Jaipur: Pointer Publisher, pp. 82–115.

Misra, M.K. & Dash, S.S. (1997). Medicinal plants used by the tibals of Koraput district, Orissa. In: Mohapatra, P.M., Mohapatra, P.C. (Eds.). Forest Management in Tribal Areas: Forest Policy and Peoples Participation. New Delhi, India: Concept Publishing Company. pp. 160–182.

Mohamed Tariq, N.P.M., Rayees Ifham, S. & Ali, A.M. (2012). Data collection methods in research for medicinal plants of Javadhu hills, Tamilnadu, India. *Int. J. Curr. Microbiol. Appl. Sci. 2*(2), 83–89.

Mohanta, R.K., Rout, S.D. & Sahu, H.K. (2006). Ethnomedicinal plant resources of Similipal Biosphere Reserve, Orissa, India. *Zoos'Print J. 21*(8), 2372–2374.

Mohapatra, S.P. & Sahoo, H.P. (2008). Some lesser known medicinal plants of the Kondha and Gond tribes of Bolangir, Orissa, India. *Ethnobotanical leaflets 12*, 1003–1006.

Mudgal, V. & Pal, D.C. (1980). Medicinal plants used by tribals of Mayurbhanj (Orissa). *Bull. Bot. Surv. India, 22*, 59–62.

Mukherjee, A. & Namhata, D. (1990). Medicinal plant lore of the tribals of Sundargarh district, Orissa. *Ethnobotany, 2*, 57–66.

Murthy, E.N. (2012). Ethnomedicinal plants used by Gonds of Adilabad District, Andhra Prades, India. *Intern. J. Pharm. Life Sci. 3(10)*, 2034–2043.

Murthy, E.N., Pattanaik, C., Reddy, C.S. & Raju, V.S. (2012). Traditional knowledge of ethnic tribes in Pranahita wildlife sanctuary, Andhra Pradesh. In: Biodiversity and Sustainable Livelihoods. Vedam Publications, pp. 35–44.

Murthy, E.N. & Raju, V.S. (2007). Ethnomedicinal plants among the Nayakpod tribe in Pranahita wildlife sanctuary, Andhra Pradesh, India. In: Abstracts: Intl. Sem. Changing Senario in Angiosperm Systematics. Shivaji Univ. Kolhapur, pp. 161–162.

Muruganandam, S., Rathakrishnan, S. & Selvaraju, A. (2012). Plants used for non-medicinal purposes by Malayali tribals in Javadhu Hills of Tamilnadu, India. *Global J. Res. Med. Pl., Indign. Med. 1*(2), 663–669.

Murugesan, P., Ganesan, R., Suresh kumar, M. & Panneer selvam, B. (2011). Ethnobotanical study of medicinal plants used by villagers in Kolli Hills of Namakkal District of Tamilnadu, India. *Int. J. Pharm. Sci. Rev. Res. 10*(1), 170–173.

Muthu, C., Ayyangar, M., Raja, A. & Ignacimuthu, S. (2006). Medicinal plants used by traditional healers in Kancheepuram district of Tamil Nadu, India. *J. Ethnobiol. Ethnomed. 43*, 1–10.

Nagalakshmi, N.V.N. (2001). Studies on crude drugs used for abortion and antifertility by the tribals of Rayalaseema, Andhra Pradesh, India. PhD Thesis, Sri Krishnadevaraya University, Anantapur,

Nagaraju, N. & Rao, K.N. (1989). Folk medicine for diabetes from Rayalaseema of A.P. *Anci. Sci. Life 9,* 31–35.

Nagaraju, N. & Rao, K.N. (1990). A survey of plant crude drugs of Rayalaseema, Andhra Pradesh, India. *J. Ethnopharmacol. 29,* 137–158.

Naidu, M.T., Babu, N.C., Kumar, O.A. & Venkaiah, M. (2012). Herbal remedies for rheumatoid arthritis used by the tribes of Vizianagaram district, Andhra Pradesh. *J. Non-Timber For. Prod. 19*(4), 303–308.

Naidu, M.T., Babu, N.C. & Venkaiah, M. (2013). Ethnic remedies against snake bite from Kotia hills of Vizianagaram district, Andhra Pradesh, India. *Indian J. Nat. Prod. Resources* 4(2), 194–196.

Natarajan, V. & Udayakumar, A. (2013). Studies on the medicinal plants used by the Malayali tribes of Kolli hill in Tamil Nadu. *International J. Basic Life Sci. 1*(1), 16–29.

Nayak, S., Behera, S.K. & Misra, M.K. (2004). Ethno-medico-botanical survey of Kalahandi district of Orissa. *Indian J. Trad. Knowl. 3,* 72–79.

Nayak, P.K. & Sahoo (2002). Trends of Tribal population in Eastern Ghats region of Orissa. *EPTRI. ENVIS News lett. 8*(2), 3–4.

Nisteswar, K. & Kumar, K.A. (1983). An enquiry in folklore medicine of Addateegala agency tract of East Godavari district of Andhra Pradesh.*Vagbhata 1,* 43–44.

Padal, S.B., Butchi Raju, J. & Chandrasekar, P. (2013a). Traditional knowledge of Konda Dora tribes, Visakhapatnam District, Andhra Pradesh, India. *IOSR J. Pharmacy. 3*(4), 22–28.

Padal, S.B., Chandrasekar, P. & Satyavathy, K. (2013b). Ethnomedicinal investigation of medicinal plants used by the tribes of Pedabayalu Mandalam, Visakhapatnam distirct, Andhra Pradesh, India. *Intern. J. Computational Engineering Res. 3*(4), 8–13.

Padal, S.B., Devender, R., Ramakrishna, H. & Prabhakar, R. (2013a). Ethnomedicinal diversity of Ananthagiri mandal of Paderu forest division in Andhra Pradesh. *Ethnobotany, 25,* 143–147

Padal, S.B., Murthy, P.P., Rao, D.S. & Venkaiah, M. (2010). Ethnobotanical studies on Paderu Division, Visakhapatnam District, Andhra Pradesh, India. *J. Phytol. 2*(8), 70–91.

Padal, S.B., Murthy, P.P., Rao, D.S. & Venkaiah, M. (2012). Ethnobotany of Paderu Division, Visakhapatnam District, AP, India. LAP LAMBERT Academic Publishing.

Padal, S.B. & Sandhyasri, B. (2013b). Ethnomedicinal investigation of medicinal plants of Sovva panchayat, Dumbriguda mandalam, Visakhapatnam district, Andhra Pradesh. *Intern. J. Engin. Sci. 2*(5), 55–61.

Padal, S.B., Sandhyasri, B. & Chandrasekar, P. (2013c). Traditional use of monocotyledons plants of Araku Valley Mandalam, Vishakapatnam District, Andhra Pradesh. *Indian J. Pharm. Biol. Sci. 6*(2), 12–16.

Padal, S.B., Vijayakumar, Y., Butchi Raju, J. & Chandrasekar, P. (2013d). Ethnomedicinal uses of shrub by tribals of Borra Panchayat, Ananthagiri Mandalam, Visakhapatnam District, Andhra Pradesh, India. *Intern. J. Pharmaceut. Sci. Invention. 2*(6), 10–12.

Padma Rao, P. & Reddy, P.R. (1999). A note on folklore treatment of bone fractures from Ranga Reddy district, Andhra Pradesh. *Ethnobotany, 11,* 107–108.

Padma Rao, P., Ramachandra Reddy, P. & Janardhana Reddy, K. (2007). Some ethnomedicines used by Chenchus in treatment of jaundice from Nagarjunasagar Srisailam Tiger Reserve (NSTR), Andhra Pradesh. *Ethnobotany, 19,* 128–130.

Panda, A. & Misra, M.M. (2011). Ethnomedicinal survey of some wetland plants of South Orissa and their conservation. *Indian J. Tradit. Knowle. 10,* 296–303.

Panda, B.K. (2007). Some ethnomedicinal plants of Karlapat Reserve Forest, district Kalahandi, Orissa. *Ethnobotany, 19,* 134–136.

Panda, S.K., Patra, N., Sahoo, G., Bastia, A.K. & Dutta, S.K. (2012). Anti-diarrheal activities of medicinal plants of Similipal Biosphere Reserve, Odisha, India. *Int. J. Med. Arom. Plants 2*(1), 123–134.

Panda, S.K., Rout, S.D., Mishra, N. & Panda, T. (2011). Phytotherapy and traditional knowledge of tribal communities of Mayurbhanj District, Orissa, India. *J. Phrmacognosy Phytotherapy 3*(7), 101–113.

Panda, S.P., Sahoo, H.K., Subudhi, H.N. & Sahu, A.K. (2014). Potential medicinal plants of Odisha used in rheumatism and conservation. *Am. J. Ethnomedicine 1*(4), 260–265.

Panda, T. & Padhy, R.N. (2008). Ethnomedicinal plants used by tribes of Kalahandi district, Orissa. *Indian J. Trad. Know. 7,* 242–249.

Panda, T., Panigrahi, S.S. & Padhy, R.N. (2005). A sustainable use of phytodiversity by the Kandha tribe of Orissa. *Indian J. Trad. Knowl. 4,* 173–178.

Pandey, A.K. & Rout, S.D. (2006). Ethnobotanical uses of plants of Simlipal Biosphere reserve (Orissa). *Ethnobotany, 18,* 102–106.

Panigrahi, N. (1999). Ethnomedicine among the 'Gondss' of Orissa. *EPTRI-ENVIS Newsletter 4*(2).

Parthipan, M., Aravindan, V. & Rajendran, A. (2011). Medicobotanical study of Yercaud hills in the Eastern Ghats of Tamilnadu, India. *Ancient Science Life. 30*(4), 104–109.

Patnaik, B.K. & Rath, S.P. (2014). Important medicinal plants used by the tribes of Sunebada plateau in Odisha. *Indian Forester 140*(1), 34–37.

Pattanaik, C., Reddy, C.S. & Dhal, N.K. (2008). Phytomedicinal study of coastal sand dune species of Orissa. *Indian J. Tradit. Knowle. 7,* 263–268.

Pattanaik, C., Reddy, C.S. & Murthy, M.S.R. (2006). Ethnomedicinal observations among the tribal people of Koraput district, Orissa, India. *Res. J. Bot. 1,* 125–128.

Pattanaik, C., Reddy, C.S. & Murthy, M.S.R. (2008). An Ethnobotanical survey of medicinal plants used by Didagi tribe of Malakangiri District of Orissa, India. *Fitoteropia 79,* 67–71.

Pattanaik, C., Reddy, C.S. & Reddy, K.N. (2009). Ethno-medicinal survey of threatened plants in Eastern Ghats, India. *Our Nature 7,* 122–128.

Prabakaran, R., Senthil Kumar, T. & Rao, M.V. (2013). Role of Non-Timber forest products in the livelihood of Malayali tribe of Chitteri Hills of Southern Eastern Ghats, Tamilnadu. *J. Appl. Pharac. Sci. 3*(5), 56–60.

Prasanthi, S., Sandhya Sri, B. & Seetharami Reddi, T.V.V. (2014). Ethnomedicine for diabetes by the Savaras of Andhra Pradesh. *J. Non-Timber Forest Products 21,* 109–112.

Prabu, M. & Kumuthakalavalli, R. (2012). Folk remedies of medicinal plants for snake bites, scorpion sting and dog bites in Eastern Ghats of Kolli hill, Tamilnadu, India. *IJRAP 3*(5), 696–700.

Prusti, A.B. (2007). Plants used as ethnomedicine by Bondo tribe of Malkangiri district, Orissa. *Ethnobotany, 19,* 105–110.

Pullaiah, T. & Kumar, T.D.C. (1996). Herbal plants of Mannanur forest. *J. Econ. Taxon. Bot. Addl. Series, 12,* 218–220.

Pullaiah, T., Sri Rama Murthy, K., Sai Prasad Goud, P., Dharmachandra Kumar, T. & Vijaya Kumar, R. (2003). Medicinal plants used by the tribals of Nallamalais, Eastern Ghats of India. *J. Trop. Med. Plants 4,* 237–243.

Rai Choudhury, H.N., Pal, D.C. & Tarafdar, C.R. (1975). Less known uses of some plants from the tribal areas of Orissa. *Bull. Bot. Surv. India 17,* 132–136.

Raja Reddy, K. (1986). Jaundice folk medicine from Chittoor District, Andhra Pradesh, India. *Ethnobotany, 3,* 5–8.

Raja Reddy, K. (1988). Folk medicine from Chittoor District, Andhra Pradesh, India used in the treatment of Jaundice. *Int. J. Crude Drug Res., 26,* 137–140.

Raja Reddy, K., Sudarsanam, G. & Gopal Rao, P. (1989). Plant drugs of Chittoor district, Andhra Pradesh, India. *Int. J. Crude Drugs, 27,* 41–54.

Rajasekaran, B. & Warren, D.M. (1994). Indigenous knowledge for socio-economic devolopment and biodiversity conservation: the Kolli hills. *Indigenous Knowledge & Devolopment Monitor 2,* 13–17.

Rajendran, A., Ramarao, N. & Henry, A.N. (1996). Hepatic stimulant plants of Andhra Pradesh, India. *J. Econ. Tax. Bot. Addl. Ser.* 12, 221–223.

Rajendran, A. & Manian, S. (2011). Herbal remedies for diabetes from Kolli Hills, Eastern Ghats, India. *Indian J. Nat. Products and Resources. 2*(3), 383–386.

Raju, D.C.S. (1998). Medicinal plants of Eastern Ghats, Andhra Pradesh. *EPTRI – ENVIS Newwsletter 3*(1), 10–13.

Raju, K., Natarajan, C., Karuppusamyy & Manian, S. (2004). New ethnomedicinal report on *Euphorbia fusiformis* from Chitteri hills of Tamil Nadu, India. *J. Swamy Bot Club 21,* 79–80.

Raju, M.P., Prasanthi, S. & Seetharami Reddi, T.V.V. (2011). Medicinal plants in folk medicine for women's diseases in use by Konda Reddis. *Indian J. Trad. Knowl. 10,* 563–567.

Raju, V.S. & Reddy, K.N. (2005). Ethnomedicine for dysentery and diarrhea from Khammam district of Andhra Pradesh. *Indian J. Trad.Knowl. 4,* 443–447.

Ramakrishna, N. (2013). Ethnobotanical studies of Adilabad district, A.P. PhD thesis, Osmania University, Hyderabad.

Ramakrishna, N., Saidulu, Ch. (2014a). Medicinal plants used in snake bite and scorpion sting by Gonds and Kolams of Adilabad dist., A.P. *Int. J. Curr. Pharmaceut. Res. 6,* 39–41.

Ramakrishna, N. & Saidulu, Ch. (2014b). Traditional herbal knowledge on reproductive disorders and sexually transmitted diseases in the Adilabad discrict of A.P. *J. Nat. Prod. Plant Resour, 4,* 1–5.

Ramana, Y.V.V.V. & Subba Reddi, C. (2002). Ethnomedicobotany of Maredumilli tribal region in the Eastern Ghats. In: *Abstracts. Proc. Natl. Sem. Conserv. Eastern Ghats.* ENVIS center, EPTRI, Hyderabad.

Ramana Naidu, B.V., Haribabu Rao, D., Subramanyam, P., Prabhakar Raju, C. & Jayasimha Rayalu, D. (2012). Ethnobotanical study of medicinal plants used by tribals in Nallamala forest area of Kurnool District, Andhra Pradesh. *Intern. J. Plant, Animal and Environmental Sci. 2*(4), 72–81.

Ramarao Naidu, B.V. & Seetharami Reddi, T.V.V. (2008). Folk herbal remedies for rheumatoid arthritis in Srikakulam district of Andhra Pradesh. *Ethnobotany, 20,* 76–79.

Ramanathan, R., Bhuvaneswari, R., Indhu, M., Subramanian, G. & Dhandapani, R. (2014). Survey of ethnomedicinal observation on wild tuberous medicinal plants of Kolli hills, Namakkal distirct, Tamilnadu. *J. Medicinal Plants Studies 2*(4), 50–58.

Ramarao, N. (1988). The Ethnobotany of Eastern Ghats in Andhra Pradesh, India. PhD thesis. Bharathiar University, Coimbatore.

Ramarao, N. & Henry, A.N. (1996). The Ethnobotany of Eastern Ghats in Andhra Pradesh, India. BSI, Kolkata.

Ramya, S., Rajasekaran, C., Sivaperumal, R., Krishnan, A. & Jeyakumararaj, R. (2008). Ethnomedicinal perspectives of botanicals used by Malayali tribes in Vattal hills of Dharmapuri (TN), India. *Ethnobotanical Leaflets 12,* 1054–1060.

Ranganathan, R., Vijayalakshmi, R. & Parameswari, P. (2012). Ethnomedicinal survey of Jawadhu hills in Tamilnadu. *Asian J. Pharm. Clinc. Res. 5*(2), 45–49.

Ranjithakani, P., Lakshmi, G., Geetha, S., Murugan, S. & Viswanathan, M.B. (1992). Ethnobotanical study on Kolli hills – a preliminary report. *J. Swamy Bot. Cl. 9*, 79–81.

Ranjithkumar, A., Chittibabu, C.V. & Renu, G. (2014). Ethnobotanical investigation on the Malayali tribes on Javadhu hills, Eastern Ghats, South India. *Ind. J. Medicine and Healthcare. 3*(1), 322–332.

Rao, B.N.S., Rajasekhar, D., Raju D.C. & Nagaraju, N. (1996). Ethno-medicinal notes on some plants of Tirumala hills for dental disorders. *Ethnobotany 8*, 88–91.

Rao, D.M. & Pullaiah, T. (2001). Ethno-medico-botanical studies in Guntur district of Andhra Pradesh, India. *Ethnobotany, 13*, 40–44.

Rao, D.M. & Pullaiah, T. (2007). Ethnobotanical studies on some rare and endemic floristic elements of Eastern Ghats- Hill ranges of South East Asia, India. *Ethnobotanical leaflets. 11*, 52–70.

Rao, D.S., Venkaiah, M., Padal, S.B. & Murthy, P.P. (2010). Ethnomedicinal plants from Paderu division of Visakhapatnam district, A.P., India. *J. Phytol. 2*(8), 70–91.

Rao, D.S., Rao, M.B., Murthy, P.P. & Venkaiah, M. (2014). Ethnobotanical uses of certain plant species from Makkauva mandal, Vizianagaram district, Andhra Pradesh. *Int. J. Curr. Res. 6*(3), 5387–5390.

Rao, K.P. & Harasreeramulu, S. (1985). Ethnobotany of selected medicinal plants of Srikakukam district, Andhra Pradesh. *Anc. Sci. Life 4*, 238–244.

Rao, M.K.V. & Prasad, O.S.V.D. (1995). Ethnomedicines of tribes of Andhra Pradesh. *J. Non Timber Forest Prod. 2*, 105–114.

Rao, N.S.B., Rajasekhar, D., Narayanaraju, K.V. & Chengalraju, D. (1995). Ethnomedicial therapy among the Chenchus of Nallamalai forest of Andhra Pradesh. *Biosci. Res. Bull. 11*, 81.

Rao, T.V.R. & Subba Reddui, C. (2002). Ethnomedicinal plants of Ratnagiri and their conservation. In: *Proc. Natl. Sem..Conserv.Eastern Ghats.* ENVIS center, EPTRI, Hyderabad, India. pp. 169–174.

Rao, V.L.N., Busi, B.R., Dharma Rao, B., Seshagiri Rao, Ch., Bharathi, K. & Venkaiah, M. (2006). Ethnomedicinal practices among Khonds of Visakhapatnam district, Andhra Pradesh. *Indian J. Trad.Knowl. 5*, 217–219.

Ravikumar, K. & Vijaya Sankar, R. (2003). Ethnobotany of Malayali tribals in Melpattu village, Javadhu Hills of Eastern Ghats, Tiruvannamalai District, Tamilnadu. *J. Econ. Taxon. Bot. 27*, 715–726.

Ravindran, K.C., Venkatesan, K., Balakrishnan, V., Chellapan, K.P. & Balsubramanian, T. (2005). Ethnomedicinal studies of Pichavaram mangroves of East coast, Tamil Nadu. *Indian J. Tradit. Knowl. 4*, 409–411.

Ravi Prasad Rao, B. & Sunita, S. (2011). Medicinal plant resources of Rudrakod sacred grove in Nallamalais, Andhra Pradesh, India. *J. Biodiversity. 2*(2), 75–89.

Ravishankar, T. (1990). Ethnobotanical studies in Adilabad and Karimnagar districts of Andhra Pradesh, India. PhD thesis, Bharathiyar University, Coimbatore.

Ravishankar, T. & Henry, A.N. (1992). Ethnobotany of Adilabad district, Andhra Pradesh, India. *Ethnobotany, 4*, 45–52.

Reddy, A.V.B. (2007). Ethnobotanical studies of a tribe inhabiting Khammam district, Andhra Pradesh. *J. Swamy Bot Club 24*, 91–94.

Reddy, C.S., Reddy, K.N., Pattanaik, C. & Raju, V.S. (2006). Ethnobotanical observations on some endemic plants of Eastern Ghats, India. *Ethnobotanical Leaflets. 10*, 82–91.

Reddy, K.N. (2002). Ethnobotany of Khammam district, Andhra Pradesh. PhD thesis, Kakatiya University, Warangal.

Reddy, K.N., Reddy, C.S. & Jadhav, S.N. (2005). Ethnobotany of certain Orchids of Eastern Ghats of Andhra Pradesh. *Indian Forester 13*(1), 90–96.

Reddy, K.N., Reddy, C.S. & Trimurthulu, G. (2006). Ethnobotanical survey on respiratory disorders in Eastern Ghats of Andhra Pradesh, India. *Ethnobotanical Leaflets.* 139–148. (http://www.siu.edu/~ebl/leaflets/ghats.html).

Reddy, K.N., Reddy, C.S. & Trimurthulu, G. (2010). Medicinal plants used by ethnic people of Medak district, Andhra Pradesh. *Indian J. Tradit. Knowle. 9*, 184–190.

Reddy, K.N. & Subbaraju, G.V. (2005). Ethnomedicine from Maredumilli region of East Godavari District, Andhra Pradesh. *J. Econ. Taxon. Bot. 29* (2), 476–48.

Reddy, K.N. & Subbaraju, G.V. (2005). Ethnobotanical medicine for rheumatic diseases from Eastern Ghats of Andhra Pradesh. In: T. Pullaiah et al. (eds.). Recent Trends in Plant Sciences. Regency Publications, New Delhi, pp. 128–138.

Reddy, M.B., Reddy, K.R. & Reddy, M.N. (1988). A survey of plants of Chenchu tribes of Andhra Pradesh, India. *Int. J. Crude Drug Res. 26*, 197–207.

Reddy, M.B., Reddy, K.R. & Reddy, M.N. (1989). A Survey of plant crude drugs from Anatapur District, Andhra Pradesh, India. *Inter. J. Crude drugs Res. 27*(3), 145–155.

Reddy, M.H., Reddy, R.V. & Raju, R.R.V. (1996). Perspective in tribal medicines with special reference to Rutaceae in Andhra Pradesh. *J. Econ. Tax. Bot. 20*, 743–744.

Reddy, M.H., Vijayalakshmi, K. & Venkataraju, R.R. (1997). Native phytotherapy for snake bite in Nllamalais Eastern Ghats, India. *J. Econ. Taxon. Bot. 12*, 214–217.

Reddy, P.R. & Padma Rao, P. (2002). A survey of plant crude drugs in folklore from Ranga Reddy district, Andhra Pradesh, India. *Indian J. Trad. Knowl. 1*, 20–25.

Reddy, P.R., Padma Rao, P. & Prabhakar, M. (2003). Ethnomedicinal practices amongst Chenchus of Nagarjunasagar Srisailam Tiger Reserve (NSTR), Andhra Pradesh – Plant remedies for cuts, wounds and boils. *Ethnobotany, 15*, 67–70.

Reddy, R.V. (1995). Ethnobootanical and phytochemical studies on medicinal plant resources of Cuddapah district, A.P., India. PhD thesis, Sri Krishnadevaraya University, Anantapur.

Reddy, R.V., Reddy, M.H. & Raju, R.R.V. (1995). Ethnobotany of *Aristolochia* L. *Acta Botanica Indica 23*, 291–292.

Reddy, R.V., Hemambara Reddy, M. & Venkataraju, R.R. (1996). Ethnobotany of less known tuber yielding plants from Andhra Pradesh, India. *J. Non-Timber Forest Prod. 3*, 60–63.

Reddy, S.R., Reddy, M.A., Philomina, N.S. & Yasodamma, N. (2014). Ethnobotanical survey of Sheshachala Hill range of Kadapa District, Andhra Pradesh, India. *Ind. J. Fundamental and Appl. Life Sci. 4*, 324–329.

Rekha, R. & Senthil Kumar, S. (2014). Ethnobotanical plants used by the Malayali tribes in Yeracaud Hills of Eastern Ghats, Salem District, Tamilnadu, India. *Global J. Res. Med. Plants & Indign. Med. 3*(5), 243–251.

Rekka, R. & Senthil Kumar, S. (2014). Indigenous knowledge on some medicinal plants among the Malayali tribals in Yercaud hills, Eastern Ghats, Salem District, Tamilnadu, India. *Intern. J. Pharma Biosci. 5*(4), 371–374.

Rout, S.D., Panda, T. & Mishra, N. (2009). Ethnomedicinal plants used to cure different diseases by tribals of Mayrbhanj district of North Orissa. *Ethnomed. 3*, 27–32.

Rout, S.D. & Pandey, A.K. (2007). Ethnomedicobiology of Simplipal Biosphere Reserve, India. In: A.P. Das & A.K. Pandey (Eds). Advances in Ethnobotany, Dehra Dun: Bishen Singh Mahendra Pal Singh. pp. 61–72.

Sabjan, G., Sundaram, G., Dharaneeshwara Reddy & Muralidhara Rao, D. (2014). Ethnobotanical crude drugs used in treatment of liver diseases by Chenchu tribes in Nallamalais, Andhra Pradesh, India. *American J. Ethnomedicine, 1*(3), 115–121.

Sahoo, A.K. & Bahali, D.D. (2003). Ethnomedicines and medico-religious beliefs in Phulbani district, Orissa. *J. Econ. Taxon. Bot. 21,* 500–504.

Sahoo, A.K. & Mudgal, V. (1995). Less known ethnobotanical uses of plants of Phulbani district, Orissa. *Ethnobotany7,* 63–67.

Sahu, C.R., Nayak, R.K. & Dhal, N.K. (2013a). Traditional herbal remedies for various diseases used by tribals of Boudh District, Odisha, India for sustainable development. *International J. Herbal Medicine 1*(1), 12–20.

Sahu, C.R., Nayak, R.K. & Dhal, N.K. (2013b). Ethnomedicinal plants used against various diseases in Boudh district of Odisha, India. *Ethnobotany, 25,* 153–159.

Sahu, C.R., Nayak, R.K. & Dhal, N.K. (2013c). The plant wealth of Boudh district of Odisha, India with reference to ethnobotany. *Intern. J. Curr. Biotech. 1*(6), 4–10.

Sandhya, B., Thomas, S., Isabel, W. & Shenbagarathi, R. (2006). Ethnomedicinal plants used by the Valaiyan community of Piramalai Hills, Tamilnadu, India: A Pilot study. *Afr. J. Trad. Compl. Alt. Med. 3*(1), 101–114.

Sandhya Sri, B. & Seetharami Reddi, T.V.V. (2011). Native herbal galactogogues used by women of Bagata tribe of Visakhapatnam district, Andhra Pradesh. *Ethnobotany, 23,* 125–128.

Sankaranarayanan, S., Bamal, P., Ramachandran, J., Kalaichelvan, T., Devvaraman, M., Vijayalakshimi, M., Dhamotharan, R., Dananjeyan, B. & Sathya Bama, S. (2010). Ethnobotanical study of medicinal plants used by traditional uses in Villupuram district of Tamilnadu, India. *J. Med. Pl. Res. 4*(2), 1089–1101.

Sankarasivaaraman, K. & Ignacimuthu, S. (2003). Medicinal plants used by tribals in the Kolli hills, Salem district, Tamil Nadu, India. *ANJAC J. Sci. 2,*55–59.

Sarangi, N. & Sahu, R.K. (2004). Ethnomedicinal plants used in venereal and gynecological disorders in Kalahandi, Orissa. *Ethnobotany 16,* 16–20.

Sarkar, N., Rudra, S. & Basu, S.K. (1999). Ethnobotany of Bangriposi, Mayurbhanj, Orissa. *J. Econ. Taxon. Bot. 23,* 509–514.

Satapathy, K.B. (1992). Medicinal uses of some plants among the tribals of Sundargarh district, Orissa. *J. Econ. Taxon. Bot. Addl. Ser. 10,* 241–249.

Satapathy, K.B. (2008). Interesting ethnobotanical uses from Juang, Koha and Munda tribes of Keojhar district, Orissa. *Ethnobotany, 20,* 99–105.

Satapathy, K.B. & Brahmam, M. (1996). Some medicinal plants used by the tribals of Sundargarh district, Orissa. In: Ethnobotany in Human Welfare, ed.: Jain, S.K., Deep publications, New Delhi, pp. 153–158.

Satapathy, K.B. & Panda, P.C. (1992). Medicinal uses of some plants among the tribals of Sundargarh district, Orissa. *J. Econ. Taxon. Bot. Addl Ser. 10,* 241–249.

Satyavathi, K., Deepika, D.S. & Padal, S.B. (2014). Ethnomedicinal plants used by the Bagata tribes of Paderu forest division, Andhra Pradesh, India. *Int. J. Adv. Res. Sci. Technol. 3,* 36–39

Savithramma, N. (2004). Diversity and Conservation of medicinal plants of Seshachalam Hill range of Andhra Pradesh. *Bull. Bot. Surv. India 46,* 438–453.

Savithramma, N., Suhrulatha, Linga Rao, M., Yugandhar, P. & Hari Babu, R. (2012). Ethnobotanical study of Penchalakona forest area of Nellore District, Andhra Pradesh, India. *Intern. J. Phytomedicine, 4,* 333–339.

Savithramma, N. & Sudharshanamma, D. (2006). Endemic medicinal plants of Eastern Ghats, India. *The Bioscan. 1,* 51–53.

Savithramma, N. & Sulochana, C. (1998). Endemic medicinal plants from Tirumala hills, Andhra Pradesh, India. *Fitoterapia 69*(3), 253–254.

Savithramma, N., Sulochana, Ch. & Rao, K.N. (2007). Ethnobotanical survey of plants used to treat asthma in Andhra Pradesh, India. *J. Ethnopharmacol. 113*(1), 54–61.

Savithramma, N., Yugandhar, P., Lingarao, M. & Venkata Ramana, D. Ch. (2013). Traditional phototherapy treatment for snake bite and scorpion sting by ethnic groups of Kadapa district, Andhra Pradesh, India. *Int. J. Pharm. Sci. Rev. Res. 20*, 64–70.

Saxena, H.O., Brahmam, M. & Dutta, P.K. (1981). Ethnobotanical studies in Orissa. In: Jain, S.K. (Ed.). Glimpses of Indian Ethnobotany. Oxford & IBH, New Delhi, pp. 232–234.

Saxena, H.O., Brahmam, M. & Dutta, P.K. (1988). Ethnobotanical studies in Simlipahar forests of Mayurbhanj district (Orissa). *Bull. Bot. Surv. India, 10*, 83–89.

Saxena, H.O. & Dutta, P.K. (1975). Studies on ethnobotany of Orissa. *Bull Bot. Surv. India, 17*, 124–131.

Sen, S.K. & Behera, L.M. (2003). Ethnomedicinal plants used against skin diseases in Bargarh district in Orissa. *Ethnobotany 15*, 90–96.

Sen, S.K. & Behera, L.M. (2007). Ethnomedicinal plants used in touch therapy at Bargarh district of Orissa. *Ethnobotany 19*, 100–104.

Senthilkumar, K., Aravindhan, V. & Rajendran, A. (2013). Ethnobotanical survey of medicinal plants used by Malayali tribes in Yercaud Hills of Eastern Ghats, India. *J. Natural Remedies. 13*(2).

Senthilkumar, S.M.S., Vaidyanathan, D., Sisubalan, N. & Basha, M.G. (2014a). Medicinal plants using traditional healers and Malayali tribes in Jawadhu hills of Eastern Ghats, Tamilnadu, India. *Adv. Appl. Sci. Res. 5*(2), 292–304.

Senthilkumar, S.M.S., Vaidyanathan, D., Sivakumar, D. & Basha, G.M. (2014b). Diversity of ethnomedicinal plants used by Malayali tribals in Yelagiri hills of Eastern Ghats, Tamilnadu, India. *Asian J. Pl. Sci. Res. 4*(1), 69–80.

Shanmukha Rao, V., Srinivasa Rao, D., Venkaiah, V.M. & Venkateswara Rao, Y. (2014). Ethnobotanical studies of some selected medicinal plants of Pathapatnam Mandalam, Srikakulam district, Andhra Pradesh, India. *Indian J. Pl. Sci. 4*(3), 22–33.

Shrivastava, S. & Kanungo, V.K. (2013). Ethnobotanical survey of Surguja District with special references to plants used by Uraon tribe in treatment of respiratory diseases. *Int. J. Herbal Medicine, 1*(3), 131–134.

Singh, H., Srivastava, S.C., Krishna, G. & Kumar, A. (2010). Comprehensive ethnobotanical study of Sundargarh district, Orissa, India. In: Ethnic Tribes and Medicinal Plants, ed. P.C. Trivedi, Pointer publishers, Jaipur, pp. 89–106.

Singh, K.K., Palvi, S.K. & Singh, H.B. (1981). Survey and biological activity of some medicinal plants of Mannanur forest, Andhra Pradesh. *Indian J. For. 4*, 115–119.

Sisubalan, N., Velmurugan, S., Malayaman, V., Thirupathy, S., Basha, M.H.G. & Kumar, R.R. (2014). Ethnomedicinal studies on villages of Thenpuranadu, Tamil Nadu, India. *Spatula DD. 4*(1), 41–47.

Sivaraj, N., Pandravada, S.R., Varaprasad, K.S., Sarath Babu, B., Sunil, K., Kamala, V., Babu Abraham & Krishnamurthy, K.V. (2006). Medicinal plant wealth of Eastern Ghats with special reference to indigenous knowledge systems. *J. Swamy Bot. Club 23*, 165–172.

Smita, S., Sangeeta, R., Kumar, S.S., Soumya, S. & Deepak, P. (2012). An ethnomedicinal survey of medicinal plants in Semiliguda of Koraput district, Odisha, India. *Bot. Res. International.v5*(4), 97–107.

Solomon Raju, A. & Jonathan, K.H. (2006). Knowledge of tribals in biodiversity conservation in the Eastern Ghats forests. *EPTRI-ENVIS Newsletter. 12*(2), 2–3.

Sri Rama Murthy, K., Ravindranath, D., Sandhya Rani, S. & Pullaiah, T. (2013). Ethnobotany and distribution of wild and cultivated genetic resources of Cucurbitaceae in the Eastern Ghats of Peninsular India. *Topclass J. Herbal Med. 2*(6), 149–158.

Subba Reddy, C. (2010). The life of hunting and gathering tribe in the Eastern Ghats.

Subramani, S.P. & Goraya, G.S. (2003). Some folklore medicinal plants of Kolli hills: Record of Natti vaidyas Sammelan. *J. Econ. Taxon. Bot. 27*(4), 665–678.

Subudhi, H.N. & Choudhury, B.P. (1985), Ethnobotanical studies in the district of Phulbani, Orissa – 1. *Bio. Sci. Res. Bull. 1*, 26–32.

Sudarsanam, G. (1987). Ethnobotanical Survey of Phyto-Pharmaco-Chemical Screening of selected medicinal plants of Chittoor district, Andhra Pradesh. PhD Thesis, S.V. University, Tirupati.

Sudarsanam, G. & Balaji Rao, N.S. (1994). Medicinal plants used by the Yanadi tribe of Nellore district, Andhra Pradesh. *Bull. Pure Appl. Sci. 13*, 65.

Sudarsanam, G. & Siva Prasad, G. (1995). Medical ethnobotany of plants used as Antidotes by Yanadi tribes in South India. *J. Herbs Spices, and Med. Pl. 3*(1), 57–66.

Sudhakar, S. & Rao, R.S. (1985). Medicinal plants of East Godavari, A.P. *J. Econ. Taxon. Bot. 7*, 399–406.

Suman Kumar, R., Venkateshwar, S., Samuel, G. & Rao, S.G. (2013). Ethnomedicinal uses of some plant barks used by Gondu tribes of Seethagondi grampanchayath, Adilabad District, Andhra Pradesh, India. *J. Nat. Prod. Plant Resour. 3*(5), 13–17.

Sundaram, G. & Balaji Rao, N.S. (1994). Medicinal plants used by the Yanadi tribe of Nellore District, Andhra Pradesh, India. *Bull. Pure and Appl. Sci. 13B*(2), 65–70.

Suneela, M. & Jyothi, K. (2014). Ethnomedicinal plants in Kolli hills of Eastern Ghats in Tamilnadu, India. *EPTRI-ENVIS Newsletter. 20*(3), 2–5.

Suneetha, J., Koteswara Rao, J. & Seetharami Reddi, T.V.V. (2013). Ethnomedicine for asthma used by the tribals of East Godavari district (Andhra Pradesh). *Ethnobotany, 25*, 120–123.

Suneetha, J., Prasanthi, S., Ramarao Naidu, B.V.A. & Seetharami Reddi, T.V.V. (2011). Indigenous phytotherapy for bone fractures from Eastern Ghats. *Indian J. Tradit.Knowle. 10*, 550–553.

Suneetha, J., Seetharami Reddi, T.V.V. & Prasanthi, S. (2009a). Herbal folk remedies for diarrhea and dysentery from East Godavari district of Andhra Pradesh. *J. Econ. Taxon. Bot. 33*, 293–299.

Suneetha, J., Seetharami Reddi, T.V.V. & Prasanthi, S. (2009b). Herbal therapy for cold and cough from East Godavari district (Andhra Pradesh). *J. Non-Timber Forest Prod. 16*, 135–138.

Suneetha, J., Seetharami Reddi, T.V.V. & Prasanthi, S. (2009c). Newly recorded ethnomedicinal plants from East Godavari district (Andhra Pradesh) for gynecological complaints. *Proc. Andhra Pradesh Akademi of Sciences, 13*, 111–119

Suneetha, J., Seetharami Reddi, T.V.V. & Prasanthi, S. (2009d). Traditional phytotherapy for bites in East Godavari district, Andhra Pradesh. *Ethnobotany, 21*, 75–79.

Sur, P.R. & Halder, A.C. (2004). Ethnobotanical study of Sambalpur district, Orissa, India. *J. Econ. Taxon. Bot. 28*, 573–584.

Suresh, K., Kottaimuthu, R., Selvin Jebaraj Norman, T., Kumuthakalavalli, R. & Sabu, M.S. (2011). Ethnobotanical study of Medicinal plants used by Malayali tribals in Kolli Hills of Tamilnadu, India. *Int. J. Res. Ayur. Phar.* 2(2), 502–508.

Suthari, S., Sreeramulu, N., Omkar, K., Reddy, C.S. & Raju, V.S. (2014). Intra cultural cognizance of medicinal plants of Warangal north Forest division, Northern Telangana. *Ethnobotany Res. Appl. 12*, 211–235.

Swamy, N.S. & Seetharami Reddy, T.V.V. (2011). Ethnomedicine for leucorrhoea from Adilabad district of Andhra Pradesh, India. *Ethnobotany, 23*, 147–149.

Taj, S.A. & Balakumar, B.S. (2014). Predominant flora of Udayagiri Hills-Eastern Ghats, Andhra Pradesh, India. *Sch. Acad. J. Biosci. 2*(5), 354–363.

Thulsi Rao, K., Reddy, K.N., Pattanaik, C. & Reddy, C.S. (2007). Ethnomedicinal importance of Pteridophytes used by Chenchus of Nallamalais, Andhra Pradesh, India. *Ethnobotanical Leaflets*: (http://www.siu.edu/~ebl/leaflets/ghats.html).

Udayan, P.S., Satheesh, G., Tuskar, K.V. & Indira, B. (2005). Ethnomedicine of Chellipale community of Namakkal District, Tamilnadu. *Indian J. Tradit. Knowl. 4*(4), 437–442.

Udayan, P.S., Satheesh, G., Tuskar, K.V. & Indira, B. (2006). Medicinal plants used by the Malayali tribe of Servarayan Hills, Yercaud, Salem District, Tamilnadu, India. *Zoo's Print. J. 21*(4), 2223–2224.

Upadhya, R. & Chauhan, S.V.S. (2000). Ethnobotanical observations of Koya tribe of Gundaala mandal of Khammam district, Andhra Pradesh. *Ethnobotany, 12*, 93–99.

Vaidyanathan, D., Salai Senthilkumar, M.S. & Ghouse Basha, M. (2013). Studies on ethno-medicinal plants used by Malayali tribal in Kolli hills of Eastern Ghats, Tamilandu, India. *Asian J. Pl. Sci. Res. 3*(6), 29–45.

Vedavathy, S. & Mrudula, V. (1996). Herbal folk medicines of Yanadis of Andhra Pradesh. *Ethnobotany, 8,* 109–111.

Vedavathy, S., Mrudala, V. & Rao, K.N. (1994). Herbal folk medicine of Tirumala and Tirupati region of Chittoor district, Andhra Pradesh. *Fitoterapia 66*(2), 167–171.

Vedavathy, S., Mrudula, V. & Sudhakar, A. (1997). Tribal Medicine Chittoor District Andhra Pradesh. India. Herbal Folk Lore Research Centre, Tirupati.

Vedavathy. S. & Rao K.N. (1989). Nephroprotectors – Folk medicine of Rayalaseema, Andhra Pradesh. *Ancient Sci. Life 9*, 164–167.

Vedavathy. S. & Rao K.N. (1994). Herbal folk medicnine of Tirumala and Tirupathi region of Chittoor district, Andhra Pradesh. *Fitoterapia, 66,* 167–171.

Vedavathy, S., Rao. K.N., Rajaiah. M. & Nagaraju, N. (1991). Folklore information from Rayalaseema region, Andhra Pradesh for family planning and birth control. *Int. J. Pharmacogn. 29*(2), 113–116.

Vedavathy, S.V., Sudhakar, A. & Mrudula, A. (1997). Tribal medicinal plants of Chittoor. *Ancient Sci. Life 16,* **307–331.**

Venkaiah, M. (1998). Ethnobotany of some plants from Vizianagaram district, Andhra Pradesh. *Flora and Fauna 4,* **90–92.**

Venkata Krishnaiah, P., Venkata Ratnam, K. & Raju, R.R.V. (2009). Multifarious uses of plants by the tribals of Rayalaseema region, Andhra Pradesh. *J. Indian Bot. Soc. 88,* 62–72.

Venkata Ratnam, K. & Raju, R.R.V. (2004). Folk medicines from Gundlabrahmeswaram Wild life Sanctuary, Andhra Pradesh, India. *Ethnobotany 16,* 33–39.

Venkata Ratnam, K. & Raju, R.R.V. (2005). Folk medicine used for common women ailments by Adivasis in the Eastern Ghats of Andhra Pradesh. *Indian J. Trad. Knowl. 4*(3), 267–270.

Venkata Ratnam, K. & Raju, R.R.V. (2008). Traditional medicine used by the Adivasis of Eastern Ghats, Andhra Pradesh for bone fractures. *Ethnobotanical Leaflets. 12*, 19–22.

Venkata Subbaiah, K.P. & Savithramma, N. (2012). Bio-Prospecting and documentation of traditional medicinal plants used to treat ringworm by ethnic groups of Kurnool District, Andhra Pradesh, India. *Intern. J. Pharmacy and Pharmaceut. Sci. 4*(1), 251–254.

Verma, R.C. (1995). Indian Tribes: Through the Ages. Director, Publication Division, Ministry of Information and Broadcasting, Government of India, New Delhi.

Vidyasagar, G.M. & Siddalinga Murthy, A.M. (2013). Medicinal plants used in the treatment of Diabetes mellitus in Bellary district, Karnataka. *Indian J. Tradit. Knowl. 12*, 747–751.

Vijayakumar, R. & Pullaiah, T. (1998a). An ethno-medicobotanical study of Prakasam district, Andhra Pradesh, India. *Fitoterapia 69*, 483–489.

Vijayakumar, R. & Pullaiah, T. (1998b). Medicobotanical plants used by the tribals of Prakasam district, Andhra Pradesh. *Ethnobotany, 10*, 97–102.

Vijayalakshmi, J. (1993). Ethno-medico-botany of antidots used by Chenchus in Ahobilam Hills of Kurnool District, Andhra Pradesh, M.Phil. Dissertation, Sri Krishnadevaraya University, Anatapur, India.

Vikneswaran, D., Viji, M. & Rajalakshmi, K. (2008). A Survey of the ethnomedicinal flora of the Sirumalai Hills, Dindigul District, India. *Ethnobotanical leaflets. 12*, 948–953.

Viswanathan, M.B. (1989). Ethnobotany of Malayalis in the Yelagiri Hills of North Arcot District, Tamilnadu. *J. Econ. Taxon. Bot. 13*(3), 667–671.

Viswanathan, M.B. (1997). Ethnobotany of Malayalis in North Arcot district, Tamil Nadu, India. *Ethnobotany 9*, 77–79.

WHO (2000). General guidelines for methodologies as research and evaluation of traditional medicine, Geneva, Switzerland, 71.

Xavier, T.F., Fred Rose, A. & Dhivyaa, M. (2011). Ethnomedicinal survey of Malayali tribes in Kolli Hills of Eastern Ghats of Tamilnadu, India. *Indian J. Trad. Knowl., 10*(3), 559–562.

CHAPTER 8

ETHNOVETERINARY MEDICINE OF EASTERN GHATS AND ADJACENT DECCAN REGION

M. HARI BABU, J. KOTESWARA RAO and
T. V. V. SEETHA RAMI REDDI

*Department of Botany, Andhra University, Visakhapatnam–530003,
India, E-mail: reddytvvs@rediffmail.com*

CONTENTS

ABSTRACT

The review deals with 271 species of plants belonging to 221 genera employed in 512 kinds of folklore prescriptions for treating 134 different types of animal diseases by the rural people and tribals of Eastern Ghats and Deccan plateau covering Odisha, Andhra Pradesh, Tamil Nadu, Telangana and Maharashtra states. Of the 512 practices 212 were found to be new along with 33 new ethnoveterinary plants. The results are compared with those of others in different parts of India. Herbal veterinary medicine offers enormous scope for further research, so all the existing information is to be correctly recorded before it is lost. Based on the age-old practices, pharmaceutical industries may formulate some novel prescriptions to cure various animal diseases. Documentation and standardization of ethnoveterinary knowledge is also important in the context of Intellectual Property Rights to check the patent claims. In future, detailed chemical and pharmacological investigations of these traditional formulations and medicinal plants will be very helpful for inventing/developing new veterinary drugs.

8.1 INTRODUCTION

Cattle and their products occupy a unique position in the national economy of India, which is predominantly dependent on agriculture. While the female progeny supplies milk, the male progeny continues to be the principal source of draft power for agriculture and rural transport. It is imperative that cattle population remains healthy and productive. Ethnomedicine is an integral part of traditional medical practices in many countries of the developing world. A large proportion of the population uses this form of treatment for primary health care and for the treatment of ailments in their livestock. Livestock is a major asset for resource poor small holding farmers and pastoralists throughout the world. It is estimated that there are 1.3 billion cattle in the world today. India is one of the very important centers of origin and domestication of domesticated animals, particularly of cattle. Such a domestication was carried out by Yadavas, who are pastoral people. India occupies first place with 281,700,000 cattle as per 2009 records. In many parts of the country, traditional veterinary practices are quite prevalent. Tribals remaining well below poverty level have their own systems of herbal medicine practiced since time immemorial. India is primarily an agricultural country with predominant rural population and hence, animals have had a very significant role in human life for food, milk, leather, fat, transport, hauling or draft, warfare, game and recreation, etc. Vedas and Puranas have many references to animals.

In India, more than 76% of total population residing in rural areas depends for their health care needs on plants. The history of medicine in India traced back to Vedic period. People in ancient India had sufficient knowledge of cattle and bird diseases and method of curing them. Techniques adopted for the diagnosis of diseased cattle and birds in ancient days may appear to be quite crude, in the modern advanced veterinary medicine. Vishnu Dharmottara Mahapuran (500–700 AD) was found well acquainted with the curing of animals although, it is not clearly known when and how the plants were used in animal health care. It might have started after the animal power was used in farming. But no evidence was found in any manuscript. Some of the traditional knowledge is vanishing very rapidly. Now-a-days, scientists in the modern era are now documenting various ethnoveterinary practices based on plant drugs. The plants are more intimately connected with the life of rural people than, perhaps, anywhere else in India. Tribal people are largely using plant crude drugs for the treatment of cattle. They use a number of plants, singly or as ingredients of medicine in the treatment of domestic animals. India has rich diversified flora providing a valuable storehouse of medicinal plants. Emphatic knowledge about vast resources of herbs and other plants having curative properties is prevailing among rural people.

According to the World Health Organization, at least 80% of people in developing countries depend largely on indigenous practices for the control and treatment of various diseases affecting both human beings and their animals. Ethnoveterinary remedies are accessible, easy to prepare and administer, at little or no cost at all to the farmer. These age-old practices cover every area of veterinary specialization and all live-stock species. Ethnoveterinary medicine differs not only from region to region but also among and within the communities. Ethnoveterinary information is in danger of extinction because of the current rapid changes in communities nowadays use a mix of local and modern practices. Promoting conservation and use of ethnoveterinary medicine does not mean downgrading or ignoring the value of modern medicine and attempting to replace one with the other. However, it does mean recognizing that both types have their strengths and limitations. In some instances, they complement each other, in others, local practices will be the better choice and again in others modern practices should be recommended.

The Eastern Ghats along the Peninsular India are divisible into three zones, the Northern Eastern Ghats, the middle Eastern Ghats and the Southern Eastern Ghats, extending over 1750 km with an average width of about 100 km and covering an area under 76° 56′ and 86° 30′ E longitudes and 11° 30′ and 22° 00′ N latitudes. The area covers parts of Orissa (South of river Mahanadi), Andhra Pradesh, Telangana and Tamil Nadu (North of river Vaigai) along the East Coast. The Eastern Ghats is forming a chain of discontinuous range of hills along the coast. The Eastern Ghats is one of the richest floristic and phyto-geographical regions of India. The rich and diversified flora provides a most valuable storehouse of medicinal plants.

8.2 REVIEW OF LITERATURE

Sudarsanam et al. (1995) explored 106 plant species used as veterinary crude drugs by the tribals *viz., Chenchus, Gadabas, Kattunayakas, Konda kapus, Koyas, Manne doras, Nakkalas, Reddy doras, Sugalis, Yanadis* and *Yerukulas* in Rayalaseema, Andhra Pradesh. Goud and Pullaiah (1996) reported 41 species of plants used by the *Chenchus, Sugalis* and *Yerukalas* of Kurnool district, Andhra Pradesh for curing veterinary ailments. Reddy et al. (1997) collected 17 species of plants belonging to 16 genera and 14 families used for treating ephemeral fevers and anthrax in cattle from the hills of Kadapa (Cuddapah) district of Andhra Pradesh. K.N. Reddy et al. (1998) presented 77 species representing 71 genera and 42 families of flowering plants used in veterinary practices by the ethnic tribes in Warangal district, Telangana. R.V. Reddy et al. (1998) recorded 48 plant species belonging to 46 genera and 29 families for folk veterinary medicine by *Chenchu, Sugali, Yanadi* and *Yerukula* tribals in Kadapaa (Cuddapah) hills of Andhra Pradesh. Girach et al. (1998) gave an account of 29 veterinary prescriptions based on 25 plant species in 20 families by *Bhumij, Kol, Munda* and *Santal* tribals of Bhadrak district of Orissa. Misra and Das (1998) reported the uses of 20 plants against 10 animal diseases by *Sabar* tribe in Ganjam district of Orissa. K.N. Reddy and Raju (1999) reported 86 plants belonging to 72 genera and 41 families used in veterinary practices by *Sugali, Yanadi* and *Yerukula* tribals of Anantapur district of Andhra Pradesh. Ramadas et al. (2000) reported 75 species of plants used in ethnoveterinary remedies by the local people in some districts of Andhra Pradesh and Maharashtra. C.S. Reddy and Raju (2000) enumerated 66 species representing 58 genera and 37 families used for common veterinary diseases by *Lambadas* and *Yerukulas* in Nalgonda district of Telangana. Naidu (2003) reported 81 species included in 78 genera of 44 families used by *Gadaba, Jatapu, Konda dora, Kuttiya, Savara* and *Yerukula* tribals for veterinary diseases in Srikakulam district, Andhra Pradesh. Patil and Merat (2003) enumerated 26 species of plants used in ethnoveterinary practices by *Bhil* and *Pawara* tribals of Nandurbar district,

Maharashtra. Ethnoveterinary practices among the *Konda reddis* of East Godavari district of Andhra Pradesh were studied by Misra and Anil Kumar (2004). Mokat and Deokule (2004) dealt with 36 plants used as veterinary medicine by *Katkari* tribals in Ratnagiri district of Maharashtra. Reddy et al. (2006) reported 35 species of 35 genera representing 28 families for treating livestock by *Chenchu, Koya, Konda reddi, Lambada, Nukadora, Porja, Savara, Valmiki* and *Yanadi* tribals in Eastern Ghats of Andhra Pradesh. A survey of ethnoveterinary medicinal plants of Cape Comerin, Tamil Nadu by Kiruba et al. (2006) yielded 34 species belonging to 30 genera and 21 families used by rural people and medical practitioners. Murthy et al. (2007) reported 21 medicinal plants used in ethnoveterinary practices by *Koyas* of Pakhal Wildlife Sanctuary, Telangana. Hari Babu (2007) recorded 39 species of crude drugs included in 37 genera and 25 families used as phytocure for veterinary diseases by *Bagata, Gadaba, Goudu, Khond, Konda dora, Konda kammara, Kotia, Mali, Mukha dora, Porja* and *Valmiki* tribals of Visakhapatnam district, Andhra Pradesh. Sanyasi Rao et al. (2008) gave information about 73 prescriptions for alleviating diseases of livestock which included 62 plant species. A total of 113 plant species belonging to 100 genera and 46 families used by rural people for the treatment of 44 veterinary health hazards in southern districts of Tamil Nadu was reported by Ganesan et al. (2008). Raju (2009) isolated 15 plants used for veterinary purpose by the *Konda reddis* of East and West Godavari districts of Andhra Pradesh and Khammam district of Telangana. Swamy (2009) reported 17 species of plants used by *Gond, Kolam, Koya, Lambada, Naikpod, Pardhan* and *Thoti* tribals of Adilabad district, Telangana, for curing veterinary ailments. Satapathy (2010) dealt with 88 plant species belonging to 46 families and 86 prescriptions for veterinary medicines in use among the tribes of Jaipur district of Orissa. Prasanthi (2010) reported 16 species of plants belonging to 16 genera with 16 practices by the *Savaras* of Andhra Pradesh. Information was provided on 29 medicinal plant species belonging to 28 genera and 21 families used by the village folk for curing various diseases in Nalgonda district of Telangana (Shashikanth et al., 2011). Salave et al. (2011) enumerated traditional ethnoveterinary knowledge of 13 plant species belonging to 11 genera and 9 families among the local inhabitants of Beed district of Maharashtra. Deshmukh et al. (2011) reported 36 plant species belonging to 33 families used for curing various animal diseases by the traditional livestock healers from Jalna district of Maharashtra. Suneetha et al. (2012) reported 69 species of plants belonging to 63 genera and 39 families used by *Konda dora, Konda kammara, Konda kapu, Konda reddi, Koya dora, Manne dora* and *Valmiki* tribals of East Godavari district, Andhra Pradesh for curing veterinary diseases. Murty and Rao (2012) identified 108 species of plants belonging to 99 genera and 51 families used by the tribals of Andhra Pradesh for ethnoveterinary purpose. Salave et al. (2012) enumerated 21 plant species belonging to 15 families used by *Dhangar, Laman* and *Vanjari* tribals for traditional ethnoveterinary practices in Ahmednagar district of Maharashtra. A study in Polasara block of Ganjam district, Orissa yielded 35 plant species used by local people for the treatment of cattle wounds (Mishra, 2013). Manikandan and Lakshmanan (2014) reported 10 species of ethnoveterinary plants belonging to 10 families used by the *Ariya gounder, Jadaya gounder* and *Kurumba gounder* tribals of Kalrayan hills, Eastern Ghats, Tamil Nadu. All the information are reviewed and a list of ethnoveterinary plants is provided by Krishnamurthy et al. (2014).

8.3 FODDER YIELDING PLANTS

Agriculture with animal husbandry is prevalent profession of rural and tribal people of India. Livestock is considered one of the main sources of livelihood and integral part of livelihood, which rely mostly on fodder extracted from forests, grasslands, agriculture and agroforestry. Majority of the fodder species are used as multipurpose and contributed to the high socioeconomic values. Poaceae is one of the largest families provide feed and fodder of animals. However grazing in forests has been often considered competitive and conflicting demands from the same land. The animals raised for milk and meat obtain their food, grazing in forests. Green grass is available usually during the rainy season in grasslands, on borders of fields and as weed of cultivation. Particularly, in winter when the green grass is not available and in summer when they are in scarce, there is more dependence on forests for fodder. During scarcity, cattle have to content themselves with inferior fodder and even unpalatable trees. Some useful economic trees, which are actually lopped for fuel and litter, are also lopped for fodder and in still higher ranges what to talk of leaves even tender stems and bark may not be spared by the hungry cattle. The damage to vegetation is greatest. The potential of trees and shrubs for green fodder production has not been fully appreciated in India. Different fodder yielding trees and shrubs differ from place to place and the tree lopped extensively for fodder in one place may not at all be lopped at another place. At the same time excessive and indiscriminate lopping of some fodder yielding trees and shrubs has resulted in destruction in some places. It was observed that shrubs are chiefly browsed by goats and sheep, whereas, the trees cater to the fodder for rest of cattle heads. The trees are generally lopped between April and December. During July and August, enough green fodder in crops is available there by giving relief to the trees and shrubs. The identification of more fodder trees is important since trees have two characteristics, which make them particularly useful during drought. Firstly they are able to draw on moisture and minerals deep in the soil, which are out of reach for grasses. Secondly the leaves of most the trees retain their nutritive value even when they are mature. Trees and shrubs provide fodder, which is of great importance during period of nutritional stress in the dry season when the nutritional value of dormant grasses and forbs is low (Chhetri, 2010). A list of fodder plants of Eastern Ghats is provided by Krishnamurthy et al. (2014).

8.4   ENUMERATION

The ethnoveterinary plants used by the tribals of Eastern Ghats and adjacent Deccan region are given in the following Table 8.1.

TABLE 8.1   Ethnoveterinary Plants of Eastern Ghats and Deccan

S. No.	Name of the plant	Part(s) used	Uses	References
1.	*Abelmoschus crinitus Wall. VN: Kona benda (Tel), Caltnet spice (E)	Tubers with seeds of jowar	Oestrum	Reddy and Raju (1999)
2.	A. esculentus (L.) Moench VN: Benda (Tel) Bhendi (O) E: Lady's finger	Fruit with stem bark of Ficus hispida	Expulsion of placenta	Sudarsanam et al. (1995)
		Root	*Urination	Satapathy (2010)
3.	A. manihot (L.) Medik	Fruit	Dysentery	Patil et al. (2010)
4.	Abrus precatorius L.	Leaf	*Anthrax	Reddy and Raju (1999)
	VN: Kuntumani (Tam), Gulivinda (Tel), Kaincha, Runjo (Oriya)			Reddy et al. (1998)
	E: Indian liquorice	Leaf with garlic and black pepper	Bronchitis	Sudarsanam et al. (1995)
		Root	*Cataract	Misra and Das (1998)

TABLE 8.1   *(Continued)*

S. No.	Name of the plant	Part(s) used	Uses	References
		Whole plant	*Dysentery	Kiruba et al. (2006) Satapathy (2010)
		Leaf	*Insect bite	Reddy and Raju (1999)
		Seed	*Liver disorders	Girach et al. (1998)
		Leaf	*Mastitis	Ganesan et al.2008
			*Retained placenta	Reddy and Raju (1999)
		Seed	*Trypanoso-miasis	Murty and Rao (2012)
		Root	Wounds	Naidu (2003); Swamy (2009)
		Seed Seed with turmeric, musk, asafoetida, castor oil	Yokegall	Goud and Pullaiah (1996) Reddy and Raju (1999)
5.	*Abutilon indicum* (L.) Sweet	Leaf	*Diarrhea	Girach et al. (1998)
	VN: Benda (Tel) E: Lady's finger	Leaf with butter milk Leaf	*Dysentery	Reddy and Raju (1999) Kiruba et al. (2006)
		Leaf	*Helminthiasis	Reddy and Raju (2000)
			Sore eyes	Naidu (2003)
		Leaf	Foot and Mouth disease	Patil et al. (2010)
		Leaves with those of *Albizia lebbeck* and *Thevetia nerifolia*	*Wounds	Mishra (2013)
6.	*Acacia auriculiformis* A. Cunn. ex Benth. VN: Thumma (Tel) E: Ear leaf acacia	Root bark with turmeric	Wounds	Suneetha et al. (2012)
7.	*A. catechu* (L.f.) Willd.	Stem bark	Diarrhea	Suneetha et al. (2012)
	VN:Podalimanu (Tel), Khaira-gachha (O) E: Cutch tree	Leaves with those of *Barleria prionitis* and neem, butter milk	Hump sores	Mishra (2013)
		Heart wood	Wounds	Satapathy (2010)

TABLE 8.1 *(Continued)*

S. No.	Name of the plant	Part(s) used	Uses	References
8.	*A. chundra* Willd. VN: Sundrafa chettu (Tel) E: Red cutch	Stem bark with garlic cloves Stem bark	*Ephemeral fever	Reddy et al. (2006) Murty and Rao (2012)
		Stem bark	*Ulcers and wounds Babesiosis	Swamy (2009) Reddy et al. (1998).
9.	*A. nilotica* (L.) Del. VN: Kaaruvelam (Tam), Nalla thumma (Tel) E: Black babul	Fruit	Galactagogue	Kiruba et al. (2006); Swamy (2009); Suneetha et al. (2012)
		Leaf	*Mouth ulcers	Salave et al. (2012)
10.	*A. pennata*(L.) Willd. VN: Tella chiki (Tel) E: Red cutch	Stem bark with garlic and black pepper	Diarrhea	Sudarsanam et al. (1995)
11.	*Acalypha indica* L.	Leaf	Wounds	Patil et al. (2010)
12.	*Acanthospermum hispidum* DC.	Leaf	Worms in wounds	Patil et al. (2010)
13.	*Achyranthes aspera* L. VN: Uttareni (Tel), Nayuruvi (Tam), Aghada (M), Apamaranga (O) E: Prickly chaff flower	Root with *Ferula asafoetida* and leaves of *Calotropis procera*	Bronchitis	Satapathy (2010)
		Whole plant	*Diuretic	Goud and Pullaiah (1996)
		Leaf with saffron Leaf	Eye problems	Reddy and Raju (1999) Kiruba et al. (2006)
		Whole plant with jaggery	*Foot and mouth disease	Swamy (2009)
		Root	Dysentery	Patil et al. (2010)
		Leaf Young tender stick	*Wounds	Ramadas et al. (2000); Murty and Rao (2012) Mokat and Deokule (2004)
14.	*Acorus calamus* L. VN: Vasa (Tel), Ghoda- vacha (O) E: The sweet flag	Rhizome	Dyspepsia	Satapathy (2010)
		Root	Eradication of lice	Suneetha et al. (2012)

TABLE 8.1   *(Continued)*

S. No.	Name of the plant	Part(s) used	Uses	References
15.	*Adhatoda vasica* Nees. VN: Addasaram (Tel), Adhatoda (Tam) E: Malabar nut	Leaf with stem of *Cissus quadrangularis*, tubers of *Withania somnifera*, black pepper and garlic	*Anthrax	Reddy and Raju (2000)
		Leaf and stem	*Fever and cough	Kiruba et al. (2006)
		Leaf	Skin diseases	Kiruba et al. (2006)
16.	*Aegle marmelos* (L.) Correa VN: Limba (O), Maredu (Tel), Vilvam (Tam) E: Bael	Leaf with that of *Datura metel* Leaf	Black-quarter Worms in wounds	Misra and Das (1998) Patil et al. (2010)
		Leaf	Bone fracture	Reddy and Raju (1999)
		Leaves with bulb of *Allium cepa* and fruits of *Cuminum cyminum*	*Enteritis	Ganesan et al. (2008)
		Stem bark with turmeric	*Mouth disease Babesiosis	Misra and Anil Kumar (2004) Reddy et al. (1998)
17.	*Aerva lanata* (L.) Juss.	Leaf	Fever	Patil et al. (2010)
18.	*Agave cantula* Roxb. VN: Kithanara (Tel) E: American aloe	Leaf with seeds of pepper	Cuts and wounds	Suneetha et al. (2012)
19.	*Ailanthus excelsa* Roxb VN: Pedda maanu, Pedda vepa (Tel) E: Tree of heaven.	Stem bark with gingelly oil Leaf	*Anorexia Worms near horns	Reddy and Raju (2000) Patil et al. (2010)
		Stem bark	*Dysentery	Suneetha et al. (2012)
		Leaf	*Malaria	Murty and Rao (2012)
		Stem bark	Stomach pain	Suneetha et al. (2012)
20.	*Alangium salvifolium* (L.f.) Wang.	Stem bark	Cough	Swamy (2009)
	VN: Udugu chettu (Tel) E: Sage-leaved alangium	Root bark with pepper seeds, mustard oil	Dog bite	Suneetha et al. (2012)

TABLE 8.1   *(Continued)*

S. No.	Name of the plant	Part(s) used	Uses	References
		Stem bark with leaves of *Hygrophila auriculata, Premna latifolia,* dry chillies and salt	*Oedema	Reddy and Raju (2000)
		Root	Snake bite	Murthy et al. (2007)
		Stem bark		Reddy et al. (1998)
21.	*Albizia amara* (Roxb.) Boivin VN: Chigara (Tel) E: Bitter albizia	Leaf	*Bone fracture	Reddy and Raju (1999)
		Stem bark	*Conjunctivitis	Reddy et al. (1998)
22.	*A.lebbeck* (L.) Benth. VN: Dirisena (Tel), Vakai	Leaf	*Conjunctivitis	Ganesan et al. (2008)
	(Tam), Shirish (M) E: Lebbeck	Stem bark with that of *Ficus religiosa,* flour of *Phaseolus mungo,* straw of *Oryza sativa,* leaves of *Coccinia grandis,* oil of *Sesamum indicum* and poultry blood	*Fracture	Ramadas et al. (2000)
		Stem bark with leaves of *Piper betel* and jaggery	Fever	Goud and Pullaiah (1996)
		Leaf and stem bark, with onion, leaves of *Mimosa pudica,* luke warm water	*Lice and wasp bite	Salave et al. (2012)
		Leaf with that of *Cleome gynandra,* stem bark of *Pongamia pinnata,* cow's urine, pepper and garlic	*Trypanosomiasis	Reddy and Raju (1999)

TABLE 8.1   *(Continued)*

S. No.	Name of the plant	Part(s) used	Uses	References
23.	*Allium cepa* L. VN: Chinnavengayam	Bulb with mustard oil	Cough	Satapathy (2010)
	(Tam), Piyaja (O) E: Onion	Bulb with fruits of *Cuminum cyminum*, leaves of *Cadaba fruticosa*, kernel of *Cocos nucifera* and *Borassus flabellifer* sugar	*Debility and general weakness	Ganesan et al. (2008)
24.	*Aloe vera* (L.) Burm. *f.* VN: Chothukathalai (Tam),	Leaf with *Areca catechu* nut	*Corneal opacity	Ganesan et al. (2008)
	Ghee-kuanri (O) E: Indian aloe	Leaf	*Inflammation	Shashikanth et al. (2011)
			Mastitis	Satapathy (2010)
			*Miscarriage	Satapathy (2010)
			*Wounds	Mishra (2013)
25.	*Alstonia scholaris* (L.) Br. VN: Edakulapala (Tel) E: Devil's tree	Latex Latex with pepper Latex	Dysentery	Girach et al. (1998); Raju (2009); Suneetha et al. (2012)
		Bark	Gastric problem	Murty and Rao (2012)
26.	**A. venenata* R.Br. VN: Pala mandhu (Tel) E: Poison devil tree	Stem bark with leaves of *Euphorbia hirta,* cow milk, small fish and castor oil	Galactagogue	Hari Babu (2007)
27.	**Alysicarpus vaginalis* DC. VN: Musaraaku (Tel) E: Alyce clover	Leaves with those of *Blepharispermum subsessile*, pepper and garlic	Anthrax	Reddy and Raju (1999)
28	*Amaranthus blitum* L.	Entire plant	To increase milk	Patil et al. (2010)
29.	*A. spinosus* L.	Whole plant	Stomachache	Patil et al. (2010)
30.	*Amorphophallus paeoniifolius* (Dennst.) Nicol. VN: Siri kand (Tel) E: Elephant foot yam	Corm	*Helminthiasis	Suneetha et al. (2012)
31.	*Ampelocissus latifolia* (Roxb.) Planch.	Powder	Flatulence	Patil et al. (2010)

TABLE 8.1 *(Continued)*

S. No.	Name of the plant	Part(s) used	Uses	References
32.	*Andrographis paniculata* (Burm. *f.*) Wall. VN: Nilavembu (Tam), Nelavemu (Tel), Bhuinimba (O), Kade-Chiraite (M) E: King of bitters	Leaf Leaveswith those of *Vitex negundo* and *Cardiospermum halicacabum* and tubers of *Curculigo orchioides* and *Urginea indica*	Fever and *cough Ephemeral fever	Mokat and Deokule (2004); Kiruba et al. (2006) Reddy et al. (1998)
		Leaf with that of *Peristrophe paniculata* and fruit of *Cuminum cyminum* and bulb of *Allium cepa* Stem. Leaf	*Foot and mouth disease	Ganesan et al. (2008) Satapathy (2010)
		Root	*Wounds	Hari Babu (2007)
33.	*A. neesiana*Wight VN: Nelavemu (Tel) E: Red English daisy	Whole plant	*Foot and mouth disease	Sudarsanam et al. (1995)
		Whole plant with jaggery, cloves and leaves of *Vitex negundo*	Indigestion, Dyspepsia	Sudarsanam et al. (1995)
34.	*Annona reticulata* L. VN: Ramaphalam (Tel) (M) E: Common custard apple	Unripe fruit	Eradication of lice	Raju (2009); Suneetha et al. (2012)
		Leaf	Wounds	Naidu (2003); Murty and Rao (2012); Mokat and Deokule (2004)
35.	*A. squamosa* L. VN: Seetha phalam (Tel), Ata (O) E: Custard apple	Seed with coconut oil Seed	Eradication of lice	Raju (2009) Satapathy (2010)
		Unripe fruit Leaf	Worms infested sores	Hari Babu (2007); Prasanthi (2010); Murty and Rao (2012); Patil et al. (2010); Mishra (2013)

TABLE 8.1   *(Continued)*

S. No.	Name of the plant	Part(s) used	Uses	References
		Leaf	Wounds	Sudarsanam et al. (1995), Goud and Pullaiah,1996; Naidu (2003); Suneetha et al. (2012)
36.	*Anogeissus latifolia* Guill.	Stem bark	*Dysentery	Salave et al. (2012)
	VN: Chirumanu, Velama (Tel), Dhamoda (M) E: Axlewood	Stem bark with leaf of *Euphorbia hirta*, onion and turmeric	*Horn cancer	Reddy and Raju (1999)
		Seed	Snake bite	Naidu (2003)
37.	*Anthocephalus chinensis*(Lamk.) Walp. VN: Kadambamu (Tel) E: Parvaty's tree	Stem bark	Dyspepsia	Suneetha et al. (2012)
38.	*Arachis hypogaea* L. VN: Verusenaga (Tel) E: Groundnut	Seed	Galactagogue	Swamy (2009)
39.	*Ardisia solanacea* Roxb. VN: Kunti (Tel) E: Spear flower	Root bark, turmeric	*Wounds	Suneetha et al. (2012)
40.	*Argemone mexicana* L. VN: Balurakkisa, Mysurapala chettu (Tel), Piwala dhotra (M) E: Mexican poppy	Root	*Anthrax	Reddy and Raju (1999)
		Leaf with young twig of *Maytenus emarginata*	*Arthritis	Deshmukh et al. (2011)
		Seed	*Eczema	Prasanthi (2005)
		Milk of the plant and oil from seeds Milk	*Wounds	Murty and Rao (2012); Mishra (2013)
41.	*Aristida setacea* Retz. VN: Puthika cheepuru (Tel) E: Hill broom	Seed with turmeric Seed	*Wounds	Hari Babu (2007); Suneetha et al. (2012); Raju (2009)

TABLE 8.1 *(Continued)*

S. No.	Name of the plant	Part(s) used	Uses	References
42.	*Aristolochia bracteo-lata* Lamk. VN: Gadidagadapa (Tel), Aaduthinnapalai (Tam), Aswali (M) E: Bracteated birthwort	Leaves with those of *Nicotiana tabacum, Zingiber officinale, Solanum xanthocarpum,* fruit of *Capsicum annum* and *Piper nigrum*seed	Anthrax	Ganesan et al. (2008)
		Leaf with that of *Anisomeles malabarica, asafoetida,* pepper and garlic Leaf	*Ephemeral fever	Reddy and Raju (1999); Kiruba et al. (2006)
		Whole plant with *Cleome gynandra* and *Enicostemma axillare*	Flatulence	Reddy et al. (1998)
		Whole plant	*Indigestion and flatulence	Naidu (2003)
		Leaf	*Infertility	Ganesan et al. (2008)
		Leaves	Worms in wounds	Patil et al. (2010)
		Leaf, neem oil	*Intestinal worms	Salave et al. (2011)
		Root Leaf with flower buds of *Madhuca longifolia*	Sores	Raju (2009); Suneetha et al. (2012) Sudarsanam et al. (1995)
43.	*A. indica* L. VN: Nalla eswari (Tel), Iswaramuli (Tam) E: Indian birth wort	Root with leaf of *Clitoria ternatea* and pepper seed	*Bloat	Ganesan et al. (2008)
		Whole plant with pepper seed	Dyspepsia	Suneetha et al. (2012)
		Leaves with those of *Acalypha indica, Leucas aspera*	*Ephemeral fever	Ganesan et al. (2008)

TABLE 8.1   *(Continued)*

S. No.	Name of the plant	Part(s) used	Uses	References
		Leaf with pepper and garlic Leaf	Insect bite	Reddy and Raju (1999); Kiruba et al. (2006)
		Leaf, root, turmeric powder and salt	Wounds	Ramadas et al. (2000)
44.	*Arundo donax* L. VN: Kakiveduru (Tel) E: Great reed	Rhizome	Dysentery	Suneetha et al. (2012)
45.	*Asparagus racemosus* Willd. VN: Chinna pillipecheri,	Tuber with leaf of *Azima tetracantha*, pepper and garlic	*Ephemeral fever	Reddy and Raju (1999)
	Chandamama gaddalu (Tel), Sathaveli (Tam), Shatavari (M) E: Wild carrot	Tuberous root Whole plant Tuberous root	Galactagogue	Kiruba et al. (2006); Patil et al. (2010); Hari Babu (2007) Prasanthi (2010); Salave et al. (2011); Deshmukh et al. (2011)
			*Inducing heat	Mokat and Deokule (2004)
		Roots with those of *Balanites roxburghii*, *Cissus quadrangularis*, leaf of *Tylophora indica*, ginger	*Tympany	Reddy and Raju (1999)
46.	*Atylosia scarabeoides* Benth. VN: Adavi ulava (Tel) E: Wild kulthi, Banakolathia (O)	Whole plant Leaf	Diarrhea	Sudarsanam et al. (1995) Prasanthi (2010); Satapathy (2010)
			*Expulsion of placenta	Sudarsanam et al. (1995)
47.	*Azadirachta indica* Juss. VN: Yapachettu (Tel), Vembu (Tam), Nimb (M) E: Margosa tree	Stem bark with that of *Ficus religiosa*, pepper, garlic	*Anthrax	Reddy and Raju (1999)
		Stem bark	*Blisters	Reddy et al. (1998)
		Leaf	*Chickenpox	Prasanthi (2010)

TABLE 8.1 *(Continued)*

S. No.	Name of the plant	Part(s) used	Uses	References
		Stem bark with leaves of *Aloe vera, Pergularia daemia* Leaf	*Ephemeral fever	Sudarsanam et al. (1995); Reddy et al. (1997)
		Leaf	Fever	Patil and Merat (2003)
		Stem bark with those of *Mangifera indica* and *Syzygium cumini*	Stomach pain	Suneetha et al. (2012)
		Leaf	*Trypanosomiasis	Murty and Rao (2012)
			Ulcers, wounds and skin diseases	Naidu (2003)
		Leaves with those of *Annona squamosa* and *Ximenia americana* Leaf	Wounds	Ramadas et al. (2000) Kiruba et al. (2006); Mishra (2013)
48.	*Azima tetracantha* Lam. VN: Uppi (Tel), Mulsangu elai (Tam) E: Needle bush	Thorny branches	*Foot and mouth disease	Swamy (2009)
		Root	*Knee pain	Murthy et al. (2007)
		Leaf with flowers of *Madhuca longifolia*	Rinderpest	Sudarsanam et al. (1995)
		Leaves with those of *Daemia extensa* and *Croton bonplandianum*	*Snake bite	Ganesan et al. (2008)
		Stem	*Tumors	Prasanthi (2010)
49.	*Balanites roxburghii* (Syn: *B. aegyptiaca*) Planch VN: Gara (Tel) Hingani (M) E: Desert date.	Stem bark	Helminthiasis	Suneetha et al. (2012)
		Seed	Flatulence	Patil et al. (2010)
		Fruit	Lice	Salave et al. (2011)

TABLE 8.1   *(Continued)*

S. No.	Name of the plant	Part(s) used	Uses	References
50.	*Barringtonia acutangula* (L.) Gaertn. VN: Hinjal (O) E: Fresh water mangrove	Fruit	Dysentery	Girach et al. (1998)
51	*Basella alba* L.	Seed	Worms in wound	Patil et al. (2010)
52.	*Bauhinia vahlii* (Wt. & Arn.) Benth. VN: Addaku (Tel) E: Camel's foot climber	Stem bark with that of *Polyalthia longifolia*	Bone fracture	Suneetha et al. (2012)
53.	**Becium filamentosum* (Forssk.) Chiov. VN: Konda rilla (Tel) E: Jumbie balsam	Plant with leaves of *Wrightia tinctoria*, pepper and garlic	Anthrax	Reddy and Raju (1999)
54	*Biophytum sensitivum* (L.) DC.	Leaf	Wounds	Patil et al. (2010)
55.	*Blepharispermum subsessile* DC. VN: Kondamamidi (Tel)	Leaf with that of *Alysicarpus vaginalis*, pepper and garlic	*Anthrax	Reddy and Raju (1999)
		Aerial part with leaves of *Ficus arnottiana, Pouzolzia zeylanica*, long pepper and garlic	*Ephemeral fever	Reddy and Raju (1999)
		Stem with that of *Vitex negundo*	*Trypanosomiasis	Reddy and Raju (1999)
56.	*Boerhavia diffusa* L. VN: Atikamamidi (Tel) E: Hogweed	Whole plant	Diuretic	Naidu (2003)
57.	*Bombax ceiba* L. VN: Buruga (Tel) E: Red silk cotton	Stem bark	Diarrhea Dysentery	Suneetha et al. (2012) Patil et al. (2010)
58.	*Brassica nigra* (L.) Koch. VN: Avalu E: Mustard	Seed with unripe fruit of mango	Anthrax, Colic, Pneumonia	Sudarsanam et al. (1995)
		Seed oil	*Kidney disorder	Murty and Rao (2012)

TABLE 8.1 *(Continued)*

S. No.	Name of the plant	Part(s) used	Uses	References
59.	*Buchanania lanzan* Spreng. VN: Chinna morlu (Tel) E: Cuddapah almond	Root bark	Diarrhea	Suneetha et al. (2012)
60.	*Butea monosperma* (Lamk.) Taub.	Stem bark	*Babesiosis	Reddy and Raju (2000)
	VN: Moduga (Tel), Palas (M) E: Flame of the forest	Seed Leaf	Intestinal worms	Prasanthi (2005); Naidu (2003)
		Flower	Stomachache	Patil et al. (2010)
		Root	*Tympany	Deshmukh et al. (2011)
61.	*Cadaba fruticosa* (L.) Druce VN: Konda mirapa (Tel) E: Indian cadaba	Leaf with ginger	Tympany	Reddy and Raju (1999)
62.	*Caesalpinia bonduc* (L.) Roxb.	Seed	Cholera	Naidu (2003)
	VN: Gachhakaya (Tel) E: Fever nut	Leaf with that of *Gardenia gummifera*, musk, *asafoetida*, long pepper and garlic	*Ephemeral fever	Reddy and Raju (1999)
		Leaf	Inflammation and intestinal worms	Naidu (2003)
		Leaf with seed of jowar, jaggery	*Oestrum	Reddy and Raju (1999)
63.	*Cajanus cajan* (L.) Millsp. VN: Kandi (Tel), Tur (M), Harada (O) E: Red gram	Leaf and seed Seed Green pods	Dysentery	Swamy (2009); Mokat et al.2010; Satapathy (2010); Patil et al. (2010)
64.	*Calotropis gigantea* (L.) Br. VN: Jilledu (Tel), Arakha (O) E: Swallow wort	Latex	Hump sores Poor vision	Mishra (2013) Patil et al. (2010)
		Root	Maggot wounds	Ramadas (2000); Satapathy (2010)
65.	*C. procera* (Ait.) R.Br. VN: Tella jilledu (Tel), Erukku (Tam), Rui (M) E: Akund	Flower with onion, turmeric and *Carum copticum*	*Ephemeral fever	Reddy et al. (1997)
		Root	*Epitaxis	Reddy and Raju (1999)

TABLE 8.1   *(Continued)*

S. No.	Name of the plant	Part(s) used	Uses	References
		Latex with red lead (vermilion) and leaf of *Sida rhombifolia*	*Pneumonia	Deshmukh et al. (2011)
		Leaf	*Scorpion sting	Reddy et al. (1998)
		Latex	Wounds	Kiruba et al. (2006); Prasanthi (2010)
		Latex	Bone fracture	Patil et al. (2010)
		Leaf	Twisting of leg	
66.	*Cannabis sativa* L. VN: Ganjayi (Tel) E: True hemp	Leaf	Diarrhea	Hari Babu (2007)
67.	*Canthium parviflorum* Lam. VN: Balusu (Tel), Kharaichedi (Tam) E: Wild dagga	Leaf	*Bone fracture	Reddy and Raju (1999); Kiruba et al. (2006)
			*Insect bite	Reddy and Raju (1999)
			Sores	Sudarsanam et al.1995
68.	*Capparis sepiaria* L. VN: Nalla uppi (Tel) E: Caper berry	Leaves with those of *Grewia damine, Chloroxylon swietenia*, pepper and garlic	Ephemeral fever	Reddy and Raju (1999)
69.	*C. zeylanica* L. VN: Thotlaku (Tel) E: Caper bush	Leaf Flower	Anti-inflammatory Germs in wounds	Reddy et al. (1998) Patil et al. (2010)
		Fruit with pepper, musk, saffron and jaggery	Oestrum	Reddy and Raju (1999)
70.	*Cardiospermum halicacabum* L. VN: Buddakakara(Tel) E: Blister creeper	Leaf with stem bark of *Carissa spinarum*, pepper and garlic	*Ephemeral fever	Reddy and Raju (1999)
		Leaf	*Night blindness	Prasanthi (2010)
		Leaf	Stomachache	Patil et al. (2010)
71.	*Careya arborea* Roxb. VN: Dhorkumbha (M), Dudippa chettu (Tel) E: Patana oak	Fruit	*Bloated stomach and fever Stomachache	Patil and Merat (2003) Patil et al. (2010)
		Stem bark	Wounds	Suneetha et al. (2012)

TABLE 8.1   *(Continued)*

S. No.	Name of the plant	Part(s) used	Uses	References
72.	*Carissa spinarum* L. VN: Kalivi (Tel)	Root	Fever	Suneetha et al. (2012)
	E: Karanda	Root with flower bud of *Madhuca longifolia*	Sores, Rheumatic arthritis	Sudarsanam et al. (1995)
		Root with turmeric Root	Wounds	Naidu (2003); Hari Babu (2007); Swamy (2009); Suneetha et al. (2012)
73.	*Casearia elliptica* Willd. VN: Giridi (Tel) E: Sword- leaf	Stem bark	*Dysentery	Murthy et al. (2007)
			Wounds and ulcers	Naidu (2003)
74.	*Cassia auriculata* L.	Leaf	Bone fracture	Swamy (2009)
	VN: Tangedu (Tel),	Tender shoot tips	Dysentery	Reddy and Raju (1999)
	Aavarai (Tam) E: Tanner's cassia	Flower with whole plant of *Enicostemma axillare* and fruits of *Cuminum cyminum*	Heart disease	Ganesan et al. (2008)
75.	*C. fistula* L. VN: Bahava (M), Rela (Tel) E: Caslia-fistula	Pods	*Asthma and pneumonia	Deshmukh et al. (2011)
		Fruit	*Bloated stomach	Patil and Merat (2003)
		Pods	Constipation	Shashikanth et al. (2011)
		Stem bark with leaf of *Mukia maderaspatana*, pepper and garlic	*Ephemeral fever	Reddy and Raju (1999)
		Leaf with that of *Piper betel*	*Diuretic	Ganesan et al. (2008)
		Leaf	*Dysentery	Sudarsanam et al. (1995)
		Fruit pulp	Intestinal worms	Pullaiah (1996)
		Seed with cumin seed and root of *Aristolochia indica*	Snake bite	Sudarsanam et al. (1995)

TABLE 8.1 *(Continued)*

S. No.	Name of the plant	Part(s) used	Uses	References
76.	*C.italica* (Mill.) Andr. VN: Nela thangedu E: Senna	Whole plant with leaf of *Calotropis gigantea*	Anthrax	Reddy and Raju (1999)
		Leaf with flower of *Calotropis gigantea* and fruit of *Terminalia chebula*	Constipation	Reddy and Raju (2000)
77.	*C. occidentalis* L. VN: Kolatapasi (Tel) E: Sicklepod India	Leaf	*Bone fracture Worms in wound	Reddy and Raju (1999) Patil et al. (2010)
78.	*C. senna* L. VN: Nela tangedu (Tel) E: Ringworm bush	Leaf	Diarrhea	Sudarsanam et al. (1995)
			*Rheumatism, Skin disease	Murty and Rao (2012)
79.	*C. siamea* L. VN: Nela tangedu (Tel) E: Kossod tree	Leaf	Bone fracture	Swamy (2009)
80.	*C. tora* L. VN: Usithagarai (Tam) E: Sickle senna	Seed	Skin disease	Manikandan and Lakshmanan (2014)
81	*Cassine albens* (Retz.) Kosterm.	Leaf	To increase milk	Patil et al. (2010)
82.	*Cassytha filiformis* L. VN: Paachi teega (Tel) E: Dodder laurel	Whole plant	Bone fracture	Reddy and Raju (2000)
83.	*Catunaregam spinosa* (Thunb.) Tirveng. VN: Manga E: Devil's walking stick	Leaf with flower of *Butea monosperma*, fruit of *Canthium parviflorum*, salt and garlic	Tympany	Reddy and Raju (1999)
84.	*Celosia argentea* L. VN: Gurugu (Tel) E: Oriental bitter sweet	Leaf with that of *Tridax procumbens*	Cuts and wounds	Shashikanth et al. (2011)
		Leaf	Honeybee bite	Shashikanth et al. (2011)
		Root	*HCN poisoning	Reddy and Raju (1999)
		Root	Emergence of placenta	Patil et al. (2010)

TABLE 8.1 *(Continued)*

S. No.	Name of the plant	Part(s) used	Uses	References
85.	*Chlorophytum tuberosum*(Roxb.) Baker VN: Buradumpa (Tel) E: Safed musili	Tuber	Dysentery	Swamy (2009)
86.	Chloroxylon swietenia DC. VN: Billu (Tel) E: Indian stain wood tree	Wood	*Neck sores	Murthy et al. (2007)
		Seed bark	*Rheumatic pain	Reddy et al. (1998)
		Leaf with turmeric Stem bark	Wounds and ulcers	Naidu (2003); Prasanthi (2010)
			*Yokegall	Reddy and Raju (1999)
87.	Cicer arietinum L.	Pulse	To expel fetus	Patil et al. (2010)
88.	*Cipadessa baccifera (Roth) Miq.VN: Brahma malika (Tel) E: Ceylon cinnamon	Stem bark	Ephemeral fever	Reddy et al. (2006)
89.	Cissampelos pareira L. VN: Visaboddi (Tel) E: Velvete-leaf pareira	Leaves with those of *Mukia maderaspatana, Pergularia daemia*, pepper and garlic	*Anthrax	Reddy and Raju (1999)
90.	Cissus pallida (Wt. & Arn.) Steud.	Tuber	*Retained placenta	Reddy and Raju (1999)
	VN: Konda gummadi (Tel) E: Snake bitters		Wounds	Ramadas et al. (2000)
91.	C. quadrangularis L. VN: Nalleru (Tel), Hadbhanga (O), Pirantai (Tam), Hadsandhi (M) E: Adamant creeper	Stem with stem bark of *Wrightia tinctoria*, leaf of *Vitex negundo*, pepper and garlic Stem bark	*Anthrax	Reddy et al. (1997); Reddy and Raju (1999)
		Stem with onion and chili powder	*Asthma	Murthy et al. (2007)
		Stem Stem and leaf	Bone fracture	Misra and Das (1998); Reddy et al. (1998); Naidu (2003); Patil and Merat (2003); Hari Babu (2007); Patil et al. (2010); Suneetha et al. (2012);
		Stem	*Dysentery	Reddy and Raju (1999)

TABLE 8.1   *(Continued)*

S. No.	Name of the plant	Part(s) used	Uses	References
		Leaf with that of *Pedalium murex*, pepper, musk and garlic Leaf	*Ephemeral fever	Reddy and Raju (1999) Manikandan and Lakshmanan (2014)
		Whole plant	*Fertility Placenta Wounds	Sudarsanam et al. (1995) Kiruba et al. (2006) Girach et al. (1998)
92.	*Citrullus colocynthis* (L.) Schr. VN: Erripuchha (Tel), Kumatti (Tam) E: Colocynthe	Root with that of *Tylophora indica*, leaf of *Securinega leucopyrus* and pepper	*Cough	Reddy and Raju (1999)
		Leaf with that of *Enicostemma axillare*, pepper, garlic and turmeric	*Horn cancer	Reddy and Raju (1999)
		Whole plant with roots of *Citrullus* sp. and *Aristolochia indica,* leaf of *Piper betel*, fruit of *Piper nigrum* and *Ferula asafoetida* Root	*Tympany Cough	Ganesan et al. (2008) Manikandan and Lakshmanan (2014)
93	*Citrus aurantifolia* (Christm.) Sw.	Fruit	Flatulence	Patil et al. (2010)
94.	*Citrus medica* L. VN: Dabba (Tel) E: Citron	Leaf with turmeric	*Scabies	Naidu (2003)
95.	*C. sinensis* (L.) Osbeck. VN: Battayi (Tel) E: Sweet orange	Leaf	Ulcers	Naidu (2003); Hari Babu (2007)
96.	*Cleistanthus collinus* (Roxb.) Benth. & Hk. *f.* VN: Karada (O), Nalla kodisha(Tel) E: Discus feather foil	Stem bark with latex of *Ficus racemosa*	*Foot and mouth disease	Misra and Das (1998)

TABLE 8.1 *(Continued)*

S. No.	Name of the plant	Part(s) used	Uses	References
		Stem bark Stem Stem bark	Sores	Naidu (2003); Hari Babu (2007); Prasanthi (2010) Raju (2009); Suneetha et al. (2012)
97.	*Clematis gouriana* Roxb. VN: Gurraputheega (Tel) E: Indian traveler's joy	Leaf	Wounds	Naidu (2003); Hari Babu (2007); Prasanthi (2010)
98.	*Cleome felina* L.f. VN: Adavi ulava (Tel) E: Wild mustard	Plant with stem bark of *Pongamia pinnata*	Impaction	Reddy and Raju (1999)
99.	*C. gynandra* L. VN: Vavtaaku (Tel) E: African spider	Whole plant with *Aristlochia bracteolata* and *Enicostemma axillare*	*Flatulence	Reddy et al. (1998)
		Leaf with that of *Albizia lebbeck*, stem bark of *Pongamia pinnata*, cow's urine, pepper and garlic	*Trypanoso-miasis	Reddy and Raju (1999)
		Leaf	Fever	Patil et al. (2010)
100.	*C. viscosa* L. VN: Kukkavaminta (Tel) E: Wild mustard	Leaf	*Cuts	Raju (2009)
		Leaves with those of *Pergularia daemia, Leonotis nepetifolia*, pepper, garlic and chilis	*Ephemeral fever	Reddy et al. (1997)
101.	*Coccinia grandis* (L.) Voigt VN: Adavi donda, Kakidonda (Tel) E: Ivy gourd	Leaf with that of *Piper betel*	*Allergy	Reddy and Raju (1999)
		Leaf with that of *Albizia amara* and stem of *Cissus quadrangularis*	Bone fracture	Shashikanth et al. (2011)
		Leaf	*Dysentery	Reddy and Raju (2000)
			*Epitaxis	Reddy and Raju (1999)
			*Opacity of cornea	Reddy and Raju (1999)

TABLE 8.1 *(Continued)*

S. No.	Name of the plant	Part(s) used	Uses	References
		Leaves with those of *Cleome viscosa, Leucas aspera* and *Pergularia daemia*	*Tympany	Reddy and Raju (1999)
		Fruit with leaf of *Trianthema portulacastrum*, onion and turmeric	*Yokegall	Reddy and Raju (1999)
		Fruit	Dizziness	Patil et al. (2010)
102.	*Cocculus hirsutus* (L.) Diels.	Leaf	Arthritis	Shashikanth et al. (2011)
	VN: Dusara teega (Tel), Musakani (O) E: Broom creeper		Blood motions	Murthy et al. (2007); Satapathy (2010)
			Lice	Goud and Pullaiah (1996)
		Leaf with poppy seeds	Urinary disorders	Sudarsanam et al. (1995)
103.	*Cocos nucifera* (L.) VN: Kobbari (Tel), Thennai (Tam) E: Coconut	Seed oil with straw of *Oryza sativa* and wood ash of *Choroxylon swietenia* Tender pods	Wounds	Armadas (2000); Kiruba et al. (2006)
104.	*Colebrookia oppositifolia* Sm. VN: Maipeeth (Tel) E: Indian squirrel tail	Leaf	Cataract	Naidu (2003)
105.	*Coleus ambonicus* Lour. VN: Vamu aaku (Tel) E: Indian borage	Leaves with those of *Ximenia americana, Azadirachta indica* and *Annona squamosa*	*Eradication of lice	Suneetha et al. (2012)
		Leaf	Infection in eyes and to kill worms	Naidu (2003)
106.	*Commelina benghalensis* L. VN: Enneddura kura (Tel), Kena (M) E: Bengal day flower	Leaf	Yoke sores	Naidu (2003); Patil and Merat (2003)

TABLE 8.1 *(Continued)*

S. No.	Name of the plant	Part(s) used	Uses	References
107	*C. forskalaei* Vahl	Entire plant	Cough in goat	Patil et al. (2010)
108	*Convolvulus arvensis* L.	Leaf	Worms in wound	Patil et al. (2010)
109.	*Corallocarpus epigaeus* Hk.*f*	Tuber	Expelling tape worms	Salave et al. (2012)
	VN: Naga donda (Tel), Mungus kand (M) E: Bitter	Tuber with pepper and garlic	Insect bite	Reddy and Raju (1999)
	apple.	Tuberous root with pepper and garlic Tuberous roots	Snake bite	Sudarsanam et al. (1995) Hari Babu (2007); Suneetha et al. (2012)
110	*Cordia gharaf* (Forsk.) Eherenb. & Asch.	Stem bark	Bone fracture	Patil et al. (2010)
111	*Coriandrum sativum* L	Whole plant	Foot and Mouth disease	Patil et al. (2010)
112.	*Crataeva magna* (Lour.) DC. VN: Ulimiri (Tel) E: Three leaved caper	Stem bark with that of *Catunaregam spinosa*, garlic cloves and black pepper	*Diphtheria	Suneetha et al. (2012)
113.	*Croton bonplandianum* Bail. VN: Kukka tulasi (Tel) E: Jungle tulsi	Leaf	Eradication of lice	Hari Babu (2007)
114.	*Cryptolepis buchananii* Roem. & Schult. VN: Adavi palatheega (Tel) E: Dude lahara	Leaf Whole plant	Galactagogue Foot and mouth disease	Naidu (2003) Patil et al. (2010)
115	*Cucumis callosus* (Rottl.) Cogn.	Leaf	Wound healing	Patil et al. (2010)
116.	*Cucurbita maxima* Lam. VN: Gummadi (Tel) E: Red pumpkin	Fruit with boiled horsegram Fruit	*Galactagogue	Hari Babu (2007) Suneetha et al. (2012)
117	*C. moschata* (Duch. ex Lamk.) Poir.	Fruit	Galactagogue	Patil et al. (2010)
118	*C. pepo* L.	Fruit	Intestinal swelling	Patil et al. (2010)
119	*Cullen corylifolia* (L.) Medic	Leaf	Worms in wound	Patil et al. (2010)

TABLE 8.1   *(Continued)*

S. No.	Name of the plant	Part(s) used	Uses	References
120.	*Curculigo orchioides-*Gaertn. VN: Nela tadi (Tel), Nilapanai (Tam) E: Golden eye grass	Tubers with stems of *Cissus quadrangularis* Dried tubers	Impaction	Reddy and Raju (1999) Kiruba et al. (2006)
		Root	Ophthalmic diseases	Goud and Pullaiah (1996); Suneetha et al. (2012)
		Tuber	*Wounds	Prasanthi (2010)
121.	*Curcuma longa* L. VN: Pasupu (Tel), Haladi (O) E: Turmeric	Rhizome with seeds of black gram and bamboo leaves	Dysentery	Satapathy (2010)
		Rhizome with leaves of *Annona squamosa*	Wounds	Ramadas et al. (2000) Patil et al. (2010)
122	*Cuscuta chinensis* Lamk.	Stem	galactagogue	Patil et al. (2010)
123.	*Cyanotis adscendens-*Dalz. VN: Yeggogulu (Tel)	Tuber with betel leaf	*Allergy	Reddy and Raju (1999)
124	*Cymbopogon martini* (Roxb.) Wats.		Diarrhoea	Patil et al. (2010)
125.	*Cynodon dactylon* (L.) Pers. VN: Dudo (O), Garika (Tel) E: Bahama grass	Whole plant	*Diarrhea	Girach et al. (1998)
		Leaf	*Trypanoso-miasis	Reddy and Raju (2000)
126.	*Dalbergia latifolia* Roxb. VN: Iridi (Tel), Shisam (M) E: Indian rose wood	Leaf	*Diarrhea	Naidu (2003)
		Stem bark	*Eradication of lice	Hari Babu (2007); Suneetha et al. (2012)
		Wood and pod of *Cassia fistula*	Fever	Patil and Merat (2003)
127.	*Datura metel* L. VN: Nallaummetta (Tel), Umathai (Tam) E: Thorn apple	Leaf	*Enteritis	Ganesan et al. (2008)
			*Horn fracture	Reddy et al. (1998)

TABLE 8.1   *(Continued)*

S. No.	Name of the plant	Part(s) used	Uses	References
128.	*Deccania pubescens* (Roth) Tirveng. VN: Bajjumanga Golden collyrium	Stem bark with those of *Carissa spinarum, Chloroxylon swietenia,* tuber of *Withania somnifera,* turmeric, long pepper and garlic	Anthrax	Reddy and Raju (1999)
129.	*Delonix regia* Raf. VN: Turayi (Tel) E: Flamboyant	Stem bark with pepper and garlic Stem bark	Ephemeral fever	Reddy et al. (2006) Murty and Rao (2012)
130.	*Dendrocalamus strictus* (Roxb.) Nees VN: Munglakarra (Tel)	Leaf with ginger and turmeric Leaf	Anthrax	Sudarsanam et al. (1995) Naidu (2003)
	E: Solid bamboo		Easy delivery	Reddy and Raju (1999)
		Leaf with that of *Caesalpinia bonduc*, flower of *Butea monosperma* and horsegram	*Panting	Reddy and Raju (1999)
		Leaf Fruit	Parturition Dysentery, Cough	Suneetha et al. (2012) Manikandan and Lakshmanan (2014)
131.	*Dendrophthoe falcata* (L.f.) Etting. VN: Bajanika, Chigara bajana (Tel) E: Mistletoe	Leaves with those of *Cissampelos pareira, Pavetta breviflora, Solanum pubescens,* stem of *Cissus quadrangularis,* stem bark of *Deccania pubescens,* pepper and garlic	*Anthrax	Reddy and Raju (1999)
		Whole plant	Bone fracture	Naidu (2003)
		Stem bark with turmeric	Wounds	Raju (2009); Suneetha et al. (2012)
132.	*Derris scandens* (Roxb.) Benth. VN: Marugaaku E: Jewel vine	Leaf with that of *Blepharispermum subsessile,* ginger, turmeric and Calcium carbonate	*Anthrax	Reddy and Raju (1999)

TABLE 8.1   *(Continued)*

S. No.	Name of the plant	Part(s) used	Uses	References
133.	*Diospyros malabarica* (Desr.) Kostel	Leaf with tuber of *Gloriosa superba*, onion and common salt Stem bark	*Impaction Diarrhea of goat	Reddy and Raju (1999) Girach (2001)
134.	*Diospyros melanoxylon* Roxb. VN: Tumiki chettu (Tel) E: Coromandel ebony	Stem bark	Diarrhea	Naidu (2003)
135.	*Dodonaea viscosa* (L.) Jacq. VN: Adivibandaru, Bandara (Tel) E: Jamaica switch sorrel	Leaves with those of *Phaseolus mungo* and *Acacia nilotica* gum Leaf	Bone fracture	Reddy and Raju (1999) Sudarsanam et al. (1995); Goud and Pullaiah (1996); Reddy et al. (1998); Naidu (2003); Swamy (2009)
136.	*Dregea volubilis* (L. *f.*) Hk. *f.*	Leaf	Sprains	Mokat et al. (2010)
	VN: Bandi gurija, Peddakadi saaku (Tel), Hirandodi (M) E: Cotton milk plant	Stem bark with fruitof *Carissa spinarum*, onion, turmeric	*Yokegall	Reddy and Raju (1999)
137.	*Drosera burmannii* Vahl. VN: Bedaku (Tel) E: Sun dew	Root	Wounds	Naidu (2003)
138.	*Eclipta prostrata* (L.) L.	Leaf Root	*Horn polish Injury	Raju (2009) Patil et al. (2010)
	VN: Guntagalaraku (Tel) E: Trailing eclipt	Leaf with turmeric	Wounds	Suneetha et al. (2012)
139.	*Ehretia laevis* Roxb. VN: Pala dantam (Tel) E: Ivory wood	Stem bark with coconut oil	Wounds	Suneetha et al. (2012)
140.	*Elephantopus scaber* L.	Leaf	Loose motions	Naidu (2003)
	VN: Yedduadugu E: Prickly leaved elephant's foot	Root	Wounds	Naidu (2003); Prasanthi (2005); Suneetha et al. (2012)
141.	*Entada pursaetha* DC.	Seed	Helminthiasis	Suneetha et al. (2012)
	VN: Gilla teega (Tel) E: Giant's rattle		Vermifuge	Naidu (2003)
			Glandular swellings	Goud and Pullaiah (1996)

TABLE 8.1 *(Continued)*

S. No.	Name of the plant	Part(s) used	Uses	References
142.	*Euphorbia antiquo-rum* L. VN: Bontha jemudu (Tel) E: Indian spurge tree	Stem	Anthrax	Sudarsanam et al. (1995)
143.	*E. hirta* L. VN: Palaku (Tel) E: Asthma plant	Leaf with groundnut oil	*Horn cancer	Reddy et al. (1998)
144	*Euphorbia nerifolia* L. VN: Aku jemudu (Tel)	Latex	Cataract	Raja Reddy et al. (1989)
145.	*E. nivulia* Buch.-Ham. VN: Boggu jilledu (Tel) E: Leafy milk hedge	Latex	*Carbuncle	Hari Babu (2007)
146	*E. tirucalli* L.	Young branch	Bone fracture	Patil et al. (2010)
147.	*Ficus benghalensis* L. VN: Vizhuthu (Tam), Marri	Prop root with honey	*Diarrhea and dysentery	Satapathy (2010)
	(Tel), Bara (O) E: Banyan tree	Tap root and flowers of *Cocos nucifera*	*Haematuria	Ganesan et al. (2008)
		Stem bark with seed oil of *Pongamia pinnata*	*Skin diseases	Goud and Pullaiah (1996)
		Stem bark with that of *Madhuca longifolia,* rhizome of *Curcuma longa*	*Wounds	Ramadas et al. (2000)
		Leaf	Bone fracture, leg swelling	Patil et al. (2010)
148.	*F. hispida* L. *f.* VN:Bhui- Umbar (M) E: Hairy fig	Fruit or leaf with that of *Atalantia racemosa* and naphthalene balls	*Wounds	Mokat et al. (2010)
149.	*F. racemosa* L. VN: Medi E: Cluster	Stem bark with pepper	Skin diseases	Hari Babu (2007)
150.	*F. semicordata* Buch.-Ham. VN: Dimri (O)	Stem bark Latex	Deworming	Suneetha et al. (2012) Prusti (2007)

TABLE 8.1 *(Continued)*

S. No.	Name of the plant	Part(s) used	Uses	References
151.	*Gardenia gummifera* L. *f.* VN: Konda jama (Tel)	Bark	*Body pains	Murthy et al. (2007)
	E: Gummy cape jasmine	Leaf bud	Wounds	Hari Babu (2007)
152.	*Glycosmis arborea* (Roxb.) DC. VN: Gulimi (Tel)	Leaf	Wounds	Naidu (2003)
153.	*Gomphrena serrata* L. VN: Gabbu aaku (Tel) E: Gomphrena	Whole plant with that of *Tephrosia purpurea* and seed of pepper	Cough	Suneetha et al. (2012)
154.	*Grewia damine* Gaertn. VN: Ulchara (Tel) E: Salvia leaved crossberry	Leaf with those of *Grewia villosa, Tephrosia purpurea, Vitex altissima*, turmeric, pepper and garlic	Anthrax	Reddy and Raju (1999)
155.	*G. hirsuta* Vahl VN: Chipri (Tel) E: Kukurbicha	Root	Bone fracture	Naidu (2003); Shashikanth et al. (2011)
156.	*G. tiliaefolia* Vahl VN: Tada chettu (Tel) E: Mulberry	Root bark	Dislocated joints	Naidu (2003)
157.	*Gymnema sylvestre* (Retz.) R. Br. ex Schultes VN: Podaptri (Tel), Sirukurinjan (Tam) E: Suger destroyer	Leaf	Diarrhea, Ephemeral fever	Sudarsanam et al. (1995); Goud and Pullaiah (1996); Kiruba et al. (2006)
158	*Hardwickia binata* Roxb.	Leaf	Galactogogue	Patil et al. (2010)
159.	*Hedyotis corymbosa* (L.) Lam. VN: Tikka chettu (Tel) E: Diamond flower	Whole plant with pepper and garlic powder	Trypanosomiasis	Reddy et al. (2006)
160	*Helicteres isora* L.	Fruits	Dysentery	Patil et al. (2010)
161.	*Hemionitis arifolia* (Burm.) Moore. VN: Ramabanam (Tel) E: Heart fern	Whole plant Fronds with long pepper	Wounds and ulcers	Naidu (2003); Hari Babu (2007)

TABLE 8.1   *(Continued)*

S. No.	Name of the plant	Part(s) used	Uses	References
162.	*Holarrhena antidysenterica* Wall. VN: Aaku paala (Tel) E: Ivory tree	Fruit with that of *Balanites roxburghii* Fruit	*Conjunctivitis Galactogogue	Shashikanth et al. (2011) Patil et al. (2010)
		Stem bark	*Helminthiasis	Hari Babu (2007)
163.	*Holoptelea integrifolia* (Roxb.) Planch. VN: Navili (Tel) E: South Indian elm	Leaf	Bronchitis	Naidu (2003)
164	*Hordeum vulgare* L.	Seed	Tumor	Patil et al. (2010)
165	*Ipomoea carnea* Jacq.	Leaf	Cuts	Patil et al. (2010)
166.	*Jatropha gossypifolia* L.	Root	Bone fracture	Naidu (2003)
	VN: Seemanepalam (Tel),	Leaf	*Eye injuries	Naidu (2003)
	Mogali erand (M) E: Belly ache bush	Stem latex and seed oil	Ringworm and dermal itching	Salave et al. (2011)
167.	*Justicia adhatoda* (L.) Nees	Leaf	Bronchitis	Satapathy (2010)
	VN: Addasaramu, Basanga (O) E: Malabar nut	Leaf, stem bark with garlic and *asafoetida*	Rheumatic pains	Sudarsanam et al. (1995)
168.	*Kalanchoe pinnata* (Lam.) Pers. VN: Gurrelamasalakura (Tel) E: Life plant	Leaf	Skin infection	Naidu (2003)
169	*Kedrostris rostrata* (Rottl.) Cogn.	Root	To increase fodder consumption	Patil et al. (2010)
170	*Lagerstroemia parviflora* Roxb.	Leaf	Bone fracture	Patil et al. (2010)
171.	*Lannea coromandelica* (Houtt.) Merr. VN: Gumpena (Tel) E: Jhingam	Stem bark	*Anthrax	Murty and Rao (2012)
		Stem bark with ginger and garlic	Fever	Manikandan and Lakshmanan (2014)
		Stem bark	*Fracture	Ramadas (2000)
172.	*Lasia spinosa* (L.) Thw. VN: Mulla bokachika (Tel)	Rhizome	*Anaemia	Prasanthi (2010)

TABLE 8.1   *(Continued)*

S. No.	Name of the plant	Part(s) used	Uses	References
173.	*Lawsonia inermis* L. VN: Gorintaku (Tel), Manjuati (O) E: Henna	Leaf	Loose motions Foot and mouth disease	Naidu (2003) Satapathy (2010)
174.	*Leea macrophylla* Roxb. ex Hornem. VN: Nela modugi (Tel) E: Hathikana	Root with axillary bud of *Bambusa arundinacea,* leaf and tuber of *Cos- tus speciosus,* rhi- zome of *Curcuma longa,* tender root of *Sterculia urens,* stem bark of *Polyalthia longifolia*	Bone fracture	Prasanthi (2010)
175.	*Leonotis nepetifolia* (L.) Br. VN: Ranabheri (Tel) E: Lion's ear	Root	Mastitis	Naidu (2003)
176	*Leucas cephalotes* (Roth.) Spr. VN: Tumbikura (Tel) E: Guma	Leaf and seed	Ulcerous wounds	Naidu (2003)
177.	*Lippia javanica* (Burm. *f.*) Spr. VN: Nagairi (O) Kampurodda (Tel)	Leaf	Diarrhea and dysentery	Girach et al. (1998)
	E:Lemon bush	Whole plant	Eradication of lice	Hari Babu (2007)
178.	*Litsea glutinosa* (Lour.) Robins VN: Chigara (Tel) E: Indian laurel	Stem bark Stem bark with that of *Cissus pallida,* root bark of *Grewia hirsuta* and tuber of *Cur- culigo orchioides*	Bone fracture Wounds	Naidu (2003) Ramadas et al. (2000)
179.	*Luffa acutangula* (L.) Roxb. VN: Beera (Tel) E: Ridged gourd	Leaf	Yoke sores	Naidu (2003)
180.	**Macaranga peltata* (Roxb.) Muel.-Arg. VN: Kulakarachettu (Tel) E: Roxburgh's lotus croton	Stem bark with that of *Pterocar- pus marsupium*	Wounds and worms	Naidu (2003)

TABLE 8.1 *(Continued)*

S. No.	Name of the plant	Part(s) used	Uses	References
181.	*Macrotyloma uniflorum* (Lam.) Verd. VN: Ulavalu (Tel) E: Horse gram	Seed	Galactagogue	Naidu (2003); Hari Babu (2007); Suneetha et al. (2012)
182.	*Madhuca longifolia* (Koen.) Mac Br.	Seed oil	*Joint pains	Naidu (2003)
	VN: Ippa (Tel) E: Mowra fat		Skin diseases	Hari Babu (2007)
183.	*Mallotus philippensis* Muell.-Arg. VN: Sindhuri (Tel) E: Kamela tree	Seed	Wounds	Hari Babu (2007); Suneetha et al. (2012)
184.	*Malvastrum coromandelianum* (L.) Garcke VN: Chiru benda (Tel)	Leaf with turmeric	Mosquito bite	Hari Babu (2007)
185.	*Mangifera indica* L.	Fruit	Constipation	Satapathy (2010)
	VN: Maavichettu (Tel), Amba (O) E: Mango	Stem bark with those of *Azadirachta indica* and *Syzygium cumini*	*Stomach pain	Suneetha et al. (2012)
		Seed with leaf of *Euphorbia nivulia*	*Vermifuge	Reddy et al. (1998)
186.	*Manilkara hexandra* (Roxb.) Dub. VN: Nimmi (Tel) E: Milk tree	Stem bark with stem of *Cissus quadrangularis* and garlic cloves	Throat infection	Naidu (2003)
187	*Manilkara zapota* (L.) P. van Royen	Leaf, seed	Bone fracture	Patil et al. (2010)
188.	*Martynia annua* L. VN: Puligoru (Tel) E: Devil's claw	Leaf with pepper and garlic Leaf	Epilepsy	Sudarsanam et al. (1995) Goud and Pullaiah (1996)
			Wounds and sores	Naidu (2003)
189.	*Maytenus senegalensis* (Lamk.) Exell. VN: Danti (Tel) E: Thorny staff tree	Leaf with turmeric	*Eradication of lice	Suneetha et al. (2012)
190.	*Melia azedarach* L. VN: Turakavepa (Tel),	Leaf	Anthelmintic	Goud and Pullaiah (1996)
	Mahanimba (O) E: Persian lilac	Leaf and flower	*Eradication of lice	Suneetha et al. (2012)

TABLE 8.1   *(Continued)*

S. No.	Name of the plant	Part(s) used	Uses	References
		Leaf	Fever, Anthelmintic	Naidu (2003)
		Leaf with root of *Smilax zeylanica,* latex of *Calotropis gigantea*	Sprain and swelling	Satapathy (2010)
191.	*Momordica charantia* L. VN: Kakara (Tel), Kalara	Leaf	Dog bite	Suneetha et al. (2012)
	(O) E: Bitter gourd		*Placenta	Misra and Das (1998)
		Whole plant with turmeric	Wounds	Naidu (2003)
192	*Momordica dioica* Roxb.	Root Leaf	Flatulence Post delivery bleeding	Patil et al. (2010)
193.	*Morinda pubescens* Sm. VN: Togaru (Tel), Bartondi (M) E: Morinda tree	Leaf	*Fertility and conception colic	Salave et al. (2011) Patil et al. (2010)
			Neck cracks	Naidu (2003)
		Stem bark with seed of *Semecarpus anacardium*, camphor and turmeric	*Rinderpest	Reddy et al. (1998)
194.	*Moringa oleifera* Lam. VN: Munaga chettu (Tel) E: Drumstick tree	Leaf and flower	Anthrax	Sudarsanam et al. (1995); Suneetha et al. (2012)
		Leaf	*Cuts and wounds	Swamy (2009)
		Seed or stem bark or root	*Helminthiasis	Murty and Rao (2012)
195.	*Mucuna utilis* Wall. ex Wight VN: Dulagondi (Tel) E: Bengal velvet-bean	Seed	Intestinal disorders and worms	Naidu (2003)
196.	*Murraya paniculata* (L.) Jack VN: Peethurumalli (Tel) E: Orange jasmine	Leaf	Bone fracture	Naidu (2003)

TABLE 8.1   *(Continued)*

S. No.	Name of the plant	Part(s) used	Uses	References
197.	*Musa paradisiaca* L. VN: Vazhai (Tam), Arati	Fruit with sesame oil	*Bronchitis	Ganesan et al. (2008)
	(Tel) E: Banana	Flower	Diarrhea	Sudarsanam et al. (1995)
		Pericarp of unripe fruit	Worms in wound	Patil et al. (2010)
198.	*M. rosacea* Jacq. VN: Adavi arati (Tel) E: Wild banana	Root	Dysentery	Suneetha et al. (2012)
199	*Nerium indicum* Mill.	Leaf	Fever	Patil et al. (2010)
200.	*Nicotiana tabacum* L. VN: Dhuanpatra (O), Pogaku (Tel) E: Tobacco	Dried seed Seed	*Foot and mouth disease	Misra and Das (1998) Swamy (2009)
		Leaves	Worms in injury	Patil et al. (2010)
201.	*Ocimum americanum* L. VN: Bhutulasi (Tel) E: Hoary basil	Whole plant	Eradication of lice	Naidu (2003); Hari Babu (2007)
		Leaf	Wounds	Naidu (2003)
202.	*O. basilicum* L. VN: Rudra jada, Sabja	Leaf with neem oil	Eczema	Sudarsanam et al. (1995)
	(Tel) E: Sweet basil	Leaf	Eradication of lice To prevent pus formation in ears	Suneetha et al. (2012) Patil et al. (2010)
203	*Ocimum tenuiflorum* L.	Leaf	Bone fracture	Patil et al. (2010)
204.	*Oroxylum indicum* (L.) Vent. VN: Bapini chettu (Tel), Phanaphania (O) E: Indian trumphet flower	Stem bark Root bark Seed	Wounds	Ramadas et al. (2000) Naidu (2003) Satapathy (2010)
205.	*Pavonia zeylanica* (L.) Cav. VN: Karrubenda (Tel) E: Ceylon swamp mallow	Leaf	Wounds	Ramadas et al. (2000)
206.	*Pedalium murex* L. VN: Enugupalleru (Tel)	Fruit	*Diuretic	Murty and Rao (2012)

TABLE 8.1   *(Continued)*

S. No.	Name of the plant	Part(s) used	Uses	References
	E: Baragokhru	Leaf with that of *Cocculus hirsutus*	*Haematuria	Reddy et al. (1998)
		Fruit with black pepper	Urinary tract infection	Sudarsanam et al. (1995)
		Leaf with ginger and common salt	Fever	Manikandan and Laksh-manan (2014)
207.	*Pergularia daemia* (Forsk.)	Leaf	*Cataract	Prasanthi (2010)
	Chiov. VN: Dushtaputheega, Jutta paala (Tel), Veliparuthi	Leaf with pepper and garlic	Ephemeral fe-ver, Rheumatic arthritis	Sudarsanam et al. (1995)
	(Tam), Utarand (M) E: Whitlow plant	Leaf with black pepper and garlic	*Eye disease	Naidu (2003)
		Leaf with pep-per, garlic and *asafoetida* Leaf	Fever	Goud and Pullaiah (1996) Reddy et al. (1998)
		Leaf with that of *Calotropis procera*	*Muscular pain	Naidu (2003)
		Leaf	*Post-natal pains	Salave et al. (2012)
		Latex	*Ring worm and skin disease	Kiruba et al. (2006)
		Root	Flatulence	Patil et al. (2010)
208.	*Phyllanthus emblica* L. VN: Pedda usiri (Tel)	Stem with leaf of *Vanda tessellata*	*Bone fracture	Swamy (2009)
	E: Emblic myrobalan	Stem bark	Dysentery	Raju (2009); Suneetha et al. (2012)
209.	*Physalis minima* L. VN: Kuppante (Tel)	Leaf	*Calf sickness Flatulence	Shashikanth et al. (2011) Patil et al. (2010)
	E: Country gooseberry		*Wounds	Prasanthi (2010)
210.	*Pimpinella tirupatien-sis* Bal. & Subr. VN: Kondadhaniyalu	Tuberous root with pepper	Colic, Rheu-matic pains	Sudarsanam et al. (1995)
211.	*Piper nigrum* L. VN: Milagu (Tam) E: Pepper	Fruit with seed of *Gossypium hirsutum,* bulb of *Allium cepa,*	Black quarter	Ganesan et al. (2008)

TABLE 8.1 *(Continued)*

S. No.	Name of the plant	Part(s) used	Uses	References
		fruit of *Cuminum cyminum*, rhizome of *Curcuma longa*, flower of *Musa paradisiaca*, leaf of *Piper betel* and sugar of *Borassus flabellifer*		
212.	*Plumbago zeylanica* L. VN: Chitramulam (Tel)	Leaf and root	Bone fracture	Sudarsanam et al. (1995)
	E: White leadwort	Root bark	Skin diseases	Goud and Pullaiah (1996)
		Whole plant	*Sores	Prasanthi (2005)
		Root	*Tumors and warts	Murthy et al. (2007)
		Plant with neem oil	*Wounds	Naidu (2003)
213.	*Plumeria alba* L. VN: Lakshmi poolu (Tel) E: Pagoda tree	Latex with lemon juice	Scabies and ulcers	Hari Babu (2007)
214.	*P. rubra* L. VN: Nuruvarahalu (Tel) E: Frangipani	Latex	Wounds, *Skin infection	Naidu (2003)
		Seed	*Stimulant	Murthy et al. (2007)
215.	*Polyalthia longifolia* (Sonner) Thw. VN:Naramamidi (Tel) E:Cemetery tree	Root bark with that of *Azadirachta indica* and stem of *Cissus quadrangularis*	Bone fracture	Suneetha et al. (2012)
216.	*Polygala arvensis* Willd. VN: Chittodi (Tel)	Leaf	Snake bite	Hari Babu (2007)
217.	*Pongamia pinnata* (L.) Pierre.	Leaf	Bronchitis, Galactagogue	Naidu (2003)
	VN: Ganuga (Tel), Pongam (Tam), Karanj (M) E: Indian beach tree	Stem bark with those of *Capparis zeylanica, Delonix regia, Wrightia tinctoria* and stems of *Cissus quadrangularis*	Oedema	Reddy et al. (1998)

TABLE 8.1 *(Continued)*

S. No.	Name of the plant	Part(s) used	Uses	References
		Seed oil Fruit oil Leaf	Skin diseases	Sudarsanam et al. (1995); Naidu (2003); Hari Babu (2007) Mokat and Deokule (2004); Kiruba et al. (2006)
218.	*Premna latifolia* Roxb. VN: Nelli (Tel) E: Headache tree	Stem	Colic, Indigestion	Sudarsanam et al. (1995)
219.	*Prosopis cineraria* L. VN: Sarkaru tumma (Tel) E: Mesquite	Leaf	Mouth ulcers	Naidu (2003)
220.	*Pterocarpus marsupium* Roxb. VN: Vegisa (Tel) E: Indian kino tree	Stem with that of *Macaranga peltata*	*Worms *Wounds	Naidu (2003)
221.	*Pterolobium hexapetalum*	Stem bark	Dyspepsia	Naidu (2003)
	(Roth) Sant. & Wagh VN: Korindakampa (Tel) E: Indian red-wing		*Wounds	Suneetha et al. (2012)
222.	*Pueraria tuberosa* DC. VN: Magasirigadda (Tel)	Tuber	*Papilloma	Sudarsanam et al. (1995)
	E: Indian kudzu	Tuber with *Cissus quadrangularis*, onion and ginger	*Motions, Tuberculosis	Murthy et al. (2007)
223.	*Rhus mysurensis* Heyne ex Wt. & Arn. VN: Vudathapooli chettu (Tel) E: Mysore sumac	Fruit	Diarrhea and dysentery	Sudarsanam et al. (1995)
224.	*Ricinus communis* L.	Seed	*Horn cancer	Reddy et al. (1998)
	VN: Jada (O), Amudamu (Tel) E: Castor	Leaf	*Placenta Bone fracture	Misra and Das (1998); Patil et al. (2010)
		Latex with lemon	Scabies	Hari Babu (2007)
		Leaf with that of *Lawsonia alba*	Wounds	Mishra (2013)

TABLE 8.1   *(Continued)*

S. No.	Name of the plant	Part(s) used	Uses	References
225.	*Rubia cordifolia* L. VN: Thamaralli (Tel) E: Indian madder	Stem bark	Postnatal problems	Naidu (2003)
226.	*Saccharum spontaneum* L. VN: Rellu gaddi (Tel) E: Thatch grass	Inflorescence with mustard oil	Wounds	Suneetha et al. (2012)
227	*Salvadora persica* L.	Leaf	Bone fracture Flatulence	Patil et al. (2010)
228.	*Sapindus emarginatus* Vahl. VN: Kunkudu E: Soapnut	Nut	*Paralysis	Sudarsanam et al. (1995)
229.	*Schleichera oleosa* (Lour.) Oken VN: Banrubai (Tel) E: Lac tree	Seed oil with root of *Derris scandens* Seed oil	Sores	Goud and Pullaiah (1996); Suneetha et al. (2012)
			Wounds and ulcers	Naidu (2003)
230.	*Semecarpus anacardium* L.f. VN: Nalla jeedi (Tel), Bibba (M) E: Marking nut	Seed oil	Foot and mouth disease	Misra and Das (1998); Deshmukh et al. (2011)
		Fruit Dried fruit	*Haemorrhagic septicemia	Misra and Anil Kumar (2004); Murty and Rao (2012)
		Seed oil	*Horn cancer Mouth ulcer	Reddy et al. (1998) Patil et al. (2010)
		Seed	Wounds	Ramadas et al. (2000) Naidu (2003)
231.	*Sesbania grandiflora* Pers. VN: Avisa (Tel) E: Sesbania	Fruit	Dysentery	Suneetha et al. (2012)
232.	*Setaria pumila* Roem. & Schult. VN: Korra (Tel) E: Yellow fox tail	Grain with pepper	Sores	Suneetha et al. (2012)

TABLE 8.1   *(Continued)*

S. No.	Name of the plant	Part(s) used	Uses	References
233.	*Shorea robusta* Gaertn. *f.* VN: Guggilamu (Tel) E:	Stem bark	Dysentery	Naidu (2003)
	Sal tree	Seed	Worms	Naidu (2003)
234.	*Solanum surattense-* Burm. *f.*	Whole plant with sesame oil	*Bone fracture	Swamy (2009)
	VN: Tikka vankaya (Tel) E: Yellow berried nightshade	Flower	*Ophthalmic diseases	Suneetha et al. (2012)
235.	*S. torvum* Sw. VN: Dhungiri (Tel) E: Turkey berry	Fruit	Diarrhea	Naidu (2003); Hari Babu (2007)
236	*S. virginianum* L.	Fruit	To improve vision	Patil et al. (2010)
237.	*Solena amplexicaulis*(Lam.) Gandhi VN: Adavi donda (Tel) E: Creeping cucumber	Tuberous root or root stock	Cuts on tongue	Naidu (2003)
238.	*Soymida febrifuga* (Roxb.) Juss. VN: Somidi chettu (Tel) E: Indian red wood tree	Stem bark	Cough and dysentery	Goud and Pullaiah (1996); Murthy et al. (2007)
			Fever	Prasanthi (2010)
		Stem bark with those of *Terminalia alata, Dichrostachys cinerea* and root of *Solanum xanthocarpum*	*Trypanosomiasis	Reddy et al. (2006)
239.	*Spermadictyon suaveolens* Roxb.	Leaf Root	Wounds Blood dysentery	Naidu (2003) Patil et al. (2010)

TABLE 8.1 *(Continued)*

S. No.	Name of the plant	Part(s) used	Uses	References
240.	*Strychnos nux-vomica* L. VN: Musini (Tel) E: Nux vomica	Fruit	*Foot and mouth disease	Raju (2009)
		Root with that of *Helianthus annuus,* petal of *Butea monosperma*	*Hump sores	Mishra (2013)
		Stem bark	Rheumatoid arthritis	Suneetha et al. (2012)
241.	*S. potatorum* L.	Seed with honey	Eye infection	Naidu (2003)
	VN: Chillapikka (Tel) E: Clearing nut	Leaf with honey and salt	*Wounds	Suneetha et al. (2012)
242.	*Syzygium cumini* (L.) Skeels	Stem bark with leaf of *Justicia adhatoda*	Diarrhea and dysentery	Satapathy (2010)
	VN: Neredu (Tel), Jammu (O) E: Jambolana	Leaf	Wounds	Suneetha et al. (2012)
243	*Tabernaemontana divaricata* L.	Leaf	shoulder injury due to yoke	Patil et al. (2010)
244.	*Tamarindus indica* L. VN: Chintha chettu (Tel)	Leaf	*Ephemeral fever Bone fracture Cramps	Misra and Anil Kumar (2004) Patil et al. (2010)
	E: Tamarind	Mesocarp	*Flatulence, Haematuria	Reddy et al. (1998)
		Seed	Fractured bones	Sudarsanam et al. (1995)
		Fruit	*Rinderpest	Misra and Anil Kumar (2004)
			*Ulcers on tongue	Hari Babu (2007)
245.	*Tectona grandis* L. f. VN: Teku (Tel), Bananila (O) E: Teak	Stem bark with turmeric Stem bark	Sores	Panduranga Raju (2009); Suneetha et al. (2012)
		Seed	Scant urination	Patil et al. (2010)

TABLE 8.1   *(Continued)*

S. No.	Name of the plant	Part(s) used	Uses	References
246.	*Tephrosia purpurea* (L.) Pers.	Leaf with green chilli and redgram	Constipation, Impaction	Sudarsanam et al. (1995)
	VN: Vempali (Tel) E: Wild indigo	Whole plant with that of *Gomphrena serrata* and seed of *Piper nigrum*	*Cough	Suneetha et al. (2012)
247	*Terminalia arjuna* (Roxb.) Wight & Arn.	Leaf	Dysentery	Patil et al. (2010)
248.	*T. bellirica* (Gaertn.) Roxb.	Stem bark	Blood dysentery	Salave et al. (2012)
	VN: Thandra (Tel), Behada	Fruit	Colic	Sudarsanam et al. (1995)
	(M) E: Belleric myrobalan		*Foot and mouth disease	Patil and Merat (2003)
		Stem bark	Tumors, Warts	Murthy et al. (2007)
		Seed	Intestinal worms	Patil et al. (2010)
249.	*T. chebula* Retz.	Fruit	Anthrax	Naidu (2003)
	VN: Karakkaya (Tel) E: Chebulic myrobalan		Dysentery	Suneetha et al. (2012)
			*Wounds	Hari Babu (2007)
250.	*Thespesia lampas* (Cav.) Dalz. & Gibs. VN: Adavibenda (Tel) E: Ran-bendi	Root	Eye disease	Naidu (2003)
251	*T. populnea* (L.) Soland ex Corr.	Fruit	To arouse sex	
252.	*Tinospora cordifolia* (Willd.) Miers ex Hook. *f.*&	Stem bark	Anthelmintic	Goud and Pullaiah (1996)
	Thoms. VN: Tippa teega (Tel), Futani-chirni (M)	Fruit with *Zingiber officinale* and *Curcuma domestica*	Fever	Salave et al. (2011)
	E: Gulancha tinospora	Dried aerial part	*Foot and mouth disease	Swamy (2009)

TABLE 8.1 *(Continued)*

S. No.	Name of the plant	Part(s) used	Uses	References
		Leaf with that of *Dodonaea viscosa*	Bone fracture	Reddy et al. (1998)
253.	*Trema orientalis* (L.) Bl. VN: Kondajana (Tel) E: Charcoal tree	Leaf	Foot and mouth disease	Naidu (2003)
254	*Tribulus lanuginosus* L.	Leaf	Worms in ears	Patil et al. (2010)
255.	*Trichosanthes tricuspidata* Lour. VN: Aguda (Tel)	Tuber with horse gram	Dysentery	Naidu (2003)
256.	*Tridax procumbens* L. VN: Gaddi chamanthi (Tel) E: Coat buttons	Leaf Whole plant	Wounds	Ramadas et al. (2000); Naidu (2003);Suneetha et al. (2012); Mishra (2013) Raju (2009)
257.	*Triticum aestivum* L. VN: Godhumalu (Tel) E: Wheat	Soup of wheat with ginger	Bone fracture	Suneetha et al. (2012)
258.	*Tylophora indica* (Burm. *f.*) Merr. VN: Nanjaruppan (Tam) E: Emetic swallow wort	Leaf	*Pyrexia	Ganesan et al. (2008)
259.	*Urena lobata* L. VN: Menda chikni (Tel),	Leaf with castor oil, eucalyptus oil	*Mastitis	Salave et al. (2011)
	Van-bhendi (M) E: Caesar weed	Leaf with castor oil Leaf	Wounds	Naidu (2003) Hari Babu (2007)
260	*Ventilago denticulata* Willd.	Leaf	Dysentery	Patil et al. (2010)
261.	*Vernonia albicans* DC. VN: Garita kami (Tel) E: Purple fleabane	Whole plant	*Worms	Prasanthi (2010)
262	*V. cinerea* (L.) Less.	Young leaf	Cough	Patil et al. (2010)
263.	*Viscum orientale* Willd. VN: Katta badanika (Tel) E: Mistletoe	Stem with coconut oil	Wounds	Suneetha et al. (2012)

TABLE 8.1 *(Continued)*

S. No.	Name of the plant	Part(s) used	Uses	References
264.	*Vitex negundo* L. VN: Tella vavili (Tel), Nirgundi (O) E: Chinese chaste tree	Leaves with those of *Leucas aspera, Zingiber officinale, Allium cepa;* seed of *Cajanus cajan* and *Piper nigrum*	*Corneal opacity	Ganesan et al. (2008)
		Leaf with that of *Acalypha indica,* pepper and garlic	*Ephemeral fever, Impaction	Reddy and Raju (1999)
		Leaf with turmeric Leaf	Wounds	Satapathy (2010) Suneetha et al. (2012)
265.	*Withania somnifera* (L.) Dunal.	Root	*Bloat	Ganesan et al. (2008)
	VN: Asvagandha (Tel), Pennerugadda (Tam) E: Aswagandha	Leaf	Eradication of lice	Suneetha et al. (2012)
266.	*Woodfordia fruticosa* (L.) Kurz. VN: Jaji (Tel) E: Fire flame bush	Flower with turmeric Flower	Wounds	Raju (2009) Suneetha et al. (2012)
267.	*Wrightia arborea* (Dennst.) Mabb. VN: Pedda pala (Tel) E: Woolly dyeing rosebay	Stem bark with those of *Strychnos nux-vomica, Azadirachta indica, Cleistanthus collinus* and fruit of *Terminalia chebula*	Rheumatoid arthritis	Suneetha et al. (2012)
268.	*Xanthium strumarium* L. VN: Peddapalleru (Tel) E: Bur weed	Plant	Swelling of glands	Naidu (2003)
269.	*Ziziphus mauritiana* Lam. VN: Regu chettu (Tel) E: Jujub	Leaf	Galactagogue	Suneetha et al. (2012)

TABLE 8.1   *(Continued)*

S. No.	Name of the plant	Part(s) used	Uses	References
270.	*Z. oenoplia* (L.) Mill. VN: Thella parimi (Tel), Kanteikoli (O) E: Jackal	Leaf with gum of *Sterculia urens,* leaf of *Acacia chundra*	*Bone fracture	Swamy (2009)
	jujube	Stem	*Sores	Hari Babu (2007)
		Stem bark with turmeric Leaf Stem bark	Wounds	Raju (2009) Ramadas et al. (2000) Suneetha et al. (2012)
		Root	*Yoke sore	Satapathy (2010)
271.	*Z. xylopyrus* (Retz.) Willd. VN: Gotika (Tel)	Root with stem bark of *Calotropis gigantea, Erythroxylum monogynum* and *Pterocarpus marsupium*	Anthrax	Murty and Rao (2012)
		Fruit with onion, pepper and turmeric	Horn cancer	Reddy and Raju (1999)

E = English, O = Oriya, Tam = Tamil, Tel = Telugu, VN = Vernacular name.

8.5 DISCUSSION

The present review deals with 271 species of plants belonging to 221 genera employed in 512 kinds of folklore prescriptions for treating over 134 different types of animal diseases by the rural people and tribals *viz., Ariya gounder, Bagata, Bhumij, Chenchu, Dhangar, Gadaba, Gond, Goudu, Jadaya gounder, Jatapu, Katkari, Kattunayakas, Khond, Kol, Kolam, Koli, Konda dora, Konda kammara, Konda kapu, Konda reddi, Konkana, Kotia, Koya, Koya dora, Kurumba gounder, Kuttiya, Laman, Lambada, Mali, Manne dora, Munda, Mukha dora, Naikpod, Nakkala, Nukadora, Pardhan, Porja, Reddy dora, Sabar, Savara, Santal, Sugali, Yanadi, Yerukula, Valmiki, Vanjari, Thakur, Thoti* and *Warl* of Eastern ghats and Deccan plateau covering Orissa, Andhra Prdesh, Tamil Nadu, Telangana and Maharashtra states. The ailments covered include allergy, anemia, anorexia, anthelmintic, anthrax, anti-inflammatory, arthritis, asthma, babesiosis, black quarter, blisters, bloated stomach, blood motions, body pains, bone fracture, bronchitis, calf sickness, carbuncle, cataract, chickenpox, cholera, colic, conjunctivitis, constipation, corneal opacity, cough, cuts and wounds, cuts on tongue, debility and general weakness, dermal itching, diarrrhea, diphtheria, dislocated joints, diuretic, dog bite, dysentery, dyspepsia, easy delivery, ectoparasites, eczema, enteritis, ephemeral fever, epilepsy, epitaxis, eradication of lice, expelling placenta, eye infection, eye injuries, eye problems, fertility and conception, fever, flatulence, foot and mouth disease, galactagogue, gastric problems, glandular swellings, haematuria, haemorrhagic septicemia, HCN poisoning, heart disease, helminthiasis, honeybee bite, horn cancer, horn fracture, horn polish, hump sores, impaction, indigestion, inducing heat, infertility, inflammation, insect bite, intestinal disorders, intestinal worms, joint pains, kidney disorder, knee pain, liver disorders, loose motions, maggot wounds, malaria, mastitis, miscarriage, mosquito bite, motions, mouth ulcers, muscular pain, neck cracks, neck sores, night blindness, oedema, oestrum, opacity of cornea, ophthalmic diseases, panting, papilloma, paralysis, parturition, pneumonia, post-natal pains, pyrexia, rheumatic arthritis, rheumatic pains, rinderpest, ring worm, scabies, scorpion bite, skin diseases, skin infection, snake bite, sore eyes, sores, sprain, stimulant, swellings, stomach pain, tape worm, throat infection, trypanosomiasis, tuberculosis, tumor, tympany, ulcers and wounds, ulcers on tongue, urinary disorders, urinary tract infection, urination, vermifuge, warts, wasp bite, worms, wounds, yokegall and yoke sores. Of the 512 practices 212 were found to be new along with 33 new ethnoveterinary plants viz., *Abelmoschus crinitus, Acacia auriculiformis, Alstonia venenata, Alysicarpus vaginalis, Arachis hypogaea, Becium filamentosum, Cadaba fruticosa, Capparis sepiaria, C.zeylanica, Cassia auriculata, C.italica, C.siamea, Cassytha filiformis, Chlorophytum tuberosum, Cipadessa baccifera, Cleome felina, Cocos nucifera, Croton bonplandianum, Deccania pubescens, Delonix regia, Gomphrena serrata, Grewia damine, Hedyotis corymbosa, Macaranga peltata, Malvastrum coromandelianum, Musa rosacea, Pavonia zeylanica, Plumeria alba, Polyalthia longifolia, Polygala arvensis, Saccharum spontaneum, Wrightia arborea,* and *Ziziphus xylopyrus* (Figures 8.1–8.18).

FIGURE 8.1 *Alstonia venenata.*

FIGURE 8.2 *Ardisia solanacea.*

FIGURE 8.3 *Asparagus racemosus.*

FIGURE 8.4 *Butea monosperma.*

FIGURE 8.5 *Careya arborea.*

FIGURE 8.6 *Curculigo orchioides.*

FIGURE 8.7 *Dodonaea viscosa.*

FIGURE 8.8 *Entada pursaetha.*

FIGURE 8.9   *Grewia tiliaefolia.*

FIGURE 8.10   *Hemionitis arifolia.*

FIGURE 8.11   *Holarrhena antidysenterica.*

FIGURE 8.12   *Macaranga peltata.*

FIGURE 8.13   *Madhuca longifolia.*

FIGURE 8.14   *Oroxylum indicum.*

FIGURE 8.15   *Pergularia daemia.*

FIGURE 8.16   *Semecarpus anacardium.*

FIGURE 8.17   *Wrightia arborea.*

FIGURE 8.18   *Ziziphus oenoplia.*

Pal (1992) reported the use of *Cissampelos pareira* for fever, *Shorea robusta* for dysentery and intestinal worms and *Casearia elliptica, Cocculus hirsutus, Madhuca longifolia, Melia azedarach, Moringa oleifera, Pongamia pinnata* for other veterinary prescriptions by *Asur, Birhore, Kondh, Mach, Munda, Oraon, Robka, Santal* tribals of West Bengal, Bihar and Orissa.

Species like *Abrus precatorius, Abutilon indicum, Acacia auriculiformis, A.catechu, A.chundra, Achyranthes aspera, Aloe vera, Andrographis paniculata, Annona reticulata, A.squamosa, Ardisia solanacea, Aristolochia indica, Azadirachta indica, Careya arborea, Carissa spinarum, Casearia elliptica, Celosia argentea, Chloroxylon swietenia, Cissus pallida, C.quadrangularis ris, Clematis gouriana, Cocos nucifera, Curculigo orchioides, Curcuma longa, Drosera burmannii, Eclipta prostrata, Elephantopus scaber, Ficus benghalensis, F. hispida, Gardenia gummifera, Glycosmis arborea, Hemionitis arifolia, Leucas cephalotes, Litsea glutinosa, Macaranga peltata, Martynia annua, Momordica charantia, Moringa oleifera, Ocimum americanum, Oroxylum indicum, Pavonia zeylanica, Plumbago zeylanica, Plumeria rubra, Pterocarpus marsupium, Pterolobium hexapetalum, Ricinus communis, Saccharum spontaneum, Schleichera oleosa, Semecarpus anacardium, Spermadictyon suaveolens, Strychnos potatorum, Syzygium cumini, Tridax procumbens, Urena lobata, Vitex negundo, Woodfordia fruticosa* and *Ziziphus oenoplia* are commonly used in the treatment of wounds. *Achyranthes aspera, Andrographis paniculata, A. neesiana, Cleistanthus collinus, Lawsonia inermis, Semecarpus anacardium, Strychnos nux-vomica, Terminalia bellirica, Tinospora cordifolia* and *Trema orientalis* are utilized for curing foot and mouth disease. *Andrographis paniculata, Areca catechu,*

Aristolochia bracteolata, A.indica, Asparagus racemosus, Azadirachta indica, Blepharispermum subsessile, Caesalpinia bonduc, Calotropis procera, Capparis sepiaria, Cardiospermum halicacabum, Cassia fistula, Cipadessa baccifera, Cissus quadrangula, Cleome viscosa, Delonix regia, Tamarindus indica and *Vitex negundo* are used for treatment of ephemeral fever. *Aegle marmelos, Albizia amara, Bauhinia vahlii, Canthium parviflorum, Cassia occidentalis, C.siamea, Cassytha filiformis, Cissus quadrangularis, Coccinia grandis, Dendrophthoe falcata, Dodonaea viscosa, Grewia hirsuta, Jatropha gossypifolia, Leea macrophylla, Murraya paniculata, Phyllanthus emblica, Plumbago zeylanica, Polyalthia longifolia, Solanum surattense, Tinospora cordifolia, Triticum aestivum* and *Ziziphus oenoplia* are used for bone setting. Herbal veterinary medicine also offers enormous scope for further research; hence all the existing information to be correctly recorded before it is lost. Basing on the age-old practices, pharmaceutical industries may formulate some novel prescriptions to cure various animal diseases.

8.6   CONCLUSION

India has got great traditional background in the field of ethnoveterinary medicine and practices, but in the process of modernization, this knowledge is vanishing very rapidly. In remote areas many herbal healers are present who have great-undocumented traditional knowledge about animal diseases, herbal treatments, herbal formulations. However, this important veterinary knowledge is in danger of extinction due to rapid modernization. Actually, this information survived by being passed from one generation to next but now-a-days younger generation does not take interest in animal husbandry practices. Due to this apathy, the valuable information about ethnoveterinary medicine is disappearing. Documentaion and standardization of ethnoveterinary knowledge are also important in the context of Intellectual Property Rights (IPR) to check the patent claims. In future, detailed chemical and pharmacological investigations of these traditional formulations and medicinal plants will be very helpful for inventing/developing the new veterinary drugs. Therefore, efforts should be made to retain this valuable information for validation and future uses. Ethnoveterinary medicine can make an economic difference, but its cost-efectiveness varies and depends on many different factors. Pharmacopoeia of ethnoveterinary medicines should be developed for its popularity and to check patenting; to develop a proper link between traditional veterinary healers and modern veterinary doctors and rare ethnoveterinary medicinal plants would be listed and preserved for posterity.

KEYWORDS

- **Deccan Plateau**
- **Eastern Ghats**
- **Ethnoveterinary Medicine**
- **Prescriptions**
- **Tribals**

REFERENCES

Deshmukh, R.R., Rathod, V.N. & Pardeshi, V.N. (2011). Ethnoveterinary medicine from Jalna district of Maharashtra state. *Indian J. Trad. Knowl. 10,* 344–348.

Galav, P.K., Ambika Nag & Katewa, S.S. (2007). Traditional herbal veterinary medicines from Mount Abu, Rajasthan. *Ethnobotany 19,* 120–123.

Ganesan, S., Chandhirasekaran, M. & Selvaraj, A. (2008). Ethnoveterinary healthcare practices in southern districts of Tamil Nadu. *Indian J. Trad. Knowl. 7,* 347–354.

Girach, R.D., Brahmam, M. & Misra, M.K. (1998). Folk veterinary herbal medicine of Bhadrak district, Orissa, India. *Ethnobotany 10,* 85–88.

Goud, P.S. & Pullaiah, T. (1996). Folk veterinary plants of Kurnool district, Andhra Pradesh, India. *Ethnobotany 8,* 71–74.

Hari Babu, M. (2007). Ethnomedicine from Visakhapatnam district, Andhra Pradesh, India. PhD Thesis, Andhra University, Visakhapatnam.

Kiruba, S., Jeeva, S. & Dhas, S.S.M. (2006). Enumeration of ethnoveterinary plants of Cape Comorin, Tamil Nadu. *Indian J. Trad. Knowl. 5,* 576–578.

Krishnamurthy, K.V., Murugan, R. & Ravi Kumar, K. (2014). Bioresources of Eastern Ghats. Dehra Dun: Bishen Singh Mahendra Pal Singh.

Manikandan, S. & Lakshmanan, G.M.A. (2014). Ethnobotanical survey of medicinal plants in Kalrayan hills, Eastern Ghats, Tamil Nadu. *Int. Letters of Natural Sciences, 12,* 111–121.

Mishra, D. (2013). Cattle wounds and ethnoveterinary medicine: A study in Polasara block, Ganjam district, Orissa, India. *Indian J. Trad. Knowl. 12,* 62–65.

Misra, K.K. & Anil Kumar, K. (2004). Ethnoveterinary practices among the *Konda Reddis* of East Godavari district of Andhra Pradesh. *Stud. Tribes Tribals 2,* 37–44.

Misra, M.K. & Das, S.S. (1998). Veterinary use of plants among tribals of Orissa. *Ancient Science of Life, 17,* 214–219.

Mokat, D.N. & Deokule S.S. (2004). Plants used as veterinary medicine in Ratnagiri district of Maharashtra. *Ethnobotany 16,* 131–135.

Mokat, D.N., Mane, A.V. & Deokule S.S. (2010). Plants used for veterinary medicine and botanical pesticide in Thane and Ratnagiri districts of Maharashtra. *J. Non-Timber Forest Products 17,* 473–476.

Murthy, E.N., Reddy, C.S., Reddy, K.N. & Raju, V.S. (2007). Plants used in ethnoveterinary practices by Koyas of Pakhal Wildlife Sanctuary, Andhra Pradesh, India. *Ethnobotanical Leaflets 11,* 1–5.

Murty, P.P. & Rao, G.M.N. (2012). Ethnoveterinary medicinal practices in tribal regions of Andhra Pradesh, India. *Bangladesh J. Plant Taxon. 19,* 7–16.

Nag, A., Praveen Galav, Katewa, S.S. & Shweta Swarnkar. (2009). Traditional herbal veterinary medicine from Kotda tehsil of Udaipur district (Rajasthan). *Ethnobotany 21,* 103–106.

Naidu, B.V.A.R. (2003). Ethnomedicine from Srikakulam district, Andhra Pradesh, India. PhD Thesis, Andhra University, Visakhapatnam.

Patil, D.A., Patil, P.S., Ahirrrao, Y.A., Aher, U.P. & Dushing, Y.A. (2010). Ethnobotany of Budhana district (Maharashtra: India): Plants used in veterinary medicine. *J. Phytol 2(12),* 22–34.

Patil, S.H. & Merat, M.M. (2003). Ethnoveterinary practices in Satpudas of Nandurbar district of Maharashtra. *Ethnobotany 15,* 103–106.

Prasanthi, S. (2010). Ethnobotany of the *Savaras.* PhD Thesis, Andhra University, Visakhapatnam.

Prusti, A.B. (2007). Plants used as ethnomedicinal by Bodo tribe of Malkangiri district, Orissa. *Ethnobotany 19,* 105–110.

Raja Reddy, K., Sudarsanam, G. & Gopal Rao, P. (1989). Plant drugs of Chittoor district, Andhra Pradesh, India. *Int. J. Crude Drugs 27,* 41–54.

Raju, M.P. (2009). Ethnobotany of the *Konda Reddis.* PhD Thesis, Andhra University, Visakhapatnam.

Ramadas, S.R., Ghotge, N.S., Ashalata, S., Mathus, N.P., Vivek Gour Broome, V.G. & Sanyasi Rao. (2000). Ethnoveterinary remedies used in common surgical conditions in some districts of Andhra Pradesh and Maharashtra, India. *Ethnobotany 12,* 100–112.

Reddy, C.S. & Raju, V.S. (2000). Folklore biomedicine for common veterinary diseases in Nalgonda district, Andhra Pradesh, India. *Ethnobotany 12,* 113–117.

Reddy, K.N., Bhanja, M.R. & Raju, V.S. (1998a). Plants used in ethnoveterinary practices in Warangal district, Andhra Pradesh, India. *Ethnobotany 10,* 75–84.

Reddy, K.N. & Raju, R.R.V. (1999). Plants in ethnoveterinary practices in Anantapur district, Andhra Pradesh. *J. Econ. Taxon. Bot. 23,* 347–357.

Reddy, K.N., Subbaraju, G.V., Reddy, C.S. & Raju, V.S. (2006). Ethnoveterinary medicine for treating livestock in Eastern Ghats of Andhra Pradesh. *Indian J. Trad. Knowl. 5,* 368–372.

Reddy, R.V., Lakshmi, N.V.N. & Raju, R.R.V. (1997). Ethnomedicine for ephemeral fevers and anthrax in cattle form hills of Cuddapah district, Andhra Pradesh, India. *Ethnobotany 9,* 94–96.

Reddy, R.V., Lakshmi, N.V.N. & Raju, R.R.V. (1998b). Folk veterinary medicinal plants in Cuddapah hills of Andhra Pradesh, India. *Fitoterapia 69,* 322–328.

Salave, A.P., Reddy, P.G. & Diwakar, P.G. (2011). Some reports on ethnoveterinary practices in Ashti areas of Beed district (M.S.) *Int. J. Appl. Biol. Pharm. Tech. 2,* 69–73.

Salave, A.P., Sonawane, B.N. & Diwakar Reddy, P.G. (2012). Traditional ethnoveterinary practices in Karanji Ghat areas of Pathardi tahasil in Ahmednagar district (M.S.), India. *Int. J. Plant, Animal Env. Sci. 2,* 64–69.

Sanyasi Rao, M.C., Varma, Y.N.R. & Vijayakumar (2008). Ethno-veterinary medical plants of the Catchments area of the river Papagni in the Chitoor and Anantapur districts of Andhra Pradesh, India. *J. Ethnobot Leaflets 12,* 217–226.

Satapathy, K.B. (2010). Ethnoveterinary practices in Jaipur district of Orissa. *Indian J. Trad. Knowl. 9,* 338–343.

Shashikanth, J., Ramachandra Reddy, P. & Padma Rao, P. (2011). Some indigenous folk-lore animal health care practices from Nalgonda district, Andhra Pradesh. *Ethnobotany 23,* 78–81.

Sudarsanam, G., Reddy, M.B. & Nagaraju, N. (1995). Veterinary crude drugs of Rayalaseema, Andhra Pradesh, India. *Int. J. Pharmacog. 33,* 1–9.

Suneetha, J., Prasanthi, S. & Seetharami Reddi, T.V.V. (2012). Plants in ethnoveterinary practices in East Godavari district, Andhra Pradesh. *J. Non-Timber Forest Products 19,* 63–68.

Swamy, N.S. (2009). Ethnobotanical Knowledge from Adilabad district, Andhra Pradesh, India, PhD Thesis, Andhra University, Visakhapatnam.

CHAPTER 9

ETHNOBOTANY OF USEFUL PLANTS IN EASTERN GHATS AND ADJACENT DECCAN REGION

M. CHANDRASEKHARA REDDY,[1] K. SRI RAMA MURTHY,[1] S. SANDHYA RANI,[2] and T. PULLAIAH[2]

[1]Department of Botany and Biotechnology, Montessori Mahila Kalasala, Vijayawada–520010, Andhra Pradesh, India, E-mail: chandra4bio@gmail.com drmurthy@gmail.com

[2]Department of Botany, Sri Krishnadevaraya University, Anantapur–515003, Andhra Pradesh, India, E-mail: pullaiah. thammineni@gmail.com; sandhyasakamuri@gmail.com

CONTENTS

ABSTRACT

Eastern Ghats is one of the richest of biodiversity centers in India. The hill ranges are discontinuous and separated by rivers, which flow through them. The *spermatophytes* of Eastern Ghats provide various useful products for human consumption, which include timber and non-timber forest products. The knowledge regarding traditional uses of various plant species is dwindling in recent days. It is very important to protect our indigenous traditional knowledge regarding the green wealth. Therefore, in the present paper the ethnobotanical uses regarding all useful plants other than ethnomedicine and ethnic food plants are given.

9.1   INTRODUCTION

The Eastern Ghats and Deccan region of Peninsular India harbors a rich diversity of ethnobotanical plants, which generate considerable benefits for social and economic perspectives. However in these days the traditional values of ethnobotanical species are difficult to reconcile with acute conflicts. There are many important sectors in developing the ethnobotany and there is no doubt that the plant kingdom is a treasure house of diverse natural products (Kala, 2007), such as medicine, food, aromatic compounds, dyes, timber, gums, resins, arrack, toddy, etc. The literature indicates that most traditional knowledge regarding the medicinal and edible plants was well documented in India, whereas the other uses of ethnoplant species have relatively limited documentation. This traditional knowledge is acquired due to the close interaction of the local communities with the forests and their products. In recent days due to rapid development, human beings are being attracted to luxurious life thus neglecting the traditional ethnic knowledge and they are adversely affecting the forests by ruthless destruction for personal use, industrialization and urbanization. Due to this the tribal people who are like gold mines of traditional knowledge regarding forest resources are also coming out of forests and attracted towards the urban culture leaving their knowledge. So it is very important to procure and document the knowledge regarding all the ethnobotanical uses of all groups of plants prior to its extinction for the benefit of future generation. Apart from this in recent days forests in Eastern Ghats region are under severe pressure for meeting growing demands for fuel, timber and other forest products from an ever increasing human and livestock population and industrial demands.

Forests of Eastern Ghats is a rich source for goods like wild food plants, honey, oils, gums, resins, gum-resins, dyes, wax, lac, fibers, fuel wood, charcoal, fencing material, brooms, wildlife products, raw materials like bamboo and cane for handicrafts, etc. besides the medicinal plants (Omkar et al., 2012). Earlier several botanists studied extensively the biodiversity and vegetation of Eastern Ghats. Rao (1998) studied vegetation and valuable plant resources that are found in the Eastern Ghats of Andhra Pradesh, India with a special note on conservation. Tree wealth and its prominent role in life and economy of tribal people living in the forests of Eastern Ghats, Andhra Pradesh were reported by Rani et al. (2003). Phytodiversity and useful plants of Eastern Ghats of Orissa with a special reference to the Koraput region was reported by Misra et al. (2009). The ethnomedicinal and ethnic food plants that are found in Eastern Ghats of India are also well reported (Ellis, 1992; Ramarao and Henry, 1996; Kadavul and Dixit, 2009; Parthipan et al., 2011; Ramasubbu et al., 2012; Krishnamurthy et al., 2014). It is equally important to document the other uses of plants (fiber, wood, dye, gums, resins, latex, arrack and toddy, etc.). An account of these different uses of plants used by the tribals in Eastern Ghats and Deccan region is given in the next section.

9.2   FIBER YIELDING PLANTS

The importance of fiber yielding plants has been considered next to the food plants in their usefulness in human society (Sahu et al., 2013). The use of the plant fibers was preferred from time immemorial due to its easy availability. The use of cotton fiber and silk is known since 5000 BC. *Boehmeria nivea, Crotalaria juncea, Corchorus capsularis, Gossypium arboretum, Hibiscus cannabinus, Linum usitatissimum* are the best-known commercial plants, which provide durable and flexible fiber. The utility of plant fibers is manifested in a diverse range of products, which includes making ropes, papers and various household materials. The fiber production also contributes significantly to the economy of the region in various ways including agricultural, clothing, small-scale industry and products for other household operations. It has been estimated that nearly 700 species yield fibers in India. However majority of the traditional fiber yielding plants remained underutilized because these uses are need based or site specific. However the plant fibers have specific qualities, such as thermal insulation, resistance to water and other desirable traits (Pandey and Gupta, 2003). Different plant parts are used for extraction of fiber like bark, leaf, stem and young shoots. It is interesting to note that there are 26 types of preparations, which were used in combinations of different parts. Use of bark as fiber is more frequent, due to the presence of long soft tissues.

9.3 DYE YIELDING PLANTS

Dyes are the natural or synthetic compounds used to add a color or to change the color of materials. Dyes are capable of being fixed to materials and do not wash out with detergents and water or fade easily on exposure to light (Rashid, 2013). In the human civilization plants have been used not only for the basic needs of life, such as food, fiber, fuel, clothing and shelter but also as sources of natural dyes for dyeing cloths, design and painting. A spectrum of beautiful natural colors ranging from yellow to black exists in the plant sources. These colors are exhibited by various organic and inorganic molecules and their mixtures (Das and Mondal, 2012a). The indigenous knowledge system associated with extraction and processing of natural dyes from plants is ancient process (Antima et al., 2012). The invention of indigo, the most important Indian natural dye, is as old as the textile marketing itself. The natural dyes are environmentally friendly, for example turmeric, the brightest of naturally occurring yellow dye, is a powerful antiseptic, which revitalizes the skin. Throughout history people have been using natural dyes for their textiles and other materials like leather, cosmetics, inks, etc. (Tiwari and Bharat, 2008), by using common locally available plants. Many natural dyestuff and stains were obtained from plants and dominated as sources of coloring producing different color like red, yellow, blue, black and a combination of these. Nature has gifted us with more than 500 dye-yielding plant species (Mahanta and Tiwari, 2005). Many of these plants have been identified as potentially rich in natural dye contents and some of them have been used in natural dyeing like Kalamkari and Lacquering toys. Almost all parts of the plants like leaves, flowers, roots, berries, bark, rhizomes, tubers, shoots, sap, wood, etc. produce dyes (Gokhale et al., 2004). Some plants have given more than one color depending upon the parts of the plant, which are used. The shade of the color a plant produces will vary according to time of the year the plant is picked, how it was grown, soil conditions, etc. The minerals in the water used in a dye bath can also alter the color. Some natural dyes contain natural mordants to hold fast the dye and to prevent them from touching the cloth (Das and Mondal, 2012b)

It is interesting to note that over 2000 pigments are synthesized by various parts of plants of which only about 150 have been commercially exploited (Siva, 2007). In India, there are more than 450 plants that can yield dyes. In addition to their dye-yielding characteristics, some of these plants also possess medicinal value (Chandramouli, 1995; Chengaiah et al., 2010). Among these more than 200 species are found in Eastern Ghats itself (S.S.C. Reddy et al., 2002; Krishnamurthy et al., 2014) of which 50 are considered to be the most important. The colors thus derived from these plants are fast colors and give strength to the cloth, making the cloth durable for years together and gives a shine as years pass on. The following plants have been extensively used in Kalamkari for coloring the fabrics on the mordanted cloth. *Albizia odoratissima* bark yields brown shades, *Rubia cordifolia* roots give red color, *Oldenlandia umbellata* roots, *Morinda citrifolia* roots, *Ventilago maderaspatana* gives red color, *Woodfordia fruiticosa* leaves are used as lavelling agent, *Punica granatum* leaves produce the olive green color on alum mordanted cloth after boiling, *Acacia catechu* bark gives rich reddish brown color after boiling on alum mordanted cloth, *Indigofera tinctoria* leaves yield blue color, *Terminalia chebula* flowers are used for producing the yellow color, *Butea monosperma* flowers give yellow color to the cotton cloth, *Curcuma longa* rhizome gives yellow, *Nyctanthus arbor-tristis* corolla tube of flowers yield orange color, *Cedrella toona* flowers yield yellow color, *Soymida febrifuga* bark yield brown shades and *Symplocos racemosa* bark yield brown shades. Few plant parts are also used as mordants and these include *Symplocos racemosa* bark and leaves, *Bixa orellana* bark, *Terminalia chebula* bark and fruit, *Curcuma longa* rhizome, *Woodfordia fruticosa* flowers, *Cassia fistula* bark, *Erythrina indica* aqueous extract, etc. (Rani et al., 2002; Pullaiah and Rani, 1999).

The wooden toys of Etikoppaka have earned the name for their exquisite craftman ship. Wood of *Wrightia tinctoria* (locally known as Ankudu, Reppala) and rarely wood of *Millingtonia hortensis* are used in toy manufacturing. Dyes (synthetic as well as plant based) have been used in lacquering of toys. Usage of vegetable dyes is very meager when compared to synthetic dyes in dyeing of toys. The following plants have been used for extracting the dyes, for lacquering of toys: *Indigofera tinctoria, Lawsonia inermis, Mallotus philippensis, Vetiveria zizanioides, Centella asiatica, Acacia catechu, Punica granatum, Terminalia arjuna, Terminalia chebula, Curcuma longa, Phyllanthus emblica,* etc. (Rani et al., 2002).

9.4   GUMS AND RESINS

Various gums, gumresins, latex, oleoresins and resins are obtained from various parts of different plants found in the forests of Eastern Ghats and

Deccan. Many of them are little known and some may prove useful and valuable articles both for medicinal (Sravani et al., 2014) and commercial purposes.

Gums obtained from plants are solids consisting of mixtures of polysaccharides (carbohydrates) which are either water-soluble or absorb water and swell up to form a gel or jelly when placed in water. They are insoluble in oils or organic solvents, such as hydrocarbons, ether and alcohol. The mixtures are often complex and on hydrolysis yield simple sugars, such as arabinose, galactose, mannose and glucuronic acid. Some gums are produced by exudation, usually from the stem bark of a tree or shrub but in a few cases from the root. The exudation is often considered to be a pathological response to injury to the plant, either accidental or caused by insect borers or by deliberate injury (tapping). Seed gums are those isolated from the endosperm portion of some seeds (Coppen, 1995).

Plant gums originating from many countries have been an important item in international trade for centuries in food, pharmaceuticals, paper, textile and other industries. Depending upon their major use, plant gums may be broadly classified as food and non-food or technological grade gums. The former can be used as food additives in various kinds of confectioneries, foods and beverages and include gum arabic, gum tragacanth, gum karaya and gum carob. The latter category finds its major use in non-food industrial applications and include 'gum ghatti,' 'gum talha' and a variety of other gums.

The term gum resin is occasionally found in the literature but it has no precise meaning (and is best avoided) although it is generally used to describe a resinous material which contains some gum. The coagulated part of some commercially important latex, such as chicle and jelutong are often referred to as non-elastic gums or masticatory (chewing) gums, but they are not gums in the proper sense of the word (Coppen, 1995).

Latex – a fluid, usually milky white in color, consists of tiny droplets of organic matter suspended or dispersed in an aqueous medium. The most well-known example is rubber latex, in which the solids content is over 50% of the weight of the latex. The solids can normally be coagulated to form a solid mass by boiling the latex. The principal components of the coagulum are cis or transpolyisoprenes and resinous material. If the polyisoprene is mainly cis, it confers elasticity to the solid and makes it rubber-like; if it is mainly trans, the solid is non-elastic and gutta-like. Latexes are usually obtained by cutting the plant to make it bleed. Latex-yielding plants occur in fewer families than those which produce gums and resins -Apocynaceae,

Euphorbiaceae and Sapotaceae are among the important ones (Coppen, 1995).

A resin, because of a high content of volatile oil, is softer than one which contains little or no oil. The term is, nevertheless, sometimes shortened to resin when describing soft resins. (The term is also used in another context to describe prepared extracts of spices or other plant materials – after evaporation of the solvent used to extract the spice a soft extract, or oleoresin, remains).

A solid or semi-solid material, usually a complex mixture of organic compounds and which is insoluble in water but soluble in certain organic solvents are called terpenes. Oil-soluble resins are soluble in oils and hydrocarbon-type solvents; spirit-soluble resins are soluble in alcohols and some other solvents. Resins are very widely distributed in the plant kingdom although a few families are notable in accounting for a large proportion of the resins of commerce (e.g., Leguminosae, Burseraceae and Pinaceae). Resins can occur in almost any organ or tissue of the plant; a few (such as lac) are produced from insects. Most resins of commerce are obtained as exudates by tapping (Coppen, 1995).

9.5   WOOD YIELDING PLANTS

Forests produce a wide range of services that are essential to human well-being and one major financial output consists of timber that can be used for a variety of manufacturing building, fuel, and other materials. Timber is harvested from forest ecosystems annually in a wide range. India is blessed with a variety of timber yielding tree species and as many as 1500 species are commercially utilized for diverse purposes. Among those it is no wonder that sandalwood which is available in India is the second most expensive wood in the world, next to the African Blackwood (*Dalbergia melanoxylon*). The carved images of gods and mythological figures have a high demand in the market. A wide variety of articles, such as boxes, cabinet panels, jewel cases, combs, picture frames, hand fans, pen holders, card cases, letter openers and bookmarks are made from sandalwood (Kumar et al., 2012). *Pterocarpus santalinus*, commonly known as Red sanders, belongs to the family Fabaceae. It is endemic to Eastern Ghats of India and considered globally endangered, with illegal harvest being a key threat. The plant is known for its characteristic timber of exquisite color, beauty and superlative technical qualities. The red wood yields a natural dye santalin, which is used in coloring pharmaceutical preparations and foodstuffs (Arunakumara et al., 2011). Next to the above-mentioned two species Teak (*Tectona grandis*) wood played a prominent role. It is widely used in preparing doors, windows and other house hold furniture (Shah et al., 2007).

Some of the other important tree species grown in India are *Azadirachta indica, Eucalyptus* spp., *Acacia* spp., *Dalbergia sissoo, Swietenia* spp., *Casuarina* spp., *Melia dubia, Ailanthus excelsa, Leucaena leucocephala,* etc. Productivity of forests in general and particularly that of commercial forest plantations is very much affected by frequent outbreak of pests and diseases, besides human interventions and various natural calamities. The total production of timber in India from forests is reported at an average 2.3 million cu.m in 2010. The wood and wood products imports to India have gradually increased since 1998 and have reached 6.3 million m³ in 2011 with a total import value of Rs. 9800 crores. Though wood is imported from about 100 countries, six countries namely Malaysia, Myanmar, New Zealand, Ghana, Ivory Coast, and Gabon constitute bulk of the timber imports to India (about 80%). Teak constitutes about 15% of total timber imports to India and the major teak exporting countries to India include Myanmar, Ivory Coast, Ghana, Ecuador, Costa Rica and Benin.

Fuel wood is the main source of energy in the developing world. The use of wood by mankind for energy purposes is as old as human civilization itself. One of the most serious problems in the developing world is shortage of fuel wood. The total fuel-wood consumption estimated in household sector is 248 million m³ and about 13 million m³ additional fuel-wood is consumed in hotels and restaurants, cottage industries and cremation of dead human bodies. This makes the total annual consumption of fuel-wood to be 261 million m³, which comes from different sources. The production of fuel-wood from forests has been estimated to be 52 million m³ (FSI, 2009) and remaining 209 million m³ from farmland, community land, homestead, roadside, canal side and other wastelands (ICFRE, 2010). India produces about 23.19 million m³ of timber log domestically and imports nearly 20% of its requirement from countries, such as Malaysia (57%) and Myanmar (18%). It is very important to be aware of the timber yielding plants and rapid growing plants which will give wood in short period to reduce timber import in future.

9.6   SACRED PLANTS

India is famous for its religious culture. During religious ceremonies a wide variety of plants are used because of their holy nature. Some plants are treated as gods and goddesses and worshipped while some species are used for doing pooja for Gods. People also believe in Kalpavriksa, for example, a tree fulfilling all human desires. In India coconut plant is known as *kalpavriksa*. One of the most common offerings in Indian temples is a coconut. It is also offered on occasions like weddings, festivals, the use of new vehicle, house, etc. The coconut is broken and placed before the Lord. It is later distributed as *prasada*. Coconut fruit is also offered as *tambulum* along with betel leaves and areca nuts. Even in *purnakumbha* a coconut is part of *kalasha*. Plants are considered sacred because of their close association with a deity (Bilva with Lord Shiva and Tulasi with Lord Krishna). Some plants are believed to have originated from bodies or limbs of Gods and hence the sanctity (*Butea* is believed to have originated from the body of Lord Brahma). Some plants became sacred owing to their association with great individuals. (Peepal under which Gouthama Budha attained enlightment is considered sacred by the Buddhists) (Reddy and Krishnaveni, 2014).

Indian women offer leaves of *Mangifera, Prosopis, Ocimum, Aegle,* etc. to God in different vratas and worship to give health and wealth. They also make *pradakshinas* (go around certain number of times) around the *Ficus benghalensis* and pray for the longevity of their husbands and for fulfillment of their wishes. In India, different sides of the houses are associated with different plants. *Ficus benghalensis* tree on the eastern side, *Ficus religiosa* in the south, *Cocos nucifera* in the east is always auspicious. *Mangifera indica* is auspicious at every place and believe that it gives wealth. Religious importance of trees can be seen from the birth to marriage. People believe that God has bestowed some specific power to certain plants, like *Ficus religiosa, Azadirachta indica, Ocimum tenuiflorum* (Syn.: *O. sanctum*), *Curcuma longa, Centella asiatica, Cynodon dactylon* which have divine qualities. Therefore these plants are used in a number of religious activities (Robinson and Cush, 1997).

9.7 ORNAMENTAL AND DECORATIVE PLANTS

Most of the present day flowers have come from the wild progenitors, a few of which still exist in natural habitat (Thomas et al., 2011). Along with the established ornamentals like rosa, primula, senecio, rhododendron, orchids, etc., many other beautiful herbs, shrubs, and trees have been introduced in the gardens. Nature has given a wealth of wild flower and ornamental plants, unfortunately many of them have been destroyed to such an extent that several have become extinct and survival of many is endangered by over exploitation by human beings (Arora, 1993). Ornamental horticulture is the functional and esthetic integration of wild flowering and beautiful plants into commercial important plants for use in landscape. The wild vegetation of Eastern Ghats and adjacent Deccan region is blessed with rich and fascinating plant species and holds a large number of curious, botanically interesting, exquisite, economically important, rare, threatened, endangered and endemic plants (Pullaiah et al., 2007). This region is a huge repository of plants of botanical importance and a reservoir of genetic variability, ecosystem diversity and species diversity.

Hundreds of wild plants from India have found their way into many European botanical gardens where they have been much appreciated. Eastern Ghats of India possess potential of raw ornamentals from wild sources. Some of the ornamental plants which are under usage in floriculture and landscaping originally hailed from Eastern Ghats and Deccan. The indigenous floristic wealth in our country proclaims its own significance and is deeply involved in our culture, literature, socio economic life, romance and poetry. It is very much interesting but a bit precarious task to incorporate such ornamental wild plants and flowers into floriculture trade. The prime source of introduction of these plants would be in the botanic gardens which can very well coordinate and exchange seeds and plant material. It is more important to collect such plants, which are not available commercially and found in nature only (Sharma and Goyal, 1991). Extensive literature survey on these wild horticultural ornamental plants was made and a fair number of wild plants that have great economic importance have been listed.

9.8   TENDU LEAF AND ECONOMY

The deciduous forest species *Diospyros melanoxylon* Roxb.is quite important as it yields valuable tendu leaves which are used for making bidis (for rolling local cigarettes). Tendu leaves make excellent wrappers, and the success of the beedi is due, in part, to this leaf (Goud et al., 1997; Lal, 2009). The leaves are in abundance shortly after the tobacco crop is cured and so are ready to be used in beedi manufacture. Collected in the summer and made into bundles, the leaves are dried in the sun for 3–6 days before being used as wrappers (Gupta, 1992). Throughout India, collection of tendu leaf (*D. melanoxylon*) generates part time employment for 7.5 million people – a majority of them tribal women (Arnold, 1995). *Bauhinia racemosa* leaves are also used for making *bidis*.

9.9   ADDITIONAL NON-TIMBER FOREST PRODUCTS

Along with the above-mentioned useful plants, some other plants were also used by the tribals living in the forests of Eastern Ghats hill ranges. Leaves, roots, bark and seeds of many plants have wider uses and help the economy of tribal's. A number of tribal families are dependent on these products for their livelihood. They collect these products and sell them in markets nearby forests or to Girijan Cooperative Corporation limited. The roots of *Decalepis hamiltonii* and *Hemidesmus indicus* are collected by the forest tribes and dried in houses, made into pieces and directly sold in tribal markets or they prepare health drink from them and sell it in bazaars. Leaves of some species like *Bauhinia vahlii* and *Butea monosperma* are collected from forests and used them to make meal plates after drying. Some species (like *Semecarpus anacardium*, *Madhuca longifolia* and *Borassus flabellifer*) are used for collection of arrack and toddy (intoxicating drink), which are used by tribes for drukenness. Several other species are used for various other purposes, like brooms and broom sticks preparation. Some species are used as pesticides and as fertilizers while some are used to protect stored grains from pests in houses and for other purposes like toothbrush, shampoos, detergents, etc.

In the present paper the details regarding some of the important traditional useful plants that are available in forests of Eastern Ghat hill ranges and adjacent Deccan region of Peninsular India are listed with their botanical names, parts used, purpose of use in Table 9.1.

TABLE 9.1   Ethnobotanical Uses of Plants

S. No.	Species	Part	Use	Reference
1	*Abutilon indicum* G. Don	Stem	Fiber	Basha et al. (2011)
			Tooth brush	Behera and Nath (2012)
2	*Abutilon hirtum* (Lam.) Sweet	Flowers	Ornamental shrub	Reddy et al. (2012)
3	*Acacia arabica* (Lam.) Willd.	Stem bark	Gum	Basha et al. (2011)
4	*Acacia chundra* (Rottl.) Willd.	Wood	Dal stirrer (*pappu gittae*), finger millet food preparation stirrer (*teddu katte*) and cots	Reddy et al. (2008)
		Stem bark	Gum	Basha et al. (2011)
		Heart wood	Brown dye	Rani et al. (2002)

TABLE 9.1 *(Continued)*

S. No.	Species	Part	Use	Reference
5	*Acacia leuco-phloea* (Roxb.) Willd.	Stem bark	Gum	Basha et al. (2011)
		Leaves, bark	Red dye	Rani et al. (2002)
6	*Acacia catechu* (L.f.) Willd.	Plant	Women worship it to remove *kujadosha*.	Reddy and Krishnaveni (2014)
7	*Acacia nilotica* (L.) Willd. ex Del.	Stem bark	Gum	Basha et al. (2011)
		Bark, fruits	Yellow-brown, black dye	Rani et al. (2002)
		Stem leaves	Agricultural implements	Behera and Nath (2012)
8	*Acacia farnesi-ana* (L.) Willd.	Bark, fruits	Yellow dye	Rani et al. (2002)
		Tree	Ornamental in gardens	Reddy et al. (2012)
9	*Acaca leuco-phloea* (Roxb.) Willd.	Leaves, bark	Red dye	Rani et al. (2002)
10	*Acacia pennata* Willd.	Bark	Brown, black dye	Rani et al. (2002)
11	*Acacia plani-frons* Wight & Arn.	Fruits	Brown, black dye	Rani et al. (2002)
12	*Acacia sinuata* (Lour.) Merr.	Bark	Brown dye	Rani et al. (2002)
13	*Achyranthus aspera* L.	Whole plant	Dye	Rani et al. (2002)
14	*Adenanthera pavonia* L.	Wood	Red dye	Rani et al. (2002)
15	*Aegle marmelos* (L.) Cor.	Bark	Red dye	Rani et al. (2002)
		Leaves	Offered to Siva in Mahasivarathri.	Reddy and Krishnaveni (2014)
16	*Agave ameri-cana* L.	Leaves	Fibers – ropes, canes, shepherd and cowherd's carrier boxes	Reddy et al. (2008)
				Basha et al. (2011); Sahu et al. (2013c)
			Fibers – Coir	Rekha and Kumar (2014)
			Laxative dye	Rani et al. (2002)
17	*Agave angusti-folia* Haw.	Leaves	Fibers – Coir	Rekha and Kumar (2014)
			Detergent	Prabakaran et al. (2013)
18	*Aglaia rox-burghiana* Hiern	Wood	Brown dye	Rani et al. (2002)

TABLE 9.1   *(Continued)*

S. No.	Species	Part	Use	Reference
19	*Alangium salvifolium* L.	Young twigs	Tooth brush	Rekha and Kumar (2014)
20	*Albizia odoratissima* (L.f.) Benth.	1. Wood 2. Stem bark	1. Agricultural tools and furniture 2. Boil with toddy for purification and doubling the activity	Naidu and Khasim (2010)
		Bark	Brown dye	Rani et al. (2002)
		Wood	Doors, cots and windows	Rekha and Kumar (2014)
21	*Albizia amara* Boivin.	Leaves powder	Shampoo	Rekha and Kumar (2014); Prabakaran et al. (2013)
22	*Albizia lebbeck* Benth.	Bark	Brown, black dye	Rani et al. (2002)
		Woody stem	Fuel	Rekha and Kumar (2014)
23	*Albizia procera* Benth.	Bark	Black dye	Rani et al. (2002)
24	*Aleuritis moluccana* Willd.	Roots	Brown dye	Rani et al. (2002)
25	*Allium cepa* L.	Bulb	Yellow to orange dye	Rani et al. (2002)
26	*Aloe barbadensis* Mill.	Whole plant	Red dye	Rani et al. (2002)
27	*Althaea rosea* Cav.	Flowers	Red dye	Rani et al. (2002)
28	*Amaranthus paniculatus* L.	Ash	Ash color dye	Rani et al. (2002)
29	*Amaranthus spinosus* L.	Plant	Auxiliary	Rani et al. (2002)
30	*Anacardium occidentale* L.	Nuts	Black dye	Rani et al. (2002)
		False fruits	Alcoholic preparation	Naidu and Khasim (2010)
		Seed	Oil	Basha et al. (2011)
31	*Andrographis echioides* Nees	Whole plant powder	Wash hair to prevent hair loss	Rekha and Kumar (2014)
32	*Annona reticulata* L.	Leaves	Black dye	Rani et al. (2002)
33	*Annona squamosa* L.	Fresh flowers paste	Wash hairs	Rekha and Kumar (2014)
34	*Anisochilus carnosus* (L.f.) Wall.	Plant	Ornamental in parks and gardens	Reddy et al. (2012)

TABLE 9.1 *(Continued)*

S. No.	Species	Part	Use	Reference
35	*Anogeissus latifolia* (Roxb. ex DC.) Wall ex Guill.	Wood	Charcoal, harrow	Sahu et al. (2013c); Rekha and Kumar (2014)
		Stem	Gums	Goud et al. (1997); Basha et al. (2011)
		Fruit	Yellow dye	Rani et al. (2002)
36	*Aphanamixis polystachya* (Wall.) Parker	Bark	Dark green dye	Rani et al. (2002)
37	*Ardisia solanacea* Roxb.	Fruits	Yellow	Rani et al. (2002)
38	*Areca catechu* L.	Nuts	Brown dye	Rani et al. (2002)
		Plant	Worship women to remove *kujadosha*.	Reddy and Krishnaveni (2014)
39	*Artemisia pallens* Wall.	Leaves	Mixed with ghee and used as incense to attract positive power	Reddy and Krishnaveni (2014)
40	*Aristida funiculata* Trin. & Rupr.	Dried Inflorescences	Brooms	Naidu and Khasim (2010)
41	*Artocarpus hirsutus* Lam.	Mature stem	Churn-staff	Rekha and Kumar (2014)
42	*Atylosia scarabaeoides* (L.) Benth.	Leaves	Oil	Basha et al. (2011)
43	*Azadirachta indica* A.Juss.	Young twigs, wood	Tooth brush, furniture and other implements	Sahu et al. (2013c); Rekha and Kumar (2014)
		Leaves, bark, twigs, fruits	As pesticides, fertilizers	Behera and Nath (2012)
		Whole plant	Worshiped for good health	Reddy and Krishnaveni (2014)
44	*Bambusa arundinacea* (Retz.) Willd.	Pounded, dried tillers, culms	Beverage (like tea) mats, baskets and supporting rods for roof	Naidu and Khasim (2010)
		Split culms	Woven into mats, baskets and fans	Rekha and Kumar (2014)
45	*Bambusa nutans* Wall. ex Munro	Leaves, culms	Construction, fencing, decorative items	Behera and Nath (2012)
46	*Bambusa vulgaris* Schrad. ex J.C. Wendl.	Leaves, young shoot	Baskets	Behera and Nath (2012)

TABLE 9.1   *(Continued)*

S. No.	Species	Part	Use	Reference
47	*Bassia longifolia* L.	Seeds	Oil	Basha et al. (2011)
48	*Barringtonia acutangula* (L.) Gaertn.	Flowers pale pink or red	Ornamental	Pullaiah and Rani (1999)
49	*Bauhinia purpurea* L.	Bark	Brown dye	Basha et al. (2011)
50	*Bauhinia tomentosa* L.	Woody stem	Fuel	Rekha and Kumar (2014)
51	*Bauhinia vahlii* Wight & Arn.	Leaves	Meal plates	Reddy et al. (2008)
		Stem	Fiber-ropes, cordage	Naidu and Khasim (2010); Sahu et al. (2013c)
52	*Bauhinia variegata* L.	Twigs	Carried with new born child in traveling to repel evil sprits	Reddy and Krishnaveni (2014)
53	*Bixa orellana* L.	Pericarp and seeds	Red, yellow dyes	Basha et al. (2011)
54	*Bombax ceiba* L.	Flux	Stuffing pillow	Reddy et al. (2008)
		Bark	Cushions	Behera and Nath (2012)
		Flowers blood red	Ornamental	Pullaiah and Rani (1999)
55	*Borassus flabellifer* L.	Leaves, male inflorescence	Umbrellas, dolls, writing pads and fans (*visana-karra*), crackers	Reddy et al. (2008)
		Cut female spadices, ramenta (browny hairs) of young leaves, stem, leaves	Toddy (intoxicating drink), cots, house construction, baskets and roofing huts	Naidu and Khasim (2010)
		Petiole	Fiber – ropes	Basha et al. (2011)
56	*Boswellia ovalifoliolata* Balak. & Henry	Stem bark	Oleogum-resin, gum – burnt and to spread fumes in home	Basha et al. (2011)
57	*Boswellia serrata* Roxb.	Stem bark	Gum	Goud et al. (1997)
58	*Bridelia crenulata* Roxb.	wood	Fire wood	Prabakaran et al. (2013)

TABLE 9.1 *(Continued)*

S. No.	Species	Part	Use	Reference
59	*Buchanania angustifolia* Roxb.	Woody stem	Fuel wood	Rekha and Kumar (2014)
60	*Butea monosperma* (Lam.) Taub.	Leaves	Meal plates	Reddy et al. (2008)
			Mulching	Behera and Nath (2012)
			Beedi (local cigarette) wrappers	Naidu and Khasim (2010)
		Stem bark, flowers	Gum, yellow and orange dyes	Basha et al. (2011)
		Red flowers	Offered to Gods in various religious activities	Reddy and Krishnaveni (2014)
		Flowers red	Ornamental	Pullaiah and Rani (1999)
61	*Calamus rotang* L.	Plants	Basket and vessels	Reddy et al. (2008)
62	*Calotropis gigantea* (L.) R.Br.	Bark	Fiber – ropes	Basha et al. (2011); Naidu and Khasim (2010)
		Leaves, flowers	Offered to Lord Shiva and Hanuman for blessings	Reddy and Krishnaveni (2014)
		Latex	Pesticide	Behera and Nath (2012)
63	*Calophyllum inophyllum* L.	Plant	Scan evil spirits, counteracting domination influence on evil spirits	Reddy and Krishnaveni (2014)
64	*Canavalia gladiata* (Jacq.) DC.	Flowers	Indoor ornamental herb	Reddy et al. (2012)
65	*Canavalia virosa* Wight & Arn.	Leaves	Oil	Basha et al. (2011)
66	*Canthium dicoccum* (Gaertn.) Teijsm. & Binn.	Wood	Wood carving, Fire wood	Prabakaran et al. (2013)
67	*Caralluma lasiantha* (Wight) N.E. Br.	Entire plant	Ornamental succulent	Reddy et al. (2012)
68	*Caralluma umbellata* Haw	Plant	Indoor plant	Reddy et al. (2012)
69	*Careya arborea* Roxb.	Wood, leaves	Furniture, fiber	Naidu and Khasim (2010); Sahu et al. (2013c)

TABLE 9.1   *(Continued)*

S. No.	Species	Part	Use	Reference
70	*Caryota urens* L.	Cut inflorescence stalk	Toddy	Naidu and Khasim (2010); Sahu et al. (2013c)
71	*Cassia fistula* L.	Leaves	Ripening of fruits	Rekha and Kumar (2014)
		Wood	Agricultural tools, fire wood	Prabakaran et al. (2013)
		Twigs	Kept in houses to keep away evil sprits	Reddy and Krishnaveni (2014)
		Flowers bright yellow	Ornamental	Pullaiah and Rani (1999)
72	*Cassia siamea* Lam.	Wood	Fire wood	Prabakaran et al. (2013)
73	*Cassine glauca* (Rottb.) Kuntze	Wood	Fire wood	Prabakaran et al. (2013)
74	*Chloroxylon swietenia* DC.	Wood	Agricultural implements and furniture	Naidu and Khasim (2010); Prabakaran et al. (2013)
		Mature stem, wood	Pounder, handle of axe	Rekha and Kumar (2014)
75	*Chomelia asiatica* Kuntze.	Woody stem	Fuel	Rekha and Kumar (2014)
76	*Chrysopogon zizanioides* L.	Roots and rachis	Mat making, brooms	Behera and Nath (2012)
77	*Citrus medica* L.	Fruits	Dandruff, kill lice	Rekha and Kumar (2014)
78	*Cleistanthus collinus* Benth.	Wood	Doors and windows	Rekha and Kumar (2014)
		Leaves	Fertilizer	Prabakaran et al. (2013)
79	*Cleome chelidonii* L.f. var. *pallai* V.S. Raju & C.S. Reddy	Seeds	Condiment	Reddy et al. (2006)
80	*Cocculus hirsutus* (L.) Diels	Plants	Pot stands, brooms, baskets, mouth baskets for bullocks	Reddy et al. (2008)
81	*Cochlospermum religiosum* (L.) Alston	Stem bark	Gum	Naidu and Khasim (2010); Basha et al. (2011); Goud et al. (1997)
82	*Cocos nucifera* L.	Veins, petiole, fiber-coir	Brooms, fish hunting instruments (*maavu*), rope making, foot mats and in beds preparation	Reddy et al. (2008); Prabakaran et al. (2013)

TABLE 9.1   *(Continued)*

S. No.	Species	Part	Use	Reference
83	*Commiphora caudata* (Wight &Arn.) Engl.	Stem	Gum	Basha et al. (2011)
84	*Corchorus aestuans* L.	Stem	Fiber	Basha et al. (2011)
85	*Corchorus olitorius* L.	Stem	Fiber – rope	Reddy et al. (2008)
86	*Corchorus trilocularis* L.	Stem	Fiber	Basha et al. (2011)
87	*Cordia macleodii* Hook.f. & Thoms.	Wood	Cots	Reddy et al. (2008)
88	*Cordia wallichii* G. Don	Wood	Doors, cots and windows	Rekha and Kumar (2014)
89	*Corypha umbraculifera* L.	Split leaves	Baskets	Rekha and Kumar (2014)
90	*Crataeva adansonii* DC.	Flowers	Ornamental tree	Reddy et al. (2012)
91	*Crotalaria juncea* L.	Bark	Fiber – gunny bags, carry bags, fishing nets	Reddy et al. (2008)
92	*Crotalaria laburnifolia* L.	Stem	Fiber	Basha et al. (2011)
93	*Crotalaria pulcherrima* Roxb.	Stem	Fiber	Basha et al. (2011)
94	*Crotalaria retusa* L.	Stem	Fiber	Basha et al. (2011)
95	*Crotalaria spectabilis*	Stem	Fiber	Sahu et al. (2013c)
96	*Crotalaria verrucosa* L.	Stem	Fiber	Basha et al. (2011)
97	*Cucurbita maxima* Duchesne	Dried fruits	Musical instrument – Tambura	Naidu and Khasim (2010); Reddy et al. (2008)
98	*Curcuma longa* L.	Rhizome	Paste is applied on face and body of the bride and groom for getting blessings on the day of marriages and other rituals.	Reddy and Krishnaveni (2014)
99	*Cyamopsis tetragonoloba* Taub.	Stem	Gum	Basha et al. (2011)

TABLE 9.1   *(Continued)*

S. No.	Species	Part	Use	Reference
100	*Cymbopogon coloratus* (Hook.f.) Stapf	Plant	Ornamental in lawns	Reddy et al. (2012)
101	*Cynodon dactylon* (L.) Pers.	Roots	Soaked in oil and used for head	Rekha and Kumar (2014)
		Grass	To pray Lord Ganesha	Reddy and Krishnaveni (2014)
102	*Cyperus rotundus* L.	Leaves and branches	Roofing and thatching	Rekha and Kumar (2014)
103	*Dalbergia lanceolaria* L.	Wood	Doors and windows	Rekha and Kumar (2014)
104	*Dalbergia latifolia* Roxb.	Mature stem	Pounder	Rekha and Kumar (2014)
105	*Dalbergia sissoo* DC.	Plant	Planted near houses and believed that it protects home from natural lighting strokes	Reddy and Krishnaveni (2014)
106	*Decaschistia crotonifolia* Wight &Arn.	Stem	Fiber	Basha et al. (2011)
107	*Decaschistia cuddapahensis* Paul & Nayar	Stem	Fiber	Rao and Pullaiah (2007)
108	*Decaschistia rufa* Craib	Stem	Fiber	Rao and Pullaiah (2007)
109	*Dendrocalamus strictus* (Roxb.) Nees	Culms	Chairs, grain store boxes, mulberry rearing trays, flower vases, water and tea cups, ladder, pipe for medicine for cattle, fans (*visanakarra*), arrows and arrow sticks	Reddy et al. (2008); Sahu et al. (2013c)
		Stem	Agricultural tools, thatching	Prabakaran et al. (2013)
110	*Desmodium oojeinensis* (Roxb.) Ohasi	Timber	Agriculturaal implements	Sahu et al. (2013c)
111	*Desmostachya bipinnata* Stapf	Leaves	Roof thatching	Behera and Nath (2012)
112	*Dillenia indica* L.	Flowers white, fragrant	Ornamental	Pullaiah and Rani (1999)

TABLE 9.1   *(Continued)*

S. No.	Species	Part	Use	Reference
113	*Dillenia pentagyna* Roxb.	Flowers yellow	Ornamental	Pullaiah and Rani (1999)
114	*Dioscorea oppositifolia* L.	Whole plant	Arched creeper	Reddy et al. (2012)
115	*Diospyros ebenum* J.Koen.	Wood	Doors and windows	Rekha and Kumar (2014)
116	*Diospyros ferrea* (Willd.) Bakh. var. *buxifolia* (Rottb.) Bakh.	Wood	Fire wood	Prabakaran et al. (2013)
117	*Diospyros melanoxylon* Roxb.	Wood ash, ripe fruits, leaves, wood	Detergent, edible, beedi wrapper, plow, insect repellant	Naidu and Khasim (2010); Sahu et al. (2013c)
118	*Diospyros montana* Roxb.	Woody stem	Fuel	Rekha and Kumar (2014)
119	*Dodonaea viscosa* L	Branches	Brooms, thatching	Reddy et al. (2008)
120	*Eleusine coracana* Gaertn.	Inflorescence	Pillows	Reddy et al. (2008)
121	*Entada pursaetha* DC.	Seeds	Paper weight	Reddy et al. (2008)
122	*Eriolaena quinquelocularis* (Wight &Arn.) Cleghorn	Stem bark	Fiber –ropes	Reddy et al. (2006)
123	*Erythrina variegata* L.	Bark leaves, flowers	Red dye	Basha et al. (2011)
124	*Erythroxylum monogynum* Roxb.	Wood	Fire wood	Prabakaran et al. (2013)
125	*Evolvulus alsinoides* (L.) L.	Whole plant	By worshipping this plant life would be lengthened	Reddy and Krishnaveni (2014)
126	*Euphorbia hirta* L.	Latex	Fodder	Naidu and Khasim (2010)
127	*Ficus benghalensis* L.	Leaves	Meal plates	Reddy et al. (2008)
		Aerial roots	Tooth brush	Rekha and Kumar (2014)
		Wood	Agricultural tools	Prabakaran et al. (2013)
		Tree	Associated with planet Saturn and women worship it for the longevity of their husbands.	Reddy and Krishnaveni (2014)

TABLE 9.1 *(Continued)*

S. No.	Species	Part	Use	Reference
128	*Ficus glomerata* Roxb.	Young twigs	Tooth brush	Rekha and Kumar (2014)
129	*Ficus racemosa* L.	Tree	Believed to have mystic powers by worshipping this	Reddy and Krishnaveni (2014)
130	*Ficus religiosa* L.	Leaves	Greeting cards	Reddy et al. (2008)
		Young twigs	Tooth brush	Rekha and Kumar (2014)
		Tree	Believed that, sitting under this tree one will get enlightenment	Reddy and Krishnaveni (2014)
131	*Gardenia gummifera* L.f.	Bark	Gum	Basha et al. (2011)
		White flowers	Ornamental	Pullaiah and Rani (1999)
132	*Gardenia resinifera* Roth	Wood	Doors and windows	Rekha and Kumar (2014)
133	*Givotia rottleriformis* Griff.	Wood	Toys	Reddy et al. (2008)
134	*Gmelina arborea* Roxb.	Mature stem	Stick for musical instruments	Rekha and Kumar (2014)
		Wood	Agricultural tools, fire wood	Prabakaran et al. (2013)
135	*Gossypium herbaceum* L.	Seed	Fiber – beds and pillows	Reddy et al. (2008)
136	*Grevillea robusta* A.Cunn. ex R. Br.	Wood	Doors and windows	Rekha and Kumar (2014)
		Wood	Agricultural tools	Prabakaran et al. (2013)
137	*Grewia hirsuta* Vahl	Stem	Fiber	Basha et al. (2011)
138	*Grewia obtusa* Wall.	Stem	Fiber	Basha et al. (2011)
139	*Grewia tiliifolia* Vahl	Stem, timber	Fiber, yoke	Basha et al. (2011); Sahu et al. (2013c)
140	*Guazuma tomentosa* Kunth	Stem	Fiber	Basha et al. (2011)
141	*Gyrocarpus americanus* Jacq.	Wood	Mulberry rearing stands, toys, cricket bats	Reddy et al. (2008)
		Woody stem	Fuel	Rekha and Kumar (2014)
142	*Haldinia cordifolia* (Roxb.) Ridsd.	Wood, fruits	Plough, hand hammer, threshing, pounding	Sahu et al. (2013c)

TABLE 9.1   *(Continued)*

S. No.	Species	Part	Use	Reference
143	*Helicteres isora* L.	Bark	Fiber	Basha et al. (2011)
144	*Hibiscus cannabinus* L.	Stem bark	Fiber – ropes, canes, hanging boxes, shepherd, cowherd carrier boxes	Reddy et al. (2008)
145	*Hibiscus rosa-sinensis* L.	Flowers	Boiled in oil and applied on hair	Rekha and Kumar (2014)
146	*Hibiscus sabdariffa* L.	Stem bark	Fiber – rope	Reddy et al. (2008)
147	*Hibiscus vitifolius* L.	Fruit	Fibers	Basha et al. (2011)
148	*Hildegardia populifolia* (Roxb.) Schott & Endl.	Stem	Fibers	Rao and Pullaiah (2007)
149	*Hiptage benghalensis* (L.) Kurz	Leaves	Oil	Basha et al. (2011)
			Narcotics	Prabakaran et al. (2013)
150	*Hiptage madablota* Gaertn.	Leaves	Oil	Basha et al. (2011)
151	*Holarrhena pubescens* (Buch.-Ham.) Wall.ex G.Don	Mature stem	All traditional religious festivals and religious ceremonies	Rekha and Kumar (2014)
152	*Holoptelea integrifolia* (Roxb.) Planch	Woody stem	Fuel	Rekha and Kumar (2014)
153	*Ipomoea carnea* L.	Stem, leaves	Fencing, mulching	Behera and Nath (2012)
154	*Ixora pavetta* Andr.	Wood	Fire wood	Prabakaran et al. (2013)
155	*Jasminum pubescens* L.	Flowers	Lakshmi puja.	Reddy and Krishnaveni (2014)
156	*Jasminum sambac* (L.) Sol.	Flowers	Used in all rituals.	Reddy and Krishnaveni (2014)
157	*Jatropha curcas* L.	Seeds	Oil	Basha et al. (2011)
		Young twigs	Toothbrush	Rekha and Kumar (2014)
		Stem	Fencing	Behera and Nath (2012)

TABLE 9.1   *(Continued)*

S. No.	Species	Part	Use	Reference
158	*Jatropha gossypifolia* L.	Seeds	Oil	Basha et al. (2011)
159	*Kydia calycina* Roxb.	Bark	Ropes	Sahu et al. (2013) c
160	*Lagenaria siceraria* (Mol.) Standl.	Fruit wall	Water bottle (carrier)	Reddy et al. (2008)
		Ripened fruits	Musical instruments	Naidu and Khasim (2010)
161	*Lannea coromandelica* (Houtt) Merr.	Wood	Furniture	Naidu and Khasim (2010)
		Stem bark	Gum	Basha et al. (2011)
162	*Lawsonia inermis* L.	Leaves	Crushed and boiled in oil and applied to hairs regularly	Rekha and Kumar (2014)
163	*Macaranga peltata* (Roxb.) Muell. – Arg.	Bark	Gum	Basha et al. (2011)
164	*Macroptilium atropurpureum* (DC.) Urban	Leaves	Oil	Basha et al. (2011)
165	*Madhuca indica* L.	Timber, leaves	Agricultural implements, thresher, used in religious ceremonies.	Sahu et al. (2013c); Reddy and Krishnaveni (2014)
			Making string for doors on all auspicious occasions to attract positive power of nature	Reddy and Krishnaveni (2014)
166	*Madhuca longifolia* (Koen.) Macbr.	Flowers	Arrack	Naidu and Khasim (2010)
167	*Madhuca latifolia* J.F. Gmel.	Wood	Fire wood	Prabakaran et al. (2013)
		Flowers	Liquor	Behera and Nath (2012)
168	*Malachra capitata* L.	Stem	Fibers	Reddy et al. (2008)
169	*Mallotus philippensis* Muell.-Arg.	Dried fruit powder	Detergent	Naidu and Khasim (2010)
		Flowers, fruits and seeds	Red dye	Basha et al. (2011)
		Woody stem	Fuel	Rekha and Kumar (2014); Prabakaran et al. (2013)

TABLE 9.1   *(Continued)*

S. No.	Species	Part	Use	Reference
170	*Mangifera indica* L.	Wood	*Teddu* (pan) for finger millet food preparation, firewood	Reddy et al. (2008); Sahu et al. (2013c)
			Wood grinder	Rekha and Kumar (2014)
171	*Melia azedarach* L.	Leaves and fruits	Fertilizer	Prabakaran et al. (2013)
172	*Melia composita* Willd.	Wood	Doors, cots and windows	Rekha and Kumar (2014)
173	*Memecylon edule* Retz.	Wood	Agricultural tools, Fire wood	Prabakaran et al. (2013)
174	*Memecylon jadhavii* K.N. Reddy et al.	Stem branches	Fuel	Reddy et al. (2006)
175	*Memecylon umbellatum* Burm.f.	Flowers bluish purple or blue	Ornamental	Pullaiah and Rani (1999)
176	*Mirabilis jalapa* L.	Rhizome powder	Tooth powder	Rekha and Kumar (2014)
177	*Moringa oleifera* Lam.	Wood	Swimming sticks	Reddy et al. (2008)
		Woody stem	Fuel	Rekha and Kumar (2014)
178	*Moringa pterygosperma* Gaertn.	Stem bark	Gum	Naidu and Khasim (2010)
179	*Musa paradisiaca* L.	Ripened fruits	Alcohol	Naidu and Khasim (2010)
		Leaves	Meal plates	Reddy et al. (2008)
		Stems with leaves and fruits	Used in entrance of houses during festivals and functions	Reddy and Krishnaveni (2014)
180	*Nymphaea nouchali* Burm.f.	Flower	Indoor and outdoor aquatic ornamental	Reddy et al. (2012)
			Used to pray for Ganesh, Siva and Lakshmi Devi	Reddy and Krishnaveni (2014)
181	*Nymphaea pubescens* Willd.	Flower	Ornamental grown in ponds and pools	Reddy et al. (2012)
182	*Ocimum basilicum* L.	Leaves	Used in functional ceremonies	Reddy and Krishnaveni (2014)
183	*Ocimum tenuiflorum* (Syn.: *O.sanctum* L.)	Plant	Liked by God Vishnu, women worship every day	Reddy and Krishnaveni (2014)

TABLE 9.1 *(Continued)*

S. No.	Species	Part	Use	Reference
184	*Origanum majorana* L.	Leaves	Used to make garlands along with other flowers for deities	Reddy and Krishnaveni (2014)
185	*Ophiuros exalatus* O.Ktz.	Leaves and branches	Roofing and thatching	Rekha and Kumar (2014)
186	*Oryza sativa* L.	Aerial parts	Roofing, thatching	Rekha and Kumar (2014)
187	*Oxalis corniculata* L.	Leaves	Blue dye	Basha et al. (2011)
188	*Oxystelma esculentum* R. Br.	Whole plant	Household creeper	Reddy et al. (2012)
189	*Pandanus odorifer* (Forssk.) Kuntze	Leaves	Offered to deities to get their blessings.	Reddy and Krishnaveni (2014)
190	*Passiflora foetida* L.	Plant	House hold creepers	Reddy et al. (2012)
191	*Passiflora incarnata* L.	Flowers	Rocky bands	Reddy et al. (2008)
192	*Peltophorum pterocarpum* (DC.) Baker	Bark, wood, leaves	Brown, black dyes	Basha et al. (2011)
193	*Phaseolus aconitifolius* Jacq.	Leaves	Oil	Basha et al. (2011)
194	*Phoenix acaulis* Roxb.	Leaves	Thatching roof	Behera and Nath (2012)
195	*Phoenix loureroi* Kunth.	Stem	Thatching	Prabakaran et al. (2013)
196	*Phoenix sylvestris* (L.) Roxb.	Leaves	Baskets, boxes, mats, brooms, liquor	Reddy et al. (2008); Sahu et al. (2013c)
		Leaf rachis, cutting inflorescence	Tooth brush, toddy	Naidu and Khasim (2010); Sahu et al. (2013c)
		Petiole	Fiber	Basha et al. (2011)
		Leaves	Broom sticks	Rekha and Kumar (2014)
197	*Phyllanthus reticulatus* Poir.	Branches	Tooth cleaners	Naidu and Khasim (2010)
198	*Pinus roxburghii* Sarg.	Flowers	Decorative items	Reddy et al. (2008)
199	*Plectronia didyma* Kurz.	Woody stem	Fuel	Rekha and Kumar (2014)

TABLE 9.1 *(Continued)*

S. No.	Species	Part	Use	Reference
200	*Pleurostylia opposita* (Wall.) Alston	Wood	Fire wood	Prabakaran et al. (2013)
201	*Pongamia pinnata* L.	Seeds	Oil	Basha et al. (2011)
		Leaves, twigs	Pesticides, tooth brush	Behera and Nath (2012); Sahu et al. (2013c)
202	*Premna tomentosa* Willd.	Wood	Fire wood	Prabakaran et al. (2013)
203	*Prosopis cineraria* (L.) Druce	Tree	The bride and bride groom go round this plant for blessings of Lord Vishnu.	Reddy and Krishnaveni (2014)
204	*Psidium guajava* L.	Young twigs, woody stem	Tooth brush, fuel	Rekha and Kumar (2014)
205	*Pterocarpus marsupium* Roxb.	Wood	Poosalu, water glasses	Reddy et al. (2008)
		Stem bark	Gums	Basha et al. (2011)
206	*Pterocarpus santalinus* L.f	Bark	Red dye	Basha et al. (2011)
		Whole plant	Ornamental tree in public gardens	Reddy et al. (2012)
207	*Pterospermum xylocarpum* (Gaertn.) Sant. &Wagh	Dried leaves, wood	Smoked just like tobacco, furniture, agricultural tools	Naidu and Khasim (2010)
		Flowers pale white	Ornamental	Pullaiah and Rani (1999)
208	*Randia malabarica* Lam	Wood	Windows	Rekha and Kumar (2014)
209	*Rhynchosia cana* DC.	Seeds	Oil	Basha et al. (2011)
210	*Rhynchosia minima* (L.) DC	Seeds	Oil	Basha et al. (2011)
211	*Salix tetrasperma* Roxb.	Woody stem	Fuel	Rekha and Kumar (2014)
212	*Sansevieria roxburghiana* Schult.f	Leaves	Fiber	Basha et al. (2011)

TABLE 9.1   *(Continued)*

S. No.	Species	Part	Use	Reference
213	*Santalum album* L.	Flowers, leaves, wood	Garland preparation, Wood carving (idols, toys, boxes)	Reddy et al. (2008)
		Wood, oil	Paste derived from wood is given as an offering to the gods and incense made of sandal wood shavings is burnt before them.	Reddy and Krishnaveni (2014)
214	*Saraca asoca* (Roxb.) Willd.	Plant	Worshiped for getting peace in their life	Reddy and Krishnaveni (2014)
215	*Schleichera oleosa* (Lour.) Oken	Seeds, timber	Cooking oil, agricultural implements, tooth brush	Behera and Nath (2012); Sahu et al. (2013c)
216	*Semecarpus anacardium* L.f	False fruits	Arrack	Naidu and Khasim (2010)
			Fertilizer	Prabakaran et al. (2013)
		Fruit, Seeds, bark	Black, gray dyes	Basha et al. (2011)
217	*Sesamum alatum* Thonn.	Seeds	Oil	Basha et al. (2011)
218	*Shorea robusta* Gaertn.	Leaves, wood, stem	Meal plates, beedi-making, furniture and agricultural implements, resin, tooth brush	Naidu and Khasim (2010); Sahu et al. (2013c)
		Leaves, stem bark	Oil and gum	Basha et al. (2011)
		Young stem, seeds	Tooth brush, cooking oil	Behera and Nath (2012)
		Plant	Offered to God to give prosperity, stability and unity among all the people.	Reddy and Krishnaveni (2014)
219	*Shorea roxburghii* Roxb.	Wood	Doors, cots, windows, fuel	Rekha and Kumar (2014)
		Wood	Thatching	Prabakaran et al. (2013)
220	*Sida acuta* Burm.	Dried plants	Brooms	Reddy et al. (2008); Naidu and Khasim (2010)
221	*Sida cordata* (Burm.f.) Borssum Walkes	Stem	Brooms	Sahu et al. (2013c)

TABLE 9.1   *(Continued)*

S. No.	Species	Part	Use	Reference
222	*Sida cordifolia* L.	Stem	Fiber	Basha et al. (2011)
223	*Smilax zeylanica* L.	Stem	Tooth brush	Behera and Nath (2012)
224	*Sorghum vulgare* L.	Leaves, branches	Roofing and thatching	Rekha and Kumar (2014)
225	*Soymida febrifuga* (Roxb.) A.Juss.	Ripe fruits	Flower vases	Reddy et al. (2008)
		Flowers and dry flowers are attractive	Ornamental	Pullaiah and Rani (1999)
226	*Sterculia urens* Roxb.	Stem, wood	Gum – skin softening, agricultural tools	Naidu and Khasim (2010)
		Stem bark	Gum	Goud et al. (1997)
227	*Streblus asper* Lour.	Stem	Tooth brush	Sahu et al. (2013c)
228	*Strychnos potatorum* L.	Wood	Fire wood	Prabakaran et al. (2013)
229	*Syzygium alternifolium* (Wt.) Walp.	Flowers cream or yellowish white, sweet scented	Ornamental	Pullaiah and Rani (1999)
230	*Syzygium cumini* (L.) Skeels	Woody stem	Fuel	Rekha and Kumar (2014)
		Leaves	Used in festivals	Reddy and Krishnaveni (2014)
231	*Tabernaemontana divaricata* (L.) R.Br	Flowers	Very much favor to Lord Shiva and offer them during Karthik mahotsavas	Reddy and Krishnaveni (2014)
232	*Tamarindus indica* L.	Wood	Fire wood	Prabakaran et al. (2013)
233	*Tarenna asiatica* (L.) Kunize	Wood	Thatching	Prabakaran et al. (2013)
234	*Tectona grandis* L.f.	Wood	Carts, chairs, tables	Reddy et al. (2008)
			Doors, cots, windows, fuel, harrow	Rekha and Kumar (2014)
235	*Tephrosia purpurea* (L.) Pers.	Leaves	Blue dye	Rani et al. (2002)
236	*Terminalia alata* Heyne ex Roth	Leaves, timber	Blue dye, agricultural implements	Rani et al. (2002); Sahu et al. (2013c)

TABLE 9.1   *(Continued)*

S. No.	Species	Part	Use	Reference
237	*Terminalia arjuna* (DC.) Wight & Arn.	Wood ash along with leaf paste, timber	Detergent, hair wash, agricultural implements	Naidu and Khasim (2010); Sahu et al. (2013c)
		Stem bark	Orange dye	Basha et al. (2011)
		Plant	Offered to God to protect crops from natural calamities	Reddy and Krishnaveni (2014)
238	*Terminalia bellerica* Roxb.	Woody stem	Fuel	Rekha and Kumar (2014)
239	*Terminalia chebula* Retz.	Wood, fruit powder, woody stem	Doors, cots, windows, firewood, tooth powder,	Sahu et al. (2013c); Rekha and Kumar (2014)
240	*Terminalia pallida* Brandis	Fruits	Brown, black dyes	Basha et al. (2011); Rani et al. (2002)
241	*Themeda cymbaria* Hackel	Wood	Stem	Prabakaran et al. (2013)
242	*Thespesia populnea* Cav.	Bark, flowers, fruits and wood	Yellow dye	Basha et al. (2011); Rani et al. (2002)
		Leaves	Detergent	Prabakaran et al. (2013)
243	*Thysanolaena maxima* (Roxb.) Kuntze	Dried inflorescence	Brooms	Reddy et al. (2008); Naidu and Khasim (2010); Sahu et al. (2013c)
244	*Tinospora cordifolia* Miers	Leaves	Oil	Basha et al. (2011)
245	*Trichosanthes bracteata* (Lamk.) Voigt	Plant	Grown as ornamental on arches in public gardens	Reddy et al. (2012)
246	*Trigonella foenum-graecum* L.	Leaves	Yellow dye	Rani et al. (2002)
247	*Urena lobata* L.	Wood	Brown dye	Rani et al. (2002)
248	*Urena sinuata* L.	Stem	Fiber	Basha et al. (2011)
249	*Ventilago madraspatana* Gaertn.	Stem bark	Oil, red dye	Basha et al. (2011)
250	*Vetiveria zizanioides* (L.) Nash	Roots	Fiber – door mats, cooler mats	Reddy et al. (2008)

TABLE 9.1 *(Continued)*

S. No.	Species	Part	Use	Reference
251	*Vitex altissima* L.f	Wood, bark	Yellow dye	Basha et al. (2011)
		Wood	Fire wood	Prabakaran et al. (2013)
252	*Vitex negundo* L.	Stem bark	Dye	Basha et al. (2011)
		Leaves, stem	Pesticide, tooth brush, fencing, tooth brush	Behera and Nath (2012); Sahu et al. (2013c)
253	*Waltheria indica* L	Stem	Fiber	Basha et al. (2011)
254	*Wedelia calendulacea* Less.	Leaves	Crushed and boiled in oil and apply hairs	Rekha and Kumar (2014)
255	*Wedelia chinensis* (Osbeck) Merril	Flowers, roots	Black dye	Rani et al. (2002)
256	*Woodfordia fruticosa* Kurz.	Wood, leaves	Firewood, fodder	Sahu et al. (2013c)
257	*Wrightia arborea* (Dennst.) Mabb.	Seeds, roots	Yellow dye	Rani et al. (2002)
258	*Wrightia tinctoria* R.Br	Wood	Toys, stirring stick for preparation of finger millet food and combs	Reddy et al. (2008)
		Leaves	Blue dye	Basha et al. (2011)
		Mature stem	Used in all traditional religious festivals and religious ceremonies	Rekha and Kumar (2014)
259	*Xylia xylocarpa* (Roxb.) Taub.	Whole plant	Fencing	Behera and Nath (2012)
260	*Yucca gloriosa* L.	Leaf	Fiber	Basha et al. (2011)
261	*Zanthoxylum budrunga* Wall.	Wood, woody stem	Doors, cots, windows, fuel	Rekha and Kumar (2014)
262	*Zizyphus mauritiana* Lamk.	Wood	Doors, cots and windows	Rekha and Kumar (2014)
263	*Ziziphus rugosa* Lam.	Leaves, wood	Fire wood	Prabakaran et al. (2013)
264	*Zizphus oenoplia* Mill.	Bark	Brown, black dye	Rani et al. (2002)
265	*Ziziphus xylopyrus* (Retz.) Willd.	Leaves, wood	Fire wood	Prabakaran et al. (2013)

ACKNOWLEDGMENTS

The receipt of financial assistance from the Council of Scientific and Industrial Research (CSIR), New Delhi, is gratefully acknowledged.

KEYWORDS

- **Dyes**
- **Ethnobotany**
- **Fibers**
- **Gums**
- **Ornamental**
- **Sacred Plants**
- **Tendu Leaf**
- **Wood**

REFERENCES

Antima, S., Dangwal L.R. & Mukta D. (2012). Dye yielding plants of the Garhwal Himalaya, India: A Case Study. I. *Res. J. Biological Sci. 1,* 69–72.

Arnold, J.E.M. (1995). Socio-economic benefits and issues in non-wood forest product use. In: Report of the International Expert Consultation of Non-Wood Forest Products. Food and Agriculture Organization of the United Nations. Rome. pp. 89–123.

Arora, J.S. (1993). Introductory Ornamental Horticulture. Ludhiana: Kalyani Publishers.

Arunakumara, K.K.I.U., Walpola, B.C., Subasinghe, S. & Yoon, M. (2011). *Pterocarpus santalinus* Linn. f. (Rathhandun): A review of its botany, uses, phytochemistry and pharmacology. *J. Korean Soc. Appl. Biol. Chem. 54,* 495–500

Basha, S.K.M., Umamaheswari, P., Rambabu, M. & Savitramma, N. (2011). Ethnobotanical study of Mamandur forest (Kadapa–Nallamalai range) in Eastern Ghats, Andhra Pradesh, India. *J. Phytology* 3, 44–47.

Behera, M.C. & Nath, M.R. (2012). Financial valuation of non-timber forest products flow from tropical dry deciduous forests in Boudh district, Orissa. *Intern. J. Farm Sciences 2,* 83–94.

Chandramouli, K.V. (1995). Sources of Natural Dyes. PPST Foundation, Madras.

Chengaiah, B., Rao, K.M., Kumar, K.M., Alagusundaram, M. & Chetty, C.M. (2010). Medicinal importance of natural dyes – review. *Intern. J. Pharm. Tech. Res. 2,* 144–154.

Coppen, J.J.W. (1995). Non wood forest products; gums, resins and latexes of plant origin. Food and Agriculture Organization of the United Nations Rome.

Das, P.K. & Mondal, A.K. (2012a). The dye yielding plants used in traditional art of 'patchitra' in pingla and mat crafts in sabang with prospecting proper medicinal value in the Paschim Midnapur district, West Bengal, India. *Int. J. Life Sc. Bt & Pharm. Res. 1,* 158–171.

Das, P.K. & Mondal, A.K. (2012b). Biodiversity and conservation of some dye yielding plants for justification of its economic status in the local areas of lateritic zone of West Bengal, India. *Advances in Bioresearch. 3,* 43–53.

Ellis, J.L. (1992). Wild Plant Resources of Nallamalais on the Eastern Ghats. Proc. Seminar on Resources, Development and Environment in the Eastern Ghats. Andhra University, Waltair. pp. 65–69.

FSI (2009). Available at www.fsi.nic.in/sfr_2009.htm. Dated 20. 01. 2015

Gokhale, S.B., Tatiya, A.U., Bakliwal, S.R. & Fursule, R.A. (2004). Natural dye yielding plants in India. *Natural Product Radiance. 3,* 228–234.

Goud, P.S.P., Murthy, K.S.R., Rani, S.S. & Pullaiah, T. (1997). Non-timber forest resources in the economy of tribals of Nallamalais, Andhra Pradesh. *J. Non-timber For. Prod. 4,* 99–102.

Gupta, P.C. (1992). Control of tobacco-related cancers and other diseases: Proceedings of an international symposium, January 15–19, 1990, TIFR, Bombay. p. 29.

ICFRE. (2010). Forest Sector Report India 2010. Indian Council of Forestry Research and Education, Dehradun. Ministry of Environment and Forests. Government of India.

Kadavul, K. & Dixit, A.E. (2009). Ethnomedicinal studies of the woody species of Kalrayan and Shervarayan Hills, Eastern Ghats, Tamil Nadu. *Indian J. Trad. Knowle. 8,* 592–597.

Kala, C.P. (2007). Local preferences of ethnobotanical species in the Indian Himalaya: implications for environmental conservation. *Curr. Sci. 93,* 1828–1834.

Krishnamurthy, K.V., Murugan, R. & Ravi Kumar, K. (2014). Bioresources of the Eastern Ghats. Bishen Singh Mahendra Pal Singh, Dehra Dun.

Krishnamurthy, K.V., Siva, R. & Senthil Kumar, T. (2002). Natural dye yielding plants of Shevaroy hills of Eastern Ghats. In: The Eastern Ghats – Proc. Natl. Sem. Conserv. Eastern Ghats. ENVIS Centre, EPTRI, Hyderabad, India.

Kumar, A.N.A., Joshi, G. & Ram, H.Y.M. (2012). Sandalwood: history, uses, present status and the future. *Curr Sci.* 103, 1408–1416.

Lal, P. (2009). Beedi – A short history. *Curr. Sci. 96 (10),* 1335–1337.

Mahanta, D. & Tiwari, S.C. (2005). Natural dye-yielding plants and indigenous knowledge on dye preparation in Arunachal Pradesh, Northeast India. *Curr. Sci. 88,* 1474–1480.

Misra, M.K., Das, P.K. & Dash, S.S. (2009). Phytodiversity and Useful Plants of Eastern Ghats of India: A Special Reference to the Koraput Region. Dehra Dun: International Book Distributors.

Naidu, K.A. & Khasim, S.M. (2010). Contribution to the Floristic Diversity and Ethnobotany of Eastern Ghats in Andhra Pradesh, India. *Ethnobotanical Leaflets 14,* 920–941.

Omkar, K., Suthari, S., Alluri, S., Ragan, A. & Raju, V.S. (2012). Diversity of NTFPs and their utilization in Adilabad district of Andhra Pradesh, India. *J. Plant Sci. 1,* 33–46.

Pandey, A. & Gupta, R. (2003). Fiber yielding plants of India. *Natural Product Radiance. 2,* 194–204.

Parthipan, M., Aravindhan, V. & Rajendran, A. (2011). Medico-botanical Study of Yercaud Hills in the Eastern Ghats of Tamil Nadu, India. *Ancient Sci Life.* 30, 104–109.

Prabakaran, R., Kumar, T.S. & Rao M.V. (2013). Role of non-timber forest products in the livelihood of Malayali tribe of Chitteri hills of Southern Eastern Ghats, Tamil Nadu, India. *J. Appl. Pharmaceut. Sci. 3,* 56–60.

Pullaiah, T. & Rani, S.S. (1999). Trees of Andhra Pradesh, India. Regency Publications, New Delhi.

Pullaiah, T, Murthy, K.S.R. & Karuppusamy, S. (2007). Flora of Eastern Ghats Hill Ranges of South East India. vol. 3., New Delhi: Regency Publications.

Pullaiah, T. & Rani, S.S. (2002). Dye yielding plants in India. In: Biodiversity in India. *vol. 1.* In: T. Pullaiah (Ed.). New Delhi: Regency Publications. pp. 24–220.

Ramarao, N. & Henry, A.N. (1996). The Ethnobotany of Eastern Ghats in Andhra Pradesh, India. BSI, Calcutta.

Ramasubbu, R., Eganathan, A. & Prabha, A.C. (2012). Medicinal and Aromatic Plants of Sirumalai Hills: Southern Eastern Ghats, India. LAP: Lambert Academic Publishing.

Rani, S.S., Murthy, K.S.M. & Pullaiah, T. (2002). Dye yielding plants of Andhra Pradesh, India. *J. Econ. Taxon. Bot. 26,* 739–749.

Rani, S.S., Murthy, K.S.M., Goud, P.S.P. & Pullaiah, T. (2003). Tree wealth in the life and economy of the tribes people of Andhra Pradesh, India. *J .Trop. For. Sci. 15,* 259–278.

Rao, M.D. & Pullaiah, T. (2007). Ethnobotanical studies on some rare and endemic floristic elements of Eastern Ghats-hill ranges of South East Asia, India. *Ethnobotanical Letters. 11,* 52–70.

Rao, R.S. (1998). Vegetation and valuable plant resources of the Eastern Ghats with specific reference to the Andhra Pradesh and their conservation. *Proc. Nation. Seminar on Conservation of Eastern Ghats,* EPTRI, Hyderabad. pp. 59–86.

Rashid, A. (2013). Dye yielding plant diversity of district Rajouri, Jammu and Kashmir state-India. *Int. J. Pharm. Biol. Sci. 4,* 263–266.

Reddy, C.S., Reddy, K.N., Pattanaik, C. & Raju, V.S. (2006). Ethnobotanical observations on some endemic plants of Eastern Ghats, India. *Ethnobotanical Leaflets 10,* 82–91.

Reddy, K.N., Pattanaik, C., Reddy, C.S., Murthy, E.N. & Raju, V.S. (2008). Plants used in traditional handicrafts in northeastern Andhra Pradesh. *Indian J. Trad. Knowle. 7,* 162–165.

Reddy, R.V., Reddy, M.H. & Raju, R.R.V. (1996). Ethnobotany of less known tuber yielding plants from Andhra Pradesh. *J. Non-Timber Forest Prod. 3,* 60–63.

Reddy, S.R. & Krishnaveni, L.H. (2014). Study of some Sacred plants of Kadapa District, Andhra Pradesh, India. *Indian J. Applied Research. 4,* 33–37.

Reddy, S.S.C., Suvarna, C. & Thummala, R. (2002). A Profile of Dye Yielding Plants. Andhra Pradesh Forest Department, Hyderabad.

Reddy, S.R., Reddy, A.M. & Yasodamma, N. (2012). Exploration of wild ornamental flora of YSR district Andhra Pradesh, India. *Indian J. Fundamental and Applied Life Sciences. 2,* 192–199.

Rekha, R. & Kumar, S.S. (2014). Ethnobotanical plants used by the Malayali tribes in Yercaud hills of Eastern Ghats, Salem district, Tamil Nadu, India. *Global J. Res. Med. Plants & Indigen. Med. 3,* 243–251.

Robinson, C. & Cush, D. (1997). The sacred cow, Hinduism and ecology. *J. Beliefs and Values. 18,* 25–37.

Sahu, C.R., Nayak, R.K. & Dhal, N.K. (2013c). The plant wealth of Boudh district of Odisha, India with reference to ethnobotany. *Int. J. Curr. Biotechnol. 1(6),* 4–10.

Sahu, S.C., Pattnaik, S.K., Dash, S.S. & Dhal, N.K. (2013). Fiber yielding plant resources of Odisha and traditional fiber preparation knowledge – an overview. *Indian J. Natural Prod. Resources. 4,* 339–347.

Samydurai, P., Jagatheshkumar, S., Aravinthan, V. & Thangapandian, V. (2012). Survey of wild aromatic ethnomedicinal plants of Velliangiri hills in the southern Western Ghats of Tamil Nadu India. *Int. J. Med. Arom. Plants. 2,* 229–234.

Shah, S.K., Bhattacharyya, A. & Chaudhary, V. (2007). Reconstruction of June-September precipitation based on tree-ring data of teak (*Tectona grandis* L.) from Hoshangabad, Madhya Pradesh, India. *Dendrochronologia. 25,* 57–64.

Sharma, S.C. & Goel, A.K. (1991). Potential of Indian wild plants as ornamentals. Horticulture—New Technologies and Applications. *Current Plant Sci. Biotechnology in Agriculture. 12,* 379–382.

Sive, R. (2007). Studies on natural dyes and dye yielding plants in India. *Curr. Sci. 92,* 916–925.

Sravani, P., Kiranmayee, Y., Narasimha, M.S., Reddy, V.S., Asha, S. & Kumar, R.B. (2014). *In-vitro* experimental studies on selected natural gums and resins for their antimicrobial activity. *Res. J. Pharmaceut. Biol. Chem. Sci. 5,* 154–172.

Thomas, B., Rajendran. A, Aravindhan, V. & Maharajan, M. (2011). Wild ornamental chasmophytic plants for rockery. *J. Mod. Biol. Tech. 1,* 20–21.

Tiwari, S.C. & Bharat, A. (2008). Natural dye yielding plants and indigenous knowledge of dye preparation in Achanakmar-Amarkantak biosphere reserve, central India. *Natural Product Radiance. 7,* 82–87.

CHAPTER 10

CONSERVATION, DOCUMENTATION AND MANAGEMENT OF ETHNIC COMMUNITIES OF EASTERN GHATS AND ADJACENT DECCAN REGION AND THEIR PLANT KNOWLEDGE

K. V. KRISHNAMURTHY,[1] BIR BAHADUR,[2] RAZIA SULTANA,[3] and S. JOHN ADAMS[4]

[1]Department of Plant Science, Bharathidasan University, Tiruchirappalli–620024, India

[2]Department of Botany, Kakatiya University, Warangal–506009, India

[3]EPTRI, Gachibowli, Hyderabad–500032, India

[4]Department of Pharmacognosy, R&D, The Himalaya Drug Company, Makali, Bangalore, India

CONTENTS

ABSTRACT

This chapter briefly summarizes the efforts so far made in the conservation, documentation and management of ethnic tribes of Eastern Ghats and the adjacent Deccan region and their ethnobotanical knowledge. Conservation efforts made by tribes themselves (like recognition of sacred temple trees and sacred groves) as well as by governmental and non-governmental organizations and institutions are detailed. Attempts made on conservation at the genetic, species and ecosystem levels involving in *situ* and *ex situ* approaches are also highlighted. Documentation efforts on traditional knowledge system as well as management details are provided. The importance of endogenous development to protect tribals and their knowledge is stressed

10.1 INTRODUCTION

'Conservation' implies the protection of environmental resources for sustainable utilization (Krishnamurthy, 2003). According to the modern concept of conservation, all these resources, whether used or not at present, should be protected. There is an ever-growing demand for these resources. Essentially three methods have been employed in the choice and use of botanical resources: (i) random method, which employs random collection of a plant for testing its value and use; (ii) phylogenetic method, which involves collection and testing of a plant whose close relatives have already been found to have some use or value; and (iii) ethno-directed method, in which attention is particularly focused on a plant whose value and use are already known to and tested by traditional ethnic communities; it is a readymade knowledge that is sure to yield the desired result, in addition to involving less research, development cost and time. In view of the above, it is all the more very vital to protect and conserve tribals, their habitats, their knowledge on useful plants and the plants themselves.

Sectorial approaches to conservation and management of traditional plant resources made sense when trade-offs among goods and services were local, more modest or unimportant. They are not sufficient today when conservation and management of services and goods have to meet conflicting goals. Hence, we are now forced to follow an integrated approach conservation and management of traditional plants. Successful conservation and management of traditional plant resources depend on two very important factors (Krishnamurthy, 2003): (i) The social, cultural, economic and political contexts of traditional societies within which management objectives are pursued; and (ii) Proper tools and methods that are to be selected to attain the aforesaid objectives. An integrated, predictive and adaptive approach to traditional phytodiversity management requires three basic types of information: (i) reliable site-specific baseline information on all aspects of plant knowledge of different ethnic societies and their documentation; (ii) knowledge on how goods and services offered by plants in specific ethnic communities will change to changing times and environments; and (iii) integrated local models that incorporate the traditional methods that incorporate traditional values, visions, need and priorities, as well as endogenous developmental changes (see later for more details). Management of ethnic botanical knowledge should also be brought about effectively through committed organizations and institutions (both governmental and non-governmental) functioning at the local, regional, national and international levels. These are required to frame policies and methodologies for execution of all conservational and managerial activities. They should also collect/collate vital data, store them and distribute them to the needy.

This chapter deals with conservation, documentation and management of ethnobotanical knowledge (in terms of both goods and services) of Eastern Ghats (E. Ghats) and adjacent Deccan region (see also Rawat, 1997; Henry Jonathan and Solomon Raju, 2007). There will be some overlap in the account given below since the three aspects mentioned above are intimately related to one another.

10.2 CONSERVATION

Conservation of ethnobotanical knowledge and plants in the study region has been attempted at three levels of plant biodiversity: genetic, species and ecosystem. Conservation at any one level automatically has helped in the conservation at the other two levels.

10.2.1 CONSERVATION OF GENETIC DIVERSITY

Conservation of genetic diversity has been planned and executed at the population level, since population is the basic unit at which genetic diversity is usually assessed. Three types of studies have been carried out on the ethnobotanical genetic diversity of the study region. The first one relates to the assessment of genetic diversity of important plant species whose populations have been collected from different provenances. This has provided the baseline data to focus conservation priority attention on the provenance showing the maximum genetic diversity as assessed through various molecular biology techniques. The second type of study relates to the estimation of minimum viable population size (MVP size) through population size estimation and population viability analysis (PVA) (see Krishnamurthy, 2003; 2012). Sridhar Reddy and Ravi Prasad Rao (2007) and Giriraj et al. (2007), for example, have made detailed observations on the distribution and population structure of *Pterocarpus santalinus*, a narrow endemic and endangered tree of E. Ghats in Kadapa (Cuddapah) hill ranges of Andhra Pradesh (Figure 10.1). The bark is used by local tribals for diarrhea and snake bite, while the wood is used to cure rheumatism, diabetes, inflammation from cuts and bleeding piles. The above authors have recorded just a very few individual trees of this species and the populations are mapped. Similarly, the distribution of *Cycas beddomei*, a critically endangered and endemic species, has been studied by Suresh Babu et al. (2007) although critical population assessments have not been made. The third type of study relates to the collection and characterization of subspecific taxa of ethnically important species, particularly land races and their relatives (Pandravada et al., 2008) and farmers varieties. On-farm and home garden conservation methodologies are being followed by some ethnic communities as well as by non-governmental organization in the tribal localities of the region studied here. The crop varieties maintained on-farm are often landraces, which are highly adapted to the local environmental conditions and invariably contain locally-adapted alleles that have been proved to be useful for specific breeding programs. Although landraces yield much less modern cultivars, they are so ancient and valuable that it is worth conserving them for posterity; they also need to be continuously monitored. For example, Pandravada et al. (2007) have suggested the landraces for some agriculturally important species and other species and pockets for on-form and *in situ* conservation. These include *Orzya officinalis* subsp. *malampuzhaensis* (in Nallamalai hills in Andhra Pradesh), *O. jeyporensis* (in Koraput in Odisha), *Cajanus cajanifolius* (in Visakhapatnam of Andhra Pradesh and Gajapati in Odisha), *Solanum erianthum*, and *Luffa acutangula* var. *amara* (both in Rayalaseema region of Andhra Pradesh), *Curcuma aromatica* and Mango varities (in Visakhapatnam), *Ensete glaucum* and *Musa ornata* (in Visakhapatanam and East Godhavari), *Cycas beddomei* (in Chittoor), *Pimpinella tirupatiensis* (in Tirupati), Minor millets (in Visakhapatnam), and Okra (in Adilabad)

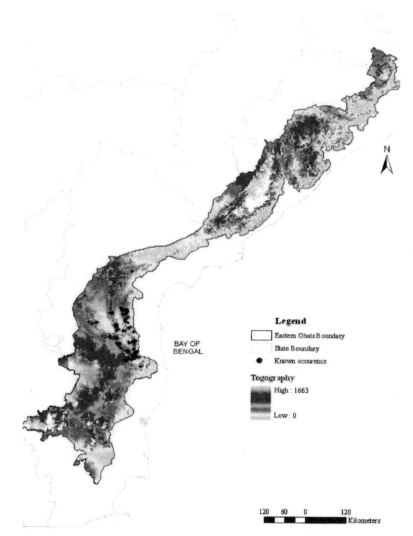

FIGURE 10.1   Locations of *Pterocarpus santalinus*, an endemic and endangered tree (Murthy et al. 2007a). (Murthy et al. 2007a; used with permission from EPTRI).

Home garden conservation is a small-level conservation effort by traditional societies. Many ethnic societies of the study region grow those plant germplasms that are needed for their use at home, kitchen or backyard gardens. Hundreds of such gardens exist in the tribal belts of all the five states through which E. Ghats run. Most of these are taxa that are either vegetables/fruits or medicinal plants; some of them are also ornamentals. These germplasms are often indigenous in the form of landraces, obsolete cultivars or rare species.

Some germplasm collections are assemblages of genotypes, populations or provenance collections that are often maintained as research materials for plant breeders and phytopathologists. They may consist of samples of domesticated plants and their wild relatives. These germplasms are maintained either in vivo in the form of plants, seeds, tubers or other propagates or in vitro in the form of single cells, embryos, buds or synseeds. There are different types of institutions involved in germplasm collections, which are called accessions (both core accessions and reserve accessions). Some of these institutions are treated subsequently in this chapter. A seed bank is a collection of seeds stored in a viable state for posterity. It is a good device for *ex situ* conservation of germplasms of sexually reproducing plants.

10.2.2 *CONSERVATION OF SPECIES DIVERSITY*

It is well-known that species is the main player in conservation conceptually, biologically and legally (Meffe and Carroll, 1994). It is because loss of species diversity is very obvious, more easily detectable and quantifiable than loss of either genetic or ecosystem diversity. The first effort in the conservation of species diversity in the study region involved the identification of those species that are in the IUCN threat list under its various categories: extinct in the wild, critically endangered, endangered, vulnerable and rare. Around 320 such taxa have been listed in Krishnamurthy et al. (2014). Biswal and Sudhakar Reddy (2007) have enlisted 127 such taxa for Odisha State. The most important threatened taxa need priority attention for conservation measures with attendant status studies on their population size, MVP size and reproductive biology. Although some attempts have been made to study the reproductive biology of some taxa belonging to E. Ghats (Sivaraj, 1991; Sivaraj and Krishnamurthy, 1989, 1992, 2002, Sivaraj et al., 2007; Solomon Raju, 2007; Solomon Raju and Henry Jonathan, 2010; Solomon Raju and Purnachandra Rao, 2002a, b; Solomon Raju et al., 2002, 2003, 2004, 2009), attention has not been focused on threatened taxa. Although effective traditional conservation measures are already in practice, more recent *in situ* and *ex situ* conservation strategies have also been taken up for execution. Both wild and domesticated taxa have been conserved by *in situ* methods, while *ex situ* methods primarily involve wild taxa and wild relatives of cultivated taxa.

One of the ways in which traditional ethnic communities of the study region had protected and conserved particular plant species was by assigning them to be the *sthalavrksha* (=temple trees). The plant species are specific to each temple and are considered as dear to the presiding deity of the concerned temple. The local ethnic community automatically considered these tree species as sacred and did not harm them wherever they are found (Krishnamurthy et al., 2014). Detailed accounts on temple plants are already given in Chapter 3.

10.2.3 CONSERVATION OF ECOSYSTEM DIVERSITY

In this method (which is the cheapest and most effective method), certain areas of ecosystems or important habitats are protected and maintained through various kinds of protected area programs or through controlled land use strategies. Invariably such areas are decided/selected by criteria, such as species richness, degree of endemism and threat index for wild taxa and land race diversity and wild relative abundance for domesticated taxa. Also taken into consideration are areas rich in tribals and in ethnic botanical knowledge. Ecosystem conservation not only protects species and their genetic diversity but also the various ecosystem services. As in species conservation efforts, here also ecosystems that need priority attention for conservation are to be given the greatest and immediate action.

Many fragile ecosystems/habitats in E. Ghats have already been identified both through extensive field studies and remote sensing and GIS technologies. Rao (1998) proposed the establishment of Biosphere reserves (see later for details on biosphere reserves) in (i) Similipal, Jeypore, Mahendragiri, and Malkangiri hill areas of Odisha; (ii) Dandakaranya forests of Koraput in Odisha; (iii) Thanjavanam-Lumbasingi-Tanjangi area of Visakhapatnam and E. Godavari districts of Andhra Pradesh; (iv) Gudem-Sapparla-Dhanukonda-Rampa-Maredumalli areas of Visakhapatnam and E. Godavari district; (v) Papi hills-Bhadrachalam areas of W. Godavari district; (vi) Nallamalai (Gundlabrahmeswaram-Ahobilam)-Seshachalem-Tirupati hills in Southern Andhra Pradesh; (vii) Shevaroy hills (Salem-Dhamapuri districts); (Viii) BR hills (near Mysore in Karnataka); and (ix) Sandur hills (in Bellary). Venkaiah (1998) also suggested the formation of Biosphere Reserves for the sal forest area near Rella, Manda and Peddakonda (all under Salur forest area) and Duggeru, Kurukutti and Thonam forests under Kurupam forest are. Sastry (2002) suggested the formation of Natural parks/ wildlife sanctuaries/Reserve forests at Similipal, Jeypore, Mahendragiri, Malkangiri, Dandakaranya,

Thanjavanam-Lumbasingi-Tanjangi, Gudem-Sapparla-Dharakonda-Rampa-
Maredumilli, Papi-Bhadrachalam hills, Nallamalais-Seshachalam, -Tirupati
hills (in Andhra Pradesh), Shervaroys (in Tamil Nadu) and Sandur and B.R.
Hills (both in Karnataka). Sudhakar Reddy et al. (2007a, b) suggested the
conservation of Mahendragiri (Gajapati district), Konda Kamuru and Upper
Sileru (Malkangiri district), Deomali and Gupteshwar (Koraput district),
Berbera (Khurda district), Niyamgiri hills (Royagada district), Kalinga Ghat
and Ranipathar (Phulbani district) and Kapilas and Saptasajya (Dhenkanal
district) based on detailed geospatial data and field studies. Murthy et al.
(2007b), based on remote sensing and GIS data on the entire E. Ghats have
identified the following nine conservation priority contiguous areas based on
biological richness value of Sections 10.1–10.4 of E. Ghats (see Chapter 1
and for more details see Krishnamurthy et al., 2014); (i) Similipal-Hadgarh-
Kuldiha (ii) Berbera (iii) Phulbani-Karlapat-Niyamgiri-Baisipalli (iv)
Lakhari valley (v) Kondakamberu-Sileru-Upper Visakha (vi) Kalrayan hills
(viii) Kolli hills and (ix) BR hills. Murthy et al. (2007b) added Shervaroy
hills to the above list. Britto and his students have identified conservation
priority zones within the following hills: Shervaroys (Balaguru et al., 2006),
Pachaimalais (Soosairaj et al., 2007) and Chitteri (Natarajan et al., 2004).

Ganeshaiah and Uma Shaanker (1998) have suggested a national approach to prioritize conservation of plants/vegetation through a hierarchical
integration of the biological and spatial elements of conservation. This approach should seek to independently map these elements of conservation and
then to integrate them to arrive at country-wide maps for conservation, for
example, contous of Conservation. A good illustration is the mapping of the
distribution of *Pterocarpus santalinus*, a threatened narrow endemic plants
species of E. Ghats (Giriraj et al. 2007). In continuation of this suggestion
the Department of Biotechnology, Government of India in association with
the department of space and under the auspices of National Bioresources
Development Board has launched two national level programs: (i) Landscape
level mapping of biodiversity of the country and its characterization using
satellite remote sensing and geographic information system, during the period 1998–2010 and (ii) Inventorying and mapping bioresources. Both these
programs covered the region studied in this volume. The major highlights
of the first program were as follows: (i) it provided a first ever baseline biodiversity database; (ii) it gave information on vegetation types, disturbance
regimes therein, biological richness, fragmentation and non-spatial database
on phytosociology; (ii) it provided GPS-tagged species archives. The entire
database is in the public domain and can be used by anyone interested (http://
bis.iirs.govt.in). The second program included bioresource inventorying of

E. Ghats. The entire stretch of E. Ghats was stratified into 2652 grids of 6.36 x 6.36 km^2 and the plant populations recorded in each grid in 2–4 transects of 1,000 x 5 m^2 size. Among the results obtained the most significant are the recording of 168 endemics (up to peninsular India level) and 28 threatened taxa; 74 taxa were recommended for inclusion in the red lists; 870 economically important species were recorded (486 medicinal, 144 edible, 98 fodder, 67 timber and 100 wild ornamental plant species).

Based on several indices a number of National Agricultural Heritage sites were identified for priority conservation in the high altitude tribal areas of Northern undivided Andhra Pradesh and Odisha (Singh and Varaprasad 2008; see also Krishnamurthy et al. 2014).

The following categories of important protected areas have already been established in the study region for the conservation of ethnic communities and their ethnobotanical resources: (i) Biosphere Resources: 18 biosphere reserves are recognized in India, two of which occur in the study region, the Similipal Biosphere Reserve (in Odisha) and Seshachalam hills (in Andhra Pradesh). (ii) Wildlife Sanctuaries: 498 wildlife sanctuaries have been established in India of which many occur in E. Ghats. The most important in Odisha occur in Badrama, Khalasundi, Hadgarh, Karlapet, Kotagarh, Lakhari valley, Sunabeda, Satkosia and Bhitarkanika. The most important in the undivided Andhra Pradesh are Etunagaram, Nagarjunasagar-Srisailam, Gundla Brahmeshwaram Metta, Salur, Sri Lankamalleswaram, Nelapattu, Rollapadu, Koringa, Kawal, Kolleru, Papikonda, Lanjamadugu, Kambalkonda, Krishna, Koundinya, Sri Lankamalleswara and Pranahita. Wildlife sancharies also exist in BR hills of Karnataka and Vedanthangal of Tamil Nadu. (iii) National parks: 92 national parks are known in India of which two occur in Odisha, four in undivided Andhra Pradesh and two in Tamil Nadu. (iv) Centre of Plant Diversity (CPD): Of the 234 CPDs recognized by IUCN in the world (up to 1998) 13 occur in India; three of these are in the study region (Nallamalais, Tirupati-Cuddapah hills and Northern Cicars. (v) Several hundreds of Reserve forests have been established throughout E. Ghats. (vi) Medicinal Plant Conservation Areas (MPCAs): these areas were identified and established by the Foundation for the Revitalization of Local Health Traditions (FRLHT), a Bangalore-based NGO in cooperation with State Forest Departments, based on wild Medicinal Plant wealth that needs conservation (Noorunnisa Begum et al., 2004). A total 55 MPCAs were established (12 in Tamil Nadu, 13 in Karnataka, 8 in undivided Andhra Pradesh, 5 in Odisha and the rest in a few other states of India). Of these, 21 are located in the study region. A number of wild medicinal plants used in folk and tribal medicine were thus conserved by these MPCAs in the wild.

Although the establishment of the above-mentioned types of protected areas is a fairly recent one in many places of the world, dating back only to about a century, the concept itself is really several centuries old (Kosambi, 1962) in India and some other parts of the world (Hughes and Chandran, 1998). There are both traditional and more recently established protected areas, since losses as well as conservation of plants have been issues of great intrinsic social and cultural concerns for the ethnic people (Shiva et al., 1991). Society and social, cultural and religion values have been called in to conserve nature and bioresources around them through sacred groves. The groves have been dedicated to the worship of the presiding deities of the concerned temples and are considered as the properties of those deities, which may be male or female. The deities are represented in the temple as a slab stone, a hero stone, *Sati* stone, or a trident, but rarely as idols. Even idols are there, they have been introduced quite late in the history of the sacred groves. Since great sanctity has been ascribed to the plants of such groves in the study area and since spiritual beings have been believed to reside in the groves, ordinary human activities are voluntarily precluded and several taboos and belief systems are associated with them. These activities included tree-felling, gathering of wood/fuel; plants, leaves, flowers, fruits and seeds, grazing by domestic animals, plowing, planting and harvesting, and dwelling (see full literature in Krishnamurthy et al., 2014). These ensured the conservation and preservation of the vegetation of sacred groves for posterity. In India, although there may be around 1,00,000 sacred groves (Swain et al., 2008), Henry Jonathan (2008) reported around 14,000 groves. According to Malhotra et al. (2001) there are 13,720 sacred groves of various sizes, diverse floristic composition and different vegetation types. Local autonomy is being given to these groves. The temple priest (who is also often the village doctor) is the only person authorized to enter the groves and that too for sustainably collecting plants/plant parts for worshipping the deity or for medicinal purposes (Krishnamurthy, 2003). There are 322 sacred groves in Odisha, 750 in undivided Andhra Pradesh, 488 in Tamil Nadu and around 75 in the E. Ghats region of Karnataka.

10.3   DOCUMENTATION

There is an enormous amount of knowledge among the ethnic tribal communities on the cultural and utilitarian uses and values of plants around them, most of which are not recorded and codified. Unfortunately, many knowledgeable tribal people have died either without divulging their knowledge on traditionally used plants or refused to part with them for reasons best known to them. This has resulted in the absence/scarcity of records on indigenous knowledge on plants. Thus, the collection and documentation of ethnobotanical data, information, knowledge and wisdom are very vital. The use of the same is often triggered by three principal categories of motivation (Krishnamurthy, 2003): (i) public policy; (ii) private sector and public interest; and (iii) culture. Documentation involves the following steps: data collection, storage of data, analysis of organized and integrated data so as to obtain useful and pertinent information, derivation of knowledge from such information through further analysis, interpretation and finally taking wise, proper and efficient management initiatives or actions through intelligent use of such knowledge.

Bibliographies related to ethnic botanical knowledge on the various ethnic communities of study region provide basic information on biology, forestry, agriculture, wildlife, conservation, food, medicine, society, culture, ethnic laws, economics and politics. These are available as published literature in journals, periodicals, books, and gray literature, such as reports, unpublished theses, manuscripts, etc. There are research papers on specific human and animal disease and the plants used by ethnic communities of the study region for both preventive and curative effects on these diseases. Most important research papers are mentioned in Chapters 7 and 8 of this volume. There are research papers on specific tribal groups of the study region and their knowledge on plants of value and use. Many of these are cited in various chapters of this volume. Research papers have also been published on the traditional botanical knowledge of different parts of the study region, like for example, Odisha, Andhra Pradesh, Tamil Nadu, and Karnataka or in different hill ranges of these states. Yet other research papers cover specific ethnic plant taxa (i.e., of particular families, genera or species). The authors of all these research papers hail from several official agencies, institutions, universities, colleges, research centers and non-governmental organizations. Several periodicals/journals have published valuable data/information on the diverse ethnic tribes of E. Ghats and the adjacent Deccan region. Journals/ periodicals that have published information on the ethnobotany of the study region are either general journals or are journals that are specifically related to ethnobotany or traditional knowledge. Most important among the latter are Indian Journal of Traditional knowledge, Asian Journal of Traditional knowledge, Journal of Economic and Taxonomic Botany, Ethnobotany, Ancient Science of Life, etc. The validation of the efficacy of traditional medicinal plants and their chemical constituents have been the subject of research papers that are being published in journals related to pharmacy, pharmacology, medicinal plants, phytochemistry, and traditional, alternate and complementary medicine, etc.

There are more than 50 books, which deal with the ethnobotany of the various traditional communities of the study region. The most important books that contain information on the ethnic tribes and their ethnobotanical knowledge on the study region are enlisted in Krishnamurthy et al. (2014). The most important gray literature relates to PhD theses from leading universities and research institutions of the study region. More than 120 PhD theses have been submitted on the ethnobotany of the various ethnic tribes/communities. Special mention must be made of the Society of Ethnobotanists, which was first established by S.K. Jain in 1980 with headquarters at the National Botanical Research Institute in Lucknow. This society has not only brought together all ethnobotanists of India but also encouraged various ethnobotanical researchers and activities. It has organized various kinds of meets and interactions between ethnobotanists. As examples of such meets can be mentioned the following: (i) Ethnobiology in Human Welfare held in November 1994 at Lucknow; it is the proceedings of this IV International Conference on Ethnobiology. It was edited by Dr. S.K. Jain and published by Deep Publications; (ii) National Workshop on Ethnobiology and Tribal Welfare organized at Coimbatore by AICRPE in Nov. 1985; (iii) National Conference on Traditional Knowledge 'Dhikshana 2008' in May 2008 at Thiruvananthapuram; (iv) the various seminars organized by EPTRI, Hyderabad on the conservation of Eastern Ghats; (v) The various conferences so far organized by the India Association for Angiosperm Taxonomy, India.

The most important effort towards documentation of ethnobotanical knowledge was made in 1982. The Ministry of Environment and Forests (MoEF), Government of India launched in 1982 the All India Co-ordinated Research Project on Ethnobiology (AICRPE). It is a multidisciplinary and multiinstitutional project involving more than 500 scientists and 27 institutions. The report of this project was submitted in 1994. It records around 10,000 medicinal plant species of which about 8,000 are wild plants. The report covers 550 tribal communities with 83.3 million people. It reports 3,900 edible wild plants, 400 fodder species, and 300 plant pesticides.

Databases form an important source of ethnobotanical data. 'Data' refer to "observation, measurements or facts referenced to some kind of accepted standard, which are subsequently integrated, processed, interpreted or otherwise manipulated to produce information." 'Information" refers to the "knowledge (product) derived from the analysis and interpretation of data" (Busby, 1997). Such data are stored, managed and readily made available for integration with other data. The absolute need for effective organization, management and use of data and information on ethnic plants is already reflected in many national and international agreements, legislations, and policy decisions, such as CBD, CITES, UNICEF, UNEP, etc. (see details in Krishnamurthy, 2003). The Biodiversity Data Management (BDM) is one such important outcome (see 'BDM UPDATE' Newsletter for relevant issues and events). The Taxonomic Databases Working Group for Plant Sciences SA 2000 and other Taxonomy Databases are another important result of this effort. For a list of Databases see Krishnamurthy (2003). The most important Databases that provide data on the ethnobotany of the study region are managed and operated by specific organizations and institutions, as detailed subsequently in this chapter. In addition, there are also Special Interest Networks, Traditional Biodiversity Application softwares, CD-ROMs and Diskettes. There are also catalogs and indexes of plant taxa known to the ethnic tribes of the study region.

The Community Biodiversity Registers (CBRs) form a 'bottoms-up system' of recorded data/information on wild and domesticated ethnodiversity in India (Gadgil et al., 1995). This is an important effort to secure community control over IKS. People of indigenous societies are encouraged to document all the known plant (and animal) species, with all available details on their uses. CBRs contain four separate sections: (i) *Background information* with two modules. Module 1 covers the total land/aquatic area, settlements and human communities, while module 2 covers the detailed local ecological history of the place; (ii) *Practical ecological knowledge* that is collected through critical surveys and then documented. Biodiversity status and utilization details are collected and recorded; (iii) *Claims of local people relate to information* provided by the local people regarding properties used and processes relating to bioresources; this is done even if such uses and properties have not yet been provided; (iv) *Scientific Knowledge* provided by the local people regarding biodiversity elements are documented. Ideally each village/traditional community should have one CBR and the preparation of the same should involve one or more local educational institutions of that area. All members outside that particular village/community are then refused normal access to the details contained in the CBR, but may be allowed on certain specific conditions. CBRs have been prepared for several villages/communities in India, including those in the study area covered by this volume.

One of the methods of documentation of ethnobotanical knowledge is the deposition of traditional plant-based materials in specially designed museums and culture collections. Although modern ethnobotany, especially of western countries, has a clear agenda centered on plant use (Cornish and Nesbitt, 2014), many traditional societies of the old world still emphasize the cultural and cognitive social approaches to the use of plants around them. Ethnobotanical knowledge, hence, should not reside only in memory, in books, manuscripts and photographs, in botanic gardens and in herbarium specimens. The focus should also be on ethnobotancial specimen collections as they form a rich source of data for contemporary ethnobotanists but also are fundamental to understanding the evolution of ethnobotany as a discipline (Cornish and Nesbitt, 2014). Such specimens have a long history and are highly effective in conveying knowledge, not only of the plant part used but also of the way in which it is processed. Ethnobotanial specimens are collected and kept in many locations in the world but specimens related to E. Ghats and the adjacent Deccan region are kept in the Economic Botany collection in Indian Museum, Kolkata, and in the Biocultural Collection Centre of Foundation for the Revitalization of Local Health Traditions (FRLHT), Bangalore. Some specimens are also kept at the Chennai Museum and the Museum of Economic Botany, Royal Botanic Gardens, Kew.

10.4 MANAGEMENT

There is an urgent need for resonance between the needs of ethnobotany and scientists on the one hand and the necessary data and information on traditional knowledge of ethnic communities on the other. In Section 10.3, the importance of collection and documentation of ethnic botanical knowledge was emphasized; also, the various kinds of documentation were discussed. The collection, documentation and distribution of these data and information are done by organizations and institutions, as mentioned earlier, in order to make effective decisions on managing the traditional plant resources and to implement these decisions.

The management of ethnobotanical knowledge in the study region was done through a number of organizations, both governmental and non-governmental. The governmental organization (both Central and State-controlled) are primarily concerned with framing laws, rules, legislations, policies and implementation methodologies for execution of conservation actions besides serving as sources of data on ethnobotany. The non-governmental organizations are involved in documentation and dissemination of data/information besides executing, on their own, conservation and management of ethnic knowledge, but under the overall control of the government.

Ethnic knowledge on plants of India, including the region studied in this volume, is managed by the following important Information Networks (see for more details Chavan and Chandramohan, 1995; Geevan, 1995): (i) Foundation for the Revitalization of Local health Traditions (FRLHT), which houses the Indian Medicinal Plants National Network of Distributed Database (INMEDPLAN) of traditional medicinal plants; (ii) Biotechnology Information System (BTIS) consisting of the Distributed Information Centres of the Department of Biotechnoliogy supported by NICNET, the computer network of the National Informatics Centre (NIC); (iii) The Indian Bioresource Information Network (IBIN) (www.ibin.co.in) is a distributed database on bio-resources and biodiversity. The Bioresources information Centres situated all over the country collaborate with IBIN. *Jeeva Sampada* is a database on Bioresources of India and it provides data on 39,000 species of plants, animals, marine organisms and microbes and comprises of more than 54,00,000 records. It also offers images, distribution maps and uses in an interactive data retrieval service (see for more details the booklet entitled Indian Bioresource information Network, Published by the Department of Biotechnology, Government of India. (iv) MEDLARS, the Medicinal Literature Analysis and Retrieval System, accessible over NICNET; (v) Biodiversity Information System (BIS) of the Indira Gandhi Conservation Monitoring Centre (IGCMC) established by the WWF and the Environmental Resources Information System (ERIS); (vi) ENVIS (Environmental Information System) India supported by the Ministry of Environment and Forests of Government of India; (vii) NBPGR (National Bureau of Plant Genetic resources) databases located at New Delhi and Hyderabad. NBPGR has the following mandates: Characterization, Evaluation, Documentation, Conservation and Bioprospecting. Collection related to the ethnobotany of the study region is deposited in NBPGR's National Gene Bank; there is also a National Herbarium of Cultivated Plants. The Botanical Survey of India also has collections of wild plants of the study region, including those used by the ethnic societies of this region. (vii) The people and plants website, an initiative of the partnership between WWF and UNESCO with the Royal Botanic Gardens, Kew taking an associate role.

10.4.1 MANAGEMENT ORGANIZATIONS AND EFFORTS

The International Union for Conservation of Nature and Natural Resources (IUCN), a federative membership organization composed essentially of governments or governmental organizations of the various countries of the world as well as scientific professional and conservation organizations, has helped and advised in the identification and preparation of a list of threatened categories plants (Red List) of India, including those of the study region. UNESCO (United National Educational, Scientific and Cultural Organization), through its Man and Biosphere Programme (MAB), has been instrumental in the establishment of protective areas in India, including those in the study area (see an earlier page of this chapter for details). It has also established cultural heritage sites. Food and Agricultural Organization (FAO) has helped in the conservation of genetic resources of food plants through its Global System for Conservation of Plant Genetic Resources and Commission on Plant Genetic Resources (PGR).

The M.S. Swaminathan Research Foundation (MSSRF) located in Chennai has carried out a project, under Indo-Canada Environment Facility, on "Coastal Wetlands: Mangroves Conservation and Management." Under this project, 14 demonstration villages falling in the East Coast Mangrove belt were targeted but only in 24 villages it was implemented. The villages were involved in the restoration of mangrove forests under their jurisdiction. The MSSRF has also established several field sites in the study region. Two of these are very important: Kolli hills in Tamil Nadu and Jeypore in Odisha. These two are involved in community-based agrobiodiversity conservation programs. The major activities are revitalization of traditional conservation, participatory plant improvement actions, and establishment of databases (Uma Ramachandran, 2002). Experts available in these two sites have formed networks with all stakeholders. After correctly identifying the resource base in these two regions and the associated problems, these are addressed through participatory meetings and training activities to strengthen their participation. This organization is also involved in *in situ*/on-form conservation of agrobiodiversity involving community seed and grain banks. In these banks seeds of traditional land races are maintained by self-help groups. These banks are linked to the Scarascia Mugnozza Community Genetic Resources Centre located at MSSRF in Chennai. The very important tribal seed banks are located in Kolli hills (Tamil Nadu where the Malayali tribes are involved by MSSRF to maintain landraces of minor millets and some legumes) and in Jeypore (Odisha) where the local tribals are involved in maintaining landraces of rice as well as its wild relatives

The Centre for Farmers' Rights at MSSRF deals with "the rights arising from the past, present and future contributions of farmers in conserving, improving and making available plant genetic resources" (Swaminathan, 1997). The concept of Farmers' Rights aims at benefit–sharing through a process of *compensation* for the use of TKS, in contrast to the system of *Internalization of benefits*. A Charter of Rights is included and some of them are mentioned by Shiva and Ramprasad (1993). A Resource Centre for Farmers' Rights has been formed by MSSRF in order to involve in the following four activities: (i) Farmers' Rights Information Service (FRIS) with a collection of several component databases, such as IPR database (with four modules respectively dealing with tribal contributions, ethnobotanical features, sacred groves, and rare angiosperms) and multimedia database on the ecological farmers of India; (ii) Community Gene Bank, meant for storing seeds of landraces, traditional cultivars and folk varieties; (iii) Community Genetic Resources Herbarium, meant to serve as a reference center for the identification of rare and threatened plants, economically important plants and traditional cultivars; and (iv) Agrobiodiversity Conservation Crops created to train young tribal and rural men and women in conservation techniques and seed science and technology.

The Environmental Protection, Training and Research Institute (EPTRI), located at Hyderabad, is one of the 25 centers established by Government of India in 1994 to protect environmental resources, particularly of ethnic communities. These centers are supported by Ministry of Environment and Forests, New Delhi under its Action Oriented Research Program for the collection, storage and dissemination of data and information. These centers are called ENVIS centers, which are centralized Networks that bring about coordination between institutions and organization. The EPTRI is meant for E. Ghats and brings about a Newsletter on E. Ghats. EPTRI has a database called EPTRI Information System, which integrates various thematic and attribute data. This is very useful for various conservation and developmental projects. It has designed, developed and updated the database on bibliography, experts, and environmental resources (flora, fauna, microbes, water, soil, geology, tribals, traditional knowledge, etc.) pertaining to E. Ghats (Srinivasa Rao, 2007). The Resource Database is designed based on UNDP Global Information Database format. This database has about 1,500 parameters. EPTRI-ENVIS is regularly conducting seminars on E. Ghats and publish their proceedings.

There are several other NGOs/organizations involved in the management of TKS/tribals of the study region. In Nallamalais 10 NGOs have been working under *Abhayaranya Samrakshan* through Holistic Resource (Array) Management (ASHRAM). The *Anantha Paryavarana Parirakshana Samithi* in Anantapur, Regional Plant Resource Centre at Bhubaneswar, the Odisha State Vanaspati Vana Society at Cuttack, the Save Eastern Ghats Organization at Chengam in Javadi hills, the Centre for Indian knowledge studies at Chennai, etc. are involved in documenting and managing tribal ethnobotanical knowledge of the study region. The Centre of Indian Knowledge System located at Chennai has been undertaking a large number of grassroot level efforts for the on-farm conservation of indigenous genetic resources, organic agriculture, integrated home gardens and *Vrksāyurveda* (ancient Indian Plant Science). Run by A.V. Balasubramanian and K.V. Vijayalakshmi, this organization has been working for the last 18 years, through field studies, on indigenous rice varieties of Tamil Nadu and adjacent regions. The organization has also helped in the establishment of Farmer's Seed Banks for seed exchange, distribution and utilization, as well as in conducting training and outreach programme's for villagers/tribals.

The Deccan Development Society (DDS) is a grassroot level organization in the Medak district of Telangana. It has adopted about 75 villages where it works with women's groups. DDS works for food security and food sovereignty for the rural people.

Another organization that manages ethnobotanical data, facilitated their use and exchange and raises important issues on TKS in India is the Society for Research and Initiatives for Sustainable Technologies and Institutions (SRISTI) initiated by Professor Anil Gupta of the Indian Institute of Management at Ahmedabad. SRISTI has already established communications with more than 300 villages of India with the main objective of capacity building at grassroot level in biodiversity conservation, protecting TKS, endogenous development (see later in this chapter for more details) and enrichment of the cultural and institution base of ethnic communities. SRISTI also provides legal counseling with reference to the protection of TKS. It has also established a sustainable development programs for villages/ethnic societies

Special mention must be made of Joint Forest Management (JFM) and Community Forest Management (CFM). These two forest management strategies were envisaged in the National Forest Policy of Government of India. Both these seek the active and real involvement of local ethnic community in forest management. In other words, the State Forest Departments and the forest dwelling people share the management activities of the forest. These two (JFM and CFM) management methods (i) promote cooperation (ii) appreciate diversity (iii) involve the collective mind of the community (iv) involve collective responsibility and accountability (v) allow equal opportunity and access to resources and benefits (vi) make everyone's voice to become important (vii) enable joint planning and management, and (viii) are not target- or product-oriented (Ravindranath et al., 2000). As a result, the process of forest management primarily promotes the active participation of the community. Local tribals are now evincing keen interest to develop degraded forests near them. In India 30 States have been selected for JFM, 1,813 Forest Protection Committees have been formed in the undivided Andhra Pradesh after its launch in 1993. In Tamil Nadu JFM was started in 1977 with the help of Village Forest Councils and with the thrust areas of forests regeneration and development of degraded forest. Both JFM and CFM have become successful with the help of Self-Help Groups, resulting not only in enhancing the total forested area but also in helping the ethnic societies to sustainably use the forest products for the their livelihood.

In 1999, the Ministry of Environment and Forests (MoEF) of Government of India prepared a National Policy and Macro-level Action Strategy on Biodiversity through a consultative process. This was called the National Biodiversity Strategy and Action Plan (NBSAP). Towards the goal of preparing NBSAP, MoEF was funded by GEF. The execution of NBSAP process was entrusted to the Technical and Policy Core Group (TPCG) coordinated by the NGO, Kalpavriksh and the Biotech Consortium India Ltd. This was operated between 2000 and 2002. The NBSAP was one of the largest environmental planning exercises carried out ever in India for the following reasons: (i) it was carried out at five different levels (local, State/Union Territory, Ecoregion, Thematic and National); (ii) It involved the entire biodiversity spectrum (both wild and domesticated/cultured) at all the levels of biodiversity (genes, species, ecosystems); (ii) the plan involves a range of aspects like scientific, technological, social, cultural, economic, political, ethical, moral and gender relations. There were 14 broad thematic areas, four cross-cutting ones and 25 sub-themes; (iv) consequent on (i) to (iii) above the plan involved all categories of people, institutions and major stakeholders. Eight interstate ecoregions were suggested for which NBSAP was to be prepared. The E. Ghats was one such ecoregion and this Ecoregion group was coordinated by Prof. T. Pullaiah and helped by Prof. K.V. Krishnamurthy and the Botanical Survey of India, Southern Circle. Based on various inputs the Draft Action Plan for E. Ghats Ecoregion was prepared and sent to the coordinating body of NBSAP (Krishnamurthy, 2001; Pullaiah, 2001). This Draft Action plan was finalized after giving details on nine aspects of E. Ghats (see details Krishnamurthy et al., 2014). This important document was very helpful in taking adequate conservation and management measures.

10.5   BIOPROSPECTING AND BIOPIRACY

As per the utilitarian approach of ethnobiology, ethnic knowledge (TKS or IKS) on various plant uses is being increasingly exploited, particularly by the western world. The process involved in this is called *bioprospecting* (see also Chapter 12 in this volume). Bioprospecting is vital for pharmaceutical, nutraceutical, cosmoceutical, biotechnological and agricultural industries (see Krishnamurthy, 2003). TKS on plants is often possessed within a tribal society by all its members or may be held by specialists, elders, shamans, clan heads or by women. The importance of TKS, especially with reference to bioprospecting has been grossly underestimated. But today the vast advancements in scientific documentation and understanding of TKS have vastly improved the situation. The sacred bondage between traditional societies and Nature has also been realized by Western Societies, culminating in the recognition of 'Cultural Landscapes' by UNESCO. Such recognition had very great implications for ownership of traditional knowledge and, hence, for *intellectual property rights (IPR)*. Loss of TKS, even if it is bits and pieces, will lead to the losing of control over vital information by the traditional societies which have been theirs for several hundreds of years. The existing mechanisms are inadequate to effective protect the genuine rights of traditional societies. Hence, *biopiracy* was rampant even until about five years back. As of today, IPRs are the most important legal measures available for protecting knowledge, discoveries, innovations, inventions and novel practices and products. Patents, which are the best known of these IPRs however, are of very limited or even no value to traditional communities as there are several difficulties in documenting their knowledge and in identifying the actual owner/inventor of that knowledge. In many cases, TKS is in the public domain and hence do not have the required 'uniqueness.' There are other difficulties as well (see details in Krishnamurthy, 2003).

Indigenous societies have recently become better informed, organized and more articulate. They have started to assert their rights forcefully through their own actions, networks and databases. Many traditional societies have entered into community forest management and participatory forest management agreements. They have got their rights to utilize forest products (on which they were all along dependent) even in protected areas in the buffer zones.

10.6 ENDOGENOUS DEVELOPMENT

Conservation of ethnic knowledge on plants requires protection of tribal peoples, who possess this knowledge. The tribal peoples are vulnerable to a range of social, cultural and economic changes that influence them and their knowledge in various ways. Often, they are forced to migrate from their habitats in search of better life, education, job and money. Many young tribals are attracted to Western ways of life and do not have interest to acquire the traditional knowledge from their elders or other members of the society. Moreover, in the last two centuries there is an increasing dominance of the Western culture, values, science and technology and these have increasingly replaced traditional knowledge systems and cultures, particularly in India due to the colonial rule. The traditional ethnic societies and people are either rejected or are marginalized and have also moved away from their rich cultural background.

In the light of the above, endogenous development is very important. It is based mainly, though not exclusively, on the locally available resources, local knowledge, culture, and leadership. It is open to integrating traditional and outside knowledge and practices in order to protect the indigenous community and its people. It helps the tribals through participatory development, local management of natural resources and ecological processes, low external input, sustainable agriculture and biodiversity, etc. Endogenous development neither implies narrowly defined development, nor romanticizes or rejects traditional knowledge. On the contrary, it is to be considered as complementary to the ongoing technological process. It addresses the local needs and contradictions, uses local potentials, enhances local economics and links them to international systems. It supports co-existence and co-evolution of a number of cultures and promotes intercultural research, exchange and dialogs. Thus, endogenous development brings together global and local knowledge's (Haverkortt et al., 2003).

The Compass program of Netherlands has supported endogenous development in south India through FRLHT, the NGO based at Bangalore. The following components have been identified by Compass for supporting endogenous development: (i) building on locally available resources; (ii) objectives based on locally felt needs and values; (iii) *in-situ* reconstruction and development of local knowledge systems through understanding, testing and improving local practices and enhancing the dynamics of local knowledge process; (iv) maximizing local control of development; (v) Identifying development niches based on the characteristic of each local situation; (vi) selective of use the external resources; (vii) retention of benefits in the locality itself; (viii) share and exchange experiences between different ethnic societies; (ix) training and capacity building; (x) developing networks and partnerships; and (xi) understanding systems of knowing, learning and experimenting.

The efforts taken by IDEA (Integrated Development through Environmental Awakening) in northern E. Ghats towards endogenous development is worth mentioning since community ownership as well as traditional leaders (Shamans) and their norms, taboos, and customary practices related to natural stability and cultural identity are breaking down. IDEA has been facilitating a process to halt degradation of cultural identity, TKS and natural resources of tribals since 1985 (in 300 villages in Odisha and northern undivided Andhra Pradesh). It has taken various measures, such as documentation, action research, formation of tribal network, traditional seed conservation (with a collection of about 285 traditional seed varieties), vegetable conservation (with about 243 wild vegetables, edible tubers, berries and nuts), traditional medicinal plants (about 500 species), besides undertaking serious endogenous development programs. These programs are related to ethnoveternary practices, biopesticides, transformation of traditions to ecofriendly processes, traditional weed management, strengthening traditional leaders and tribal identity, developing organic markets, ecotourism, popularizing traditional knowledge, protecting tribal cosmovisions and belief systems, etc. (Gowtham Shankar, 2003).

There is great variability of endogenous development options if one considers seriously the diversity of worldviews, values, practices, knowledge concepts and ecological and political context. Hence, there are a number of opportunities and constraints related to endogenous development of ethnic societies in E. Ghats and the adjacent Deccan region, as in any other traditional societies of the world. In order to meet these opportunities and constraints we must first create the enabling environment (Haverkort et al., 2003) at the local, regional, national and international levels.

KEYWORDS

- **Conservation**
- **Databases**
- **Documentation**
- **Endogenous Development**
- **Traditional Botanical Knowledge**

REFERENCES

Balaguru, B., John Britto, S., Nagamurugan, N., Natarajan, D. & Soosairaj, S. (2006). Identifying conservation priority zones for effective management of tropical forests in Eastern Ghats of India. *Biodiv. Conserv. 15*, 1529–1543.

Biswal, A.K. & Sudhakar Reddy, C. (2007). Prioritization of taxa for conservation in Orissa. In: Proc. Natl. Sem. Conserv. Eastern Ghats. ENVIS Centre, EPTRI, Hyderabad, India. pp. 73–74.

Busy, J.R. (1997). Management of information to support conservation decision making. In: D.L. Hawksworth, P.M. Kirk & S. Dextre Clarke (Eds.). Biodiversity Information: Needs and Options. Wallingford, UK: CAB International.

Chavan, V. & Chandramohan, D. (1995). Databases in Indian Biology: the state of art and prospects. *Curr. Sci. 68*, 273–279.

Cornish, C. & Nesbitt, M. (2014). Historical Perspective on western Ethnobotanical Collections. In: J. Salick, K. Konchar & Nesbitt, M. (Eds.). Curating Biocultural Collections, Kew: Kew Publishing Royal Botanical gardens. pp. 271–293.

Gadgil, M., Devasia, P. & Seshagiri Rao, P.R. (1995). A comprehensive framework for nurturing practice ecological knowledge. Centre for Ecological Science, Indian Institute of Science, Bangalore, Karnataka. pp. 1–74.

Ganeshaiah, K.N. & Uma Shaanker, A. (1998). Contours of Conservation – A national agenda for mapping biodiversity. *Curr. Sci. 75*, 292–298.

Geevan, C.P. (1995). Biodiversity Conservation Information Networks: A Concept Plan. *Cur. Sci. 69*, 906–914.

Giriraj, A., Shilpa, G., Sudhakar Reddy, C., Sudhakar, S., Beierkuhnlein, C. & Murthy, M.S.R. (2007). Mapping the geographical distribution of *Pterocarpus santalinus* L.f. (Fabaceae) – an endemic and threatened plant species using ecological niche modeling. In: Proc. Natl. Sem. Conserv. Eastern Ghats. ENVIS Centre, EPTRI, Hyderabad, India. pp. 446–457.

Gowtham Shankar (2003). Building on Tribal resources. Endogenous Development in the Northern Eastern Ghats. pp. 115–128. In: B. Haverkort, K. Van't Hooft & W. Hiemstra (Eds.). Ancient Roots, New Shoots-Endogenous Development in Practice. London: Compas, Zed Books.

Haverkort, B., Van't hooft, K. & Hiemstra, W. (Eds.). (2003). Ancient Roots, New Shoots-Endogenous Development in Practice., London: Compas, Zed Books.

Henry Jonathan, K. (2008). Sacred groves – the need for their conservation in the Eastern Ghats region. *EPTRI-ENVIS Newsletter 14*(3), 3–5.

Henry Jonathan, K. & Solomon Raju, A.J. (2007). Threats to biodiversity of Eastern Ghats: The need for conservation and management measures. *EPTRI-ENVIS Newsletter 13*(2), 9–10.

Hughes, J.D. & Chandran, M.D.S. (1998). Sacred groves around the earth: an overview. In: P.S. Ramakrishnan (Ed.). Conserving the Sacred for Biodiverisity Management, New Delhi: Oxford & IBH Publishing Co. Pvt. Ltd., pp. 69–85.

Kosambi, D.D. (1962). Myth and Reality: Studies in the Formation of Indian Culture. Bombay: Popular Press.

Krishnamurthy, K.V. (2001). Report on National Biodiversity Strategy and Action Plan (NBSAP) for Part of the Tamil Nadu Sector of Eastern Ghats Ecoregion. Tiruchirappalli, India.

Krishnamurthy, K.V. (2003). Text Book on Biodiversity. Science Publishers, New Hampshire, USA.

Krishnamurthy, K.V. (2012). The need for estimation of minimum viable population size for the threatened plants of India. In: G. Maiti & S.K. Mukherjee (Eds.) Multidisciplinary Approaches in Angiosperm Systematics. Vol. II. Publication Cell, Univ. Kalyani, W. Bengal, India. pp. 141–148.

Krishnamurthy, K.V., Murugan, R. & Ravikumar, K. (2014). Bioresources of the Eastern Ghats: Their Conservation and Management. Dehra Dun: Bishen Singh Mahendra Pal Singh.

Malhotra, K.C., Gokhale, Y., Chatterjee, S. & Srivastava, S. (2001). Cultural and Ecological Dimensions of Sacred Groves in India. IWSA, New Delhi and Indira Gandhi Rashtriya Manav Sangrahalaya, Bhopal, India.

Meffe, G.K. & Carroll, C.R. (1994). Principles of Conservation Biology. Sunderland, Mass, USA: Sinauer Associates.

Murthy, M.S.R., Sudhakar, S., Jha, C.S., Sudhakar Reddy, C., Pujar, G.S., Roy, A., Gharai, B., Rajasekhar, G., Trivedi, S., Pattanaik, C., Babar, S., Sudha, K., Ambastha, K., Joseph, S., Karnatak, H., Roy, P.S., Brahmam, M., Dhal, N.K., Biswal, A.K., Mohapatra, A., Mohapatra, U.B., Misra, M.K., Mohapatra, P.K., Mishra, R., Raju, V.S., Murthy, E.N., Venkaiah, M., Venkata Raju, R.R., Bhakshu, L.M., Britto, S.J., Kannan, L., Rout, D.K., Behera, G. & Tripathi, S. (2007a). Biodiversity Characterization at landscape level using Satellite Remote Sensing and Geographic Information System in Eastern Ghats. In: Proc. Natl. Sem. Conserv. Eastern Ghats. ENVIS Centre, EPTRI, Hyderabad, India. pp. 394–405.

Murthy, M.S.R., Sudhakar, S., Jha, C.S., Sudhakar Reddy, C., Pujar, G.S., Roy, A., Gharai, B., Rajasekhar, G., Trivedi, S., Pattanaik, C., Babar, S., Sudha, K., Ambastha, K., Joseph, S., Karnatak, H., Roy, P.S., Brahmam, M., Dhal, N.K., Biswal, A.K., Mohapatra, A., Mohapatra, U.B., Misra, M.K., Mohapatra, P.K., Mishra, R., Raju, V.S., Murthy, E.N., Venkaiah, M., Venkata Raju, R.R., Bhakshu, L.M., Britto, S.J., Kannan, L., Rout, D.K., Behera, G. & Tripathi, S. (2007b). Vegetation land cover and Phytodiversity Charaterization at Landscape Level using Satellite Remote Sensing and Geographic information system in Eastern Ghats, India. *EPTRI-ENVIS Newsletter 13*(1), 2–12.

Natarajan, D., John Britto, S., Balaguru, B., Nagamurugan, N., Soosairaj, S. & Arockiasamy, D.I. (2004). Identification and conservation priority sites using remote sensing and GIS – A case study from *Chitteri* hills, Eastern Ghats, Tamil Nadu. *Curr. Sci. 86*, 1316–1323.

Noorunnisa Begum, S., Ravikumar, K., Vijaya Kumar, R. & Ved, D.K. (2004). Profile of medicinal plants diversity in Tamil Nadu MPCAs. In: *Abstracts – Natl. Sem. New Frontiers in Plant Taxonomy and Biodiversity Conserv.* TBGRI, Tiruvananthapuram, India. pp. 97–98.

Pandravada, S.R., Sivaraj, N., Kamala, V., Sunil, N. & Varaprasad, K.S. (2008). Genetic Resources of wild relatives of crop plants in Andhra Pradesh – Diversity, Distribution and Conservation. *Proc. A.P. Akademi of Science, Special Issue on Plant Wealth of Andhra Pradesh 12*, 101–119.

Pandravada, S.R., Sivaraj, N., Kamala, V., Sunil, N., Sarath Babu, B. & Varaprasad, K.S. (2007). Agri-Biodiversity of Eastern Ghats-exploration, collection and conservation of crop genetic resources. In: *Proc. Natl. Sem. Conserv. Eastern Ghats*. ENVIS Centre, EPTRI, Hyderabad, India. pp. 19–27.

Pullaiah, T. (2001). *Draft Action Plan – Report on National Biodiversity Strategy and Action Plan: Eastern Ghats Eco-Region*. Anantapur, India.

Rao, R.S. (1998). Vegetation and valuable plant resources of the Eastern Ghats with specific reference to the Andhra Pradesh and their conservation. In: The Eastern Ghats – Proc. Natl. Sem. Conserv. Eastern Ghats. ENVIS Centre, EPTRI, Hyderabad, India. pp. 59–86.

Ravindranath, N.H., Murali, I. & Malhotra, K.C. (2000). Community Forest Management and Joint Forest Management in India. New Delhi, India: Oxford & IBH.

Rawat, G.S. (1997). Conservation status of forests and wildlife in the Eastern Ghats, India. *Environ. Conserv. 24*, 307–315.

Sastry, A.R.K. (2002). Hotspots concept: Its application to the Eastern Ghats for biodiversity conservation. In: *Proc. Natl. Sem. Conserv. Eastern Ghats.* ENVIS Centre, EPTRI, Hyderabad, India. pp. 184–186.

Shiva, V. & Ramprasad, V. (1993). *Cultivating Diversity.* Research Foundation for Science, Technology and National Resources Policy. Dehradun, India.

Shiva, V., Anderson, P., Schiicking, H., Gray, A., Lohman, L. & Cooper, D. (1991). Biodiversity: Social and Ecological Perspectives. New Jersey, USA: Zed Books.

Singh, A.K. & Varaprasad, K.S. (2008). Criteria for identification and assessment of agrobiodiversity heritage sites: Evolving sustainable agriculture. *Curr. Sci. 94*, 1131–1138.

Sivaraj, N. (1991). Phenology and reproductive ecology of angiosperm taxa of Shervaroy hills (Eastern Ghats, South India). PhD Thesis, Bharatidasan University, Tiruchirapallu, India.

Sivaraj, N. & Krishnamurthy, K.V. (1989). Flowering phenology in the vegetation of Shervaroys (Eastern Ghats: South India). *Vegetatio 79*, 85–88.

Sivaraj, N. & Krishnamurthy, K.V. (1992). Fruiting behavior of herbaceous and wood flora of Shervaroy hills. *Trop. Eco. 33*(2), 55–63.

Sivaraj, N. & Krishnamurthy, K.V. (2002). Phenology and reproductive ecology of tree taxa of Shervaroy hills (Eastern Ghats, South India). In: Proc. Natl. Sem. Conserv. Eastern Ghats. ENVIS Centre, EPTRI, Hyderabad, India. pp. 69–87.

Sivaraj, N., Pandravada, S.R. & Krishnamurthy, K.V. (2007). Gender distribution and breeding systems among medicinal plants of Eastern Ghats. In: Proc. Natl. Sem. Conserv. Eastern Ghats. ENVIS Centre, EPTRI, Hyderabad, India. pp. 148–153.

Solomon Raju, A.J. (2007). Bird-flower interactions in the Eastern Ghats. In: *Proc. Natl. Sem. Conserv. Eastern Ghats.* ENVIS Centre, EPTRI, Hyderabad, India. pp. 281–285.

Solomon Raju, A.J. & Henry Jonathan, K. (2010). Reproductive ecology of *Cycas beddomei* Dyer (Cycadaceae), an endemic and critically endangered species of southern Eastern Ghats. *Curr. Sci. 99*, 1833–1840.

Solomon Raju, A.J. & Purnachandra Rao, S. (2002a). Pollination ecology and fruiting behavior in *Acacia sinuata* (Lour.) Merr. (Mimosaceae), a valuable non-timber forest plant species. *Curr. Sci. 82*, 1467–1471.

Solomon Raju, A.J. & Purnachandra Rao, S. (2002b). Reproductive ecology of *Acacia concinna* (shikakai) and *Semecarpus anacardium* (marking nut), with a note on pollinator conservation in the Eastern Ghats of Visakhapatnam district, Andhra Pradesh. In: Proc. Natl. Sem. Conserv. Eastern Ghats. ENVIS Centre, EPTRI, Hyderabad, India. pp. 316–326.

Solomon Raju, A.J., Puranchandra Rao, S., Zafar, R. & Roopkalpana, P. (2003). Bird-flower interactions in the Eastern Ghats forests. *EPTRI-ENVIS Newsletter 9*(3), 2–5.

Solomon Raju, A.J., Purnachandra Rao, S. & Ezradanam, V. (2002). Bird-pollination in *Helicteres isora* and *Spathodea campanulata* with a note on their conservation aspects. In: Proc. Natl. Sem. Conserv. Eastern Ghats. ENVIS Centre, EPTRI, Hyderabad, India. pp. 308–315.

Solomon Raju, A.J., Purnachandra Rao, S. & Ezradanam, V. (2004). Pollination by bats and passerine birds in a dry season blooming tree species, *Careya arborea*, in the Eastern Ghats. *Curr. Sci. 86*, 509–511.

Solomon Raju, A.J., Venkata Ramana, K. & Henry Jonathan, K. (2009). Anemophily, anemo-chory, seed predation and seedling ecology of *Shorea tumbuggaia* Roxb. (Dipterocarpaceae), an endemic and globally endangered red-listed semi-evergreen tree species. *Curr. Sci. 96*, 827–833.

Soosairaj, S., John Britto, S., Balaguru, B., Nagamurugan, N. & Natarajan, D. (2007). Zonation of conservation priority sites for effective management of tropical forests in India: a value-based conservation approach. *Appl. Ecol. Environ. Res. 5(2)*, 37–48.

Sridhar Reddy, M. & Ravi Prasad Rao, B. (2007). Observations on distribution and popula-tion structure of *Pterocarpus santalinus* L.f. in Kadapa hill ranges. In: *Abstracts – Internatl. Sem. Changing Scenario in Angiosperm Systematics.* Shivaji University, Kolhapur, India. pp. 46–47.

Srinivasa Rao, A. (2007). Networking of Digital Resources on the Environment of Eastern Ghats – the EPTRI-ENVIS Perspective. In: *Proc. Natl. Sem. Conserv. Eastern Ghats.* ENVIS Centre, EPTRI, Hyderabad, India. pp. 602–607.

Sudhakar Reddy, C., Rout, D.K., Pattanaik, C. & Murthy, M.S.R. (2007a). Mapping of prior-ity areas of conservation significance in Eastern Ghats: A case study of Similipal Biosphere Reserve using Satellite Remote Sensing. In: Proc. Natl. Sem. Conserv. Eastern Ghats. ENVIS Centre, EPTRI, Hyderabad, India. pp. 435–438.

Sudhakar Reddy, C., Varma, Y.N.R., Brahmam, M. & Raju, V.S. (2007b). Prioritization of endemic plants of Eastern Ghats for biological conservation. In: Proc. Natl. Sem. Conserv. Eastern Ghats. ENVIS Centre, EPTRI, Hyderabad, India. pp. 3–13.

Suresh Babu, M.V., Srinivasa Rao, V. & Ravi Prasad Rao, B. (2007). *Cycas beddomei* Dyer (Cycadaceae). A global endemic of Tirupati and Kadapa hills. In: *Abstracts – Internatl. Sem. Changing Scenario in Angiosperm Systematics.* Shivaji University, Kolhapur, India. p. 54.

Swain, P.K., Siva Rama Krishna, I. & Murty, B.L.N. (2008). Sacred Groves – Their Distribution in Eastern Ghats Region. *EPTRI-ENVIS Newsletter 14*(4), 3–5.

Swaminathan, M.S. (1997). Implementing the global biodiversity Convention: IPR for public good. In: P. Pushpangadan, K. Ravi & Santhosh, V. (Eds.). Conservation and Economic Evaluation of Biodiveristy. Vol.2. New Delhi: Oxford & IBH Publishing Co.Pvt. Ltd. pp. 399–412.

Uma Ramachandran. (2002). Bettering Agriculture to Preserve Mountain Ecosystems. Examples from Eastern Ghats. In: Proc. Natl. Sem. Conserv. Eastern Ghats. ENVIS Centre, EPTRI, Hyderabad, India. pp. 548–557.

Venkaiah, M. (1998). Plant biodiversity in Vizianagaram district, Andhra Pradesh. In: The Eastern Ghats – Proc. Natl. Sem. Conserv. Eastern Ghats. ENVIS Centre, EPTRI, Hyderabad, India. pp. 1–5.

CHAPTER 11

COMPUTER APPLICATIONS IN ETHNOBOTANY

ASHISH KUMAR PAL[1] and BIR BAHADUR[2]

[1]Formulation Analytical Research Department, Aurobindo Pharma Limited and Research Centre, Survey No. 313, Bachupally Village, Quthubullapur Mandal, R.R. District 500090, Telangana, India, E-mail: ashishkumarhyd@gmail.com

[2]Department of Botany, Kakatiya University, Warangal 506008, Telangana, India, E-mail: birbahadur5april@gmail.com

CONTENTS

ABSTRACT

To date, many researchers have had to rely on was their memory of the data they collected and the meaning of those data in the context of their ethnobotanic study. However, the working of memory create two potential problems for researchers analyzing the field data. First, researchers may use those data that were most dramatic in the fieldwork and erroneously present them as being the most significant; second, they may use more data from the later stages of fieldwork and less of what happened in the middle or beginning because the later data are fresher and clearer in their minds. Computer-assisted qualitative data analysis (CAQDA) can help the careful analyst avoid both these problems. Also, the volume of data and its compilation also presents a big challenge to modern ethnobotanist. Quantitative Computer Soft wares can help reduce these issues. The use of computers could stimulate team approaches in research that would generate a wealth of data and make important analytic contributions. Also in teams, the data are gathered by a number of researchers who in many cases have different degrees of training as well as different degrees of insight, there is no way to assure consistency in what each researcher thinks it important to record. Thus computer are of great help in ethnobotanical research. The present paper is an effort to bring forth several qualitative and quantitative softwares now available in market that offers the researcher the choice the optimum software depending upon both its advantages and limitation so that a suitable conclusion can be drawn for ethnobotanical research work.

## 11.1	INTRODUCTION

Computer-assisted data analysis (CADAS) is associated with the analysis of aggregate data according to the tenets of logical positivism. There are more than twenty computer programs designed to assist researchers analyzing ethnographic data, and these programs may be used by researchers with a variety of epistemological orientations.

Computer-assisted analysis in sociology is currently associated with both categories of researches namely quantitative and qualitative research. Statistical procedures available in main stream software packages, such as statistical analysis system (SAS) and statistical package for the social sciences (SPSS) helps the analysis of aggregate data, and thus computer-assisted analysis carries suggestions of hard data, computation, and objectivity.

Qualitative research aggregate data analysis using statistical procedures has many setbacks, as it either misses important sociological causes of social action or emphasizes at the expense of understanding. Software for the analysis of qualitative data has appeared relatively recently, because qualitative sociologists have been slow to adopt to softwares.

11.2 UTILITIES OF COMPUTERS IN ETHNOBOTANY

Computer-assisted qualitative data analysis (CAQDA) programs automate analysis procedures that have been used by ethnobotanists. This opens up new avenues through the use of linked coding schemes, hypertext, and case-based hypothesis testing. Just as it can with aggregate data, computer assistance can facilitate systematic computational research with qualitative data. In addition, CAQDA technologies can be useful to researchers who place themselves outside the positivistic research tradition.

CAQDA does not differ fundamentally, for the most part, from the no mechanical qualitative analysis traditions from which it has developed. Most computers ease the labor burden and broaden the scope of common analysis tasks, such as typing up field data and memos, searching for text, coding data, sorting and comparing codes. Hypermedia is an unique contribution of computer technology to the analysis of qualitative data. Linking text, analysis, and non-text materials (graphics, sound, and video) in a single analytical space outside the mind's eye is not possible manually.

Such software helps to organize, manage and analyze information. The advantages of using this software include being freed from manual and clerical tasks, saving time, managing huge amounts of qualitative data, having increased flexibility, and having improved validity and auditability of qualitative research. Concerns include increasingly deterministic and rigid processes, privileging of coding, and retrieval methods, rectification of data, increased pressure on researchers to focus on volume and breadth rather than on depth and meaning, time and energy spent learning to use computer packages, increased commercialism, and distraction from the real work of analysis.

11.3 RELEVANCE OF COMPUTERS IN CURRENT ETHNOBOTANY PRACTICES

The use of computer assisted data analysis is comparatively a newer concept in ethnobotany. Field notes have been transcribed into word processors and many ethnobotanist now carry portable lap-top computers to the field. The use of computers for the analysis (rather than the gathering) of ethnobotanical data is a comparatively recent development.

11.4   TYPES OF SOFTWARES

Computers can be programmed to accomplish four different kinds of analysis: numerical/arithmetic analysis, writing and document processing, data organization, and symbolic manipulation. Ethnobotanists use computers for all these kinds of analysis.

11.5   TYPES OF COMPUTER-ASSISTED QUALITATIVE DATA ANALYSIS (CAQDA) BASED ON THEIR FUNCTION

CAQDA softwares are classified into the following types based on their function and their ability to analyze the data gathered (Weitzman and Miles, 1995).

11.5.1   DOCUMENT PROCESSING: SEARCHING AND RETRIEVING

Word processing is the major feature of computer assistance for the ethnographer. Many ethnobotanist require computer assistance for searching with a word processor. Basic searches retrieve a text string from a single computer file. More advanced searches count the occurrences of "a string" and stand-alone search engines can search multiple files and produce extracts of search "hits" in context. Specialized programs developed for both CAQDA and commercial uses enhance the search and retrieval process. Many of these programs are designed for what Tesch (1990) called descriptive-interpretive work rather than theory building, see Tesch (1991). For searching and retrieving, packages including GOFER, Metamorph Orbis, Sonar Professional, The Text Collector, WordCruncher, ZyINDEX, and FYI3000PLUS expand on the capacities of word processors in several ways (Weaver and Atkinson, 1994).

First, these packages create and manage the ethnographic database. Some of these packages manage files off-line (data remain in separate, unaltered text files); others manipulate the data directly. Usually, document processors work on documents that have already been produced in a word-processing package. Orbis manages files produced in XyWrite or NotaBene; MetaMorph and WordCruncher are particularly adept with WordPerfect documents. Others read files produced by a variety of word-processing, database, spreadsheet, and even drawing programs. Nearly all can manage plain text files, and some packages require files to be in this format before they can work with them.

The second value-added feature of document processors is their search features. As part of their management of the qualitative dataset, document processors allow the analyst to specify a variety of computer files in which to conduct a single search. ZyIndex, for example, searches documents that remain in their native format off-line, allows the analyst to keep track of changes to documents through several revisions, and indexes files so they can be readily included or excluded from particular searches. Document processors can mount complex searches: combinations or sequences of text strings; strings within specified proximity of each other; word synonyms, stems, and roots; and searches defined through Boolean, fuzzy, or set logic. Some display the results of searches interactively so that analysts can see how the addition or deletion of certain search terms in a complex search affects the number of hits produced.

Document processors are designed to make it easy for ethnographers to investigate data they have collected. Compared to word processors, document processors do a better job of placing the complete ethnographic dataset in the hands of the analyst. They allow the ethnographer to search more easily for desired pieces of text and to investigate how the text can be arranged in the dataset.

11.5.2 DATA ORGANIZATION

Searching and retrieving allows the analyst to inspect but not alter the ethnographic database. However, CAQDA packages, such as askSam, Folio Views, MAX, Tabletop, HyperQual2, Kwalitan, Martin, QUALPRO, and The Ethnograph allow the analyst to alter the form of the ethnographic database by organizing its text.

Organizers expand on document processors in two ways: (1) Organizers allow the ethnobotanist to attach a structure to the database. Some document processors can retrieve text chunks in context. Organizers create context by giving analysts control over the structure of the database, and this structure can be manipulated and analyzed by the researcher. Organizers can also structure the database by adding database fields for factual information and for memos that are produced during analysis; and (2) Addition of organizers is the ability to code data according to a theoretical scheme developed by the analyst. Organizers are designed to tag chunks of text with analytical codes and to retrieve codes and tagged text. Retrieval of codes frequently includes the ability to search for multiple codes, to retrieve the text associated with codes, or to count codes.

11.5.3   ORGANIZING AND ANNOTATING

Organizing and annotating are two basic tasks of qualitative data analysis. Some computer applications are designed to translate these activities with fidelity from hard copy to electronic form. For example, HyperQual2 and Martin use note cards as an organizing metaphor. Like their hard-copy counterpart, the note cards of these CAQDA packages each contain a single chunk of text. Electronic cards can be replicated and sorted into stacks, and these stacks then provide the raw materials to write up memos, annotations, and the ethnographic report. Another way to organize a hard-copy database is to use database-like fields. Fields can contain a wide variety of information including factual information that situates the ethnographic text to which it is attached (data collector, date of interview or observation, information about the subject of the note) or analytical information about the text itself. CAQDA software, such as askSam facilitates the creation, insertion, and organization of these fields. Once organized, these CAQDA programs can quickly search and retrieve information from database fields and quickly count and tabulate the results of these searches.

CAQDA packages that accommodate organizing and annotating the database are useful in a variety of situations, but they are particularly useful in research projects as they expand in size and scope. Multisite or multiyear ethnobotanical projects generate a plethora of notes that beg for efficient organization. Flexible annotations are particularly valuable in multiresearcher projects in that each researcher provides his/her own analysis and commentary.

11.5.4   CODING, RETRIEVING AND COUNTING

Coding and retrieving is one of the central tasks of CAQDA software packages. Many of the software packages discussed above can code textual data, retrieve text based on applied codes, and tabulate which codes have been applied to which text. Most packages discussed in this section and below use coding and retrieving as their primary method of analysis or as a preface to other kinds of analysis. There are many ways to apply codes to text. Software, such as Kwalitan, QUALPRO, or The ethnograph number each line in the ethnographic database and apply codes to specific lines. Some packages encourage coding on the computer screen, whereas others encourage the analyst to code a numbered print-out of the text for later entry.

Once the codes are applied to the database, CAQDA software greatly accelerates analysis based on retrieving codes. Code-and-retrievers find codes using the same powerful features that document processors applied to the raw database. Multiple codes may be searched for at once. Hierarchies of codes can be established so that searches for higher-order terms also retrieve instances of lower-order terms. Complex searches can be formulated using Boolean, sequential, and proximity logic. Retrieval may yield a display of text associated with a code or a union of codes, or it may yield counts where those codes were applied. A number of CAQDA packages support cross-tabular displays of counts.

Organizing with CAQDA alters the ethnographic database in two ways. First, the database can be organized using database fields, hierarchical levels, or annotations so that the analyst has an easier time placing data in context and moving about in large ethnographic databases. Second, the database can be organized by applying codes to the text of the database so that the analyst can retrieve information from the database based on a theoretical mark-up of the text. CAQDA software facilitates the administration of both of these activities, but it does little to guide the intellectual work involved.

11.5.5 SYMBOLIC MANIPULATION

There are three kinds of CAQDA software for symbol manipulation. Some symbol manipulators begin where code-and-retrievers leave off. These packages focus analysts' attention on the coding process, encouraging them to create positive links between codes and to develop theory as they create a coding scheme. A second form of symbol manipulation is done by theory-building software. These packages take material that has been abstracted from the database through coding or other means and analyze relationships between codes or concepts. The final kind of CAQDA software that facilitates symbolic manipulation is hypothesis testers. These packages facilitate the advancement and testing of causal statements about relationships between codes or concepts in multiple cases in the database.

11.5.6 VALUE-ADDED CODERS

The coding process already contains the seeds of symbol manipulation. Value-added coders add additional coding and analysis features to allow the analyst to move closer to the manipulation of concepts usually by moving further from the ethnographic text. Software packages, such as AQUAD allow the analyst to search purposefully through the ethnographic database for combinations of codes. The analyst can look for theoretically significant combinations of codes, tabulate the number of instances, and compare them to counts for combinations of codes that represent competing theories. Value-added coders consider the ethnographic database on a case-by-case basis so the counts and cross-tabulations they produce are a case-based numerical summary in contrast to the variable-based summaries provided by quantitative analysis.

A second way of transforming coding into symbol manipulation is to involve the computer in the construction of the coding scheme. Other value-added coders involve the computer in the coding process without imposing hierarchical constraints on the coding scheme. In ATLAS.ti, for example, the coding scheme is not constrained by the software but is retained to manipulate and analyze on its own. Text, codes, and memos can be linked in the program and these links later inspected and manipulated in conjunction with the original ethnographic text. Maps of relationships between elements in the database provide an analytical metaphor distinct from quantitative summary statistics or cross-tabulations.

11.5.7 THEORY BUILDERS

Compared to value-added coders, theory-building CAQDA software moves the analyst a step further from the ethnographic text. Software packages, such as ETHNO, Inspiration, MECA, and MetaDesign are designed to facilitate the conceptual manipulation of ethnographic data. Theory-building CAQDA software packages do not actually construct theory, of course. They construct a graphical map (node and links) of data. Nodes represent data (field notes, memos, codes, etc.), and links represent relationships between data. Maps may help the analyst picture the project's theoretical shape, the concepts in use, the relationship between those concepts, and the ethnographic data that have been collected regarding each of those concepts and links. Theory-building software facilitates experiments with different concepts and links within the research project.

But theory-building CAQDA packages need not be reserved for the armchair ethnographer idly speculating on abstract relationships in field data. Theory builders can also incorporate links to the original text that encourage grounding in the original data and checks on concept validity. In addition, theory builders need not be reserved for analyzes of a nearly finished research project (nor need they be the exclusive province of the principal investigator). Theory builders can aid researchers who are mapping complex empirical concepts or events during the course of fieldwork.

11.5.8 HYPOTHESIS TESTING

Some value-added coders, such as HyperRESEARCH and AQUAD as well as stand-alone packages, such as QCA use hypothesis testing, the third form of symbol manipulation. Hypothesis-testing software bridges the gap between qualitative and quantitative analysis by facilitating case-based analysis of qualitative data. These packages allow the analyst to specify hypotheses based on codes applied to text (in HyperRESEARCH and AQUAD) or based on a descriptive matrix of cases (in QCA). Hypothesis testers determine how causally antecedent features of cases are related to outcomes. Boolean algebra is used to define the antecedent conditions for each case in the database. CAQDA software reduces large numbers of cases into statements that identify under what conditions the outcome of interest prevails.

Qualitative hypothesis testing determines what qualities of cases are crucial for a specified outcome. In contrast, quantitative hypothesis testing focuses on the contribution of different variables to the outcome. Apart from this difference, CAQDA packages that include hypothesis-testing features are similar to statistics software that dominates computer-assisted analysis of quantitative data. Hypothesis testers encourage the analyst to develop ideas in the form of equations (Boolean rather than arithmetic) and to investigate how different terms (binary codes rather than multivalve variables) in the equation affect its ability to accurately explain outcomes.

Stand-alone hypothesis testers remove the analyst from the original database. These software packages are useful in the analysis of data from a variety of sources and not only from ethnobotanical field studies. Hypothesis testers that include search-and-retrievers or data organizers may encourage the analyst to remain in contact with the database even as analysis proceeds along more abstract and quasi-quantitative avenues. Ideally, hypothesis-testing software allows the analyst to ensure reliability through hypothesis checking and to maintain validity by returning frequently to re-examine the original database and the codes, memos, and annotations that have accumulated over the course of the research project.

Symbol manipulation includes a variety of techniques for analyzing ethnographic data in ways that take advantage of microcomputers. Value-added coders encourage the analyst to develop explicit links between codes and data as the analysis proceeds. The software keeps track of the relationships between codes as they develop and then makes them available for later re-inspection and analysis. Theory builders facilitate exploration of concepts in research projects through graphical displays and the ability to quickly move between different levels of detail. Finally, hypothesis testers move CAQDA closer to the practices of quantitative research by embracing the goals of reliability and explanation. Hypothesis-testing packages may even allow analysts to strive for reliability and causal explanation without losing the traditional advantages of qualitative data with respect to validity.

11.6   QUALITATIVE DATA ANALYSIS (QDA) GENERAL APPROACHES

Many CAQDA software packages facilitate data analysis from the grounded-theory perspective. Grounded theorists advocate close contact with raw data, the emergence of analytical categories from the data through memo writing, and comparison as the primary analytical tool. Elements of grounded theory are common in CAQDA in part because Glaser and Strauss are explicit about the principles and procedures involved in this kind of analysis as stated by Straus (1987).

11.7   SYSTEMIZATION OF ETHNOBOTANICAL METHODS

The use of CAQDA software makes explicit the methods of analysis used in converting ethnobotanical data into meaningful ethnobotanical reports. The explicit discussion of methods of analysis in the grounded-theory school midwifed the development of much CAQDA software. Computer-assisted analysis goes beyond discussion, however, by allowing researchers to share details of their analysis process. Even when ethical concerns prevent the sharing of raw data, the use of CAQDA may increase reliability by making explicit the concrete steps taken in moving from data to conclusion.

Secondly, the use of computers fosters increased reliability and generalizability by expanding the amount of data that can be managed and exhaustively analyzed within a single ethnographic project. Data expand rapidly in ethnographies involving multiple sites or multiple researchers.

CAQDA software may allow researchers to access large ethnobotanical databases directly without the theoretical intermediary of a single intellectual vision or research goal. The computer can accommodate data collected by multiple fieldworkers and facilitate coding, re-coding, linking, and re-linking by multiple investigators. Within this analytical space, differing understandings of the same database can be produced and compared, and analysts can examine the procedures undertaken to produce each account.

QDA that is consistent with grounded theory uses a sequential style of analysis that is highly data-intensive. Advocates of these methods urge the analyst to begin data analysis while collection is under way, to reduce the data using codes or categories, to shuttle between data and codes, and to compare coded and raw data to make tentative and ultimate conclusions. This analytical strategy returns the analyst to the database over and over again, and each step of analysis is readily translated into computer modules and procedures.

11.8   STEPS AND BASIC CAPABILITIES OF CAQDA

11.8.1   ENTERING DATA

Entering data into the computer is an important decision with enormous consequences. Which data are entered into the computer, how they are entered, and which remain outside the computer shape all further analyzes of the data. Data can enter a computer in a myriad of forms, from the "beginning" methods of text processing on a word processor to "advanced" methods of digital signal processing of videotape. Primary consideration for researchers entering text data into the personal computer is the size of the textual unit of analysis. Notes entered into a dedicated CAQDA package are divided into analysis "key words" which can be single words, lines of text, paragraphs, hypertext note cards, or larger files.

Especially important is the size of these key words as they are de-contextualized and re-contextualized during the analysis process. Larger and more elaborate key words of text are more likely to contain data falling into several analytical categories, and this may complicate analysis.

On the contrary for those interested in context-dependence, smaller key words may prove worthless unless to the CAQDA software as it contains elaborate coding or linking procedures. Practical issues also arise at the data-entry stage. Specialized and non-relevant field notes, interview transcripts, and memos can create many user-based confusions leading to greater errors in the final outcome.

11.8.2   ORGANIZING DATA

The organization of data depends upon the research and its objectives. The number of researchers involved in the project, the number of field sites, the variety of data types, and the theoretical orientation of the researchers all influence how the dataset is organized. The end users, for example, the researchers have to familiarize themselves with basic and software use.

11.8.3   SEARCHING FOR AND RETRIEVING DATA

Another axillary use of CAQDA is that it enhances the researcher's ability to search for and retrieve text. Search and retrieval surely does not mean the end of the computer's usefulness as a qualitative data analyst though several CAQDA packages are designed for this kind of analysis. CAQDA software usually allows researchers to search for root forms of words or synonyms, use combination searches, such as those based on word proximity or word order. Retrieval of searched-for items is usually dependent on the keywords.

11.8.4   CODING OR INDEXING DATA

Coding is a most important feature of CAQDA. The use of the computer does not affect the fundamentals of data coding. The field notes has to be manually coded before entering the data into their CAQDA package. Since coding is more of a mechanical process and require less intellect of the researcher, CAQDA are best useful in this step.

The disadvantage of computer assistance is that it can impose limitations on the coding process creating problems for ethnographers. Analysts must be confident that using the computer facilitates their work.

11.8.5 ANALYZING CODES

Analysis of codes starts simultaneously as the first data are coded. Codes are defined in relationship to each other, so their application to a set of data implies theory. CAQDA software can make this implicit theory explicit by generating a list or map of codes and their relationships. Some packages constrain the development of a coding scheme to encourage the analyst to make positive connections between codes, such as hierarchical connections between more and less inclusive ones or sequential connections between coded events.

Once sufficient data have been coded, other analytical possibilities develop. In most CAQDA packages, analysts search for codes as easily as they explore raw data. Packages that retrieve text associated with particular codes or conjunctions of code may be useful for analysts interested in interpretational analysis.

Apart from data entry, the analysis of codes is the area of the computer's greatest influence on theory and methods. Software design may force the analyst to consider the previously unexamined relationship between concepts in the research project. The flip side of the coin is that software may limit the ability of the analyst to develop theory in desired directions. The ability to mount comprehensive searches for codes /sets of codes means that the ethnographic analysis may benefit from less bias. But large-scale searches can also bury the analyst in chaotic results. In short, the computer-assisted analysis of codes has theoretical and methodological implications surpassed only by those taken during the first steps of data entry.

11.8.6 LINKING DATA

Software available during the last decade permit analysts to create hypertext links between combinations of data, codes, memos, and research reports. Graphics, sound, and video may also be incorporated into " hyperspace" databases as sited by Weaver and Atkinson (1994).

Analysis based on data linking is most useful for ethnobotanists who collect non-textual data, especially if hypertext moves out of the researcher's office and becomes a medium for the distribution of research reports. For researchers working outside of the positivistic tradition, linking data may be particularly valuable. Hyperlinks concretize nonlinear data-analysis techniques and free the researcher from reliance on computation. Reports that incorporate graphics sound, and video can more readily make the case for the significance of context.

But hypertext technology also imposes special limitations on analysts. At present, the incorporation of text into hypertext "spaces" is inevitably fraught with more burdensome formatting limitations than those imposed by traditional text databases. Integrating sound or video into an ethnographic database involves technological expertise beyond the use of the word processor. In addition, the publication of materials using sound or video technology may introduce new ethical considerations, such as the protection of research subjects' confidentiality.

11.8.7   ANALYZING LINKS

Analyzing links within the database is a more general form of analyzing codes. Links may be analyzed only after a certain number have been established in the data. Once established, the links may be abstracted from the original data and analyzed as a system or network of their own. Compared to the analysis of codes, the analysis of links is more flexible and general. Greater complexity is possible in hypertext links than in coding schemes, so the representation of linked data may consequently be more complicated.

11.9   TYPES OF SOFTWARES USED IN ETHNOBOTANY

11.9.1   AQUAD

The first version of Analysis of Qualitative Data (AQUAD) was developed in Germany in 1987 to compensate for a shortage of manpower in research projects. At this time several programs for qualitative analysis had been in existence for a number of years. Simpler types of software tools were using the search function of available word processors or database programs. Some were already designed for the special demands of qualitative analysis, although they did not offer more than simple counting and retrieval functions. This program made use of the rich potential of so called " logical programming", and it was the stimulus and model for AQUAD.

In all qualitative analyzes following the "coding paradigm" the main task is to reduce the usually wordy and redundant descriptions, explanations, justifications, field notes, protocols of observations, etc., that comprise the researcher's data texts to some kind of systematic description of the meaning of the data.

AQUAD is a program for the generation of theory on the basis of qualitative data. Since theory-building and hypothesis-testing have been traditionally the domain of researchers who work with quantitative data, theoretical notions based on qualitative data are easily distrusted. Although we acknowledge (and have no desire to claim otherwise) that qualitatively developed statements do not achieve the same degree of generalization as statistically tested statements, it is important to make sure qualitatively developed conclusions are based on as rigorous a verification process as possible. Therefore, special emphasis is placed in AQUAD on objectivity, reliability, and validity. The researcher is encouraged to use procedures, such as the repeated interpretation of the same text by the same analyst or by different analysts over time, or to pay attention to issues of internal validity, such as whether the categories have been used consistently, whether their defined range of meaning has been maintained, whether the meanings represented by specific categories indeed correspond to the content of the text passages that are sorted into them, etc.

More information about these issues will be presented later in this paper. Finally, two more attributes of AQUAD deserve attention. Unlike many other qualitative analysis programs, AQUAD supports some versions of conventional content analysis or linguistic analysis by allowing the user not only searching for words and phrases that occur in the data text and examine their frequency, but the program can extract words together with their context (key word in context – KWIC – indexes).

Furthermore, AQUAD provides for the attachment of researcher memos to text segments. Principally empirical research follows a path towards discovery that starts from descriptive or categorical analyzes and leads via postulating or observing regularities to statements, which explain these connections at least tentatively. No matter whether an analysis results in a taxonomic, correlative or causal order of the phenomena under study, the process of research is focused on reduction that is reducing concrete details and moving to higher levels of abstraction and generalization. That will enable one to see the essentials. AQUAD tries to contribute to this goal, but at the same time able to keep open the way back to the manifold, concrete, and colorful details of the original database.

11.9.1.1 UTILITIES OF AQUAD (SEE, HTTP://WWW.AQUAD.DE/EN/)

AQUAD facilitates the tools of content analysis as under:

- Text searching: AQUADs basic utility is looking for segments in texts. Keywords can be entered and the AQUAD search engine can be used to search related results.
- Retrieving by file name, code, keyword, or parts of memo texts: Retrieving segments according to criteria.
- Coding: Labeling segments in texts, audios, photos, or videos can also be done in AQUAD
- Annotation: Inserting annotations linked to parts or whole texts, audios, photos, or videos. This facilitates in easy analysis of the data and thus developing a theory.
- Construction of linkage hypotheses: Linking data to reach a conclusion is most important feature for any analysis tool. AQUAD looks for relationships among codes.
- Comparison of cases/files: contrasting coding among files.
- Word analysis: Counting words according to criteria can also be accomplished using AQUAD.

FIGURE 11.1 Constructing tables combining criteria and arranging data in rows (Source: http://www.aquad.de/en/; used with permission.)

11.9.2 CODING ANALYSIS TOOLKIT

Coding Analysis Toolkit (CAT) (see, http://cat.ucsur.pitt.edu/) is a free service of the Qualitative Data Analysis Program (QDAP), and hosted by the University Centre for Social and Urban Research, at the University of Pittsburgh, USA and QDAP-UMass, in the College of Social and Behavioral Sciences, at the University of Massachusetts Amherst, USA. CAT was the 2008 winner of the "Best Research Software" award from the organized section on Information Technology and Politics in the American Political Science Association. It is a web-based suite of CAQDAS tools.

CAT is able to import ATLAS.ti, which is a computer program used mostly, but not exclusively, in qualitative research or qualitative data analysis.data, but also has an internal coding module. It was designed to use keystrokes and automation as opposed to mouse clicks, to speed up CAQDAS tasks.

11.9.3. COMPENDIUM

Compendium is open source like those latter products but it is driven by an academic institute rather than commercial or hobby interest. From a practical point of view Compendium is more than a mind mapping product, something larger and more ambitious a framework for capturing the connections between information and ideas.

Compendium (see, http://www.graphic.org/mind-mapping-software/compendium-review.html) is developed in KMi, in collaboration with various US partners, and funded from externally secured research grants.

The builders of Compendium have provided a number of screen cast movies on their website which take us through the initial concepts contained within the program. They are well worth viewing as a quick way to get to know the software.

11.9.3.1 DATABASE DRIVEN (SEE, HTTP://COMPENDIUM.OPEN. AC.UK/INDEX.HTML)

The program starts up with an invitation to create a user account and then an initial mind map. This is the first sign that Compendium is built from the ground up using a database to provide a multi-faceted collaborative tool. Most mind mapping software has had collaboration capability bolted on as the internet has become faster and more reliable but this is not the case with Compendium.

11.9.3.2   MAP CREATION

After following the initial screencasts one will be able to create maps. These are completely freeform and there are a number of different types of nodes (node appears to be the chosen term for an item in most mind mapping programs). As Compendium is a tool for mapping ideas, discussions and problems the nodes are more specialist than in pure mind mapping tools. You can add a node that asks a question, answers it or ascribes positive or negative attributes to another node. There are also decision making and argument nodes.

11.9.3.3   CONTEXT CLICK CONSISTENCY

There is some inconsistency as to when an area or a node is in focus (in terms of the mouse) This means one sometimes fail to get response to a mouse click and have to repeat it after waiting to realize that nothing's going to happen. This is the sort of minor irritation that open source users have to live with which wouldn't happen with a commercial product. Having said that there is a full and comprehensive set of keyboard shortcuts for almost any task and they are all listed in sections in the Help file.

11.9.3.4   CONNECTING LINKS

One difference with Compendium over other mind mapping products is that lines connecting the nodes aren't drawn in automatically as nodes are created. They are dragged into place with a right-click (ALT+mouse button for Mac users) once the nodes are in place. This makes map creation slightly slower than with a package like XMind, which connects nodes at creation, but allows complete and absolute freedom as to where lines should go, with multiple end and starting points no problem at all.

11.9.3.5 NODE PROPERTIES

The nodes can have all properties ascribed to them, comments, tags, etc. Tags show some of the power that Compendium has behind the maps that can be drawn. There is a group of pre-assigned tags provided with the package but also the ability to add your own. Once items have been assigned relevant tags, or perhaps updated after a project meeting, you can search on those tags (as well as other properties) to find all nodes with that tag, not only in the current mind map but across however many maps you can get access to.

11.9.3.6 BELLS AND WHISTLES

Compendium has some neat tricks as well. One can overlay a map on top of an image to add notes and graphics to it. This allows one to, for example, annotate satellite maps or highlight items on a picture of a circuit board. These tasks normally need knowledge of a complex graphics program but with Compendium it is straightforward. Setting Text on mouse over is very useful too. If care is taken over the text chosen this can be used effectively so that people looking at a mind map can get the overall picture before delving down into the layers that will specifically interest them.

11.9.3.7 COMPENDIUM CONCLUSIONS

Compendium has the mind mapping concept at its heart is much more than a tool to draw and update mind maps. Many people use Compendium as an organizer for research projects or even their lives and its ability to link to almost any object certainly lends itself to this. The downside is that one need to use Compendium as the link between all activities. Compendium visually represents thoughts and illustrates the various interconnections between different ideas and arguments. The creation of "issue maps" graphically represents the relations between issues and questions and facilitates the understanding of interconnected topics through pictorial representation.

It can be used by a group of people in a collaborative manner to convey ideas to each other using visual images.

11.9.4 ELAN AND ANNEX

ELAN is written in the Java programming language as a local tool and stores the transcription data in a specialized XML format, EAF (ELAN Annotation Format). It is available for Windows, MacOS X and Linux. On Windows and Mac it dele- gates the media playback to an available high performance native media framework: DirectX/DirectShow or QuickTime on Windows and QuickTime on Mac. On Linux JMF is used. Due to its reliance on native media solutions it can provide frame-accuracy in particular on Windows systems.

ANNEX is written as an ELAN compliant server-based tool relying on streaming media via the combination Darwin media streamer and Quicktime client. Due to the used Internet protocol ANNEX cannot provide the same high timing accuracy as ELAN. While ELAN creates a local index for content searches on physical directory structures, ANNEX works with an index created for a whole Language Achieve Management and Upload System (LAMUS) archive using the Postgres database system. ELAN and ANNEX have the following major features (see, https://tla.mpi.nl/tools/tla-tools/ annex/attachment/annex-elan_flyer_2006-05-11/):

- Association of up to 4 videos to a transcription document and synchronized viewing;
- Accurate alignment of annotations to the media, with maximum precision of 1 millisecond;
- Creation of unlimited number of tiers (layers) and unlimited number of annotations;
- Free selection by the researcher of the tier structure and the values to be used on these tiers;
- Tiers can be grouped hierarchically; different types of dependencies between tiers can be defined;
- Multiple, customizable views on tiers and annotations;
- Specialized viewer for time series data, such as from cyber gloves;
- Creation and use of controlled vocabularies;
- Import modules for Shoe-box/Toolbox, Transcriber, and CHAT, export to Toolbox and CHAT, tab-delimited text and interlinear text;
- Highly customizable, inter-linear-style printing of the transcript;
- Versatile structured and unstructured search options;
- Productivity enhancements, such as semi-automatic segmentation, to kenizing, copying of tiers, keyboard shortcuts;
- Multiple undo and redo;
- Support for templates;

- Input methods for different character sets, such as IPA, Chinese, Hebrew, Arabic, etc.
- Multilingual user interface;
- Adheres to standards like XML and Unicode.

11.9.5   TAMS ANALYZER

TAMS stand for Text Analysis Markup System. It is a convention for identifying themes in texts (web pages, interviews, field notes). It was designed for use in ethnographic and discourse research.TAMS Analyzer is a program that works with TAMS assigns ethnographic codes to passages of a text just by selecting the relevant text and double clicking the name of the code on a list. It then allows extracting, analyzing and saving coded information. TAMS Analyzer is open source; it is released under GPL v2. The Macintosh version of the program also includes full support for transcription (back space, insert time code, jump to time code, etc.) when working with data on sound files (Figure 11.2).

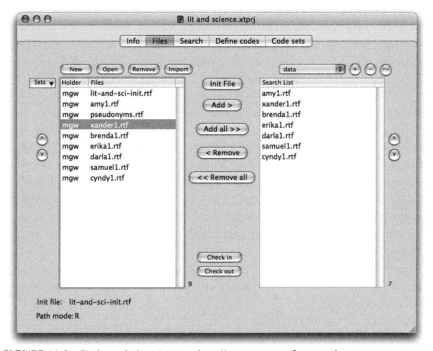

FIGURE 11.2   Project window (source: http://tamsys.sourceforge.net/).

11.9.6   QDA MINER

QDA Miner is a mixed methods and qualitative data analysis software developed by Provalis Research. The program was designed to assist researchers in managing, coding and analyzing qualitative data QDA Miner was first released in 2004 after being developed by Normand Peladeau. The latest version-4 was released in December 2011. QDA Miner is widely used software for qualitative research and used by market researchers, survey companies, government, education researchers, crime, fraud detection experts, journalists, etc.

Structure of work in QDA Miner QDA Miner functions using an 'Internal Database Structure' so that the documents are imported into the database and all the workings are held in about six project files. This will vary according to the types of data being stored. Being designed for the management of mixed-methods projects, QDA Miner and its sister programs can handle large amounts of data in a variety of formats. Data can be imported or a project file can originate and be structured directly from spreadsheet applications, when for example analyzing open-ended survey questions. The ability to archive and compress the project file facilitates the backing up and movement of projects. The compression reduces file size to between 15% and 20% of the original. The QDA Miner interface is divided up into resizable areas, with lists of variables, documents and codes and the main document window (with coded margin) are viewable simultaneously throughout.

11.9.6.1   VARIABLES STRUCTURE

QDA Miner operates using a cases and variables structure which is unusual amongst CAQDAS packages. The variable values of the currently selected case are displayed adjacent to the data along with the case from which the data derives and the main codes list. A new project can be created in number of different ways, for example, from a list of documents; from a database/spreadsheet containing quantitative variables; using a document converter, etc.

Variables act as holders for different types of qualitative/categorical and quantitative information. New document type variables can be added at any time to handle additional types of data, for example, notes for each case.

Variables of various types (e.g., numeric, nominal, ordinal, Boolean, string) are used for case filtering and comparison. This information can be imported from a spreadsheet.

Structure of work in QDA Miner QDA Miner functions using an 'internal database structure' so documents are imported into the database and all the workings are held in about six project files. This will vary according to the types of data being stored. Being designed for the management of mixed-methods projects, QDA Miner and its sister programs can handle large amounts of data in a variety of formats

Data can be imported or a project file can originate and be structured directly from spreadsheet applications, when for example analyzing open-ended survey questions. The ability to archive and compress the project file facilitates the backing up and movement of projects. The compression reduces file size to between 15% and 20% of the original. The QDA Miner interface is divided up into resizable areas, with lists of variables, documents and codes and the main document window (with coded margin) are viewable simultaneously throughout.

Data types and format in QDA Miner Textual formats: QDA Miner directly stores text saved as 'rich text format' (rtf) and ASCII (txt). However, the Document Conversion Wizard can convert various file formats, such as MS Word, WordPerfect, HTML, Adobe Acrobat to work with inside the software. Textual data is fully editable using standard Windows formatting toolbars. Objects, such as tables and graphic elements can be embedded into rtf documents

QDA Miner can also convert projects created in other CAQDAS packages, including ATLAS.ti, HyperRESEARCH and NVivo.

11.9.6.2 DATABASE FORMATS

Several database and spreadsheet file formats (including MS Access, Excel, dBase, Paradox) and any data file with an ODBC driver (Oracle, MS SQL, etc.) can also be imported.

11.9.6.3 CODING PROCESSES IN QDA

Miner Codes can be assigned to any segment of text, to one or several table cells, or a whole graphic or another embedded object. Drag and drop codes on to text or double click on a code to assign it to selection. Codes can be applied to whole paragraphs without manual text selection by drag and drop.

11.9.6.4   DATA ORGANIZATION IN QDA MINER

Documents are organized by Cases for filtering purposes. Cases can be grouped and ordered according to selected variable values, and these orderings follow through to generated outputs. Any numerical, categorical, logical and date data may be used to categorize cases, which can be automatically assigned upon data import. QDA Miner can handle more than 2000 variables and several million cases.

11.9.6.5   OUTPUT IN QDA MINER

Tabular outputs can be printed or exported to Excel, HTML, comma or tab-delimited files. Coding retrieval results may also be exported as a new QDA Miner project. Text reports may be saved to disk in Rich Text, MS Word™, ASCII or HTML format. Various graphical displays can be generated and exported, for example, cluster plots, heat maps, etc. Export to BMP, WMF, PNG and JPG format. Whole project export to spreadsheet and database file formats, for example, Quattro Pro, Lotus, Excel, Paradox, dBase.

11.9.7   HyperRESEARCH

HyperRESEARCH is used by qualitative researchers in areas, such as health care, legal, sociology, anthropology, music, geography, geology, education, theology, philosophy, history, market research, focus group analysis and most other fields using qualitative research approaches. This is designed to assist with any research project involving analysis of qualitative data. It's easy to use and works with both Mac and Windows computers. So when collaborating with multiple researchers, everyone gets to use their preferred computer. HyperRESEARCH is powerful, and flexible, which means that no matter how the data is approached, the software allows one to "do it your way."

HyperRESEARCH helps analyze almost any kind of qualitative data, whether it's audio, video, graphical or textual. The intuitive interface and well-written documentation – and especially the step-by-step tutorials – help get you up and running with your own data quickly and easily (see, http://www.researchware.com/).

HyperRESEARCH enables coding and retrieval of source material, theory building, and analyzes of data. With its multimedia capabilities, HyperRESEARCH allows to work with text, graphics, audio, and video sources. HyperRESEARCH has been in use by qualitative researchers since it was first introduced in 1991 by Researchware, Inc./rHyperRESEARCH is fully cross-platform. The overviews of various features are given below (see, http://www.researchware.com/):

- Easy-to-use interface: offers the same intuitive case-based interface on all supported platforms. It provides a turnkey solution by employing a built-in help system and several tutorials with step-by-step instructions.
- Flexible methodologies: supports Case-base or Source-based qualitative methodologies or combinations. With flexible organization of codes from any source to any case, support for code frequencies and other code statistics, HyperRESEARCH is well suited for mix-method approaches to qualitative research.
- Fully Cross-Platform: designed from the ground up to work well on Microsoft Windows and Apple's OS X. The HyperRESEARCH study file format and all of the various media types supported will work across operating systems.
- Multi-Media capabilities: allows you to work with text, graphics, audio, and video source material in many popular formats.

11.9.8 MAXQDA

MAXQDA is professional software for qualitative and mixed methods data analysis for Windows and Mac, which are used by thousands of people worldwide (see, http://www.maxqda.com/products/maxqda).

Released in 1989 it has a long history of providing researchers with powerful, innovative and easy to use analytical tools that help make a research project successful.

11.9.8.1 ORGANIZE AND CATEGORIZE YOUR DATA

The clearly structured user interface of MAXQDA is divided into four windows, which reflect essential work areas in the process of qualitative data analysis and allow intuitive handling.

Data can be imported from interviews, focus groups, online surveys, web pages, images, audio and video files, spreadsheets, and RIS data easily. Attach post-it like notes (memos) and sort your data into groups.

a. Code and Retrieve

Mark important information in your data with different codes by using regular codes, colors, symbols, or emoticons.

Organize your thoughts and theories in memos and stick them to any element of your project.

Retrieve coded segments quickly and efficiently and make use of powerful search tools with automatic coding options.

b Code and Transcribe Audio and Video Files

The MAXQDA Multimedia Browser enables to code audio and video files directly without having to create a transcript. The coded segments are treated like any other segments in MAXQDA. You can retrieve, comment and assign a weight to these segments in the same way as with other segments. Advantage: MAXQDA-11 has extended transcription functions with which you can adapt the speed or the sound volume of your audio and video files.

c. Mixed Methods

MAXQDA offers a lot of helpful Mixed Methods functions to complete your data analysis Integrate quantitative methods or data into your project. MAXQDA offers excellent Mixed Methods features to include variables or to quantify the results of your qualitative analysis (Figure 11.3).

11.9.9 NVIVO

NVivo is software that helps you easily organize and analyze unstructured information, so that you can ultimately make better decisions (see, http://www.qsrinternational.com/products_nvivo.aspx).

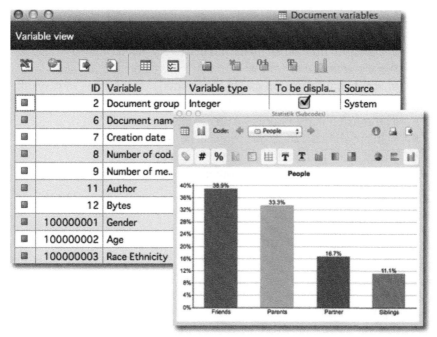

FIGURE 11.3 Database interface of MAXQDA (Source: http://www.maxqda.com/products/maxqda; used with permission.)

Whatever the materials, whatever the field, whatever the approach, NVivo provides a workspace to help at every stage of project – from organizing your material, through to analysis, and then sharing and reporting.

NVivo is a qualitative data analysis (QDA) computer software package produced by QSR International. It has been designed for qualitative researchers working with very rich text-based and/or multimedia information, where deep levels of analysis on small or large volumes of data are required.

NVivo is used predominantly by academic, government, health and commercial researchers across a diverse range of fields, including social sciences, such as anthropology, psychology, communication, sociology, as well as fields, such as forensics, tourism, criminology and marketing uses (see, http://www.qsrinternational.com/products_nvivo.aspx).

From health research and program evaluation, to customer care, human resources and product development – NVivo is used in virtually every field. Yale University, USA, World Vision Australia, the UK Policy Studies Institute and Progressive Sports Technologies all use NVivo to harness information and insight.

For individuals use NVivo to:

- Spend more time on analysis and discovery, not administrative tasks.
- Work systematically and ensure don't miss anything in data.
- Interrogate information and uncover subtle connections in ways that simply aren't possible manually.
- Rigorously justify findings with evidence.
- Manage all material in one project file.
- Easily work with material in own language.
- Effortlessly share work with others.

For organizations use NVivo to:

- Get the most out of your data – from customer and employee feedback to information about product performance – to make new discoveries and ultimately, better decisions.
- Easily manage your information and enhance internal workflow and reporting processes.
- Deliver quality outputs backed by a transparent discovery and analysis process.
- Justify decision making with sound findings and evidence-based recommendations.
- Revisit data easily. Build up the big picture over time.
- Increase productivity and reduce project timeframes.

NVivo is intended to help users organize and analyze non-numerical or unstructured data. The software allows users to classify, sort and arrange information; examine relationships in the data; and combine analysis with linking, shaping, searching and modeling. The researcher or analyst can test theories, identify trends and cross-examine information in a multitude of ways using its search engine and query functions. They can make observations in the software and build a body of evidence to support their case or project.

NVivo accommodates a wide range of research methods, including network and organizational analysis, action or evidence-based research, discourse analysis, grounded theory, conversation analysis, ethnography, literature reviews, phenomenology, mixed methods research and the Framework methodology. NVivo supports data formats, such as audio files, videos, digital photos, Word, PDF, spreadsheets, rich text, plain text and web and social media data. Users can interchange data with applications like Microsoft

Excel, Microsoft Word, IBM SPSS Statistics, EndNote, Microsoft OneNote, SurveyMonkey and Evernote; and order transcripts from within NVivo projects, using TranscribeMe.

9.9.10 QIQQA

Qiqqa (pronounced "Quicker") is a freeware and freemium reference management software that allows researchers to work with thousands of PDFs. It combines PDF reference management tools, a citation manager and a mind map-brainstorming tool. It integrates with Microsoft Word XP, 2003, 2007 and 2010 and BibTeX/LaTeX to automatically produce citations and bibliographies in thousands of styles. Researchers and research groups can store, synchronize and collaborate on their PDF documents, annotations, tags and comments using the internet cloud-based Qiqqa Web Libraries.

Key Features (see, http://www.qiqqa.com/)

- Rich PDF viewer supporting annotating, tagging, notes, searching and cross-referencing.
- Filtering and reporting against your tags, autoTags and aiTags.
- A full-text search across your entire PDF library.
- The tagging of text annotations and the associated annotation report features enables Qualitative research and Grounded theory methodologies against your PDF documents and scans.
- Automatic extraction of paper metadata and integration with GoogleScholar to automatically build the rest of your citation metadata.
- Sync your documents, metadata and annotations across multiple computers and to a private online Web Library.
- Integration with web browsers to support searching of the internet for new PDFs to add to your document library.
- Optical character recognition (OCR) of PDF documents to support text searching of scanned PDFs.
- Text and image export text export for PDFs (including scanned PDFs)
- Automatically generate your citations and bibliographies in Microsoft Word XP, 2003, 2007 and 2010 with support of thousands of Citation Style Language (CSL) styles.
- BibTeX export to allow researchers using LaTeX to format their lists of references in a consistent manner, for example, using LyX.

- Integrates with the built-in Microsoft Word 2007 and 2010 reference management systems.
- A brainstorming tool allowing you to incorporate your ideas, PDF documents, annotations and information on the internet.
- Import your PDFs and metadata from other reference managers.

11.9.11   f4ANALYZE

f4ANALYZE supports you in analyzing your textual data. You can develop codes, write memos, code (manually or automatically), and you can analyze cases and topics. The program is slim and easy to learn – you'll greatly speed up your work. f4analyze is an inexpensive, easy-to-learn QDA software for research projects with up to 30 texts. It supports your analysis of word processing files by providing functions for coding, memoing, retrievals, and frequency analyzes. Our reference manual is only 12 pages long because f4analyze focuses on core functions. This is a guarantee for an easy start. We have put emphasis on user-friendliness on all operating systems. (see, http://www.audiotranskription.de/english/f4-analyze)

Coding Text

Codes help to structure your texts. No matter if codes are developed inductively or deductively, you can effortlessly create and organize your code system in f4analyze. Different colors can be assigned to the codes – if a segment of text is coded, it will appear in the same color as the respective code. Clicking on a code in the code list retrieves all coded passages in a well-arranged list.

Text filter and retrieval

With a few clicks one can start a word search in both texts and comments. To retrieve certain statements or segments in texts, simply click on the desired codes and the texts that should be searched. One can further bundle your texts in text groups – for example, if one specifically wants to search through statements by academics from Wisconsin (as opposed to searching through statements from academics all over the US). Everything can be exported to RTF.

Code distribution

In the code distribution table, you can display the coding frequency in total numbers. The segments of text assigned to the respective codes can also be directly displayed in the table. Frequency numbers as well as text segments can be exported as a CSV file.

Open software tool

Your data is stored in a single project file – you can export everything from this file in case you want to work in Word, MAXQDA or ATLAS.ti at a later point. f4analyze supports export as text file and as a table for parts of the project. The whole project can be exported as an XML-file. f4analyze runs on Windows & Linux machines, as well as on Macs, and is available in English and German. In f4analyze, YOU decide where and how you want to work.

11.9.12 XSIGHT

XSight was released in 2006 and supported until January 2014. Developed by QSR International for qualitative data analysis (QDA), it is a tool for researchers or individuals who are undertaking short-term qualitative research analysis on projects involving non-numerical data. Qualitative research can encompass business intelligence, marketing research or data analysis. (see, http://students.pugh.co.uk/index.php?nID=productDetail&manu=63&prod ID=1079).

This was superseded by NVivo 10 for Windows, which offers equivalent functionality with greater flexibility and enables researchers to work with more data types likes PDFs, surveys, images, video, audio, web and social media content. XSight software assists researchers or other professionals working with non-numerical or unstructured data to compile, compare and make sense of their information. It provides a range of analysis frameworks for importing, classifying and arranging data; tools for testing theories and relationships between items; and the ability to visually map and report thoughts and findings.

Designed for rapid analysis, XSight can handle small or large volumes of data and search and query tools support the review and reflection process and users can look for patterns, make comparisons, and interrogate the data in seconds. XSight is used predominantly by commercial market researchers, but also by professionals and students in a diverse range of areas, from health and law to telecommunications and tourism. It is useful for evaluating a variety of information to review results and draw conclusions. Some examples of uses include tracking customer satisfaction, testing an advertising campaign, researching new packaging or even evaluating information garnered in research, such as community consultation projects.

11.9.13   ATLAS.TI

ATLAS.ti is a powerful software package for the visual qualitative analysis of large bodies of textual, graphical and audio or video material. It offers a variety of tools for accomplishing the tasks associated with any systematic approach to fieldwork material. ATLAS.ti helps you to uncover the complex phenomena hidden in your qualitative data, offer a powerful and intuitive environment for coping with the inherent complexity of tasks and data, and keeps you focused on the data under analysis. For a comparison between ATLAS.ti and N6, the reader may refer the paper by Barry (1998).

11.9.14   THE ETHNOGRAPH

The Ethnograph v5.0 is a versatile computer program designed to make the analysis of fieldwork material easier, more efficient, and effective. It is possible to import text-based qualitative material, typed up in any word processor, straight into the program. The Ethnograph helps to search and note segments of interest within fieldwork material mark them with code words and run analyzes which can be retrieved for inclusion in reports or for further analysis.

11.9.15   N6 BY QSR

N6 is the latest version of the NUDIST software. It combines efficient management of fieldwork material with powerful processes of Indexing Searching and Theorizing.

Designed for researchers making sense of complex material, N6 offers a complete toolkit for rapid coding, thorough exploration and rigorous management and analysis. With a full command language for automating coding and searching, and a Command Assistant that formats the command for you, N6 powerfully supports a wide range of methods. Its command files and import procedures make project set up very rapid, and link qualitative and quantitative data. Tolerant of large data sets, it ships with QSR Merge, which seamlessly merges two or more projects (N4, N5 or N6) for teams or multi-site research.

For a comparison between ATLAS.ti and N6, the reader may refer the work of Barry (1998).

11.10   RECENT DEVELOPMENTS IN COMPUTER APPLICATIONS IN ETHNOBOTANY

Many ethnobotanists are using more unconventional methods to create a database. The FLAAR staff utilizing a special 3D scanner training course at Z Corp learned how to handle the portable ZScanner 800 and how to process the data from it.

The report shows the processing of two pataxte pods. The pataxte is what is left after removing the cacao, which is the fruit from where the chocolate is made. Using the leftover pataxte pods, they tried to reconstruct the entire fruit using 3D scanning (Figures 11.4 and 11.5).

FIGURE 11.4   Figure showing the morphological closeness of the 3D scan generated and real Cacao shells. (Photos courtesy of Nicholas Hellmuth, FLAAR Reports 3D research programs. http://www.maya-ethnobotany.org.)

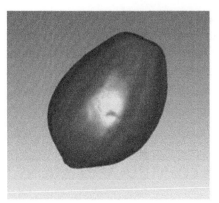 

FIGURE 11.5   Figure showing the morphological closeness of the 3D scan generated and real Papaya fruit. (Photos courtesy of Nicholas Hellmuth, FLAAR Reports 3D research programs. http://www.maya-ethnobotany.org.)

Similarly, FLAAR staff use the same 3D scanning for papaya. McGrath (2011) in collaboration with the University of Nebraska Medical Center (UNMC), developed the paper work book with inexpensive Augmented Reality, to create a "magic book" effect through which the plants can be seen to "pop up" in three dimensions off the page. Augmented Reality (AR) is a technology that allows content creators to merge two media for the purpose of an enhanced reading experience: traditional printed material and 3D computer graphics. The physical book's location and orientation are tracked by a low cost camera, such as a webcam or the camera of a cell phone, so the graphics move as the physical object is manipulated.

For the ethnobotanical work book, Rober McGrath created 3D renditions of the 15 plant species in the book, along with a software application that recognize 15 unique markers, which displays one of the plants floating atop each marker. These markers can be printed out (e.g., on sticky paper), and affixed to the appropriate pages of the workbook. The workbook will look almost the same with the addition of the markers, and can be used just as before. But with AR application and an inexpensive web camera, the plants will pop off the page in 3D on the computer screen. The software package will be available for inexpensive computers that can be used in a classroom, nature center or at home.

11.11   QUANTITATIVE RESEARCH IN ETHNOBOTANY

The concept of quantitative ethnobotany is relatively new and the term itself was coined only in 1987 by Prance and coworkers (Prance, 1991).

Quantitative ethnobotany may be defined as "the application of quantitative techniques to the direct analysis of contemporary plant use data". Quantification and associated hypothesis-testing help to generate quality information, which in turn contributes substantially to resource conservation and development. Further, the application of quantitative techniques to data analysis necessitates refinement of methodologies for data collection.

Quantitative research is often contrasted with qualitative research, which is the examination, analysis and interpretation of observations for the purpose of discovering underlying meanings and patterns of relationships, including classifications of types of phenomena and entities not involving mathematical models.

Quantitative research is generally made using scientific methods, which includes:

1. the generation of models, theories and hypotheses;
2. the development of instruments and methods for measurement;
3. experimental control and manipulation of variables;
4. collection of empirical data;
5. modeling and analysis of data.

Generally, multivariate and statistical methods aim at making large data sets mentally accessible, structures recognizable and patterns explicable, if not predictable. Johnson and Wichern give five basic applications for these methods (Johnson and Wichern, 1988):

1. Data reduction or structural simplification: The phenomenon being studied is represented as simply as possible with reduced number of dimensions but without sacrificing valuable information. This makes interpretation easier.
2. Sorting and grouping: Groups of similar objects or variables are created.
3. Examining relationships among variables: Variables are investigated for mutual interdependency. If interdependencies are found the pattern of dependency is determined.
4. Prediction: Relationships between variables are determined for predicting the values of one or more variables on the basis of observations on the other variables.
5. Testing of hypothesis: Specific statistical hypotheses formulated in terms of the parameters of multivariate populations are tested. This may be done to validate or reject assumptions.

11.12 TYPES OF QUANTITATIVE ANALYSIS SOFTWARES

There are many computer packages available for analysis of multivariate data sets. Some of them are discussed as follows:

11.12.1 BIOMEDICAL PACKAGE (BMDP) (see, http://www. statsols.com/?pageid=6)

A statistical language and library of over forty statistical routines developed in 1961 at UCLA, Health Sciences Computing Facility under Dr.Wilford Dixon. BMDP was first implemented in Fortran for the IBM 7090. Tapes of the original source were distributed for free all over the world.

BMDP is the second iteration of the original BIMED programs. It was developed at UCLA Health Sciences Computing facility, with NIH funding. The "P" in BMDP originally stood for "parameter" but was later changed to "package". BMDP used keyword parameters to define what was to be done rather than the fixed card format used by original BIMED programs.

BMDP supports many statistical functions like:

- simple data description,
- survival analysis,
- ANOVA,
- multivariate analyzes,
- regression analysis, and
- time series analysis.

11.12.2 CANOCO

Canoco is one of the most popular programs for multivariate statistical analysis using ordination methods in the field of ecology and related fields. User's Guides of the recent Canoco versions (4.0 and 4.5) were cited more than 6900 times in the past 16 years (1999–2014, ISI Web of Knowledge).

Canoco 5 is the new, much re-worked version of the Canoco software, released in October 2012. This site offers one access to additional resources for the effective use of the software, as well as a brief overview of Canoco 5 new features. Use the menu at the right side of this page to access these resources. (see, http://www.wageningenur.nl/en/Expertise-Services/

Research-Institutes/plant-research-international/show/Canoco-for-visualization-of-multivariate-data.htm).

In Canoco 5, data import, analyzes and making graphs are integrated in a single Canoco 5 project. The Canoco Adviser helps in choosing data transformations and methods of analysis. Numerical analyzes that used to take many runs, are now available through a single analysis template and the Analysis Notebook concisely summarizes the results and allows access to the full results. All analyzes done on a set of data tables are now collected within Canoco 5 projects, sharing the analytical and graphing settings. Canoco 5 helps to make even better publication-quality ordination diagrams. The manual has been largely re-written and the large set of real-life examples is updated and extended to show new ways of working with multivariate data. (see, http://www.canoco5.com/index.php/canoco5-overview).

All statistical methods offered by Canoco for Windows 4.5 are available, such as DCCA method – including their partial variants, with Monte Carlo permutation tests for constrained ordination methods, offering appropriate permutation setup for data coming from non-trivial sampling designs

Principal coordinates of neighbor matrices (PCNM) method is available within the variation partitioning framework. Present implementation matches the suggestions described in Legendre and Legendre (2012) under an alternative method name (dbMEM).

11.12.3 NTSYSpc

NTSYSpc can be used to discover pattern and structure in multivariate data. For example, one may wish to discover that a sample of data points suggests that the samples may have come from two or more distinct populations or to estimate a phylogenetic tree using the neighbor-joining or UPGMA methods for constructing dendrograms. Of equal interest is the discovery that the variations in some subsets of variables are highly inter-correlated (clustered). The program originated as NTSYS in the 1960s but over the years is has been completely redesigned and greatly extended for use on PCs.

The input can be descriptive information about collections of objects or directly measured similarities or dissimilarities between all pairs of objects. The kinds of descriptors and objects used depend upon the applicational morphological characters, abundances of species, presence and absence of properties, etc. NTSYSpc can transform data, estimate dis/similarities among objects, and prepare summaries of the relationships using cluster analysis, ordination, and multiple factor analyzes. Many of the results can be shown both numerically and graphically. The software is designed for both classroom and research.

Version 2.2 (see: http://www.exetersoftware.com/cat/ntsyspc/ntsyspc. html) for Windows is easy to use yet still has the speed and functionality of the previous versions. There is an interactive mode and a batch mode with a simple command language (useful for analysis of simulations and multiple datasets). The program takes advantage of the Windows environment and allows long file names and the processing of large datasets. Plot options windows allows you to customize the plots (specify titles, fonts, sizes, colors, scales, line widths, background colors, margins, and many other aspects of what is plotted). There is also a print preview mode. NTS data files are ASCII files that can be shared with other programs. Long input lines are supported. A spreadsheet-like data editor is included that makes it easy to create and edit data files. It can be also used as an ASCII text editor for very large files. Matrices can be read from Excel XLS and CSV files and trees can be read from one type of nexus files. An option is provided to output to MATLAB M files.

11.12.4   TWINSPAN FOR WINDOWS (see, http://www. canodraw.com/wintwins.htm)

This software, written by Mark O. Hill and Petr Smilauer (with contribution of CajoTerBraak and John Birks), is available free of charge for any non-commercial or commercial purposes and can be installed and used even if you do not have a license for Canoco for Windows. It implements divisive classification method named TWINSPAN (Two-way Indicator Species Analysis) and analyzes datasets provided in one of the data formats accepted by Canoco (Cornell condensed format, full format or free format). TWINSPAN for Windows is distributed using an independent installation program.

11.13   FUTURE RELEVANCE OF CAQDA IN ETHNOBOTANY

Ethnobotanists interested in computer assistance in their work must acquaint themselves with the variety of capabilities and programs available because no one program dominates the CAQDA field.

The analytical principles of these context-dependent methods are more difficult to codify than those of grounded theory. So, while grounded theorists may find themselves able to take advantage of a wide variety of computer resources as they move from QDA to CAQDA, researchers working in other traditions may find that computer assistance limits their analyzes unless they limit the extent to which they make use of computers.

The large number of CAQDA software programs available suggests that these software packages are in a preliminary stage of computer entry into the qualitative field. With time, the computer will do for qualitative data analysis what it has done in the quantitative realm: reduce labor, regularize procedures for data gathering and analysis, and establish conventions for the reporting of results. Moreover, the diversity of program options will allow these advances to occur along parallel methodological lines so that regularizing data-handling procedures will not require homogeneous epistemological stances. On the other hand, the still infrequent mention of CAQDA in ethnobotanical writing means that the expansion of software choices has not yet influenced the course of ethnographic research. CAQDA may be a significant advance for positivist researchers, but its potential for regularizing analysis in the qualitative field has not been reached.

The computer offers three ways of facilitating qualitative analysis that may lead to, but are no guarantee of, the enthronement of a CAQDA killer app.

First, CAQDA packages reduce the administrative burdens of analysis. Administrative assistance is a strong reason to climb learning curves in some research projects, such as those that use grounded-theory methods or those large projects that involve multiple sites or multiple researchers. But given the diversity of techniques for ethnographic analysis, administrative reduction is compellingly attractive to only a fraction of qualitative researchers.

Second, many CAQDA programs allow the user to analyze materials that are difficult to access without the computer. These packages integrate text, graphics, sound, and video; they encourage analysis based on the creation of links between distinct pieces of the ethnographic database; and they open up new possibilities for the presentation of research. However, not only do many work exclusively with text, but also text and graphics are the dominant form of the report. Multimedia capability alone does not create a killer app.

Third, some of the features of symbol-manipulation software are not easily replicated without a computer. Like symbol-manipulation software, these software must offer analysts the ability to perform analyzes that are unmanageable without a computer. To be compelling, the CAQDA software package will have to constitute its own best marketing device. In addition, the methodological and epistemological diversity of data analysis means that CAQDA software will have to offer different analytical facilities to different analysts. At present, popular software packages meet the challenges of one group or another, but no killer app appears to be on the horizon.

To date, many researchers have had to rely on was their memory of the data they collected and the meaning of those data in the context of their study. However, the workings of memory create two potential problems for researchers analyzing field data.

First, researchers may use those data that were most dramatic in the field-work and erroneously present them as being the most significant; second, they may use more data from the later stages of fieldwork and less of what happened in the middle or beginning because the later data are fresher and clearer in their minds. CAQDA can help the careful analyst avoid both of these problems.

Researchers who use CAQDA still face issues related to representation. Data quality is directly tied to the ability of the researcher to observe significant phenomena in the course of fieldwork and to recognize what he or she has seen. While CAQDA can compensate for small failures of detailed observation or sharp insight, it is no substitute for either.

The use of CAQDA could stimulate team approaches in research that would generate a wealth of data and make important analytic contributions, but CAQDA does not eliminate the validity problems inherent to team ethnobotanist. Because data in teams are gathered by a number of researchers who in many cases have different degrees of training (as well as different degrees of insight), there is no way to assure consistency in what each researcher thinks it important to record. Thus, there are validity problems for which CAQDA cannot compensate. Ethnobotanist spend much of their time engaged in filework rather than fieldwork, but quality analysis that has a high degree of validity and reliability remains dependent on the competence and consistency of fieldworkers.

Quantitative data analysis is not inconsistent with the qualitative analysis. Statistics and mathematics help in reaching to a decisive conclusion by building theories. It is desirable to do a joint study where statistical regularities associating variables are sought, through Quantitative analysis and seek to identify patterns of interaction that characterize the set of observations. Combining the two leads to a "mixed-method approach" that can take various forms: data collection and analysis can be either separated or addressed together, and each of them can be used in service of the other.

ACKNOWLEDGEMENTS

We wish to keep on record our grateful thanks to all concerned and various websites for kindly permitting to use the data reported in this paper. We also thank Prof. K.V. Krishnamurthy for reading the manuscript critically and useful suggestions.

KEYWORDS

- **Computer-Assisted Qualitative Data Analysis (CAQDA)**
- **Ethnobotanic Study**
- **Qualitative and Quantitative Softwares**

REFERENCES

Berg, B.L. (1995). Qualitative Research Methods for the Social Sciences. Boston: Allyn & Bacon.

Barry, C.A. (1998). Choosing Qualitative Data Analysis Software: Atlas/ti and Nudist Compared Sociological Research Online, 3, p. 3 http://www.socresonline.org.uk/3/3/4.html

Dohan, D. & Snchez-Jankowski, M. (2003). Robert Wood Johnson Foundation Scholars in Health Policy Research Program, School of Public Health, University of California, Berkeley, California.

Glaser, B.G. & Strauss, A.L. (1967). The Discovery of Grounded Theory: Strategies for Qualitative Research. New York: Aldine.

Johnson, R.A. & Wichern, D.W. (1988). Applied Multivariate Statistical Analysis. Prentice Hall, Inc.: Englewood Cliffs, New Jersey.

Moore, J.W. & Garcia, R. (1978). Homeboys: Gangs, Drugs, and Prison in the Barrios of Los Angeles. Philadelphia, PA: Temple University Press.

Prance, G.T. (1991). What is Ethnobotany Today? *J. Ethnopharmacol. 32*, 209–216.

Robert, E., McGrath, Alan Craig, Dave Bock & Ryan Rocha (2011). Augmented Reality for an Ethnobotany Workbook Institute for Computing in the Humanities, Social Sciences and Arts and National Center for Supercomputing Applications University of Illinois, Urbana—Champaign 1201 W. Clark Street Urbana, Illinois 61808.

Strauss, A.L. (1987). Qualitative Analysis for Social Scientists. New York: Cambridge University Press.

Tesch, R. (1990). Qualitative Research: Analysis Types and Software Tools. New York: Falmer.

Tsesch, R. (Ed.) (1991). Computer and Qualitative Data II, Special Issue, Pts. 1.2, *Qualitative Sociology, 14*(3, 4).

Weitzman, E.A. & Miles, M.B. (1995). Computer Programs for Qualitative Data Analysis: A Software Sourcebook. Thousand Oaks, CA: Sage

Weaver, A. & Atkinson, P. (1994). Microcomputing and Qualitative Data Analysis. Aldershot, UK: Avebury.

3D Software for Mayan Ethnobotany Cacao (Pataxte species), FLAAR Reports March (2010).

3D Software for Mayan Ethnobotany Papaya, FLAAR Reports March (2010).

http://www.aquad.de/en/ (accessed on 21 November 2014).

http://cat.ucsur.pitt.edu/ (accessed on 21 November 2014).

http://www.graphic.org/mind-mapping-software/compendium-review.html (accessed on 21 November 2014).

http://compendium.open.ac.uk/index.html (accessed on 21 November 2014).

https://tla.mpi.nl/tools/tla-tools/annex/attachment/annex-elan_flyer_2006–05–11/ (accessed on 21 November 2014).

http://www.researchware.com/ (accessed on 21 November 2014).

http://www.maxqda.com/products/maxqda (accessed on 21 November 2014).

http://www.qsrinternational.com/products_nvivo.aspx (accessed on 21 November 2014).

http://www.qiqqa.com/ (accessed on 21 November 2014).

http://www.audiotranskription.de/english/f4-analyze (accessed on 21 November 2014).

http://students.pugh.co.uk/index.php?nID=productDetail&manu=63&prodID=1079 (accessed on 21 November 2014).

http://www.statsols.com/?pageID=6 (accessed on 2 December 2014).

http://www.wageningenur.nl/en/Expertise-Services/Research-Institutes/plant-research-international/show/Canoco-for-visualization-of-multivariate-data.htm (accessed on 2 December 2014).

http://www.canoco5.com/index.php/canoco5-overview (accessed on 2 December 2014).

http://www.exetersoftware.com/cat/ntsyspc/ntsyspc.html (accessed on 2 December 2014).

http://www.canodraw.com/wintwins.htm (accessed on 2 December 2014).

CHAPTER 12

ETHNOBOTANY, ETHNOPHARMACOLOGY, BIOPROSPECTING, AND PATENTING

P. PUSPANGADAN, V. GEORGE and T. P. IJINU

Amity Institute for Herbal and Biotech Products Development, 3 Ravi Nagar, Peroorkada P.O., Thiruvananthapuram–695005, Kerala, India. E-mail: palpuprakualm@yahoo.co.in; georgedrv@yahoo.co.in; ijinutp@gmail.com

CONTENTS

ABSTRACT

Genetic resources particularly those related to ethnobotany, associated with Traditional Knowledge (TK) have great potentials and their contributions to global economy and global intellectual property regimes are enormous. Ranging from subsistence uses by indigenous and local communities for their livelihood security to the high-tech research and development programs on bioprospecting, ethnobotanical resources and associated TK find an ever increasing demand and utility in a diverse array of sectors, such as biopharmaceuticals, biotechnology (including agricultural biotechnology and health care), crop protection, agricultural seed production, horticulture, phytomedicines, cosmetics and a myriad of other areas of products and processes development based on wild and domesticated genetic resources and their derivatives extracted from both *in situ* and *ex situ* sources. Bioprospecting helps mobilizing funds to conserve biodiversity in both protected and unprotected wilderness areas. Value addition and assigning economic value to biodiversity enhance human resource development, capacity building in chemical and gene prospecting and other relevant biotechnologies; protection of IPRs, farmers rights, etc.

12.1   INTRODUCTION

Chemical and gene prospecting of wild biological resources of actual or potential values will have significant application in agriculture, medicine and industry. Bioprospecting can bring forth substantial economic returns from the products and processes to be derived from biodiversity and biotechnology. Bioprospecting helps mobilizing funds to conserve biodiversity in both protected and unprotected wilderness areas. Value addition and assigning economic value to biodiversity enhance human resource development, capacity building in chemical and gene prospecting and other relevant biotechnologies, protection of IPRs, Farmers Rights, etc. Traditional resources rights of local and ethnic communities; economic development of the country – particularly the rural and tribal communities by improving their source of income and living standards through location specific production and processing technologies based on local biogenetic resources evolve environmentally friendly policies and programs on biodiversity conservation and bioprospecting. Application of modern biotechnology has many critical areas of agriculture including aquaculture; healthcare; medicine, particularly in developing vaccine; diagnostics; gene/protein; cosmetics; environmental protection and bio-energy.

12.2 TRADITIONAL KNOWLEDGE

Traditional Knowledge (TK) is a community based system of knowledge that has been developed, preserved and maintained over many generations by the local and indigenous communities through their continuous interactions, observations and experimentations with their surrounding environment. It is unique to a given culture or society and is developed as a result of the co-evolution and co-existence of both the indigenous cultures and their traditional practices of resource use and ecosystem management. TK is a general term, which refers to the collective knowledge, beliefs and practices of indigenous/local people on sustainable use and management of their ambient resources. Through years of observations and analysis, trial, error or experimentations, the traditional communities have been able to identify useful as well as harmful elements of their ambient flora and fauna. Such knowledge (acquired through ages) has always remained as part of their life, culture, traditions, beliefs, folklores, arts, music, dance, etc. TK covers a broad spectrum of the local and indigenous people's traditional life and culture, art, music, architecture, agriculture, medicine, engineering and a host of other spheres of human activity. TK thus can be of direct or indirect benefit to society as it is often developed, in part as an intellectual response to the necessities of their life. Protection and maintenance of TK of local and indigenous communities is vital for their well-being and sustainable development and for their intellectual and cultural vitality.

12.3 ALL INDIA CO-ORDINATED RESEARCH PROJECT ON ETHNOBIOLOGY (1982–1998)

The Indian Council of Agricultural Research convened a meeting of its inter organizational panel for food and agriculture on September 21, 1976 under the Chairmanship of Prof. M.S. Swaminathan, the then Director General, ICAR. Prof. Swaminathan felt the urgent need to undertake an ethnobiological study of the tribals of the country to tap and document the fast disappearing life style, knowledge system and wisdom of these people. This panel decided to form a team of experts to examine the current status of ethnobiological studies of the tribal areas and to submit a report as to how the

biological resources found in these communes could be conserved and utilized for socio-economic improvement of tribals on one hand and country on
the other. Dr. T.N. Khoshoo along with Dr. E.K. Janaki Ammal prepared the
All India Co-ordinated Research Project on Ethnobiology (AICRPE) project
proposal which was considered by the high level committee of Science and
Technology, Govt. of India. Department of Science and Technology (DST)
formerly launched the project in July 1982 under the Man and Biosphere
Program (MAB) of UNESCO. When the Ministry of Environment and
Forest (MoEF) came in to being the MAB program along with AICRPE was
transferred to MoEF. In September 1983, MoEF set up a co-ordination unit
at RRL, Jammu (now known as Indian Institute of Integrative Medicine,
CSIR-IIIM) with Dr. P. Pushpangadan as the Chief Coordinator of this
project for overall supervision, co-ordination and implementation of various programs included in the AICRPE (AICRPE, 1998; Pushpangadan and
Pradeep, 2008).

From the deliberations it emerged that the biological resources in the
tribal and other backward areas were affected due to the indiscriminate and
unplanned management. Initially the focus was given on the botanical aspect
and the zoological part was completely neglected. But later the incorporation of the zoological aspect became inevitable as the tribals use a big range
of animal products. Ethnobiology brings together diverse disciplines like
botany, zoology, anthropology, linguistics, sociology, archeology and others.
Of late, with the renewed interest in traditional medicine, ethnobiology is
gaining prime importance.

This multi-institutional and multi-disciplinary project was operated in
about 27 centers by over 500 scientific personnel located in the different institutions spread over the length and breadth of the country. AICRPE during
the course of its operation (1982–1998) recorded information on the multidi-
mensional perspectives of the life, culture, tradition and knowledge system
associated with biotic and abiotic resources of the 550 tribal communities
comprising over 83.3 million people belonging to the diverse ethnic group.
In India there are 550 communities of 227 ethnic groups. There are 116
different dialects of 227 subsidiary dialects spoken by tribals of India. The
knowledge of these communities on the use of wild plants for food, medicine and for meeting many other material requirements are now considered
to be potential information for appropriate S&T intervention for developing
value added commercially marketable products. The TK are oral in tradition and not qualified for the formal IPR system. The vast information collected by the AICPPE team is locked up as unattended reports for want of
proper resources. Traditional knowledge on about 10,000 plants have been

collected during the course of the project. It may be mentioned here that the classical systems of medicine (Ayurved, Siddha, Unani, Amchi, etc.) makes use of only 2500 plants where as we have a database on 10,000 plants which requires further scientific validation. Out of this 8000 wild plant species used by the tribals for medicinal purposes, about 950 are found to be new claims and worthy of scientific scrutiny. Out of 3900 or more wild plant species used as edible as subsidiary food/vegetable by tribals. About 8000 are new informations and at least 250 of them are worthy of investigation. Out of 400 plant species are used as fodder 100 are worth recommending for wider use and out of 300 wild species used by tribals as piscides or pesticides, atleast 175 are quite promising to be developed as safe pesticides (Pushpangadan, 1995; Pushpangadan and Pradeep, 2008).

12.4 THE KANI TRIBE AND *AROGYAPPACHA* (=ELIXIR OF HEALTH)

The Kanis, a semi nomadic community, is the predominant tribe inhabiting the forests of the southern most parts of the Western Ghats in Kerala (in the districts of Thiruvananthapuram, Kollam and Pathanamthitta), India. Traditional occupation of Kanis includes craftwork like basket-making, mat-making using *Ochlandra* stem, cane works, etc. They are also engaged in collection of non-timber forest produces (NTFPs) like honey, bee wax, medicinal plants, python fat, etc. The Kanis are well known for their rich knowledge on medicinal plants of the region (Pushpangadan and Pradeep, 2008).

In one of the field expeditions in the mountainous forests of the Southern Western Ghats in Kerala, a few young Kani men accompanied the AICRPE team led by its Chief Coordinator. During the arduous trekking across the forests the scientists noticed that the tribals frequently ate some fruits, which kept them energetic and agile. The Chief Coordinator and the accompanying scientist (Dr. S. Rajasekharan, an Ayurvedic specialist) were almost exhausted at one time when they were offered these strange seeds. After consuming the same the scientists also felt a 'sudden flush of energy and strength.' When asked about the source of the fruits, the Kani young men were first reluctant to reveal their secret. The team convinced the Kani men that if they passed on the information to them they would not misuse it and that they would conduct scientific investigation and, if found promising, a drug would be developed for the welfare of the humanity. The team leader Dr. Pushpangadan also assured the Kanis that if any marketable drugs/products were developed from this plant, the financial benefits accrued from the same would be equally shared with them and their community. The Kani men then showed the plant to the scientists from which the fruit was obtained. The scientists identified the plant as *Trichopus zeylanicus* subsp. *travancoricus* Burkill *ex* Narayanan. The Kanis call it as 'Arogyapacha,' meaning ever-greener of health or elixir of health (Pushpangadan and Pradeep, 2008).

12.5   SCIENTIFIC INVESTIGATION ON *AROGYAPPACHA* AND DEVELOPMENT OF THE HERBAL DRUG *JEEVANI*

Dr. Pushpangadan collected samples of this plant and took it to his eth-nopharmacology laboratory at IIIM, Jammu where he was working at the time. He and his team at IIIM carried out phytochemical and pharmaco-logical evaluation of this plant. The study revealed that the plant contained various biodynamic compounds notably certain glycolipids and non-steroid compounds with profound adaptogenic and immuno-enhancing proper-ties. IIIM has filed two patents on the same. In the meantime in 1990, Dr. Pushpangadan moved to Thiruvananthapuram to assume the position of Director, Jawaharlal Nehru Tropical Botanical Garden and Research Institute (JNTBGRI). At JNTBGRI, he organized an ethnopharmacology division and recruited a multidisciplinary team of scientists drawn from Ayurveda, ethnobiology, biochemistry, phytochemistry, pharmacy and pharmacology. Dr. S. Rajasekharan also joined this team as an Ayurvedic scientist and Dr. V. George joined as Head of Phytochemistry group. This Division developed a scientifically validated and standardized herbal formulation – *Jeevani* with *Arogyappacha* as one of the constituents. This drug after necessary clini-cal trials was transferred to Arya Vaidya Pharmacy (AVP) Coimbatore Ltd. against a license fee of Rs. 10 lakhs (US$ 25,000) and a royalty of 2% at ex-factory sale rate. While transferring the technology, JNTBGRI with the ap-proval of its competent authority agreed to share the license fee and royalty received from AVP with the Kani tribe (Pushpangadan and Pradeep, 2008).

Kani tribe is an unorganized forest dwelling semi-nomadic tribe. Our prime concern in the beginning was therefore to evolve a viable mecha-nism for receiving such funds and utilizing the same for the welfare of the community. Several ways of sharing the benefits were discussed at many levels and it was finally decided to set up a trust fund of the tribe. The very idea of the trust fund had originated from very useful and protracted discus-sion Dr. Pushpangadan had with Prof. Anil K. Gupta, the founder and co-ordinator of SRISTI and Honey Bee Network. It took however, almost two years to transfer the benefits to the tribe. With the help of some local NGOs, JNTBGRI scientists and some motivated government officials, the tribes were encouraged to form a registered trust, Kerala Kani Samudaya Kshema Trust (KKSKT) with Kani adults as its members. The trust was fully owned and managed by the Kani tribe. About of 60% of the Kani families of Kerala are now members of this Trust. In February 1999, the amount due to them (Rs. 6.5 lakhs) which was till then kept by JNTBGRI in a separate account was transferred to the Trust. As per the rules of the Trust the license fee and royalty received on account of the sale of 'Jeevani' drug will be in a fixed deposit and only the interest accrued from this amount will be utilized for the benefits/welfare of the members of the Kani tribe (Anand, 1998; Anuradha, 1998; Bagla, 1999; Pushpangadan and Pradeep, 2008).

This model was thus developed and perfected over a period of about 12 years starting from 1987 to 1999 in full consultation with the Kani tribe. In fact, the whole process of this benefit sharing started much before the CBD evolved. It took almost 3 years for the Kani tribe to receive this benefit. The delay was mainly due to the inherent inability and absence of any organized mechanism for Kani tribe to receive such benefit. The secretary of Scheduled Caste and Scheduled Tribe Department of Government of Kerala played a crucial role in the formation of the Kani Trust and also in effecting a smooth transfer of the amount due to the Kanis from JNTBGRI to the Kani Trust. The Kani Trust now continues to receive the royalty.

In addition to the license fee and royalty that Kani Trust is receiving, a large number of Kani families are now getting benefit from the cultivation of *Arogyapacha* and supply of the raw-material (i.e., the leaves of the plant) to the pharmaceutical company for the production of the drug. JNTBGRI has trained many tribal families for the cultivation of *Arogyapacha* in and around their dwellings in the forest. All along these years, starting from 1987, it was the mutual trust, respect, transparency and frequent interaction and communication between JNTBGRI and the Kani tribe that contributed to the success of this benefit-sharing model.

India has the distinction of the first country in the world in experimenting a benefit-sharing model that implemented the Article 8(j) of CBD, in letter and spirit. It was the Jawaharlal Nehru Tropical Botanic Garden and Research Institute (JNTBGRI) in Kerala (where the author was director) that demonstrated indigenous knowledge system merits support, recognition and fair and adequate compensation. The model, which later on came to be known as 'JNTBGRI Model' or 'Pushpangadan Model' or 'Kani Model,' relates to the sharing of benefits with a tribal community in Kerala, the Kanis, from whom a vital lead for developing a scientifically validated herbal drug

(*Jeevani*) was obtained by scientists of JNTBGRI. The 'JNTBGRI Model' has got wider acclaims, acceptance and popularity the world over, because it was the first of its kind that recognized the resource rights and IPR of a traditional community by way of sharing equitably the benefits derived out of the use of a knowledge that has been developed, preserved and maintained by that community for many generations. Further, it demonstrates the vast and as yet under-explored or untapped potentials of the Indian traditional knowledge systems, particularly the traditional health care practices of the local and indigenous people in India (Anand, 1998; Anuradha, 1998; Bagla, 1999; Pushpangadan and Pradeep, 2008). Considering the significant outcome of this model in community empowerment and income generation and poverty eradication of a tribal community, Dr Pushpangadan was awarded with the UN-Equator Initiative Prize (under individual category) at the World Summit on Sustainable Development held in Johannesburg in August, 2002.

Thus, 'JNTBGRI Model' is perhaps a unique experiment ever done, wherein the benefits accrued from the development of a product based on an ethnobotanical lead were shared with the holders of that traditional knowledge. It would perhaps require further refinement and modification if it has to be applied for other cases of equitable benefit sharing in India or elsewhere. In the absence of any internationally agreed laws or guidelines, the type of benefit sharing and their *modus operandi* would differ from case to case, depending upon the stakeholder's contributions, the type of products developed and their processes of development, and several other factors, including the self-reliance and capability of the stakeholder communities, and the political and bureaucratic set up prevailing in a particular country.

12.6   ETHNOBOTANY RESEARCH IN 20ᵀᴴ AND 21ˢᵀ CENTURY

Very little organized work had been done in ethnobotany in India till about 50 years ago. Organized fieldwork and other studies in the subject were started in the Botanical Survey of India. Also there has been a resurgence of interest developed in ethnobotanical research in various institutions. Dr. E.K. Janaki Ammal initiated researches on ethnobotany in BSI. She studied food plants of certain tribals of South India. When the senior author of this paper joined Regional Research Laboratory, Jammu under Dr. Janaki Ammal for about 8 months she has fondly told the senior author about the importance of doing ethnobotanical studies in South India. However, Pushpangadan could do it only after 1984 when he started research in Ethnobotany (Pushpangadan and Atal, 1984, 1986). From 1960, Dr. S.K. Jain from BSI started intensive fieldwork among the tribals of Central India. He devised methodology for ethnobotany particularly in the Indian context. The publications from this group in the early sixties triggered the ethnobotanical activity in many other centers, particularly among botanists, anthropologists and medical practitioners in India (Bondya et al., 2006; Bora and Pandey, 1996; Borthakur, 1981; Borthakur, 1996; Borthakur and Gogoi, 1994; Hajra, 1981; Hajra et al., 1991; Jain, 1987, 1991, 2002, 2005, 2006, 2010; Jain and Goel, 1987, 2005, Jain and Sikarwar, 1998; Jain et al., 1994, 1997; Janaki Ammal, 1956; Joshi, 1995, Joseph and Kharkongor, 1981; Manilal, 1978, 1980a, b, c, 1981, 1996, 2005, 2012; Manilal et al., 2003; Mohanty, 2003, 2010; Mohanty and Rout, 2001; Patil, 2000, 2001, Pushpangadan, 1986, 1990; Pushpangadan et al., 1995, 2012; Pushpangadan and Dan, 2011; Pushpangadan and George, 2010; Mitra, 1998a, 1998b; Singh et al., 2011; Subramoniam et al., 1997, 1998; Vartak, 1981; Vartak and Gadgil, 1980, 1981). During the last four decades similar work has been initiated at various centers, such as National Botanical Research Institute (NBRI), Lucknow, National Bureau of Plant Genetic Resources (NBPGR), Delhi, Jawaharlal Nehru Tropical Garden and Research Institute (JNTBGRI), Palode, Central Council of Research in Unani Medicines (CCRUS), Central Council of Research in Ayurveda and Siddha (CCRAS) and in some Universities.

Human strategies for survival have long depended on the ability to identify and utilize plants. Generations of experience- success, failure, intuition, accidental discovery, error, trial, or empirical reasoning, etc. might have contributed to building a broad base knowledge on individual plant species and its value/utility and this was transmitted to subsequent generations. Incremental improvements in managing and utilizing such resources were made by successive generations.

12.7 TRADITIONAL KNOWLEDGE DIGITAL LIBRARY

India has initiated two important task force programs relating to creation of a Traditional Knowledge Digital Library (TKDL) and designing a Traditional Knowledge Resource Classification (TKRC). The Department of Indian Systems of Medicine and Homeopathy (ISMH) spearheaded the initiative. The ISMH set up the TKDL task force, by drawing experts from Central Council of Research of Ayurveda and Siddha, Banaras Hindu University, National Informatics Centre, Council of Scientific and Industrial Research, and Controller General of Patents and Trade Marks. The Indian TKRC has information on 5000 subgroups and the structure of TKRC is compatible with the International Patent Classification (IPC). TKRC would help enhance the quality of patent examinations by facilitating the patent examiners to access pertinent information on traditional knowledge in an appropriately classified form (Mashelkar, 2001).

All unauthorized access to biological resources and associated TK and TM and patenting without compensation and knowledge of the knowledge holder is generally referred as biopiracy. Many instances of such biopiracies have been detected in recent times. Several examples are known where patents have been granted and applications, which are based on codified TM like the Indian System of TM like Ayurveda, Siddha, Unani and Amchi. It was against the above said background more particularly after the successful revoking of the US patent on Turmeric by CSIR that the Ministry of Health and Family Welfare jointly with CSIR established the Traditional Knowledge Digital Library (TKDL) in 1999. The TKDL project involve translations of the formulations involving medicinal plants included in the various versus/ prose contained in the classical text of Indian System of Medicine like the Ayurveda, Siddha and Unani, into a digitized database in several international languages (English, Spanish, German, French, Japanese and Hindi). TKDL will give legitimacy to existing TK/TM and by ensuring case of retrieval of TM related information by patent examiners and thereby prevent granting of patents.

TKDL classifieds that entire TM related information contained in Ayurveda in a modern system as per the format of International Patent Classification (IPC) into sections, classes, subclasses, main groups and subgroups (Gupta, 2005). This classification system evolved by India is known as Traditional Knowledge Resource Classification (TKRC). TKRC has been developed for Ayurveda, Unani and Siddha system of medicine where about 8000 subgroups have been created for classifying the codified (published) TK/TM informations particularly with respect to Indian Systems of Medicine. TKRC is now recognized by the experts of IPC Union and WIPO. TKDL is a proprietary database made available to patent offices for preventing misappropriation and for collaborative research positive protection of IP of oral tradition.

12.8   THIRD WORLD NATIONS AND INTELLECTUAL PROPERTY RIGHTS

Many third world nations have no capability and legal expertise to develop suitable measures needed to control, protect and maintain their bioresources and associated cultural expressions, including the traditional wisdom on the use of bioresources. Many traditional communities in these countries fear that they are losing control on their knowledge systems and that outsider are appropriating their knowledge and resources without their consent and approval. There are several cases of such illegal plundering of indigenous people's knowledge and resources have been reported from third world nations in Africa, South America and Asia. The recent revocation of the US Patents on turmeric (use of turmeric in wound healing) and on a new variety of *Banisteriopsis caapi* by USPTO revealed that the claims made on these two patents were drawn from the traditional knowledge base of the developing countries. These are just two examples of the patent cases that the developing countries have won. Several other ongoing cases of challenging and re-examination of patents filed in developed countries, especially in USA, based on the age-old knowledge system (both undocumented and public domain knowledge) and resources of local and indigenous communities living in the developing countries. In such a situation the third world nations should strive to develop appropriate policies and procedures to bring out legislations to recognize the values and the rights of indigenous and local communities over their knowledge, innovations and practices, particularly those associated with genetic resources. In order to guarantee the implementation of such policies, the third world nations need to pass legislation to facilitate access to genetic resources and benefit sharing, based on prior informed consent, access agreements (based on mutually agreed terms) and material transfer agreements. In most third world nations, these rights refer to a right to use resources for subsistence purposes, which are considerably different from the interests centered on bioprospecting activities. Many countries are presently discussing the possibility of developing a national register for documenting the knowledge, innovations and practices of indigenous and local communities. Such community registers could assist the states in ensuring controlled access and use of traditional knowledge systems, and equitable sharing of benefits as well. This register could function as a legal document that certifies the claim of the community about the knowledge and it could also form the basis for evolving a licensing method based on *sui generis* form of Intellectual Property Rights (IPR). While undertaking such an exercise the countries need to consider the international regimes on IPR, particularly the Trade Related Intellectual Property Rights of the World Trade Organization (WTO-TRIPs) and the World Intellectual Property Organization (WIPO).

Indian Government has passed three revolutionary bills to protect the national Intellectual Property Rights (IPR), viz., Patents (second and third)

Amendment Bill, Biological Diversity Act of 2002 and Plant Variety Protection and Farmers' Rights (PVPFR) Act of 2001. These Acts and Rules are now in position. It contains adequate provisions that would help safeguard the sovereign rights of the country over its biological resources, protect the indigenous knowledge systems associated with biological diversity, and recognize the farmer's rights to save, use, exchange, share or sell the plant varieties which they have developed, improved and maintained over many generations through indigenous practices of selection, domestication and conservation. The specific clauses pertaining to the above provisions include: controlled access and use of biological material and associated knowledge for commercial utilization, bio-survey or bio-utilization with the prior approval of the National Biodiversity Authority; disclosure of source of material and knowledge; and availability of indicative traditional knowledge as ground for opposition or revocation of a patent or a plant variety registered under the PVPFR Act. The national legislation on biological diversity would necessitate registration of local and indigenous knowledge throughout the country, and strict implementation of Prior Informed Consent (PIC) (of the owners of the biological resources and associated knowledge as well as the Government), so that access to and transfer of resources and knowledge and resultant benefit sharing could be regulated and monitored smoothly and efficiently. It also charges Government with monitoring and opposing IPR infringement of Indian resources and knowledge. Efforts of several NGOs (e.g., SRISTI) and the government sponsored National Innovations Foundation (NIF) provide the platform to build the registration and benefit sharing system at the grassroots. Encouraging such measures internationally through the World Intellectual Property Organization (WIPO) would be advisable.

The Intellectual property is a class of property emanating from the activities of the human brain/intellect and just like any other property can have monopolistic ownership rights and are enforceable by legal rights developed in various jurisdictions. The IPR is essentially a concept and practice developed by industrialized western countries during the last two centuries to allow monopolistic ownership or control over intangible products. Such concept and practices are quite alien to third world nations particularly to the traditional communities. To them all the natural resources and the associated knowledge systems, innovation and practices of the indigenous and local communities as characterized as a body of knowledge integrated in a holistic view of world. It is built by the members of the group/communities through generation and is considered to be a continuous process of development and refinement and thus considered to achievements of the entire community and considered to be sacred common property to be shared and cared and not to be traded as marketable commodity. But in modern world where the technologically advanced countries of the world are marching fast and indiscriminately converting the resources and the associated knowledge into marketable commodities and in this process no communities in this world can escape its impact. Taking lead/s from traditional societies about the specific use of natural resources, the industrialized modern societies who are increasingly dominated by powerful corporate bodies employs researchers, inventors and technologists and manages to access the resources including and make marketable commodities but fail to acknowledge the traditional communities from whom they got the initial lead.

Historically the process of commoditization and trading of bioresources started about 3 centuries back when the Europeans reached the Biodiversity rich nations, colonizing these countries and exploring the bioresources. The colonial traders collected both the natural resources and the associate knowledge systems and with S&T intervention made value added commodities and marketed them even in the countries from where they collected the resources/raw materials. The industrial scale of production of value added products based on the raw materials and the associated knowledge of the traditional societies led to a destructive extraction of the bioresources, which undermined the ecological security and stability of biodiversity rich third world nations as well as the livelihood security of the traditional communities of the Third world nations. The attractive attributes of the mass produced value added products from the factories of the developed nations attracted even the traditional communities and the efforts to procure them have driven such communities and countries into debt traps. This was the process by which the colonial powers exploited the national resources of the biodiversity rich third world nations during the past 300 years and that made them rich pushing the 3rd world countries in to poverty.

Just like any property, movable or immovable has to be protected in order to prevent from stealing, the rights of the property created from the intellectual efforts also has to be protected from infringement. The result of such intellectual efforts has come to be known as Intellectual Property and the rights over them denote Intellectual Property Rights (IPR). IPR are of different types, such as patents, designs, trademarks, copyrights, etc. Out of these, patent is the most important and it embodies the creative strength through innovations. Medicinal knowledge can be protected by patents. For granting a patent the invention must be: (1) new and useful (novelty) (2) involve an inventive step (non-obviousness) and (3) capable of industrial application (utility). Most of TM/TK may not strictly qualify these parameters for patent filing. At time TM could offer excellent leads based on which scientists could develop novel inventions and development of novel therapeutics. Obtaining patents on medicinal plants used in TM need to satisfy the following. (1) It must be confidential information known to one person/family/community. (2) Specification about the combination of drugs. (3) Mode of preparation—part used, pretreatment, if any and method of preparation and mode of presentation and application.

12.9 GENESIS OF THE SUBJECT ETHNOPHARMACOLOGY

Ethnopharmacology as a scientific term was first introduced at an international symposium held at San Francisco in 1967 (Efron et al., 1962). This was used while discussing the theme 'Traditional Psychoactive drugs' in this Symposium. But later Rivier and Bruhn (1979) made an attempt to define Ethnopharmacology as "a multidisciplinary area of research concerned with observation, description and experimental investigation of indigenous drugs and their biological activities. It was later redefined by Bruhn and Holmstedt (1983) as "The interdisciplinary scientific exploration of biologically active agents traditionally employed or observed by man". In its entirety, pharmacology embraces the knowledge of the history, source, chemical and physical properties, compounding, biochemical and physiological effects, mechanism of action, absorption, distribution, biotransformation, excretion and therapeutic and other uses of drugs. A drug is broadly defined as any substance (chemical agent) that affects life processes. Therefore, briefly, the main component of ethnopharmacology may be defined as pharmacology of drugs used in ethnomedicine. However none of the above said definitions captures the true spirit of this interdisciplinary subject. Ethno- (Gr., culture or people) pharmacology (Gr., drug) is about the intersection of medical ethnography and the biology of therapeutic action, for example, a transdisciplinary exploration that spans the biological and social sciences. This suggests that ethnopharmacologists are professionally cross-trained – for example, in pharmacology and anthropology – or that ethnopharmacological research is the product of collaborations among individuals whose formal training includes two or more traditional disciplines. In fact, very little of what is published as ethnopharmacology meets these criteria.

Hansen et al. (1995) has suggested that the objectives of Ethnopharmacology should focus on: (i) the basic research aiming at giving rational explanation to how a traditional medicine works, and (ii) the applied research aiming at developing a traditional medicine into a modern medicine

(Pharmacotherapy) or to develop its original usage by modern methods (Phytotherapy).

The scientific evaluation and standardization of traditional remedies using exclusively the parameters of the modern medicine is both conceptually wrong and unethical. Evaluation of traditional remedies particularly those of the classical traditions has to be based on the theoretical and conceptual foundation of these classical systems of medicine, but may utilize the advancements made in modern scientific knowledge, tools and technology. In fact it is important to combine the best of elements of concept and practice from traditional medicines and modern medicines with the objective to improve the health care system of humankind. Such an integrated approach to study and develop holistic health care system is termed as the Ethnopharmacological approach. The concept of Ethnopharmacology research in India evolved in 1980s independently of this international initiative.

Ethnopharmacology research in India was initiated at Regional Research Laboratory (RRL), Jammu in 1985 by the then Director Dr. C.K. Atal along with his student Dr. P. Pushpangadan, the then chief coordinator of All India Co-ordinated Research Project on Ethnobiology (AICRPE) and the senior author of this communication. Dr. Atal, however left RRL in mid-80s. But Dr. Pushpangadan and his students, colleagues and a few other enthusiasts, notably Dr. A.K. Sharma, Dr. S. Rajasekharan, Dr. V. George, Dr. P.G. Latha, Dr. K. Narayanan Nair, Dr. B.G. Nagavi, Shri. P.R. Krishna Kumar, etc. continued their effort to develop ethnopharmacology research. They observed that subjecting the traditional herbal remedies including the remedies of the classical systems like Ayurveda, Siddha and Unani to the parameters of modern medicine is not only foolish, but suicidal. Both these systems are conceptually quite different. The concept of disease, its etiology, manifestation and approach to treatment, etc. are all viewed on a holistic basis contrary to the reductionistic approach of modern medicine. Only an integrated approach that combines the best of theory, concepts and methods of the classical systems of medicine, such as Ayurveda, Siddha and Unani with the modern scientific knowledge (Phytochemistry and Pharmacology), tools and technology can bring in the desired results.

The concept and methods of Ethnopharmacology research thus developed by the authors contain experts from diverse disciplines like Ayurveda, Siddha, scholars of Sanskrit and Tamil languages (who can correctly interpret the classical texts of Ayurveda and also its theoretical basis like 'Sankhya' and 'Vaiseshika' philosophy), ethnobotany/ethnomedicine, chemistry, pharmacognosy, pharmacology, biochemistry, molecular biology, pharmacy, etc. The main objective of this approach was to develop appropriate

techniques to evaluate the traditional remedies in line with the classical con-
cepts of Ayurvedic pharmacy and pharmacology, such as the 'Rasa,' 'Guna,'
'Veerya,' 'Vipaka' and 'Prabhava,' in other words 'Samagrah Guna' of the
'Draya Guna' concept of Ayurveda. The senior author was successful in con-
vincing Prof. M.G.K. Menon way back in 1985 who then agreed to be the
Chief Patron of the newly formed National Society of Ethnopharmacology.
This society was formally registered in 1986 with the senior author as its
first founder president. The first ethnopharmacology laboratory started
functioning at Regional Research Laboratory, Jammu under the All India
Coordinated Research Project on Ethnobiology (AICRPE) funded by the
Ministry of Environment and Forest, Govt. of India. However, the first full-
fledged Ethnopharmacology Division was started in 1992 at Jawaharlal
Nehru Tropical Botanic Garden and Research Institute (JNTBGRI) where
the senior author joined in 1990 as its Director. At JNTBGRI the team could
successfully demonstrate the integrated approach and could develop novel
scientifically verified standardized herbal drugs. Some herbal drugs developed at JNTBGRI after filing patents were released for commercial production. The Ethnopharmacology Society in association with JNTBGRI
and with the financial assistance of DANIDA organized the first National
Conference on Ethnopharmacology in Trivandrum, Kerala from 24th to
26th May 1993. Selected papers in this conference were compiled and
published as 'Glimpses of Indian Ethnopharmacology' in 1995. The 2nd national conference of Ethnopharmacology was organized at J.S.S. College of
Pharmacy, Mysore in 1997 and the 3rd at Pankaj Kasthuri Ayurveda College,
Trivandrum in 2004 and the 4th at Amala Cancer Research Institute, Thrissur
in 2006. In 1999 Feb. the senior author moved from JNTBGRI, Trivandrum
to National Botanical Research Institute (NBRI) Lucknow, a pioneer plant
research institute under the umbrella of Council of Scientific and Industrial
Research (CSIR). International Society of Ethnopharmacology in association with National Society of Ethnopharmacology and National Botanical
Research Institute (NBRI) have organized the Vth International Congress on
Ethnopharmacology in November, 1999 at NBRI, Lucknow. At NBRI, the
senior author has established a state of the art Ethnopharmacology laboratory and Herbal Product Development Division where the latest analytical
techniques, such as HPTLC, High-Through put analysis, activity guided isolation techniques and similar other innovative new techniques in validating,
formulating and standardizing the herbal products, etc. were introduced.

Ethnopharmacological impulse to modern medicine can lead to many
novel useful drugs. Traditional medicine in general is a powerful source of

biologically active compounds. Ethnopharmacology has become a scientific backbone in the development of active therapeutics based upon traditional medicine of various ethnic groups. The ultimate aim of ethnopharmacology is validation of these traditional preparations, either through pharmacological findings or through the isolation of active substances. Harmful practices can be discouraged, such as the use of plants containing tumor-producing pyrrolizidine alkaloids. Selection of plant for serious study depends basically on two approaches. One approach is random screening of plants for their medicinal value. Another approach is that Ethnopharmacological survey of plants of a particular region or cultural group depending on their use in traditional system by choosing a specific therapeutic target. Screening program based on ethnoparmacological information has more success rate than random screening (George and John, 2008). The first and most important stage in a drug development program using plants is collection and analysis of information on the use(s) of the plant(s) by various indigenous cultures. Ethnobotany, Ethnomedicine, folk medicine and traditional medicine can provide information that is useful as a pre-screen to select plants for experimental pharmacological studies (Bigoniya, 2008).

12.10 WORKSHOP ON ETHNOBIOLOGY AND TRIBAL WELFARE

A National Workshop on Ethnobiology and Tribal Welfare was organized on behalf of the Ministry of Environment, Govt. of India in association with the International Institute of Ayurveda (IAA), Coimbatore, Tamil Nadu by the AICRPE Co-ordination Unit. The workshop was held from 1st to 3rd November, 1985 at Patanjilipuri Campus of the IAA, Coimbatore. The aim of this work shop was to bring together the senior administrators, planners, scientists, voluntary agencies associated with tribal welfare programs as well as the tribal representatives in order to interact and evolve ways and means by which the information generated from AICRPE could immediately be translated into action. The workshop was attended by 185 participants consisting of 30 administrators at the level of Secretary, Directors and Forest Conservators, 35 Scientists, 60 representatives from leading voluntary organizations and 60 tribal representatives. The three days deliberation emerged in the context of the fact that destruction of the material resource base due to deforestation caused great hardship and economic misery to tribals. The workshop after discussion on the various issues and problems of the tribals and also keeping in view of the AICRPE project findings made specific recommendation for improving the socio economic status and quality of life of the tribal people. The conference made some recommendations and submitted to the Ministry of Environment and Forests, Govt. of India (Pushpangadan, 1993).

Another National Conference as part of the AICRPE, to streamline the traditional knowledge towards a sui generic regime in the post WTO scenario named 'Dhishana 2008' was organized during May 23 to 25, 2008 at Thiruvananthapuram, Kerala. This Conference was organized in association with the Ministry of Environment and Forests, Government of India. This was also supported by the other ministries and agencies of the Govt. of India viz. NMPB, CAPART and DST. The major objective of the conference was to evolve appropriate *sui generis* mechanisms in the context of CBD, WTO and TRIPS requirements. Scientists, legal luminaries, policy makers and activists together with representatives of TK holders from tribal and non-tribal backgrounds came together for the purpose. The conference came out with the Thiruvananthapuram Declaration on Traditional Knowledge (TDTK), a landmark document on TK and biodiversity, with focus on tribal communities of Kerala (Pushpangadan and Pradeep, 2008).

12.11   CONCLUSION

Ethnobotanical research can provide a wealth of information regarding both past and present relationships between plants and the traditional societies. Investigations into traditional use and management of local flora have demonstrated the existence of extensive local knowledge of not only about the physical and chemical properties of many plant species, but also the phenological and ecological features in the case of domesticated species. In addition to its traditional roles in economic botany and exploration of human cognition, ethnobotanical research has been applied to current areas of study, such as biodiversity prospecting and vegetation management. It is hoped that, in the future, ethnobotany may play an increasingly important role in sustainable development and biodiversity conservation. In interaction with the traditional areas of science, ethnobotany gives out several interrelated and interdisciplinary subjects and link ethnomedicine, ethnoarchaeology, ethnobryology, ethnoecology, ethnoagriculture, ethnonarcotics, ethnopharmacology, etc.

KEYWORDS

- **Biopiracy**
- **Ethnobotany**
- **Ethnopharmacology**
- **IPR**
- **Jeevani**
- **Traditional Knowledge**
- *Trichopus zeylanicus*

REFERENCES

AICRPE (All India Coordinated Research Project on Ethnobiology). (1998). Final Technical Report 1992–1998, Ministry of Environment and Forest, Govt. of India.

Anand, U. (1998). The Wonder Drug. New Delhi: UBS Publishers Ltd.

Anuradha, R.V. (1998). Sharing with the Kanis; a case study from Kerala, India, Benefit-Sharing case studies, Fourth Meeting of the Conference of Parties to the CBD, Bratislava, May, Secretariat of the Convention on Biological Diversity, Montreal.

Bagla, P. (1999). Model Indian deal generates payments. *Science, 283,* 1614–1615.

Bigoniya, P. (2008). Ethnopharmacological Approach in Herbal Drug Development. *The Pharma Review,* pp. 109–116.

Bondya, S.L., Khanna, K.K. & Singh, K.P. (2006). Ethnomedicinal uses of leafy vegetables from the tribal folklore of Achnakmar-Amarkantak Biosphere Research, Madhya Pradesh and Chhattisgarh, *Ethnobotany, 18,* 145–148.

Bora, H.R. & Pandey, A.K. Less known wild food plants of Assam. In: J.K. Maheshwari (Ed.), Ethnobotany in South Asia. Jodhpur, India: Scientific Publisher, pp. 357–358.

Borthakur, S.K. (1981). Studies in ethnobotany of the Karbis (Mikirs) of Assam: Plant masticatories and dyestuffs. In: S.K. Jain (Ed.). Glimpses of Indian Ethnobotany. New Delhi, India: Oxford and IBH Publishing Co. pp. 182–190.

Borthakur, S.K. (1996). Wild edible plants in markets of Assam. India: An ethnobotanical investigation. In: S.K. Jain (Ed.). Ethnobiology in Human Welfare. New Delhi, India: Deep Publications. pp. 31–34.

Borthakur, S.K. & Gogoi, P. (1994). Indigenous technology of making writing materials among the Tai Khamtis, *Ethnobotany, 6,* 5–8.

Efron, D.H., Holmstedt, B. & Kline, N.S. (Eds.) Ethnopharmacological Search for Psychoactive Drugs. U.S. Public Health Service Publication #1645, U.S.Government Printing Office. 1967.

George, V. & John, A.J. (2008). The role of ethnomedical leads in drug discovery In: P. Pushpangadan, V. George, K. K. Janardhanan (Eds.), Ethnopharmacology Recent Advances. Delhi: Daya Publishing House, pp. 79–85.

Gupta, V.K. (2005). Traditional Knowledge Digital Library, Paper presented at Sub-regional experts meeting in Asia on Intangible Cultural Heritage: Safeguarding and Inventory-Making Methodologies, Bankok, Thailand.

Hajra, P.K. (1981). Nature conservation in Khasi folk beliefs and taboos, In: S.K. Jain (Ed.). Glimpses of Indian Ethnobotany, Oxford & IBH Publishing Co.: New Delhi, pp. 149–152.

Hajra, P.K., Baishya, A.K., Nayar M.P. & Sastry, A.R.K. (1991). Ethnobotanical Notes on the Miris (Mishings) of Assam Plains In: Contribution to Indian Ethnobotany, S.K. Jain (Ed.). Jodhpur: Scientific Publishers, pp. 161–169.

Hansen, K., Nyman, U., Smitt, U.W., Adsersen, A., Gudiksen, L., Rajasekharan, S. & Pushpangadan, P. (1995). In vitro screening of traditional medicines for anti-hypertensive effect based on inhibition of the angiotensin converting enzyme (ACE). *J. Ethnopharmacol.* *48*(1), 43–51.

Holmstedt, B. & Bruhn, J.G. (1983). Ethnopharmacology—A Challenge. *J. Ethnopharmacol.,* *8,* 251–256.

Jain, S.K. (1987), A Manual of Ethnobotany. Scientific Publishers, Jodhpur, Rajasthan, India.

Jain, S.K. (1991). Dictionary of Indian Folk Medicine and Ethnobotany, Deep Publications, New Delhi, India.

Jain, S.K. (2002). Bibliography of Indian Ethnobotany. Scientific Publications, Jodhpur, India. 2002.

Jain, S.K. (2005). Dynamism an ethnobotany, *Ethnobotany, 17,* 20–23.

Jain, S.K. (2006), Ethnobotany in the new millennium-some through on future direction in Indian Ethnobotany, *Ethnobotany, 18,* 1–3.

Jain, S.K. (2010). Ethnobotany in India: Some thoughts on future work, *Ethnobotany, 22,* 1–4.

Jain, S.K. & Goel, A.K. (1987). Workshop Exercise-I: Proforma For Field Work. *In:* A *Manual of Ethnobotany* (1st Edition), Scientific Publishers, Jodhpur, India. pp. 142–147.

Jain, S.K. & Goel, A.K. (2005). Some Indian plants in Tibetan traditional medicine. *Ethnobotany, 17,* 127–136.

Jain, S.K. & Sikarwar, R.L.S. (1998). Some Indian plants used in Latin American ethnomedicine, *Ethnobotany, 10,* 61–65.

Jain, S.K., Vinay, R., Sikarwar, R.L.S. & Saklani, A. (1994). Botanical distribution of psycho-active plants of India, *Ethnobotany, 6,* 65–75.

Jain, S.K., Sikarwar R.L.S. & Pathak, V. (1997). Ethnobotanical aspects of some plants in Latin America, *Ethnobotany, 9,* 16–23.

Janaki Ammal, E.K. (1956). Introduction to the subsistence economy of India. In: L.T.Jr. William, (Ed.). Man's Role in Changing the Face of the Earth. University of Chicago Press, Chicago, USA. pp. 324–335.

Joshi, P. (1995). Ethnobotany of the Primitive Tribes in Rajasthan. Jaipur, India: Printwell. pp. 168–183.

Joseph, J. & Kharkongor, P. (1981). A preliminary ethnobotanical Survey in the Khasi and Jaintia hills, Meghalaya, In: S.K. Jain, (Ed.), Glimpses of Indian Ethnobotany. New Delhi: Oxford & IBH Publishing Co., pp. 115–123.

Manilal, K.S. (1978). Hortus Malabaricus of Van Rheede and Floristic and Ethnobotanical studies in India, Madras Herbarium (MH) 125th Anniv. Sem. Proc., Coimbatore, p.45.

Manilal, K.S. (1980a). Malayalam plant names from Hortus Malabaricus, *Bot. Hist. Hort. Malab.*, pp. 113–120.

Manilal, K.S. (1980b). Malayalam plant names from Hortus Malabaricus in modern botanical nomenclature, *Bot. Hist. Hort. Malab.*, pp. 70–77.

Manilal, K.S. (1980c). The implication of Hortus Malabaricus with the botany and history of peninsular India, *Bot. Hist. Hort. Malab.*, pp. 1–5.

Manilal, K.S. (1981). Ethnobotany of the Nagari Script in Hortus Malabaricus. In: Recent Researches in Indian Ethnobotany. Society of Ethnobotanists, Lucknow (Abst.), p.1.

Manilal, K.S. (1996). Hortoos Malabarikoosum – Itty Achudanum: A study on the role of Itty Achudan in the compilation of the Hortus Malabaricus (in Malayalam), Mentor Books/P. K. Brothers, Calicut.

Manilal, K.S. (2005). Hortus Malabaricus, a book on the plants of Malabar, and its impact on the religious of Christianity and Hinduism in the 17th century Kerala. *Indian J. Bot. Res.*, *1*(1), 13–28.

Manilal, K.S. (2012). Hortus Malabaricus and the Socio-Cultural Heritage of India. Indian Association for Angiosperm Taxonomy (IAAR), Deptartment of Botany, University of Calicut.

Manilal, K.S., Sathish Kumar, C. & Ramesh, M. (2003). Carl Linnaeus and Hortus Malabaricus: a 250th Anniversary Tribute to Species Plantarum, *Rheedea*, *13*, 3–18.

Mashelkar, R.A. (2001). Intellectual Property Rights and the Third World, *Curr. Sci.*, 81, 955–965.

Mitra, R. (1988a). Ethnoeconomic significance of the common Myrtle – A plant sacred to Greeks and Romans, *Ethnobotany*, *10*, 1–5.

Mitra, R. (1988b). Selected references on Ethnobotany 1996–1998, *Ethnobotany*, *10*, 142–144.

Mohanty, R.B. (2003). Oral and dental health care in folklores of Orissa: an ethnobotanical observation, *Ethnobotany*, *15*, 125–126.

Mohanty, R.B. (2010). New wild edible plants from some tribal pockets of Dhenkanal district, Odisha, *Ethnobotany*, *22*, 113–113.

Mohanty, R.B., Rout, M.K. (2001). Indigenous rice germplasm and their cultivation technique in folklores of Orissa: an ethnobotanical Study, *Ethnobotany*, *13*, 24–28.

Patil, D.A. (2000). Sanskrit plant names in an ethnobotanical perspective, *Ethnobotany*, *12*, 60–64.

Patil, D.A. (2001). Ethnography of the drug safed-musali in India. *Ancient Science of Life*, *21*, 51–65.

Pushpangadan, P. (1986). Search for new sources of biodynamic compounds form Tribal Medicine. *Crude Drugs*, *7*, 40–43.

Pushpangadan, P. (1990). Health status of tribals of India. Key note address, Proceedings of National Seminar on 'Health of Tribal People,' Indian Society for Health Administrators (ISHA), pp. 57–61.

Pushpangadan, P. & Atal, C.K. (1984). Ethno-medico botanical investigation in Kerala. 1. Some primitive tribals of Western Ghats and their herbal medicine, *J. Ethnopharmacol.*, *11*, 59–77.

Pushpangadan, P. & Atal, C.K. (1986). Ethnomedical and ethnobotanical investigations among some scheduled caste communities of Travancore, Kerala, India. *J. Ethnopharmacol.*, 16, 175–190.

Pushpangadan, P., Sharma, A.K. & Rajasekharan, S. (1995). Ethnopharmacology of *Trichopus zeylanicus* – The Ginseng of Kerala: A Review. In: P. Pushpangadan, U. Nyman & V.

George, (Eds.), Glimpses of Indian Ethnopharmacology. Thiruvananthapuram, Kerala, India: TBGRI Publication, pp. 137–145.

Pushpangadan, P. & Pradeep, P.R.J. (2008). A Glimpse at Tribal India—An Ethnobiological Enquiry. Amity Institute for Herbal and Biotech products Development, Thiruvananthapuram, Kerala, India.

Pushpangadan, P. & George, V. (2010). Ethnomedical practices of rural and tribal populations of India with special reference to the Mother and childcare. *Indian J. Tradit. Knowle., 9,* 9–17.

Pushpangadan, P. & Dan, V.M. (2011). Modern methods and strategies of bioprospecting of traditional knowledge and the issues related to benefit sharing. *Ethnobotany 23,* 1–20.

Pushpangadan, P., Dan, V.M., Ijinu, T.P. & George, V. (2012). Food, Nutrition and Beverage. *Indian J. Tradit. Knowle., 11,* 26–34.

Rivier, J. & Bruhn, J.G. (1979). Editorial. *J. Ethnopharmacol., 1,* 1.

Singh, V.N., Ibemhal C.L., Chiru Community & Baruah, M.K. (2011). An ethnobotanical study of Chirus-A less known tribe of Assam. *Indian J. Tradit. Knowle., 10,* 572–574.

Subramoniam, A., Madhavachandran, V., Rajasekharan, S. & Pushpangadan, P. (1997). Aphrodisiac property of *Trichopus zeylanicus* extract in small mice. *J Ethnopharmacol., 57,* 21–27.

Subramoniam, A., Evans, D.A., Rajasekharan, S. & Pushpangadan, P. (1998). Hepatoprotective activity of *Trichopus zeylanicus* extract against paracetamol induced hepatic damage in rats. *Indian J. Exp. Biol., 36,* 385–389.

Vartak, V.D. (1981). Observations on wild edible plants from Hilly Regions of Maharashtra and Goa: Resume and Future Prospects. In: S.K. Jain (Ed.), Glimpses of Indian Ethnobotany. New Delhi, India: Oxford and IBH Publishing Co., pp. 261–271.

Vartak, V.D. & Gadgil, M. (1980). Studies in ethnobotany: A new vistas in botanical Science, *Biovigyanam, 6,* 151–156.

Vartak, V.D. & Gadgil, M. (1981). Studies on sacred groves along the Western Ghats from Maharashtra and Goa: Role of beliefs and folklores. In: S.K. Jain, (Ed.). Glimpses of Indian Ethnobotany. New Delhi, India: Oxford and IBH Publishing Co., pp. 272–278.

INDEX

Printed and bound by CPI Group (UK) Ltd, Croydon, CR0 4YY

23/10/2024

01777704-0015